Robots, Drones, UAVs and UGVs for Operation and Maintenance

A Call for Authors

Introducing the New CRC Press

ICT in Asset Management Series

CRC Press is proud to announce the formation of the **ICT in Asset Management** series, edited by Professor Diego Galar. This new series is dedicated to publishing high-quality texts and research monographs pertaining to cutting-edge developments in information and communications technology (ICT) in asset management and the closely allied discipline of maintenance. Due to the growing need for complex data handling, these areas of study and practice are poised to become the spark that ignites the fourth industrial revolution.

We are looking for authored books or edited volumes that:

- Describe the self-learning and smart systems that predict failure, make diagnoses, and trigger maintenance
- Discuss sensor networks, sensor fusion, cloud computing, data mining, big data analytics, and similar technologies
- Explore the capture and mining of information to accurately assess the health of the whole system, including infrastructure, factories, facilities, and vehicles
- Demonstrate how to determine the probability of a shutdown or slowdown
- Present the latest advances in manufacturing, reliability, maintenance, operations, quality management, quality control, design, automation, and other areas related to data collection across different assets

We welcome all related project and concept ideas.

Proposals for the series should be submitted directly to:

Professor Diego Galar
Luleå University of Technology
971 87 Luleå, Sweden
Email: diego.galar@ltu.se
Phone: +46 920 49 24 37

About the Series Editor:

Diego Galar holds an M.Sc in telecommunications and a Ph.D in manufacturing from the University of Zaragoza, Spain. He has been a professor at several universities, including the University of Zaragoza and the European University of Madrid, Spain. At the University of Zaragoza, he also has served as a senior researcher in the Aragon Institute of Engineering Research (i3A), director of academic innovation, and pro-vice-chancellor. In industry, he has been a technological director and a certified business manager. Currently, he is a professor of condition monitoring in the Division of Operation and Maintenance at Luleå University of Technology (LTU), Sweden, where he is coordinating several EU-FP7 projects related to different maintenance aspects and is involved with the LTU-SKF University Technology Centre (UTC) focused on SMART bearings. He is also a visiting professor at the University of Valencia, Spain; Polytechnic Institute of Braganza, Portugal; University of the Valley of Mexico; University of Sunderland, UK; and Northern Illinois University, USA. Professor Galar has authored more than 100 journal and conference papers, books, and technical reports in the field of maintenance.

WWW.CRCPRESS.COM

CRC Press
Taylor & Francis Group

Robots, Drones, UAVs and UGVs for Operation and Maintenance

Diego Galar, Uday Kumar, and Dammika Seneviratne

CRC Press
Taylor & Francis Group
Boca Raton London New York

CRC Press is an imprint of the
Taylor & Francis Group, an **informa** business

MATLAB® is a trademark of The MathWorks, Inc. and is used with permission. The MathWorks does not warrant the accuracy of the text or exercises in this book. This book's use or discussion of MATLAB® software or related products does not constitute endorsement or sponsorship by The MathWorks of a particular pedagogical approach or particular use of the MATLAB® software

CRC Press
Taylor & Francis Group
52 Vanderbilt Avenue,
New York, NY 10017

© 2020 by Taylor & Francis Group, LLC
CRC Press is an imprint of Taylor & Francis Group, an Informa business

No claim to original U.S. Government works

Printed on acid-free paper

International Standard Book Number-13: 978-1-138-32211-0 (Hardback)

This book contains information obtained from authentic and highly regarded sources. Reasonable efforts have been made to publish reliable data and information, but the author and publisher cannot assume responsibility for the validity of all materials or the consequences of their use. The authors and publishers have attempted to trace the copyright holders of all material reproduced in this publication and apologize to copyright holders if permission to publish in this form has not been obtained. If any copyright material has not been acknowledged please write and let us know so we may rectify in any future reprint.

Except as permitted under U.S. Copyright Law, no part of this book may be reprinted, reproduced, transmitted, or utilized in any form by any electronic, mechanical, or other means, now known or hereafter invented, including photocopying, microfilming, and recording, or in any information storage or retrieval system, without written permission from the publishers.

For permission to photocopy or use material electronically from this work, please access www.copyright.com (http://www.copyright.com/) or contact the Copyright Clearance Center, Inc. (CCC), 222 Rosewood Drive, Danvers, MA 01923, 978-750-8400. CCC is a not-for-profit organization that provides licenses and registration for a variety of users. For organizations that have been granted a photocopy license by the CCC, a separate system of payment has been arranged.

Trademark Notice: Product or corporate names may be trademarks or registered trademarks, and are used only for identification and explanation without intent to infringe.

Visit the Taylor & Francis Web site at
http://www.taylorandfrancis.com

and the CRC Press Web site at
http://www.crcpress.com

Contents

Preface vii
Authors ix

1. Introduction 1
   1.1   Autonomous Vehicles 1
   1.2   Industrial Assets 21
   1.3   Inspection of Industrial Assets 26
   1.4   Maintenance of Industrial Assets 28
   References 33

2. Development of Autonomous Vehicles 37
   2.1   History of Development of Autonomous Robots 37
   2.2   Dynamics and Machine Architectures 48
   2.3   Robots and Machine Intelligence 52
   2.4   Programming Autonomous Robots 57
   2.5   Adaptive Algorithms and Their Use 62
   References 70

3. Autonomous Inspection for Industrial Assets 73
   3.1   Autonomous Vehicle Inspection Platform 73
   3.2   Inspection Communications and Transport Security 80
   3.3   Obstacle Avoidance 83
   3.4   Inspection Modes and Content 87
   3.5   Inspection Methods 93
   References 106

4. Sensors for Autonomous Vehicles in Infrastructure Inspection Applications 111
   4.1   Sensors and Sensing Strategies 111
   4.2   Sensor Types: Introduction 115
   4.3   Sensors for Military Missions 130
   4.4   Sensor-Based Localization and Mapping 133
   4.5   Sensor Fusion, Sensor Platforms, and Global Positioning System 141
   References 150

5. Data Acquisition and Intelligent Diagnosis 155
   5.1   Data Acquisition Principle and Process for Laser Scanning, Visual Imaging,
         Infrared Imaging, UV Imaging 155
   5.2   Cloud Data Post-Processing Technology 174
   5.3   Cloud Data Intelligent Diagnosis 179
   References 183

6. Three-Dimensional Visualization 187
   6.1   Overview 187
   6.2   Line Security Diagnosis for Multisource Data Fusion 206
   6.3   Three-Dimensional Visualization Applications 216
   References 220

7. Communications ... 227
- 7.1 Communication Methods ... 227
- 7.2 Radio Communication ... 230
- 7.3 Mid-Air Collision (MAC) Avoidance ... 235
- 7.4 Communications Data Rate and Bandwidth Usage ... 241
- 7.5 Antenna Types ... 244
- 7.6 Tracking with Multiple Autonomous Vehicles ... 265
- References ... 268

8. Autonomous Vehicles for Infrastructure Inspection Applications ... 271
- 8.1 Power Line Inspection ... 271
- 8.2 Building Monitoring ... 280
- 8.3 Railway Infrastructure Inspection ... 288
- 8.4 Waterways and Other Infrastructures ... 294
- References ... 303

9. Failure Detection Application in Autonomous Vehicles ... 307
- 9.1 Repeated Inspections and Failure Identification ... 307
- 9.2 Autonomous Vehicle Emergency Inspection Applications ... 313
- 9.3 Autonomous Vehicle Navigation Security ... 322
- References ... 337

10. Autonomous Inspection and Maintenance with Artificial Intelligence Infiltration ... 341
- 10.1 Artificial Intelligence Techniques Used in AVs ... 341
- 10.2 Artificial Intelligence Approaches for Inspection and Maintenance ... 346
- 10.3 Current Developments of AVs with AI ... 353
- References ... 361

11. Big Data Analytics for AV Inspection and Maintenance ... 365
- 11.1 Big Data Analytics and Cyber-Physical Systems ... 365
- 11.2 Big Data Analytics in Inspection and Maintenance ... 376
- 11.3 Integration of Big Data Analytics in AV Inspection and Maintenance ... 381
- 11.4 Utilization of AVs in Industry 4.0 Environment ... 384
- References ... 388

Index ... 393

Preface

Autonomous vehicle (AV) technology offers the possibility of fundamentally changing our lives, perhaps most visibly in the transportation sector. Equipping cars and light vehicles with this technology will likely reduce crashes, energy consumption, the costs of congestion, and pollution. In a wider sense, AV technology has the potential to substantially affect safety, energy use, and, ultimately, land use and the environment. This book begins by explaining what autonomous robots are and noting their potential use in many different applications, not just on the road driving people from place to place but in the workplace for inspections, maintenance, and many other tasks.

The main difficulty is training robots to effectively carry out tasks in different and little-known environments. Autonomous robots, i.e., freely moving robots that operate without direct human supervision, are expected to function in complex, unstructured environments and make their own decisions on what action to take in any given situation. In such cases, systems based only on classical control theory are insufficient. This requires major improvements in the control architectures. Other architectural needs include the ease and quality of a robot's specification and programming. Chapter 2 tackles this issue by looking at the programing of autonomous robots and other related details.

Autonomous robots gain information on their surroundings via sensors. The information is processed in the robot's "brain," consisting of one or more microcontrollers; after processing, motor signals are sent to the actuators (motors) of the robot and it can take action. Thus, the "brain" is the system that provides an autonomous robot, however simple, with the ability to process information and decide which actions to take.

The emphasis in the development of autonomous robots is currently on speed, energy efficiency, sensors for guidance, guidance accuracy, and enabling technologies, such as wireless communication and the global positioning system (GPS). Chapter 7 goes into more detail on the communication aspect, discussing communication methods in general before turning to radio communication specifically. It defines communication as a process of transmitting and receiving meaningful information or intelligence. Electronic communication involves converting speech or intelligence into electrical signals using transducers. The signals are processed and transmitted to a receiver. The receiver processes the received signals and drives the transducer which converts the processed signals into speech or intelligence. Put otherwise, transducers convert energy from mechanical to electrical and vice versa. The chapter also defines and explains communications data rate (CDR) and bandwidth usage.

Chapters 2–4 and Chapter 8 give an overview of the development of AVs and distant inspection operations for industrial assets using AVs and unmanned aerial vehicles (UAVs). The chapters discuss the use of AVs in infrastructure inspection and explain the types of sensors used for these applications. Autonomous robots, including UAVs, pipe inspection gauges, and remotely operated vehicles, are currently used in various industrial settings for inspection and maintenance. As autonomous robots can be programmed for repetitive and specific tasks, they can fruitfully be used for the inspection and maintenance of industrial assets.

Chapter 5 discusses laser scanning technology, from data acquisition from sensors and intelligent diagnosis to data processing and visualization. It provides a general overview of laser scanning technologies and mentions some specific applications. The chapter also discusses artificial intelligence (AI), defining it as the science and engineering of making intelligent machines, especially intelligent computer programs. AI is useful for such activities as searching, recognizing patterns, and making logical inferences. As the chapter points out, AI techniques have potential for system diagnostics.

The inspection of critical structural components, machinery, and spots that are hard to reach is technically complex and mostly done by specially trained staff and/or specialized equipment. Remotely controlled UAVs equipped with high definition photo and video cameras can simplify these inspection tasks. Chapter 6 presents an inspection method and three-dimensional visualizations that lead to critical failure detection by AVs and UAVs (developed in Chapter 9). It also analyzes autonomous inspection and maintenance with AI infiltration (developed in Chapter 10) and Big Data analytics for AV inspection and maintenance. Essentially, the use of Big Data analytics in maintenance represents the fourth level of maturity in predictive maintenance. The fourth level, Predictive Maintenance 4.0, or PM 4.0, is described at length in Chapter 11.

The final chapter, Chapter 11, analyzes maintenance under a Big Data platform and shows how AI can be applied to classify the likely failure patterns and estimate equipment condition. It explains the important position of AVs in an Industry 4.0 environment and anticipates their further development. Industry 4.0 includes the complete manufacturing value chain—from raw materials, to unfinished goods, to the production shop floor, to the warehouse, storage, and delivery. Automation is at its core, and robots are an essential part. Inevitably, as smarter environments give rise to more information, robots will become more intelligent and easier to operate. Indeed, the ultimate goal of Industry 4.0 is an autonomous smart factory that can produce customizable products.

MATLAB® is a registered trademark of The MathWorks, Inc. For product information, please contact:
The MathWorks, Inc.
3 Apple Hill Drive
Natick, MA 01760-2098 USA
Tel: 508-647-7000
Fax: 508-647-7001
E-mail: info@mathworks.com
Web: www.mathworks.com

Authors

Dr. Diego Galar is Full Professor of Condition Monitoring in the Division of Operation and Maintenance Engineering at LTU, Luleå University of Technology, Lulea, Sweden, where he is coordinating several H2020 projects related to different aspects of cyber-physical systems, Industry 4.0, IoT or Industrial AI, and Big Data. He was also involved in the SKF UTC located in Luleå focused on SMART bearings and also actively involved in national projects with the Swedish industry or funded by Swedish national agencies like Vinnova.

He is also principal researcher in Tecnalia (Spain), heading the Maintenance and Reliability research group within the Division of Industry and Transport.

He has authored more than five hundred journal and conference papers, books, and technical reports in the field of maintenance, working also as member of editorial boards, scientific committees and chairing international journals and conferences and actively participating in national and international committees for standardization and R&D in the topics of reliability and maintenance.

In the international arena, he has been visiting Professor in the Polytechnic Institute of Bragança (Portugal), University of Valencia and NIU (USA), and the Universidad Pontificia Católica de Chile. Currently, he is visiting professor in University of Sunderland (UK), University of Maryland (USA), and Chongqing University in China.

Uday Kumar, the Chaired Professor of Operation and Maintenance Engineering, is Director of Luleå Railway Research Center and Scientific Director of the Strategic Area of Research and Innovation-Sustainable Transport at Luleå University of Technology, Luleå, Sweden. Before joining Luleå University of Technology, Dr. Kumar was Professor of Offshore Technology (Operation and Maintenance Engineering) at Stavanger University, Norway. Professor Kumar has research interest in the subject area of reliability and maintainability engineering, maintenance modeling, condition monitoring, LCC and risk analysis, etc. He has published more than 300 papers in international journals and peer-reviewed conferences and has made contributions to many edited books. He has supervised more than 25 PhD theses related to the area of reliability and maintenance. Professor Kumar has been a keynote and invited speaker at numerous congresses, conferences, seminars, industrial forums, workshops, and academic institutions. He is an elected member of the Swedish Royal Academy of Engineering Sciences.

Dammika Seneviratne currently works as post-doctoral researcher in the Division of Operation and Maintenance—Luleå University of Technology, Luleå, Sweden, and senior researcher in Tecnalia, Spain. He holds a BSc degree in Mechanical Engineering from the University of Peradeniya, Sri Lanka, specialized in Production Engineering. He received his MSc degree in Mechatronics Engineering from the Asian Institute of Technology, Thailand. After working for a number of years as a Mechanical Maintenance Engineer in various organizations, he attained a PhD degree in Offshore Technology from the University of Stavanger. His research interests include condition monitoring, operation, and maintenance engineering in railway systems; risk-based inspection planning in offshore oil and gas facilities; reliability and risk analysis and managements; and risk-based maintenance.

1
Introduction

1.1 Autonomous Vehicles

Autonomous vehicle (AV) technology will fundamentally change transportation. Technological advancements are creating a continuum between conventional, fully human-driven vehicles and AVs which partially or fully drive themselves and may ultimately require no driver at all. Within this continuum are technologies that enable a vehicle to assist and make decisions for a human driver. Such technologies include crash warning systems, adaptive cruise control (ACC), lane keeping systems, and self-parking technology.

Equipping cars and light vehicles with this technology will reduce crashes, energy consumption, and pollution—and reduce the costs of congestion.

This technology is most easily conceptualized using a five-level hierarchy suggested by the National Highway Traffic Safety Administration (NHTSA) with different benefits realized at different levels of automation:

- Level 0 (no automation): The driver is in complete and sole control of the primary vehicle functions (brake, steering, throttle, and motive power) at all times and is solely responsible for monitoring the roadway and for safe vehicle operation.
- Level 1 (function-specific automation): Automation at this level involves one or more specific control functions. If multiple functions are automated, they can operate independently of each other. In this case, the driver has overall control and is solely responsible for safe operation but can choose to cede limited authority over a primary control (as in ACC). Alternatively, the vehicle can automatically assume limited authority over a primary control (as in electronic stability control), or the automated system can provide added control to aid the driver in certain normal driving or crash-imminent situations (e.g., dynamic brake support in emergencies).
- Level 2 (combined function automation): This level involves automation of at least two primary control functions designed to work in unison to relieve the driver of controlling those functions. Vehicles at this level of automation can share authority when the driver cedes active primary control in certain limited driving situations. The driver is still responsible for monitoring the roadway and safe operation and is expected to be available for control at all times and on short notice. The system can relinquish control with no advance warning, and the driver must be ready to control the vehicle safely.
- Level 3 (limited self-driving automation): At this level of automation, the driver can cede full control of all safety-critical functions under certain traffic or environmental conditions and rely heavily on the vehicle to monitor changes in those conditions requiring transition back to driver control. The driver is expected to be available for occasional control but with sufficiently comfortable transition time.
- Level 4 (full self-driving automation): The vehicle is designed to perform all safety-critical driving functions and monitor roadway conditions for an entire trip. Such a design anticipates that the driver will provide destination or navigation input but is not expected to be available for control at any time during the trip. This includes both occupied and unoccupied vehicles.

The type and magnitude of the potential benefits of AV technology will depend on the level of automation achieved. For example, some of the safety benefits of AV technology may be achieved from function-specific automation (e.g., automatic braking), while the land use and environmental benefits are likely to be realized only by full automation (Level 4) (Anderson et al., 2016).

1.1.1 Brief History and Current State of Autonomous Vehicles

For decades, futurists have envisioned vehicles that drive themselves, and research into AV technology can be divided into three phases.

1.1.1.1 Phase 1: Foundational Research

From approximately 1980 to 2003, university research centers worked on two visions of vehicle automation. As one thrust, researchers pursued the development of automated highway systems (AHS), in which vehicles depended significantly on the highway infrastructure to guide them (Anderson et al., 2016).

One of the first major demonstrations of such a system took place in 1997 over a 7.6-mile stretch of California's I-15 highway near San Diego. Led by the California Partners for Advanced Transit and Highways (PATH) program, the "DEMO 97" program demonstrated the platooning of eight AVs guided by magnets embedded in the highway and coordinated with vehicle-to-vehicle (V2V) communication (Ioannou, 1998).

A second research thrust was to develop both semiautonomous and AVs that depended little, if at all, on highway infrastructure. In the early 1980s, a team led by Ernst Dickmanns at Bundeswehr University Munich in Germany developed a vision-guided vehicle that navigated at speeds of 100 km/h without traffic (Bâela & Mâarton, 2011). Carnegie Mellon University's NavLab developed a series of vehicles, named NavLab 1 through NavLab 11, from the mid-1980s to the early 2000s. In July 1995, NavLab 5 drove across the country in a "No Hands Across America" tour, in which the vehicle steered autonomously 98% of the time, while human operators controlled the throttle and brakes. Other similar efforts around the world sought to develop and advance initial AV and highway concepts (Anderson et al., 2016).

1.1.1.2 Phase 2: Grand Challenges

From 2003 to 2007, the US Defense Advanced Research Projects Agency (DARPA) held three "Grand Challenges" that markedly accelerated advancements in AV technology. The first two Grand Challenges charged research teams with developing fully autonomous vehicles for competition in a 150-mile off-road race for $1 million and $2 million prizes, respectively. No vehicle completed the 2004 Grand Challenge—the best competitor completed less than 8 miles of the course (BBC News, 2004). However, five teams completed the 2005 Grand Challenge course, held only 18 months later. The fastest team completed the course in just under 7 h, with the next three fastest finishing within the next 35 min (DARPA, undated).

In 2007, DARPA held its third and final AV challenge, dubbed the "Urban Challenge." As the name suggests, vehicles raced through a 60-mile urban course, obeying traffic laws and navigating alongside other autonomous and human-driven vehicles. Six teams finished the course, and three completed the race within a time of 4.5 h, including time penalties for violating traffic and safety rules. This Grand Challenge spearheaded advancements in sensor systems and computing algorithms to detect and react to the behavior of other vehicles, to navigate marked roads, and to obey traffic rules and signals (Anderson et al., 2016).

1.1.1.3 Phase 3: Commercial Development

The DARPA Challenges solidified partnerships between auto manufacturers and the education sector and mobilized a number of endeavors in the automotive sector to advance AVs. These included the Autonomous Driving Collaborative Research Lab, a partnership between GM and Carnegie Mellon University (Carnegie Mellon University, undated) and a partnership between Volkswagen and Stanford University (Stanford University, undated).

Introduction 3

Google's Driverless Car initiative has brought autonomous cars from the university laboratory into commercial research. The program began shortly after the DARPA Urban Challenge and drew on the talents of engineers and researchers from several teams who participated in that competition. In the years since, Google has developed and tested a fleet of cars and initiated campaigns to demonstrate the applications of the technology through, for example, videos highlighting mobility offered to the blind (Google, 2012). Google is not alone. In 2013, Audi and Toyota both unveiled their AV visions and research programs at the International Consumer Electronics Show, an annual event held every January in Las Vegas (Hsu, 2013). Nissan has also recently announced plans to sell an AV by 2020 (Anderson et al., 2016).

1.1.2 Current State of Autonomous Vehicle Technology

Google's vehicles, operating fully autonomously, have driven more than 500,000 miles without a crash attributable to the automation. Advanced sensors to gather information about the world, increasingly sophisticated algorithms to process sensor data and control the vehicle, and computational power to run them in real time have permitted this level of development (Anderson et al., 2016).

1.1.2.1 Making Sense of the World

In the most general terms, AVs employ a "sense-plan-act" design that is the foundation of many robotic systems. A suite of sensors on the vehicle gathers raw data about the outside world and the vehicle's relations to its environment. Software algorithms interpret the sensor data—for example, lane markings from images of the road or behavior of other vehicles from radar data. They use these data to make plans about the vehicle's own actions—its overall trajectory down the road and immediate decisions such as accelerating and changing directions. These plans are converted into actionable commands to the vehicle's control system; i.e., steering, throttle, brakes.

Many "sense-plan-act" loops run in parallel on an AV. One loop may run at extremely high frequency to initiate rapid emergency braking, while another runs less frequently to plan and execute complex behaviors such as changing lanes. In some cases, the planning component of the loop is extremely short and resembles a sense-act cycle instead of a sense-plan-act cycle. For instance, a vehicle may gather data about obstacles immediately in front of it at very high frequency and initiate emergency braking if any obstacle is detected within a short distance. In this case, the sensor data may directly trigger a vehicle action.

With perfect perception (a combination of sensor data gathering and interpretation of those data), AVs could plan and act perfectly, achieving ultra-reliability. Vehicles never tire; their planning algorithms can choose provably optimal behaviors, and their execution can be fast and flawless. For example, if a deer were to leap into the path of a human-driven vehicle, the driver may make mistakes in choosing whether to swerve, brake, or take another course of action. The driver may also make mistakes in executing the action; for example, oversteering a swerve. AVs need never make these mistakes. Computer algorithms can rapidly evaluate, compare, select, and execute the best action from among a number of maneuvers, taking into account the vehicle's speed, the animal's trajectory, the position and behavior of other vehicles, and the utility of various outcomes (Anderson et al., 2016).

One of the more difficult challenges for AVs is making sense of the complex and dynamic driving environment—for example, perceiving the deer. The driving environment includes many elements (Anderson et al., 2016):

- Other vehicles on the road, each of which operates dynamically and independently.
- Other road users or on-road obstacles, such as pedestrians, cyclists, wildlife, and debris.
- Weather conditions, from sunny days to severe storms.
- Infrastructure conditions, including construction, rough road surfaces, poorly marked roads, and detours.
- Traffic events, such as congestion or crashes.

It is in making sense of the world that humans often outperform robots. Human eyes are sophisticated and provide nearly all of the sensory data we use to drive. We are also adept at interpreting what we see. Although our eyes are passive sensors, only receiving information from reflected light, we can judge distances, recognize shapes, and classify objects such as cars, cycles, and pedestrians and see in a tremendous range of conditions. Of course, we are far from perfect. Our sight and our cognition of visual information vary and can be dangerously limited in several situations: adverse ambient conditions such as darkness, rain, and fog, when we are tired or distracted, or when we are impaired through the use of drugs or alcohol (Olson, Dewar, & Farber, 2010).

A second limitation is that, like human eyes, camera systems are better able to gather data in some ambient conditions (e.g., clear sunny days) than others (e.g., fog or rainstorms). Changes in ambient conditions also pose challenges, as camera systems calibrated to certain conditions may have difficulty interpreting data in others. This problem of autonomous camera calibration is also a fundamental robotics research problem (Furukawa & Ponce, 2009).

Camera-based systems, i.e., computer vision systems, are analogous to human eyes and visual cognition. They can "see" very long distances and provide rich information about everything in their field of view. Cameras are also inexpensive, making them important components for cost-effective autonomy. However, they have two important limitations. First, the underlying algorithms are not nearly as sophisticated as humans at interpreting visual data. The Solutions in Perception Challenge is an annual competition that embodies this difference, challenging engineering teams to develop computer vision and other sensor algorithms that can detect, recognize, and locate objects. In the 2011 competition, for example, the objects included a number of items that would be found on supermarket shelves. None of the competing teams reached the goal of 80% accuracy (Markoff, 2011).

AVs have a critical advantage over humans: they can draw upon a much wider array of sensor technologies than cameras alone. While many major advances have been made in the last decade, however, the interpretation of visual data (and sensor data more generally) remains a fundamental research problem in the field of computer vision. We can expect advances in both sensor technology and perception algorithms, but matching human perception under best conditions is a long-term research challenge (Anderson et al., 2016).

1.1.2.1.1 Light Detection and Ranging Sensor Systems

Light detection and ranging (LiDAR) systems feature prominently in robotic systems, including AVs. LiDAR systems determine distances to obstacles by using laser range finders, which emit light beams and calculate the time of flight until a reflection is returned by objects in the environment. Many sophisticated LiDAR systems couple multiple laser range finders with rapidly rotating mirrors to generate 3D point clouds of the environment. Developed during the DARPA Grand Challenges and used by teams in the Urban Challenge and by Google, the Velodyne HDL-64E LIDAR uses 64 lasers that provide 1.3 million data points per second and offer a 360° field of view. LIDAR is typically useful over a shorter range than other sensors—the Velodyne provides data up to 120m away, depending on the reflectivity of the object.

LiDAR systems' two key limitations are range (less useful at long ranges) and reflectivity (poor reflection from certain kinds of materials). The Velodyne's specifications state that it detects black asphalt, which has low reflectivity, to a range of just 50m (Velodyne, 2010). The costs of LiDAR systems range widely but are expected to decline in the near future (Anderson et al., 2016).

1.1.2.1.2 Sensor Suites

Each sensor provides different kinds of data and has its own limitations related to field of view, ambient operating conditions, and the elements in the environment that it can sense. Because the limitations are fairly well understood, the usual practice is to construct suites of complementary sensors that are positioned around the vehicle to prevent blind spots—both visual blind spots (i.e., due to occluded views) and material blind spots (i.e., the inability to detect certain kinds of objects or certain properties of objects in the environment).

Sensors can be integrated to perceive more about the environment than can be learned purely from the sum of individual sensors' data. As one example, vision systems can detect colors of surfaces in the distance, while LiDAR can be used to determine the material as that surface approaches. When these two types of sensors are coupled, a system can learn that green surfaces in the distance correspond to grass, allowing the vehicle to make greater sense of the far away environment (Thrun et al., 2007).

Vehicles also use sensor suites for localization, i.e., determining their own position in the world. The use of the Global Positioning System (GPS) is essential for localization. Vehicle GPSs receive signals from orbiting satellites to triangulate their global coordinates. These coordinates are cross-referenced with maps of the road network to enable vehicles to identify their position on roads.

The accuracy of GPS has improved significantly since 2000, when the US government made it fully available to civilian users. However, GPS error can still be large—several meters, even under ideal conditions. The errors grow rapidly when obstacles or terrain occlude the sky, preventing GPS receivers from obtaining signals from a sufficient number of satellites. This is a significant concern in urban areas, where skyscrapers create "urban canyons" in which GPS availability is severely limited (Anderson et al., 2016).

GPS is typically coupled with inertial navigation systems (INSs), which consist of gyroscopes and accelerometers, to continuously calculate the position, orientation, and velocity of a vehicle without the need for external references. INSs are used to improve the accuracy of GPS and to fill in "gaps" such as those caused by urban canyons. The key challenge with INS is drift; even over very short periods, small errors can aggregate into large differences between calculated and true positions. For example, a 10 s period during which the system relies on INS because the GPS signal is unavailable can result in more than a meter of drift in calculated position, even with some of the most sophisticated systems (Applanix, 2012).

1.1.2.1.3 Environmental Challenges

Certain ambient conditions (e.g., severe precipitation, dense fog) may pose problems for multiple sensors simultaneously. Common failure conditions such as these limit the extent to which sensor combinations can compensate for individual sensor limitations. It must be noted, however, that these same conditions pose problems for humans. Indeed, robotic sensors such as radar may prove more effective than human vision, and the rapid reaction of planning algorithms may be particularly valuable, making autonomous systems imperfect but potentially safer than human drivers in these adverse conditions (Anderson et al., 2016).

Terrain poses challenges as well. A sensor configuration appropriate for a flat environment may be inappropriate for steep hills, where sensors must look "up" or "down" the slopes. Different terrains can require different sensor configurations, which may not be readily changeable. While sensors can be put on adjustable mounts to accommodate this problem, this adds complexity and cost (Urmson et al., 2006).

Road materials also change from region to region. They are typically concrete and asphalt but can be dirt, cobblestone, and other materials. Different materials have different reflectivity, and sensors calibrated to certain materials may have difficulty detecting other materials with equal fidelity.

Construction projects and roadwork are particularly difficult to negotiate, as there may be little consistency in signage and alerts, roadway materials may change suddenly, and the maneuvers needed to navigate through construction zones may be complex and poorly marked. Moreover, these areas often involve deviations from preconstructed maps, so vehicle localization may be particularly difficult.

Each of these factors can have implications for where AVs can successfully operate. For example, in the United States, weather and terrain vary significantly, as do the road materials and signage practices. A vehicle that operates easily on flat terrain in Louisiana may have significant performance challenges on Colorado's snowy and steep roads or in New York City's congested urban canyons (Anderson et al., 2016).

1.1.2.1.4 Graceful Degradation

Sensor failure (as opposed to external environmental conditions) can pose serious performance threats (Hwang, Kim, Kim, & Eng, 2010). Sensors may fail because of electrical failures, physical damage, or age. It will be critical for AVs to have internal sensing and algorithms that can detect when internal components are not performing adequately. This is not easy. A sensor that fails to provide any data is easily detected as nonfunctioning, but a sensor that occasionally sends spurious data may be much harder to detect.

These and other failures will require a system that degrades gracefully (Berger & Rumpe, 2012). AVs will likely need to have an ultrareliable and simple low-level system that uses minimal sensor data to perform basic functions in the event of main system degradation or failure. The backup system must also be able to detect degradation and failure and override control rapidly and safely. The task of graceful degradation may be complicated by traffic conditions and roadways. If a system fails in the middle of a curve in dense traffic, it may need to be able to navigate to a safe area to pull over (Anderson et al., 2016).

1.1.2.1.5 V2V and V2I Communication

The role of vehicle-to-vehicle (V2V) and vehicle-to-infrastructure (V2I) communication in enabling AV operation remains unclear. While this technology could ease the task of automated driving in many circumstances, it is not clear that it is necessary. Moreover, V2I might require substantial infrastructure investments—for example, if every traffic signal must be equipped with a radio for communicating with cars.

1.1.2.1.6 Sharing the Drive

Partly as a result of all of these challenges, most stakeholders anticipate a "shared driving" concept will be used on the first commercially available AVs. Vehicles will drive autonomously in certain operating conditions, for example, below a particular speed, only on certain kinds of roads, in certain driving conditions, and will revert to traditional, manual driving outside those boundaries or at the request of a human driver.

To experience the greatest benefits of the technology, human drivers will need to be able to engage in other tasks while the vehicle is driving autonomously. For safety, however, they will need to quickly reengage (in a matter of seconds or less) at the vehicle's request. Cognitive science research on distracted driving suggests this may be a significant safety challenge. Similarly, developing the appropriate mental models for human–machine collaboration may be a challenge in creating a technology usable by the general public.

1.1.2.1.7 Integrity, Security, and Verification

Software upgrades might need to be backward-compatible with earlier models of vehicles and sensor systems. Moreover, as more vehicle models offer autonomous driving features, software and other system upgrades will have to perform on increasingly diverse platforms, making reliability and quality assurance even more challenging. System security is also a concern; viruses or malware must be prevented from subverting proper functioning of vehicles' systems.

State transportation departments may need to anticipate the use of vastly different kinds of AVs operating on roadways. This may pose challenges for the registration and requirements necessary for the vehicles to operate and for the level of training particular operators must have. One short-term action that might improve safety is requiring stricter conformance to road signage requirements, particularly those involving construction or some alteration to the roadway. This would aid human drivers and ease some of the perception requirements for AVs (Anderson et al., 2016).

1.1.3 Why Is Autonomous Vehicle Technology Important Now?

AV technology merits the immediate attention of policymakers for several reasons. First, the technology appears close to maturity and commercial introduction. Google's efforts—which involve a fleet of cars that collectively have logged hundreds of thousands of autonomous miles—have received widespread media attention and demonstrate this technology has advanced considerably. Every major commercial automaker is engaged in research in this area, and the full-scale commercial introduction of truly autonomous (including driverless) vehicles is predicted to occur within 5–20 years. Several states in the United States have passed laws to regulate the use of AVs, and many more laws have been proposed. As these technologies trickle (or flood) into the marketplace, it is important for policymakers to understand the effects that existing policy (or lack thereof) is likely to have on the development and adoption of this technology (Anderson et al., 2016).

Second, the stakes are high. In the United States alone, more than 30,000 people are killed each year in crashes, approximately 2.5 million are injured, and the vast majority of these crashes are the result of human error (Choi, Zhang, Young, Singh, & Chen, 2008). By reducing the opportunity for human error, AV technologies have the potential to reduce the number of crashes (Anderson et al., 2016).

AVs may reduce congestion and its associated costs; estimates suggest the effective road capacity (vehicles per lane per hour) can be doubled or tripled. The costs of congestion can be greatly reduced if vehicle operators can productively conduct other work. AV technology promises to reduce energy use as well; automobiles have become increasingly heavy over the past 20 years partly to meet more rigorous crash test standards. If crashes become exceedingly rare events, it may be possible to dramatically lighten automobiles.

In the long run, AVs may improve land use. Quite apart from the environmental toll of fuel generation and consumption, the existing automobile shapes much of our built environment. Its centrality to our lives accounts for the acres of parking in even our most densely occupied cities. With the ability to drive and park themselves at some distance from their users, AVs may obviate the need for nearby parking for commercial, residential, or work establishments, and this, in turn, may enable a reshaping of the urban environment and permit new in-fill development as adjacent parking lots become unnecessary.

Along with these benefits, however, AVs could have many negative effects. By reducing the time cost of driving, AVs may encourage greater travel and increase the total vehicle miles traveled (VMT), leading to more congestion. Urban sprawl may increase if commuters move ever farther away from workplaces. Similarly, AVs may eventually shift users' preferences towards larger vehicles to permit other activities. In theory, this could even include beds, showers, kitchens, or offices. If AV software becomes standardized, a single flaw might lead to many accidents. Internet-connected systems might be hacked by the malicious. And perhaps the biggest risks are simply unknowable (Anderson et al., 2016).

From seatbelts, to air bags, to antilock brakes, automakers have often been reluctant to incorporate expensive new technology, even if it can save many lives (Mashaw & Harfst, 1990). Navigating the AV landscape makes implementation of these earlier safety improvements appear simple by comparison. Negotiating the risks to reach the opportunities will require careful policymaking (Anderson et al., 2016).

1.1.4 Components of Autonomous Vehicles

The AV system can be divided into four main components, as shown in Figure 1.1. The vehicle senses the world using many different sensors mounted on it. These hardware components gather data about the environment. The information from the sensors is processed in a perception block and turned into meaningful information. A planning subsystem uses the output from the perception block to plan behavior and create short- and long-range plans. A control module ensures the vehicle follows the path provided by the planning subsystem and sends control commands to the vehicle (Kocić, Jovičić, & Drndarević, 2018).

The first fully autonomous vehicles were developed in 1984 at Carnegie Mellon University and in 1987 by Mercedes-Benz and the University of Munich. Since then, many companies and research organizations have developed prototypes of AVs and are intensively working on the development of full vehicle autonomy (Kocić, Jovičić, & Drndarević, 2018).

As mentioned previously, a significant event in AV development was the Defence Advanced Research Project Agency's (DARPA) Grand Challenge events in 2004 and 2005 (Thrun, 2006; Montemerlo, 2006) and Urban Challenge event in 2007 (Buehler, Iagnemma, & Singh, 2009). These events demonstrated machines could independently perform the complex human task of driving. In the 2007 DARPA Urban Challenge, six of 11 AVs in the finals successfully navigated an urban environment to reach the finish line, a landmark achievement in robotics (Kocić, Jovičić, & Drndarević, 2018).

Current challenges in AVs' development are scene perception, localization, mapping, vehicle control, trajectory optimization, and higher-level planning decisions. New trends in autonomous driving include end-to-end learning and reinforcement learning (Kocić, Jovičić, & Drndarević, 2018).

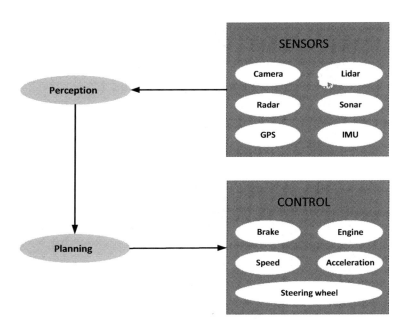

FIGURE 1.1 Block diagram of the AV system (Kocić, Jovičić, & Drndarević, 2018).

1.1.5 Autonomous Vehicle Applications

There has been a continuous and gradually increasing interest in developing autonomous ground vehicles for different applications. We classify the types of applications as:

1. Automated highway systems (AHS).
2. In-city and urban driving.
3. Off-road driving.
4. Specialty applications.

The fourth class includes a series of diverse applications and problems, for example, closed deployment environments, tasks requiring special motion, docking, and convoying (Ozguner & Redmill, 2008).

An example of an AHS is shown in Figure 1.2. An example of an off-road vehicle and ION, the intelligent off-road navigator, are shown in Figures 1.3a and b, respectively (Chen & Ozguner, 2006; Redmill, Martin, & Ozguner, 2006). Figure 1.3c shows a vehicle designed for city urban driving and developed for DARPA Urban Challenge 2007 (Ozguner & Redmill, 2008).

Studies on the development of AHS usually advocate an ingress–egress pairing, during which the car will follow the assigned lane on the highway. Early testing and demonstration implementations assumed cars would follow specialized technological aids indicating the precise location of the car with respect to the lane.

Figure 1.2 shows a technology advocated for location information with respect to the lane, a radar-reflecting stripe indicating the distance from the center of roadway, and the relative orientation of the car. When this technology was developed, GPS and precision maps were not commonly available. Today, it is assumed that precision maps will identify individual lanes and GPS reception will provide precise location information in real time (Ozguner & Redmill, 2008).

The cars shown in Figure 1.2 are from Demo'97, a test held on a 7.5-mile segment of Highway I-15 in San Diego. This segment was a segregated two-lane highway normally used for rush-hour high-occupancy vehicle traffic. Traffic flowed in the same direction in both lanes, and there were no intermediate entry and exit points. The curvature of the highway lanes was benign and suited for high-speed (70 mph) driving; other traffic was minimal to nonexistent. A general AHS would presumably have merge and exit lanes, but the single entry–exit aspect of Demo'97 made it a single activity: drive down the lane and possibly handle

Introduction

FIGURE 1.2 Two AVs in Demo'97 following a radar-reflecting stripe and undertaking a pass (Ozguner & Redmill, 2008).

FIGURE 1.3 (a) Off-road TerraMax at Grand Challenge 2004, (b) Off-road ION at Grand Challenge 2005, (c) Urban vehicle ACT at Urban Challenge 2007 (Ozguner & Redmill, 2008).

simple interactions with other vehicles. We call this behavior a meta-state. Dealing with interchanges produced by entry and exit lanes would require other meta-states (Ozguner & Redmill, 2008).

The DARPA Grand Challenges of 2004 and 2005 (mentioned above and shown in Figure 1.3a and b) were both off-road races. As such, the only behavior and, thus, the only meta-state required was following the path with obstacle avoidance from point A to point B. However, since there was no "path" or "lane" that could be discerned from a roadway, the only method of navigation was to rely on GPS and INS-based vehicle localization and a series of predefined "waypoints". Obstacle avoidance was needed, as in an AHS, although in the less structured off-road scenario, greater freedom of movement and deviations from the defined path are allowed. The Grand Challenge race rules ensured there were no moving obstacles, and different vehicles would not encounter each other in motion. General off-road driving would not have this constraint (Ozguner & Redmill, 2008).

Finally, fully autonomous urban driving introduces a significant number of meta-states, situations where different behavior is required, and different classes of decisions need to be made. The DARPA Urban Challenge (mentioned above), although quite complex, had fairly low speed limits, careful drivers, and no traffic lights. Visual lane markings were unreliable, and, thus, true to life, and the terrain was fairly flat, although some areas were unpaved, generating an unusual amount of dust and creating problems for some sensors.

Although "lanes" are obvious in highway systems and urban routes, it is reasonable to assume that off-road environments also present a set of constraints that indicate the drivability of different areas and, thus, provide the possibility of defining lanes. A sketch of a vehicle on a path with waypoints and lanes is shown in Figure 1.4 (Ozguner & Redmill, 2008).

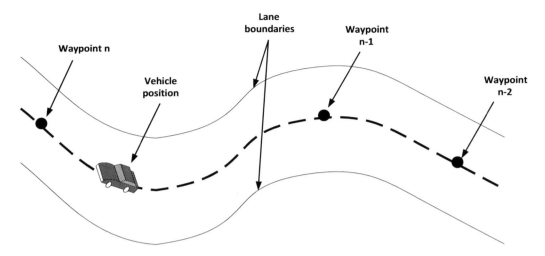

FIGURE 1.4 Roadway with waypoints and lane markings (Ozguner & Redmill K, 2008).

1.1.6 Architecture and Hierarchy Issues of Autonomous Vehicles

A generic functional architecture for an AV is given in Figure 1.5 (Ozguner & Redmill, 2008).

The "plan" can be simple or complex, but the overall configuration will cover all three application areas under consideration. Figures 1.6 and 1.7 show the details of the hardware architecture for AVs developed in 1996 and 2007, more than 10 years apart. Although some technologies have changed, and in spite of one being for AHS and the other for autonomous urban driving, the similarities between the two configurations are obvious (Ozguner & Redmill, 2008).

Note that the Demo'97 car does not have a GPS system and relies totally on infrastructure-based queues to find its position with respect to the roadway. The car has a special (stereo) radar system that senses the radar-reflective stripe it straddles on the lane and a vision system that senses the white lane markers on both sides of the lane. It senses other cars on the roadway via radar and a separate LiDAR unit.

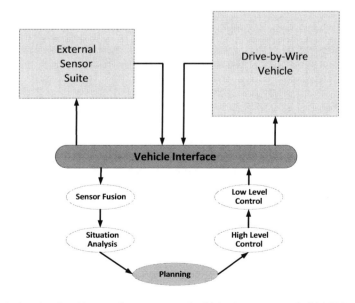

FIGURE 1.5 Generic functional architecture for an automated vehicle (Ozguner & Redmill, 2008).

Introduction

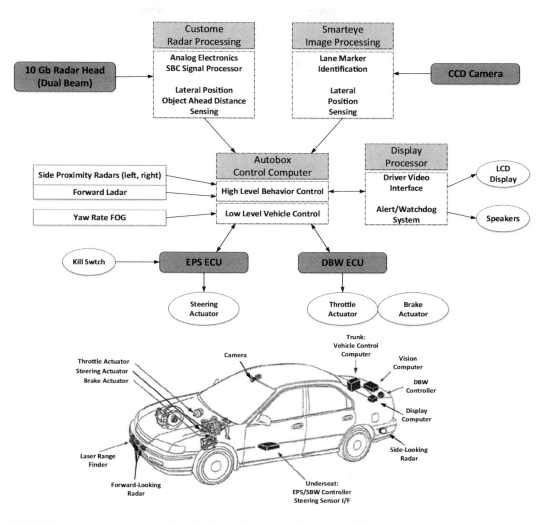

FIGURE 1.6 Architecture for an AV used in Demo'97 (Ozguner & Redmill, 2008).

In the Urban Challenge car developed by OSU, shown in Figure 1.7, the overall architecture is very similar. In this case, direct sensor-based lane detection is not fully integrated into the vehicle, although the sensor suite utilized would have allowed this. The vehicle relied on high-precision GPS signals and inertial and dead-reckoning positioning technologies (Ozguner & Redmill, 2008).

1.1.7 Potential Impacts of Autonomous Vehicles

AV operations are inherently different from human-driven vehicles. AVs can be programmed to not break traffic laws. They do not drink and drive. Their reaction times are quicker, and they can be optimized to smooth traffic flows, improve fuel economy, and reduce emissions. They can deliver freight and unlicensed travelers to their destinations. This section examines some of the largest potential benefits (Fagnant & Kockelman, 2015).

1.1.7.1 Safety

AVs have the potential to dramatically reduce crashes. Table 1.1 highlights the magnitude of automobile crashes in the United States and indicates sources of driver error that may disappear as vehicles become increasingly automated (Fagnant & Kockelman, 2015).

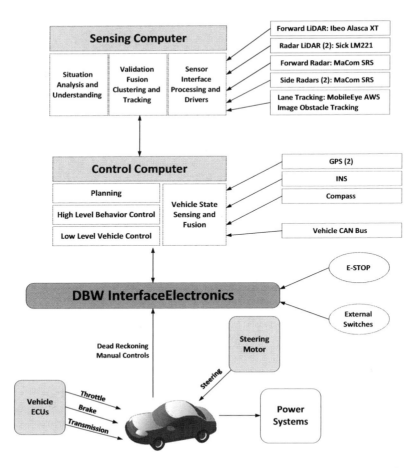

FIGURE 1.7 Architecture of ACT, an urban driving car used in DARPA's Urban Challenge (Ozguner & Redmill, 2008).

TABLE 1.1

US Crash Scope and Selected Human and Environmental Factor Involvement (Fagnant & Kockelman, 2015).

Total crashes per year in the United States	5.5 million
• % human cause as primary factor	93%
Economic costs of US crashes	$277 billion
• % of US GDP	2%
Total fatal and injurious crashes per year in the United States	2.22 million
Fatal crashes per year in the United States	32,367 million
• % of fatal crashes involving alcohol.	31%
• % involving speeding.	30%
• % involving distracted driver.	21%
• % involving failure to keep in proper lane.	14%
• % involving failure to yield right of way.	11%
• % involving wet road surface.	11%
• % involving erratic vehicle operation.	9%
• % involving inexperience or overcorrecting.	8%
• % involving drugs.	7%
• % involving ice, snow, debris, or other slippery surface.	3.7%
• % involving fatigued or sleeping driver.	2.5%
• % involving other prohibited driver errors (e.g., improper following, driving on shoulder, wrong side of road, improper turn, improper passing)	21%

Over 40% of these fatal crashes involve some combination of alcohol, distraction, drug involvement, and/or fatigue. Self-driven vehicles would not fall prey to human failings, suggesting the potential for at least a 40% fatal crash-rate reduction, assuming automated malfunctions are minimal and everything else remains constant (such as the levels of long-distance, night-time, and poor-weather driving). Such reductions do not reflect crashes due to speeding, aggressive driving, overcompensation, inexperience, slow reaction times, inattention, and various other driver shortcomings. Driver error is believed to be the main reason for over 90% of all crashes (National Highway Traffic Safety Administration, 2008). Even when the critical reason for a crash is attributed to the vehicle, roadway, or environment, additional human factors such as inattention, distraction, or speeding are regularly found to have contributed to the crash occurrence and/or injury severity (Fagnant & Kockelman, 2015).

The scope of potential benefits is substantial, both economically and politically. Over 30,000 persons die each year in the United States in automobile collisions (National Highway Traffic Safety Administration, 2012), with 2.2 million crashes resulting in injury (Traffic Safety Facts, 2013). Traffic crashes remain the primary reason for the death of Americans between 15 and 24 years of age (CDC, 2011).

At $277 billion, the annual economic cost of crashes is over double that of congestion and is highlighted as the number one transportation goal in the nation's federal legislation, Moving Ahead for Progress in the 21st century (MAP-21). These issues have long been top priorities of the US Department of Transportation's Strategic Plan.

While many driving situations are relatively easy for an AV to handle, designing a system that can perform safely in nearly every situation is challenging (Campbell, Egerstedt, How, & Murray, 2010). For example, recognition of humans and other objects in the roadway is critical but more difficult for AVs than human drivers (Dalal & Triggs, 2005; Economist Technology Quarterly, 2012; Farhadi, Endres, Hoiem, & Forsyth 2009). A person in a roadway may be small or large, standing, walking, sitting, lying down, riding a bike, and/or partly obscured – all of which complicate AV sensor recognition. Poor weather, such as fog and snow, and reflective road surfaces from rain and ice create other challenges for sensors and driving operations. Evasive decisions should depend on whether an object in the vehicle's path is a large cardboard box or a large concrete block, and computer vision has much greater difficulty than human vision in identifying material composition. When a crash is unavoidable, it is crucial that AVs recognize the objects in their path so they may act accordingly. Liability for these incidents is a major concern and could be a substantial impediment to implementation (Fagnant & Kockelman, 2015).

Ultimately, some analysts predict that AVs will overcome many of the obstacles that inhibit them from accurately responding in complex environments. Hayes (2011) suggests motor-vehicle fatality rates (per person-mile traveled) could eventually approach those seen in aviation and rail, about 1% of current rates. Some foresee the creation of "crash-less cars" (KPMG & CAR, 2012). However, drivers could take their vehicles out of self-driving mode and take control. Google's only reported AV crash occurred when a human driver was operating the vehicle. The rate at which human control is needed will be a substantial factor in the safety of these vehicles (Fagnant & Kockelman, 2015).

1.1.7.2 Congestion and Traffic Operations

Aside from making automobiles safer, researchers are developing ways for AV technology to reduce congestion and fuel consumption. For example, AVs can sense and possibly anticipate lead vehicles' braking and acceleration decisions. Such technology allows smoother braking and fine speed adjustments in following vehicles, leading to fuel savings, less brake wear, and reductions in traffic-destabilizing shockwave propagation. AVs are also expected to use existing lanes and intersections more efficiently with shorter gaps between vehicles, more coordinated platoons, and more efficient route choices. Many of these features, such as ACC, are already being integrated into automobiles, and some of the benefits will be realized before AVs are fully operational (Fagnant & Kockelman, 2015).

These benefits will not happen automatically. Many congestion-saving improvements depend not only on automated driving capabilities but also on cooperative abilities through vehicle-to-vehicle (V2V) and vehicle-to-infrastructure (V2I) communication. Vehicle communication is likely to become standard on most vehicles even before the significant proliferation of AV capabilities, with the US NHTSA announcing its intention to mandate all new light-duty vehicles to be equipped with V2X capabilities (National Highway Traffic Safety Administration, 2014). Even without V2X communication, significant congestion reduction could occur if the safety benefits alone are realized. The US Federal Highway Administration (FHWA) estimates that 25% of congestion is attributable to traffic incidents, around half of which are crashes (Federal Highway Administration, 2005).

Multiple studies have investigated the potential for AVs to reduce congestion under differing scenarios. Under various levels of AV adoption, congestion savings due to ACC measures and traffic monitoring systems could smooth traffic flows by minimizing accelerations and braking in freeway traffic. This could increase fuel economy and congested traffic speeds by 23%–39% and 8%–13%, respectively, for all vehicles in the freeway travel stream, depending on V2V communication and how traffic smoothing algorithms are implemented (Atiyeh, 2012). If vehicles are enabled to travel closer together, the system's fuel and congestion savings rise further, and some expect a significant increase in highway capacity on existing lanes (Tientrakool, 2011). Shladover and colleagues estimate cooperative ACC (CACC) deployed at 10%, 50%, and 90% market penetration levels will increase lanes' effective capacities by around 1%, 21%, and 80%, respectively (Shladover, Su, & Lu, 2012). Gap reductions, coupled with near-constant velocities, will produce more reliable travel times—an important factor in trip generation, timing, and routing decisions. Similarly, shorter headways between vehicles at traffic signals (and shorter startup times) mean more AVs could more effectively utilize green time at signals, considerably improving intersection capacities (Fagnant & Kockelman, 2015).

Over the long term, new paradigms for signal control, such as autonomous intersection management, could use AVs' powerful capabilities. Some evidence shows advanced systems could nearly eliminate intersection delays while reducing fuel consumption, though this concept is only theoretical and certainly a long way off. To implement such technologies, Dresner and Stone estimate a 95% or more AV market penetration may be required, leaving many years before deployment (Dresner & Stone, 2008).

Of course, many such benefits may not be realized until high numbers of AVs are present on the roads. For example, if 10% of all vehicles on a given freeway segment are AVs, there will likely be an AV in every lane at regular spacing during congested times, and this could smooth traffic for all travelers (Bose & Ioannou, 2003). However, if just 1 out of 200 vehicles is an AV, the impact would be nonexistent or greatly lessened. If one AV is following another, the following AV can reduce the gap between the two vehicles, increasing effective roadway capacity. This efficiency benefit is also contingent on higher AV shares. Technical and implementation challenges must be met to realize the full potential of high usage, including the implementation of cloud-based systems and city or region-wide coordinated vehicle-routing paradigms and protocols. Finally, while AVs have the potential to increase roadway capacity with higher market penetration, the induced demand resulting from more automobile use might require additional capacity needs (Fagnant & Kockelman, 2015).

1.1.7.3 Travel Behavior Impacts

The safety and congestion-reducing impacts of AVs have the potential to create significant changes in travel behavior. For example, AVs may provide mobility for those too young to drive, the elderly, and the disabled, thus generating new roadway capacity demands. Parking patterns could change as AVs self-park in less expensive areas. Car- and ride-sharing programs could expand, as AVs serve multiple persons on demand.

Most of these ideas point towards more VMT and automobile-oriented development, though perhaps with fewer vehicles and parking spaces. Added VMT may bring other problems related to high automobile use, such as increased emissions, greater gasoline consumption and oil dependence, and higher obesity rates (Fagnant & Kockelman, 2015).

As of July 2014, state legislation in California, Florida, Michigan, Nevada, and Washington, D.C. mandated that all drivers pursuing AV testing on public roadways be licensed and prepared to take over vehicle operation, if required. As AV experience increases, this requirement could be relaxed, and AVs may be permitted to legally chauffeur children and persons who otherwise would be unable to safely drive. Such mobility may be increasingly beneficial, as the US population ages, with 40 million Americans presently over the age of 65 and this demographic growing at a 50% faster rate than the nation's overall population (US Census Bureau, 2011). Wood observes that many drivers attempt to cope with physical limitations through self-regulation, avoiding heavy traffic, unfamiliar roads, night-time driving, and poor weather, while others stop driving altogether (Wood, 2002). AVs could facilitate personal independence and mobility, while enhancing safety, thus increasing the demand for automobile travel (Fagnant & Kockelman, 2015).

With increased mobility among the elderly and others, as well as lowered travel effort and congestion delays, the United States can expect VMT increases, along with associated congestion, emissions, and crash rates, unless demand management strategies are thoughtfully implemented (Kockelman & Kalmanje, 2006; Litman, 2013).

However, AV benefits could exceed the negative impacts of added VMT. For example, if VMT were to double, a reduction in crash rates per mile traveled by 90% yields a reduction in the total number of crashes and their associated injuries and traffic delays by 80%. Likewise, unless new travel from AV use is significantly underestimated, the existing infrastructure capacity on roadways should be adequate to accommodate the new/induced demand, thanks to AVs' congestion-mitigating features, like traffic smoothing algorithms (Atiyeh, 2012) and effective capacity increases through CACC (Shladover, Su, & Lu, 2012), as well as public infrastructure investments, like V2I communication systems with traffic signals (KPMG & CAR, 2012), designed to support these capabilities. However, other negative impacts, such as sprawl, emissions, and health concerns, may not be readily mitigated (Fagnant & Kockelman, 2015).

It is possible that already congested traffic patterns and other roadway infrastructure will be negatively affected, because of increased trip making. Indeed, Smith argues, "Highways may carry significantly more vehicles, but average delay during the peak period may not decrease appreciably. Similarly, emissions per vehicle mile traveled may decrease, but total emissions (throughout the day) may actually increase" (Smith, 2013). However, AVs could enable smarter routing in coordination with intelligent infrastructure, quicker reaction times, and closer spacing between vehicles to counteract increased demand.

Whether arterial congestion improves or degrades ultimately depends on how much induced VMT is realized, the relative magnitude of AV benefits, and the use of demand management strategies, such as road pricing. Emissions are predicted to fall when travel is smooth, rather than forced, with Berry saying a 20% reduction in accelerations and decelerations should lead to a 5% reduction in fuel consumption and associated emissions (Berry, 2010). Thus, while AVs may increase VMT, emissions per mile could be reduced (Fagnant & Kockelman, 2015).

Additional fuel savings may accrue through AVs' smart parking decisions (Bullis, 2011; Shoup, 2005), helping avoid "cruising for parking." For example, in-vehicle systems could communicate with parking infrastructure to enable driverless drop-offs and pickups. This same technology could improve and expand car sharing and dynamic ride-sharing by allowing nearby, real-time rentals on a per-minute or per-mile basis. If successful, this has great promise for program expansions, since users could simply order a vehicle online or using mobile devices, much like an on-demand taxi, to take them to their destinations. Preliminary results (Fagnant & Kockelman, 2016) for Austin, Texas, using an agent-based model for assigning vehicles around a core region indicate that each shared AV (SAV) could replace around ten privately owned or household-owned vehicles. These simulations assumed that the SAVs operated within a prescribed 12 miles by 24 miles geofence, where trip intensity is relatively high; longer trips to or from destinations outside the geofence were not considered (Fagnant & Kockelman, 2015).

As shown in Figure 1.8, even in Seattle where vehicle use is more intense than national averages (Puget Sound Regional Council, 2006), just under 11% of vehicles are "in use" throughout the day, even at peak times, though usage rises to 16% if only newer vehicles are monitored (Fagnant & Kockelman, 2015).

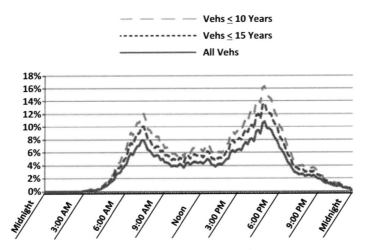

FIGURE 1.8 Vehicle use by time of day and by vehicle age (Puget Sound Regional Council, 2006).

1.1.7.4 Freight Transportation

Freight transport on and off the road will be impacted by autonomous driving. As one example, mining company Rio Tinto is already using 53 self-driving ore trucks; these trucks have driven 2.4 million miles and carried 200 million tons of materials (Rio Tinto, 2014). The same technologies that apply to autonomous cars can apply to the trucking industry, increasing fuel economy and lowering the need for truck drivers. While workers would likely still need to load and unload cargo, long-distance journeys may be made without drivers, with warehousing employees handling container contents at either end. Autonomously operated trucks may face significant resistance from labor groups, like the Teamsters, and competing industries, such as the freight railroad industry (Fagnant & Kockelman, 2015).

Additional benefits can emerge through higher fuel economies when using tightly coupled road-train platoons, thanks to reduced air resistance of shared slipstreams, not to mention lowered travel times from higher capacity networks (a result of shorter headways and less incident-prone traffic conditions). Bullis estimates that 4 m inter-truck spacings could reduce fuel consumption by 10%–15%, and road-train platoons would facilitate adaptive braking, potentially enabling further fuel savings (Bullis, 2011). Kunze and colleagues did a successful trial run using 10 m headways between multiple trucks on public German motorways (Kunze, Ramakers, Henning, & Jeschke, 2009), and a variety of autonomously platooned Volvo trucks recently logged approximately 10,000 km along Spanish highways (Newcomb, 2012). However, tight vehicle spacing on roads could cause problems for other motorists trying to exit or enter highways, possibly resulting in the need for new or modified infrastructure with dedicated platoon lanes and thicker pavements to handle high truck volumes (Fagnant & Kockelman, 2015).

1.1.7.5 Anticipating AV Impacts

Since AVs are only in the testing phase, it is difficult to precisely anticipate actual outcomes. Nevertheless, it can be useful to roughly estimate likely magnitudes of impact. Based on research estimates for the potential impacts discussed above, this section quantifies crash, congestion, and other impacts for the US transportation system (including changes in parking provision, VMT, and vehicle counts). To understand how AVs' assimilation into the road network might work, multiple assumptions are needed and explained below. To further understand the impact, we assume three AV market penetration shares: 10%, 50%, and 90%. These not only represent market shares but also technological improvements over time, since it could take many years for the United States to see high penetration rates. The analysis is inherently imprecise, as it provides an order-of-magnitude estimate of the broad economic and safety impacts this technology may have.

Introduction   17

We assume the primary benefits for AV use will include safety benefits, congestion reduction (comprised of travel time savings and fuel savings), and savings realized from reduced parking demands, particularly in areas with high parking costs. The assumptions driving these estimated impacts are discussed in this section, as are the assumptions used to estimate changes in VMT, to estimate AV technology costs, and to select an appropriate discount rate for net present value (NPV) calculations (Fagnant & Kockelman, 2015).

1.1.7.6 Changes in VMT and Vehicle Ownership

VMT per AV is assumed to be 20% higher than that of non-AV at the 10% market penetration rate and 10% higher at the 90% market penetration rate. This reflects the notion that early adopters will have more pent-up demand for such vehicles than later buyers.

Fagnant and Kockelman's preliminary agent-based simulations (Fagnant & Kockelman, 2016) underscore this idea. For the Austin, Texas, market, a fleet of SAVs serving over 56,000 trips a day was found to travel 8.7% of its mileage unoccupied (empty). This figure fell to 4.5% when ride-sharing was permitted, and minor (less than 1%) net VMT reductions were realized when demand rose by a factor of 5 and ride-sharing was permitted. Analysis of the various simulation results suggests each SAV could serve the same number of trips as 10 household-owned vehicles (if all replaced travel were to lie within a 12 mile × 24 mile geofence). While the 10^{-1} replacement rate may be too high for mass adoption settings, especially in locations with lots of long-distance trip making and low-density development, 10 household vehicles are assumed to be replaced here for every SAV operated (10% of the fleet), resulting in the implicit assumption that around half of all AV trips will be served by SAVs and the other half by personally owned AVs (Fagnant & Kockelman, 2015).

Additional VMT increases may be realized from induced demand, as travel costs and congestion fall. In his review of literature spanning 30 years in California and across the United States, Cervero shows that the long-term (6 years or more) urban area elasticity of VMT (demand for road travel) with respect to the number of highway lane-miles supplied ranges from around 0.47 to 1.0, averaging 0.74 (Cervero, 2001). This suggests that if a region's lane-miles increase by 1%, regional VMT may rise by 0.74% over the long term, after controlling for population, income, and other factors. If tolls and/or other traffic management policies are put in place to stem excessive demand, demand elasticity should be lower. Of course, a 0.74 elasticity value is likely high, since AVs' capacity effects are probably uniform, rather than targeted. Many road segments in a region are not currently congested and do not exhibit latent or elastic demand. Therefore, if we assume a 0.37 elasticity, system-wide VMT may be expected to rise 26% under 90% AV market penetration assumptions (i.e., 60% freeway congestion reduction and 15% arterial congestion reduction, due to an increase in effective capacity) (Fagnant & Kockelman, 2015).

While the congestion-relieving impacts of AVs may be similar to those of adding lane-miles, they differ in another crucial respect beyond their uniform versus targeted capacity increases, as noted above. Personal values of travel time (VOTT) may also fall because of drivers' increased productivity gains as they are freed for purposes other than driving. Gucwa attempted to estimate the joint implications of increased travel due to capacity and value of travel time changes using simulations of the San Francisco Bay Area. When increasing roadway capacity between 10% and 100%, and simultaneously reducing the VOTT from current levels to somewhere between high quality rail and half of current (in-car) values of time, his model results produced a 4%–8% increase in VMT region-wide, because of changes in destination and mode choices (Gucwa, 2014). Of course, AVs may also travel while unoccupied, and long-term housing and employment shifts may generate extra VMT (Fagnant & Kockelman, 2015).

Cervero's (2001) framework (with halved elasticity values) and Gucwa's (2014) simulations produced two different VMT outcomes that may represent the respective high and low ranges of reasonable VMT growth scenarios. Therefore, we select 20% and 10% increases in VMT per AV as assumptions for the 10% and 90% AV market penetration rates, respectively, reflecting reasonable estimates within these bounds. These VMT increases are expected to apply system-wide, across personally owned AVs, SAVs and AVs used for shipping and freight (Fagnant & Kockelman, 2015).

1.1.7.7 Discount Rate and Technology Costs

For NPV calculations, a 10% discount rate is assumed; this is higher than the 7% rate required by the US Office of Management and Budget (OMB) for federal projects and TIGER grant applications (LaHood, 2011). We do so to reflect the uncertainty of this emerging technology. Early-introduction costs (perhaps seven years after initial rollout) at the 10% market penetration level are assumed to add $10,000 to the purchase price of a new vehicle, falling to $3,000 by the 90% market penetration share. Internal rates of return for initial costs are included at the $37,500 level, and this may be closer to the added price of AV technologies, a couple of years after they are first introduced (Fagnant & Kockelman, 2015).

1.1.7.8 Safety Impacts

US crash rates for non-AVs are assumed constant, based on NHTSA's 2011 values, and the severity distribution of all crashes remains unchanged from the present. As noted previously, over 90% of the primary factors in crashes are human errors (National Highway Traffic Safety Administration, 2008), and 40% of fatal crashes involve driver alcohol or drug use, driver distraction, and/or fatigue (National Highway Traffic Safety Administration, 2012).

Therefore, AVs may be assumed to reduce crash and injury rates by 50%, versus non-AVs at the early 10% market penetration rate (reflecting savings from eliminating the aforementioned factors, as well as fewer legal violations like running red lights), and 90% safer at the 90% market penetration rate (reflecting the near elimination of human error as a primary crash cause, thanks to improved vehicle automation technology).

Pedestrian and bicycle crashes (with motor vehicles) are assumed to enjoy half of the AV safety benefits, since just one of the two crash parties (the driver) relies on the AV technology. Similarly, motorcycles may not enjoy autonomous status for a long time (and their riders may be reluctant to relinquish control), and around half of all fatal motorcycle crashes do not involve another vehicle. Therefore, motorcycles are assumed to experience just a 25% decline in their crash rates, relative to the declines experienced by other motor vehicles.

While safety improvements will likely be greater than new safety risks, it is possible that new risks will be greater for some system users under certain circumstances, particularly at early technology stages. Lin argues that increased safety to some users at the expense of others is not necessarily a clear-cut benefit, even if net safety risks to the whole population are lower (Lin, 2013).

In the following calculations, crash costs are estimated first based on their economic consequences, using National Safety Council (2012) guidance, and then on higher comprehensive costs, as recommended by the USDOT (Trottenberg, 2011), to reflect pain and suffering and the full value of a statistical life (Fagnant & Kockelman, 2015).

1.1.7.9 Congestion Reduction

Shrank and Lomax's congestion impact projections (Schrank, Eisele, & Lomax, 2012) for 2020 are used in what follows as a baseline. They assume a $17 per person-hour value of travel time, $87 per truck-hour value of travel time, and state-wide average gas prices in 2010. They estimate 40% of the nation's roadway congestion occurs on freeway facilities (with the remainder on other streets). By 2020, US travelers will experience around 8.4 billion hours of delay while wasting 4.5 billion gallons' of fuel (due to congestion), for an annual economic cost of $199 billion (Fagnant & Kockelman, 2015).

This analysis assumes AVs are equipped with CACC and traffic-flow-smoothing capabilities. At the 10% AV market penetration level, freeway congestion delays for all vehicles are estimated to fall 15%, mostly due to smoothed flow and bottleneck reductions. This is lower than Atiyeh suggests so that induced travel, though additional congestion benefits, may be realized (fewer crashes, a small degree of increased capacity from CACC, and smarter vehicle routings) (Atiyeh, 2012). At the 50% market penetration level, a cloud-based system is assumed to be active; Atiyeh suggests 39% congestion improvements from smoothed flow. Further capacity enhancements of 20% may also be realized. With crashes falling because of safety improvements, another 4.5% in congestion reduction may be obtained. Again, induced travel will counteract some of these benefits, and a 35% delay reduction on freeways is estimated in this analysis. Finally, at the 90% level, freeway congestion is assumed to fall by 60%, with the near doubling of roadway capacity (Shladover, Su, & Lu, 2012) and dramatic crash reductions. However, capacity and delay are not linearly related, and congestion abatement may be even greater than these predictions at 90% market penetration (Fagnant & Kockelman, 2015).

At the arterial-roadway level, congestion is assumed to realize fewer benefits from AVs (without near-complete market penetration and automated intersection management (Dresner & Stone, 2008), as delays emerge largely from conflicting turning movements, pedestrians, and other transportation features that AV technologies cannot address so easily. Therefore, arterial congestion benefits are assumed to be just 5% at the 10% market-penetration level, 10% at 50% penetration, and 15% at 90% penetration. AV fuel efficiency benefits are assumed to begin at 13%, increasing to 25% with 90% market penetration, because of better route choices, less congestion, road-train drag reductions (from drafting), and more optimal drive cycles. Non-AVs on freeways are assumed to experience 8% fuel economy benefits during congested times of day under a 10% market penetration, and 13% at the 50% and 90% penetration levels. For simplicity, this analysis assumes all induced travel's added fuel consumption will be fully offset by AVs' fuel savings during non-congested times of day (Fagnant & Kockelman, 2015).

1.1.7.10 Parking

Parking savings comprise the final monetized component. Litman estimates that comprehensive (land, construction, maintenance and operation) annual parking costs are roughly $3,300 to $5,600 per parking space in central business districts (CBDs), $1,400 to $3,700 per parking space in other central/urban areas, and $680 to $2,400 per space in suburban locations (Litman, 2012). Simply moving a parking space outside the CBD may save nearly $2,000 in annualized costs, while moving one to a suburban location may save another $1,000. In addition, fewer overall spaces should be needed, thanks to car sharing. Therefore, while not every AV will result in a moved or eliminated parking space, $250 in parking savings will be realized per new AV, following the earlier assumption of 10% of AVs being publicly shared (Fagnant & Kockelman, 2015).

1.1.7.11 Summary of Economic Impacts

Table 1.2 summarizes all these estimated impacts. It suggests economic benefits will reach $196 billion ($442 billion, comprehensive) with a 90% AV market penetration rate. Meaningful congestion benefits are estimated to accrue to all travelers early on, while the magnitude of crash benefits grows over time (and accrues largely to AV owners/users). For example, congestion savings represent 66% of benefits, and crash savings represent 21% of benefits—at the 10% market penetration level, versus 31% and 54% of benefits, respectively, at the 90% penetration rate. When comprehensive crash costs are included, overall crash savings jump by more than a factor of three (Fagnant & Kockelman, 2015).

These results are consistent with the findings of Manyika, Chui, Bughin, Dobbs, Bisson, and Marrs (2013). These authors estimate global AV impacts of $200 billion to $1.9 trillion by 2025, assuming 5%–20% of all driving is either autonomous or semiautonomous and valuing the lowered burdens of in-vehicle travel time, at least for drivers, who can now perform other activities in route. If the 10% market penetration estimates used here are scaled globally (at least within the developed world), and the lowered burden of in-vehicle time is added, overall economic benefits are likely to fall to within Manyika et al.'s range (Fagnant & Kockelman, 2015).

Additional monetized congestion benefits may be realized beyond the values shown in Table 1.2, with falling VOTT. For example, an hour stuck driving in traffic may be perceived as more onerous than an hour spent being driven by an AV (Fagnant & Kockelman, 2015).

1.1.7.12 Privately Realized Benefits

While Table 1.2 illuminates AVs' social benefits, it is also important to anticipate the privately realized benefits of AV ownership and use. These benefits are assessed using the assumptions in Table 1.2 at the 10% market penetration and a $10,000 added purchase price, taking into account monetary savings from

TABLE 1.2

Estimates of Annual Economic Benefits from AVs in the United States (Fagnant & Kockelman, 2015)

	Assumed Market Shares		
	10%	50%	90%
Crash Cost Savings from AVs			
Lives saved (per year)	1,100	9,600	21,700
Fewer crashes	211,000	1,880,000	4,220,000
Economic cost savings	$5.5 B	$48.8 B	$109.7 B
Comprehensive cost savings	$17.7 B	$158.1 B	$355.4 B
Economic cost savings per AV	$430	$770	$960
Comprehensive cost savings per AV	$1,390	$2,480	$3,100
Congestion Benefits			
Travel time savings (M hours)	756	1,680	2,772
Fuel savings (M gallons)	102	224	724
Total savings	$16.8	$37.4	$63.0
Savings per AV	$1,320	$590	$550
Other AV Impacts			
Parking savings	$3.2	$15.9	$28.7
Savings per AV	$250	$250	$250
VMT increase	2.0%	7.5%	9.0%
Change in total # vehicles	−4.7%	−217%	−42.6%
Annual savings: Economic costs only	$25.5 B	$102.2 B	$201.4 B
Annual savings: Comprehensive costs	$37.7 B	$211.5 B	$447.1 B
Annual savings per AV: Economic costs only	$2,000	$1,610	$1,760
Annual savings per AV: Comprehensive costs	$2,960	$3,320	$3,900
NPV of AV benefits minus added purchase price: Economic costs only	$5,210	$7,250	$10,390
NPV of AV benefits minus added purchase price: Comprehensive costs	$12,510	$20,250	$26,660
Assumptions			
Number of AVs operating in the United States	12.0 M	45.1 M	65.1 M
Crash reduction fraction per AV	0.5	0.75	0.9
Freeway congestion benefit (delay reduction)	15%	35%	60%
Arterial congestion benefit	5%	10%	15%
Fuel savings	13%	18%	25%
Non-AV foil owing vehicle fuel efficiency benefit (freeway)	8%	13%	13%
VMT increase per AV	20%	15%	10%
% of AVs shared across users	10%	10%	10%
Added purchase price for AV capabilities	$10,000	$5,000	$3,000
Discount rate	10%	10%	10%
Vehicle lifetime (years)	15	15	15

reduced fuel use and insurance, along with several levels of daily parking savings and (hourly) travel time savings.

Privately realized benefits are estimated using assumptions of a $10,000 purchase price provided in Table 1.2. These are first compared to 50% insurance cost savings from a base of $1,000 per year and 13% fuel savings from a base of $2,400 per year (American Automobile Association, 2012) over a 15-year vehicle life. Parking costs of $250 are added, representing about $1 per work day. Finally, driven time under autonomous operation is added under $1 per hour and $5 per hour assumptions, with total annual vehicle hours traveled estimated based on US average VMT (10,600 miles per year) divided by an assumed

Introduction

TABLE 1.3

AV Owners' Privately Realized Internal Rates of Return (From 0% to 10% Market Share) (Fagnant & Kockelman, 2015)

Development Stage	Estimated Added Costs	\$0 and \$0	\$0 and \$1	\$1 and \$1	\$5 and \$1	\$1 and \$5	\$5 and \$5	\$5 and \$10	\$10 and \$10
Current	\$100k+	−19	−17	−15	−11	−9	−6	−2	0
Initial price	\$37.5k	−12	−8	−6	0	2	6	12	16
Mass production	\$10k	3	8	11	23	28	38	56	68

Benefits (Daily Parking and Hourly Value of Travel Time Savings) (%)

average speed of 30 mph (Federal Highway Administration, 2013). Privately realized internal rates of return are also compared to a higher added-technology price, $37,500 (Fagnant & Kockelman, 2015).

This results in the range of benefits shown in Table 1.3, across various purchase prices, values of time, and parking costs. At current high-technology costs of $100,000 or more, benefits are mostly small compared to purchase prices, except for individuals with very high values of time. Once prices come down to $37,500, persons with high VOTT and/or parking costs may find the technology a worthwhile investment. Only at the $10,000 added price does the technology become a realistic investment for many, with even the $1 per hour time value savings and $1 daily parking cost savings generating an 11% rate of return for AV owners (Fagnant & Kockelman, 2015).

We are not attempting to quantify or monetize several other possible impacts. For example, many of the nation's 240,000 taxi drivers and 1.6 million truck drivers (Bureau of Labor Statistics, 2012) could be displaced by AV technologies, while greenhouse gas emissions, infrastructure needs, and rates of walking may fall or rise, depending on the induced VMT. Increased sprawl or automobile-style development could also result, as projected by Laberteaux (2014). Such impacts are not included in the analysis (Fagnant & Kockelman, 2015).

While exact magnitudes of all impacts remain uncertain, this analysis illustrates the potential for AVs to deliver substantial benefits to many, if not all, Americans, thanks to sizable safety and congestion savings. Even at 10% market penetration, this technology has the potential to save over 1,000 lives per year and offer tens of billions of dollars in economic gains, once added vehicle costs and possible roadside hardware and system administration costs are covered (Fagnant & Kockelman, 2015).

1.2 Industrial Assets

Industrial assets are defined in a narrow sense as equipment deployed by the industry to convert inputs to outputs which can be then marketed by the industry as goods or services.

Industries in the fields of mining (minerals, diamonds etc.), transportation (space, airlines, bus, railways, shipping, automobiles, trucking etc.), manufacturing (automobiles, tools, equipment, instruments, electronics etc.), construction (roads, buildings, stadiums, dams, bridges etc.), and production (chemicals, metals, pharmaceuticals, petroleum etc.) can be considered to have industrial assets.

Industrial assets can be specialized, tailor-made for specific operations, or used in multiple industries. Table 1.4 gives some examples of industrial assets in different types of industries for both specific and general purposes (Syamsundar, 2017).

The table includes mechanical, electrical, structural, civil, electronic, computer, instrumentation, and telecommunication systems. A single system can contain subsystems, assemblies, subassemblies, and components. A typical example of industrial equipment along with its subsystems, assemblies, and components is given in Table 1.5 (Syamsundar, 2017).

TABLE 1.4

Examples of Industrial Assets (Syamsundar, 2017)

Si. No.	Industry	Some Equipment
1	Mining	Excavators, loaders, dumpers, draglines, shovels, rigs
2	Manufacturing	Lathes, milling, drilling, boring, shaping machines
3	Power plant	Generators, boilers, turbines, coal handling, transportation, and disposal equipment
4	Power transmission	Cabling, transmission line towers, transformers
5	Iron and steel industry	Raw materials handling, coke ovens, sintering machines, pelletizing machines, blast furnaces, steel making facilities, casting facilities, rolling facilities
6	Transport	Aircraft, ships, trains, buses, automobiles, bikes, trucks
7	Defense	Armor, tanks, battleships, aircraft
8	Communication networks	Cables, nodes, hardware, software

TABLE 1.5

Industrial Equipment for Electric Overhead Traveling Crane (Syamsundar, 2017)

Si. No.	Systems	Subsystems	Assemblies/Components
1	Mechanical	Long Travel Assembly	Gearbox, Brake, Drive Wheel Assembly, Non-Drive Wheel Assembly, Floating Shaft Assembly
		Cross Travel Assembly	Gearbox, Brake, Drive Wheel Assembly, Non-Drive Wheel Assembly, Floating Shaft Assembly
		Main Hoist Assembly	Drum Assembly, Brake Assembly, Hook Assembly, Ropes
		Auxiliary Hoist Assembly	Drum Assembly, Brake Assembly, Hook Assembly, Ropes
2	Electrical	LT Electrics, CT Electrics, MH Electrics, Current Collector Assembly	Motors, Cables, Electrical Panels, Switches
3	Structural	LT Girder Assembly, CT Trolley Assembly, MH Platform Assembly	Long Travel Girder, Cross Travel Frame, End Carriage, Balancer Body

1.2.1 Current Industrial Practice

Engineering assets can be divided into three categories (Seneviratne, Ciani, Catelani, & Galar, 2018):

1. Nonlinear assets: These can be further classified as component-based assets, which include production assets such as machines, mobile assets such as vehicles, and fixed physical assets such as manufacturing facilities (De Francesco & Leccese, 2012).
2. Linear assets: These can be defined as engineering structures or infrastructures that often cross a long distance and can be divided into different segments that perform the same function but are subjected to different loads and conditions.
3. Hybrid systems: These comprise a combination of linear and nonlinear assets.

The differences between the asset categories are summarized in Table 1.6.

1.2.2 Types of Assets

An asset is a resource owned or controlled by an individual, corporation, or government with the expectation that it will generate future cash flows. Common asset categories include: current, noncurrent, physical, intangible, operating, and nonoperating. Correctly identifying and classifying assets is critical to the survival of a company, specifically its solvency and associated risks (Corporate Finance Institute (CFI)).

The International Financial Reporting Standards (IFRS) framework defines an asset as follows: "An asset is a resource controlled by the enterprise as a result of past events and from which future economic benefits are expected to flow to the enterprise" (CFI). The main types are explained below.

Introduction 23

TABLE 1.6

Three Asset Categories (Seneviratne, Ciani, Catelani, & Galar, 2018)

Category	Example	System Configuration	Characteristics
Linear assets	Roads, railway tracks, pipelines, power cables, canals, and waterways	Tree structure or networks	• Maintained and renewed in place and in segments • No clear physical boundary for segments • All segments normally perform the same function • Maintenance costs, failures, and maintenance and operational events are associated with segments • Segments can be dynamic and are usually long lived
Nonlinear assets	Pumps, cars, machine tools	Complex physical structure	• Installed and replaced as a whole but maintained at base maintainable unit (BMU) level, often in workshops • BMU has clear physical boundaries • BMUs often play different functions • Maintenance cost, failures, and maintenance and operational events are associated with BMUs • BMUs are usually static • Lifetime varies
Hybrid assets	Power plant boilers, refrigerators, refineries	Complex physical structure, but linear and nonlinear parts have a clear boundary	• Installed, maintained, and replaced in parts for large assets and as a whole for small assets • Linear subsystems and nonlinear subsystems can often be separated

1.2.2.1 Inventory

Inventory is a current asset account found on the balance sheet. It consists of all raw materials, work in progress, and finished goods that a company has accumulated. It is often deemed the most illiquid of all current assets and, thus, is excluded from the numerator in the quick ratio calculation.

1.2.2.2 Property, Plant, and Equipment (PP&E)

Property, plant, and equipment (PP&E) is a noncurrent, tangible capital asset shown on the balance sheet of a business and used to generate revenues and profits. PP&E plays a key part in the financial planning and analysis of a company's operations and future expenditures, especially with regard to capital expenditures.

The PP&E account is often denoted as net of accumulated depreciation. This means that if a company does not purchase additional new equipment (therefore, its capital expenditures are zero), its net PP&E should slowly decrease in value every year because of depreciation. This can be better determined by a depreciation schedule.

PP&E is a tangible fixed-asset account item and is generally very illiquid. A company can sell its equipment but not as easily as it can sell its inventory or investments such as bonds or stock shares. The value of PP&E for different companies will vary with the operations. For example, a construction company will generally have a significantly higher PP&E balance than an accounting firm.

1.2.2.2.1 What Classifies as Property, Plant, and Equipment?

PP&E basically includes any of a company's long-term, fixed assets. PP&E assets are tangible, identifiable, and expected to generate an economic return for the company for more than one year or one operating cycle (whichever is longer). The account can include machinery, equipment, vehicles, buildings, land, office space, office equipment, and furnishings, among other things. Of all these asset classes, land does not typically depreciate over time.

If a company produces machinery (for sale), that machinery does not classify as PP&E. The machinery used to produce the machinery for sales is PP&E, but the machinery manufactured for sale is classified as inventory. The same goes for real-estate companies that hold building and land under their assets. Their office buildings and land are PP&E, but the houses they sell are inventory.

There is an interplay between the inventory account and the cost of goods sold in the income statement.

1.2.3 Properties of an Asset

There are three key properties of an asset (CFI):

- Ownership: Assets represent ownership and can be eventually turned into cash and cash equivalents for those owners.
- Economic value: Assets have economic value and can be exchanged or sold.
- Resource: Assets are resources that can be used to generate future economic benefits.

1.2.4 Classification of Assets

Assets are generally classified in three ways (CFI):

- Convertibility: Classifying assets based on how easy it is to convert them into cash.
- Physical existence: Classifying assets based on their physical existence.
- Usage: Classifying assets based on their business operation usage (Figure 1.9) (Corporate Finance Institute (CFI)).

1.2.5 Classification of Assets: Convertibility

If assets are classified based on their convertibility into cash, they are classified as either current or fixed assets (CFI).

1. Current assets

    Current assets are assets that can be easily converted into cash and cash equivalents (typically within a year). Current assets are also termed liquid assets. Examples include:
    - Cash
    - Cash equivalents
    - Short-term deposits
    - Stock
    - Marketable securities
    - Office supplies.

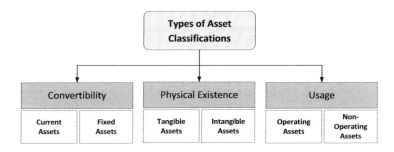

FIGURE 1.9 Types of asset classifications (CFI).

2. Fixed or noncurrent assets
Noncurrent assets are assets that cannot be easily and readily converted into cash and cash equivalents. Noncurrent assets are also termed fixed assets, long-term assets, or hard assets. Examples include:
- Land
- Building
- Machinery
- Equipment
- Patents
- Trademarks.

1.2.6 Classification of Assets: Physical Existence

If assets are classified based on their physical existence, they are classified as either tangible or intangible assets (CFI).

1. Tangible assets
Tangible assets are assets that have a physical existence (we can touch, feel, and see them). Examples of tangible assets include:
- Land
- Building
- Machinery
- Equipment
- Cash
- Office supplies
- Stock
- Marketable securities.

2. Intangible assets
Intangible assets are assets that do not have a physical existence. Examples of intangible assets include:
- Goodwill
- Patents
- Brand
- Copyrights
- Trademarks
- Trade secrets
- Permits
- Corporate intellectual property.

1.2.7 Classification of Assets: Usage

If assets are classified based on their operational usage, they are classified as either operating or nonoperating assets (CFI).

1. Operating assets
Operating assets are assets that are required in the daily operation of a business. In other words, operating assets are used to generate revenue. Examples of operating assets include:
- Cash
- Stock

- Building
- Machinery
- Equipment
- Patents
- Copyrights
- Goodwill.

2. Nonoperating assets

Nonoperating assets are assets that are not required for daily business operations but can still generate revenue. Examples of nonoperating assets include:
- Short-term investments
- Marketable securities
- Vacant land
- Interest income from a fixed deposit.

1.2.8 Importance of Asset Classification

Classifying assets is important to a business. For example, understanding which assets are current assets and which are fixed assets is important in understanding the net working capital of a company. In the scenario of a company in a high-risk industry, understanding which assets are tangible and intangible will help it determine its solvency and risk. Determining which assets are operating and which are nonoperating is important to understand the contribution of revenue from each asset. Therefore, knowing how to classify assets is integral to business success (CFI).

1.3 Inspection of Industrial Assets

Inspections can be targeted to need, and interventions can be timed to preempt expensive and often distressing asset failure, rather than dictated by routine alone. Inspections are driven by a considered balance of investment and risk, offering the greatest impact on risk reduction at least cost.

The processes mentioned here are established and tested best practice. They are used to illustrate an asset inspection guidance principle. They should not necessarily be adopted by all organizations. It is the principle of the approach that is important (Bown, Chatterton, & Purcell, 2014).

1.3.1 Inspection of All Types of Industrial Assets

1.3.1.1 Understanding the Role of the Asset

For effective management of assets, an appropriate level of understanding of the following is required:

- The consequences of failure of the asset.
- The expectations of the performance of the asset in fulfilling its role. For example, what is the expected annual probability of an asset overtopping, and is that different from the "as built" design standard?
- The actual loading on an asset.

Without this understanding from a performance and risk assessment perspective, it will be difficult to target resources for inspections of those assets where risks and consequences of failure are higher. This applies to all tiers of inspection and is discussed in more detail in the following section (Bown, Chatterton, & Purcell, 2014).

Introduction

1.3.1.2 Targeting Inspections through Tiering

- A tier 1 inspection is the default level, routine inspection.
- Tier 2 and 3 inspections seek more detailed information than is routinely collected in tier 1 inspections. Tier 2 inspections are nonintrusive investigations carried out by an appropriate expert. Tier 3 involves intrusive investigations into the fabric of the asset. Both require notable investment (tier 2 less so than tier 3) and need to be justified in terms of efficiency gains, performance, and risk. Thus, each can only be triggered after proper consideration of performance and risk.

The flow chart in Figure 1.10 shows these basic principles and the integrated link to performance and risk-related activities (Bown, Chatterton, & Purcell, 2014).

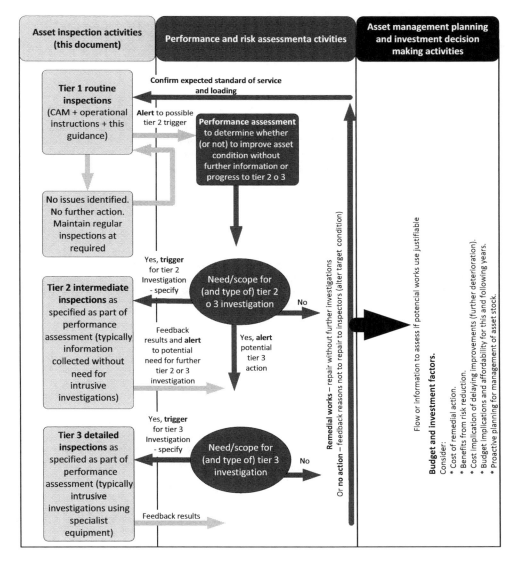

FIGURE 1.10 Flowchart showing general process of tiered inspections and assessments (Bown, Chatterton, & Purcell, 2014).

The following principles are embedded in the flow chart.

- There is a distinction between roles:
  - Asset inspection—those arranging and reporting the results of inspections or more detailed assessments, i.e., gathering physical data.
  - Performance and risk assessment—those making decisions as a result of the inspection findings and alerts, i.e., desk-based analysis of inspection data and taking account of other wider considerations.

  The roles are distinct but could be carried out by the same member/s of staff depending on the needs of the organization (Bown, Chatterton, & Purcell, 2014).
- The expected Standard of Protection (SoP) and loading for all assets should be well known by all. This will improve the efficiency of the inspection processes; for example, the inspection frequency can be reduced when risks are low and the need for further investigations can be given a lower priority.
- There will be a range of observations which could raise an alert. Each may trigger more detailed investigations if assessed as required and justifiable.

1.4 Maintenance of Industrial Assets

Most industrial assets are complex repairable systems. Maintenance decisions are still made based on the experience and understanding of the maintenance engineers/managers. This leads to sub-optimal decisions and inefficient and ineffective maintenance. Reliability is a key attribute of such assets. Reliability-centered maintenance (RCM) as a maintenance philosophy and reliability engineering as an engineering discipline have been well developed over the past several decades.

1.4.1 Asset Maintenance

A company must perform maintenance on its assets to maximize its return on investment (ROI). Choosing an appropriate asset maintenance policy for each asset is important to create an effective maintenance environment. A maintenance policy defines a company's approach to maintaining a particular asset.

It is important to consider the impact of asset failure, as well as asset performance, reliability, age, and condition, when deciding what type of maintenance policy to use. There are three different maintenance policies to choose from:

1. Use-based maintenance
2. Failure-based maintenance
3. Condition-based maintenance (CBM).

Use-based maintenance is maintenance triggered by time, distance, or event. Changing the oil in a car every 3,000 miles is an example of use-based maintenance. We simply replace the oil at a given interval.

Failure-based maintenance refers to running an asset to failure. An example of a failure-based approach to vehicle maintenance would be to not change the oil at all and run the engine to failure.

CBM is a maintenance policy that assesses the actual condition of an asset to determine whether maintenance needs to be done. Going back to the example of an oil change, we would have to inspect the oil and do a lubrication analysis to determine the level of particulates in the oil and decide whether it's time to change it.

The goal of CBM is to predict equipment failure so that maintenance can be scheduled only when it is needed. CBM is performed before the asset fails or performance is hindered. It lowers disruptions to normal operations, reduces the costs of asset failure, minimizes unscheduled downtime, and improves the reliability of equipment (DPSI Remarkable Technology, 2015).

1.4.1.1 How to Choose the Right Asset Maintenance Policy

There are several factors to consider before choosing a maintenance policy for an asset:

- Risks of failure
- Maintenance costs
- Asset performance
- Asset reliability.

For some assets, the risk of failure is too huge from a liability perspective, so a failure-based maintenance policy would not be an appropriate choice. The costs of a maintenance policy are also important to consider. CBM is the most expensive because it requires regular inspection and maintenance work, as well as a one-time investment in equipment, such as sensors and monitoring equipment. Avoiding asset failure with a CBM policy is recommended if an asset's failure could lead to safety risks or a loss of reputation and sales. While it is less costly upfront to set up a use-based maintenance policy, it could be more expensive in the long run, as use-based maintenance is performed regularly, regardless of an asset's condition.

When deciding on an asset maintenance policy, a company needs to consider what an asset does, what would happen if it fails, and the probability of it failing. It should consider each maintenance policy's one-time and ongoing costs, risks, and benefits. Once it determines an appropriate maintenance policy for each asset, it should create work plans in its computerized maintenance management software (CMMS). The company can use the software to schedule CBM inspections and use-based maintenance tasks, as well as create work orders for failure-based maintenance (DPSI Remarkable Technology, 2015).

1.4.2 Using Predictive Maintenance for Industrial Assets

For asset-intensive companies, such as metals and paper mills, mines, and chemical plants, predictive maintenance often represents the highest value. Predictive maintenance addresses key business challenges on the factory floor—unplanned machine breakdowns or a lack of asset visibility—and delivers the highest returns. If successful, predictive maintenance creates substantial infrastructure and expertise for widening the digital footprint across the organization (Capgemini, 2017).

1.4.2.1 Challenges of Predictive Maintenance

The major challenge of predictive maintenance is the breakdown of equipment because of machine degradation, component wear, and other factors invisible to operators. These machine breakdowns have direct implications on the organization's finances, productivity, and reputation.

Other challenges include a difficulty in manual monitoring of equipment health, a lack of asset visibility, and a lack of standardization in efficient maintenance planning processes.

Such unpredicted machine failures can be avoided at stage 1 (see Figure 1.11) by adopting advanced maintenance techniques that systematically process the continuously generated equipment data. This provides insight into the operational condition of the equipment, allowing maintenance supervisors and process supervisors to make more informed decisions (Capgemini, 2017).

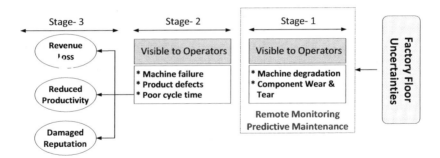

FIGURE 1.11 Remote monitoring/predictive maintenance (Capgemini, 2017).

1.4.2.2 How Does Predictive Maintenance Work?

The predictive maintenance approach helps to monitor and assess equipment health based on the analysis of various parameters, including temperature, pressure, vibration, RPM, and flow rate. Various data processing systems capture these data using sources such as OPC, Historian, and SQL. They then perform data analysis, where the streamed equipment data are compared with pre-identified failure patterns, captured and stored using historical data. Machine learning processes enable the system to analyze and store machine failure patterns that iteratively learn from data, allowing systems to find hidden signs without any explicit conditions. A match in the streamed equipment data with pre-identified failure patterns triggers alarms and notifications indicating a deterioration of machine health and the potential for equipment failure (see Figure 1.12) (Capgemini, 2017).

1.4.2.3 How to Approach Predictive Maintenance Design

The predictive maintenance approach varies across industries and depends on the types of industry-specific machines and equipment in use. A structured process for organizations to design the predictive maintenance strategy is detailed below (see Figure 1.13) (Capgemini, 2017).

1.4.2.3.1 Create List of Plant Equipment

It is essential to have a master list of all the equipment on the plant floor to ascertain which machines require continuous monitoring for seamless operations. The equipment list needs to be reviewed for completeness to ensure no critical equipment is ignored.

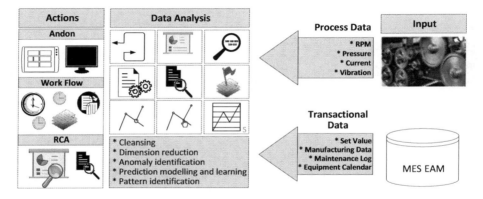

FIGURE 1.12 Predictive maintenance work (Capgemini, 2017).

Introduction

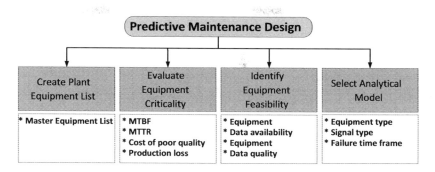

FIGURE 1.13 Predictive maintenance design (Capgemini, 2017).

1.4.2.3.2 Evaluate Equipment Criticality

Once the equipment list has been created, the next step is to evaluate how indispensable the equipment is with respect to overall operations and maintainability. Various factors influence the equipment criticality evaluation, and a few critical factors are shown in the graphs below. All factors should be considered collectively when evaluating criticality.

1.4.2.3.3 Identify Equipment Feasible for Monitoring

The next step is to determine if the equipment can be monitored. Monitoring equipment with high criticality will be less useful if the equipment doesn't produce reliable or quality data or the equipment is not compatible with the required retrofits.

A few critical factors determining whether equipment can usefully be monitored are the following:

- Availability of equipment data

    The key factor that decides if the equipment is feasible to be monitored is the availability of equipment data. Either historical equipment data should be available or the equipment should have the ability to stream data to derive insights for monitoring. The following characteristics are essential:
    - Reliability: Stable and consistent results are produced over subsequent iterations.
    - Resolution: Gathered data accurately depict the exact health of the equipment.
    - Networking: Data can be collected from equipment and shared with other sources for analysis.
- Possibility of retrofitting equipment with required data sources

    If there is no default mechanism to capture required data from the equipment, it is essential to retrofit the equipment with external data sources. The following factors need to be considered to evaluate the feasibility of doing so:
    - Compliance and regulation complications.
    - Cost to retrofit external data source.
    - Technology compatibility.

1.4.2.3.4 Selecting the Appropriate Analytical Model

Once the critical equipment is feasible for remote monitoring and preventive maintenance, the next step is to select an appropriate analytical model. The selection of the model depends on the type of equipment, failure mode and cause, and the signal type and failure time frame.

- Equipment and signal type

    Defining an analytical model depends on the type of equipment, failure mode and cause, and the signal type generated or captured. Capgemini's analytical model framework is shown in Table 1.7.

TABLE 1.7
Capgemini's Analytical Model Framework (Capgemini, 2017)

Equipment Category	Equipment Types	Failure Mode	Failure Cause	Signal Type	Signal Captured	Statistical Model	Visualization
Rotating machinery Reciprocating machinery	Pumps, turbines…	Mechanical, electrical…	Corrosion, fatigue, cavitation, electrical discharge, seal leak/rupture…	Vibration, acoustic, temperature…	Vibration, acoustic, temperature…	Normal distribution, Fourier series, residual distribution	Data visualization: • Histograms • Density • Plot • Scatter • Binscatter
Thermal equipment	Heat exchangers…	Mechanical, electrical…	Leak rupture, tube side plugged, shell side plugged, too much flow through valve	Vibration, temperature…	Vibration, temperature…	Normal distribution, Fourier series, residual distribution	Data visualization: • Histograms • Density • Plot • Scatter • Binscatter
Repetitive process	Specific equipment that allows the installation of assembly lines…	Mechanical, electrical…	Corrosion, fatigue, electrical discharge…	Vibration, acoustic, temperature…	Vibration, acoustic, temperature…	Normal distribution, Fourier series, residual distribution	Data visualization: • Histograms • Density • Plot • Scatter • Binscatter
Containment and transfer equipment	Containment equipment: Laminar flows, equipment particulate airborne concentration, and others.	Mechanical, electrical…	Corrosion, fatigue, electrical discharge…	Vibration, acoustic, temperature…	Vibration, acoustic, temperature…	Normal distribution, Fourier series, residual distribution	Data visualization: • Histograms • Density • Plot • Scatter • Binscatter

- Failure time frame

    Another key factor to be considered when selecting an analytical model is the time frame between the occurrence of a symptom and the actual failure. Certain equipment tends to fail quickly and, as such, requires highly agile analytical models for monitoring.

After the selection of the appropriate analytical model to monitor critical equipment, the next challenge is the implementation of a predictive solution.

REFERENCES

American Automobile Association, 2012. *Your Driving Costs: How Much Are You Really Paying to Drive?* Heathrow, FL: American Automobile Association.

Anderson J. M., Kalra N., Stanley K. D., Sorensen P., Samaras C., Oluwatola O. A., 2016. *Autonomous Vehicle Technology. A Guide for Policymakers.* Santa Monica, CA: RAND Corporation. ©Copyright 2016 RAND Corporation. Library of Congress Cataloging-in-Publication Data is available for this publication. ISBN: 978-0-8330-8398-2.

Applanix, 2012. POS LV 120 specifications. October 2012.

Atiyeh C., 2012. Predicting traffic patterns, one Honda at a time. MSN Auto, June 25.

Bâela L., Mâarton L., 2011. *Nonlinear Control of Vehicles and Robots.* London: Springer-Verlag.

BBC News, 2004. Desert race too tough for robots. March 15, 2004. http://news.bbc.co.uk/2/hi/technology/3512270.stm. Viewed: February 17, 2019.

Berger C., Rumpe B., 2012. Autonomous driving—Five years after the urban challenge: The anticipatory vehicle as a cyber-physical system. In *10th Workshop on Automotive Software Engineering*, Braunschweig, Germany, September 2012, pp. 789–798.

Berry I., 2010. *The Effects of Driving Style and Vehicle Performance on the Real-World Fuel Consumption of U.S. Light-Duty Vehicles.* Cambridge, MA: Massachusetts Institute of Technology.

Bose A., Ioannou P., 2003. Analysis of traffic flow with mixed manual and semi-automated vehicles. *IEEE Transactions on Intelligent Transportation Systems*, 4, 173–188.

Bown C., Chatterton J., Purcell A., 2014. Asset performance tools – Asset inspection guidance. Report – SC110008/R2. Environment Agency, Horizon House, Bristol, UK, July 2014.

Buehler M., Iagnemma K., Singh S., 2009. *The DARPA Urban Challenge: Autonomous Vehicles in City Traffic.* Springer Tracts in Advanced Robotics. Heidelberg: Springer.

Bullis K., 2011. How vehicle automation will cut fuel consumption. MIT's Technology Review. October 24.

Bureau of Labor Statistics, 2012. *Occupational Outlook Handbook: Transportation and Moving Occupations.* Washington, DC: Bureau of Labor Statistics.

Campbell M., Egerstedt M., How J., Murray R., 2010. Autonomous driving in urban environments: Approaches, lessons and challenges. *Philosophical Transactions of the Royal Society A Mathematical Physical and Engineering Sciences*, 368, 4649–4672.

Capgemini, 2017. Using predictive maintenance of industrial assets: Your starting point to the digital manufacturing journey. Consulting Technology Outsourcing.

CDC, 2011. *Injury Prevention and Control: Data and Statistics.* Atlanta, GA: Center for Disease Control.

Cervero R., 2001. Induced demand: An urban and metropolitan perspective. In *Prepared for Policy Forum: Working Together to Address Induced Demand*, Berkeley, CA, USA.

Chen Q., Ozguner U., 2006. Intelligent off-road navigation algorithms and strategies of team desert buckeyes in the DARPA grand challenge '05. *Journal of Field Robotics*, 23(9), 729–743.

Choi E.-H., Zhang F., Young Noh E., Singh S., Chen C. L., 2008. *Sampling Design Used in the National Motor Vehicle Crash Causation Survey.* Washington, DC: National Highway Traffic Safety Administration's National Center for Statistics and Analysis. DOT HS 810 930.

Corporate Finance Institute (CFI). Types of assets: Classifying assets based on convertibility, physical existence and usage. CFI Education Inc. https://corporatefinanceinstitute.com/resources/knowledge/accounting/types-of-assets/. Viewed: April 14, 2019.

Dalal N., Triggs B., 2005. Histogram of oriented gradients for human detection. In *IEEE International Conference on Computer Vision and Pattern Recognition, CVPR'05*; June 2005, San Diego, USA. Vol. 1, pp. 886–893.

DARPA—See U.S. Defense Advanced Research Projects Agency.

De Francesco E., Leccese F., 2012. Risks analysis for already existent electric lifelines in case of seismic disaster. In *Proceedings of the 11th International Conference on Environment and Electrical Engineering, EEEIC*, Art. No. 6221490; Rome, Italy. pp. 830–834. doi: 10.1109/EEEIC.2012.6221490. Viewed: February 14, 2019.

DPSI Remarkable Technology, 2015. Realistic price. EAM & CMMS Software, March 19, 2015. www.dpsi.com/blog/how-to-select-the-best-maintenance-policy-for-your-assets/.

Dresner K., Stone P., 2008. A multiagent approach to autonomous intersection management. *Journal of Artificial Intelligence Research*, 31, 591–656.

Economist Technology Quarterly, 2012. Inside story: Look, no hands. September 1 Issue: 17–19.

Fagnant D. J., Kockelman K., 2015. Preparing a nation for autonomous vehicles: Opportunities, barriers and policy recommendations. *Transportation Research Part A*, 77(2015), 167–181.

Fagnant D. J., Kockelman K., 2016. Dynamic ride-sharing and optimal fleet sizing for a system of shared autonomous vehicles. In *Presented at the Annual Meeting of the Transportation Research Board and Under Review for Publication in Transportation*. August 22, 2016. Austin, Texas. DOI: 10.1007/s11116-016-9729-z.

Farhadi A., Endres I., Hoiem D., Forsyth D., 2009. Describing objects by their attributes. IEEE *Conference on Computer Vision and Pattern Recognition*. University of Illinois at Urbana-Champaign, Miami, FL, USA. DOI: 10.1109/CVPR.2009.5206772

Federal Highway Administration, 2005. *Traffic Congestion and Reliability: Linking Solutions to Problems*. Washington, DC: General Books.

Federal Highway Administration, 2013. *Public Data for Highway Statistics*. Washington, DC: Office of Highway Policy Information.

Furukawa Y., Ponce J., 2009. Accurate camera calibration from multi-view stereo and bundle adjustment. *International Journal of Computer Vision*, 84(3), 257–268.

Google, 2012. Self-driving car test: Steve Mahan. March 28, 2012. www.youtube.com/watch?v=cdgQpa1pUUE. Viewed: February 14, 2019.

Gucwa M., 2014. Mobility and energy impacts of automated cars. In *Presentation to the Automated Vehicles Symposium*, Burlingame, CA, USA, July 16.

Hayes B., 2011. Leave the driving to it. *American Science*, 99, 362–366.

Hsu T., 2013. "CES 2013: Lexus driverless car: 'Technology alone is not the answer,'" *Los Angeles Times*, January 7, 2013.

Hwang I., Kim S., Kim Y., Eng Seah C., 2010. A survey of fault detection, isolation, and reconfiguration methods. *IEEE Transactions on Control Systems Technology*, 18(3), 636–653.

Ioannou P., 1998. Development and experimental evaluation of autonomous vehicles for roadway/vehicle cooperative driving. California PATH Research Report UCB-ITS-PRR-98-9, Berkeley, CA, USA.

Kocić J., Jovičić N., Drndarević V., 2018. Sensors and sensor fusion in autonomous vehicles. School of Electrical Engineering, University of Belgrade, Belgrade, Serbia, November 2018.

Kockelman K., Kalmanje S., 2006. Road pricing simulations: Traffic, land use and welfare impacts for Austin, Texas. *Transportation Planning and Technology* 29(1), 1–23.

KPMG & CAR, 2012. *Self-Driving Cars: The Next Revolution*. Ann Arbor, MI: KPMG & CAR.

Kunze R., Ramakers R., Henning K., Jeschke S., 2009. Organization of electronically coupled truck platoons on German motorways. In *Intelligent Robotics and Applications: Second International Conference; Vol. 5928, pp. 135–146*. Berlin, Heidelberg: Springer. DOI: https://doi.org/10.1007/978-3-642-10817-4_13

Laberteaux K., 2014. How might automated driving impact US land use? In *Presentation to the Automated Vehicles Symposium*, Burlingame, CA, USA, July 16. Toyota Research Institute-North America, and Toyota Motor Engineering & Manufacturing North America.

LaHood R., 2011. Notice of funding availability for the department of transportation's national infrastructure investments under the full-year continuing appropriations. *U.S. Department of Transportation Federal Register*, 76(156), 50310.

Lin P., 2013. The ethics of saving lives with autonomous cars are far Murkier than you think, wired. July 30.

Litman T., 2012. *Parking Management: Strategies, Evaluation and Planning*. Victoria, BC: Victoria Transport Policy Institute.

Litman T., 2013. *Online Transportation Demand Management Encyclopedia*. Victoria, BC: Victoria Transport Policy Institute.

Manyika J., Chui M., Bughin J., Dobbs R., Bisson P., Marrs A., 2013. *Disruptive Technologies: Advances that Will Transform Life, Business and the Global Economy*. McKinsey Global Institute: New York, USA.

Markoff J., 2011. In search of a robot more like us. *New York Times*, July 11, 2011.

Mashaw J. L., Harfst D. L., 1990. *The Struggle for Auto Safety*. Cambridge, MA: Harvard.

Montemerlo M., 2006. Winning the DARPA grand challenge with an AI robot. In *Proceedings of the 21st National Conference on Artificial Intelligence and the 18th Innovative Applications of Artificial Intelligence Conference*, July 2006; pp. 982–987. Boston, Massachusetts, USA.

National Highway Traffic Safety Administration, 2008. National motor vehicle crash causation survey. Report DOT HS 811 059, U.S. Department of Transportation.

National Highway Traffic Safety Administration, 2012. *Fatal Analysis Reporting System*. Washington, DC: U.S. Department of Transportation.

National Highway Traffic Safety Administration, 2014. U.S. Department of Transportation announces decision to move forward with vehicle-to-vehicle communication technology for light vehicles. NHTSA 05-14, Washington, DC, USA

National Safety Council, 2012. *Estimating the Costs of Unintentional Injuries*. Washington, DC: National Safety Council.

Newcomb D., 2012. Road-train test keeps cars in line, wired.

Olson P. L., Dewar R., Farber E., 2010. *Forensic Aspects of Driver Perception and Response*, Second Edition. Tucson, AZ: Lawyers and Judges Publishing Company.

Ozguner U., Redmill K., 2008. Sensing, control, and system integration for autonomous vehicles: A series of challenges. *SICE Journal of Control, Measurement, and System Integration*, 1(2), 129–136.

Puget Sound Regional Council, 2006. *2006 Household Activity Survey*. Seattle, WA: Puget Sound Regional Council.

Redmill K., Martin J. I., Ozguner U., 2006. Sensing and sensor fusion for the 2005 desert buckeyes DARPA grand challenge off-road autonomous vehicle. In *2006 IEEE Intelligent Vehicles Symposium*, pp. 528–533, June 2006. Tokyo, Japan: IEEE.

Rio Tinto, 2014. Rio Tinto improves productivity through the world's largest fleet of owned and operated autonomous trucks.

Schrank D., Eisele B., Lomax B., 2012. 2012 urban mobility report. Texas Transportation Institute. College Station, TX.

Seneviratne D., Ciani L., Catelani M., Galar D., 2018. Smart maintenance and inspection of linear assets: An Industry 4.0 approach. *ACTA IMEKO*, 7(1), 50–56.

Shladover S., Su D., Lu X.-Y., 2012. Impacts of cooperative adaptive cruise control on freeway traffic flow. In *Proceedings of the 91st Annual Meeting of the Transportation Research Board*, Washington, DC, USA.

Shoup D., 2005. *The High Cost of Free Parking*. Chicago, IL: APA Planners Press. ISBN 978-1-884829-98-7.

Smith B. W., 2013. Managing autonomous transportation demand. *Santa Clara Law Review*, 52, 1413.

Thrun S., 2006. Stanley: The robot that won the DARPA grand challenge. *Journal of Robotic Systems - Special Issue on the DARPA Grand Challenge*, 23(9), 661–692.

Thrun S., Montemerlo M., Dahlkamp H., Stavens D., Aron A., Diebel J., Fong P., Gale J., Halpenny M., Hoffmann G., Lau K., Oakley C., Palatucci M., Pratt V., Stang P., Strohband S., Dupont C., Jendrossek L. E., Koelen C., Markey C., Rummel C., Niekerk J. V., Jensen E., Alessandrini P., Bradski G., Davies B., Ettinger S., Kaehler A., Nefian A., Mahoney P., 2007. Stanley: The robot that won the DARPA grand challenge. In M. Buehler, K. Iagnemma, & S. Singh (Eds.), *The 2005 DARPA Grand Challenge: The Great Robot Race* (pp. 1–43). Berlin: Springer.

Tientrakool P., 2011. Highway capacity benefits from using vehicle-to-vehicle communication and sensors for collision avoidance. *Proceedings of the 74th IEEE Vehicular Technology Conference, VTC Fall 2011*. September 2011, San Francisco, CA, USA.

Traffic Safety Facts, 2013. Report DOT HS 811 753, U.S. Department of Transportation, Washington, DC.

Trottenberg P., 2011. *Treatment of the Value of Preventing Fatalities and Injuries in Preparing Economic Analysis – 2011 Revision*. Washington, DC: U.S. Department of Transportation.

Urmson C., Ragusa C., Ray D., Anhalt J., Bartz D., Galatali T., Gutierrez A., Johnston J., Harbaugh S., "Yu" Kato H., Messner W., Smith B., Snider J., Spiker S., Ziglar J., "Red" Whittaker W., Clark M., Koon P., Mosher A., Struble J., 2006. A robust approach to high-speed navigation for unrehearsed desert terrain. *Journal of Field Robotics*, 23(8), 467–508.

U.S. Census Bureau, 2011. Age and sex composition: 2010, C2010BR-03.

Velodyne, 2010. *High Definition Lidar HDL-64E S2*. Morgan Hill, CA: Velodyne Lidar, Inc.

Wood J., 2002. Aging driving and vision. *Clinical and Experimental Optometry*, 85, 214–220.

2
Development of Autonomous Vehicles

2.1 History of Development of Autonomous Robots

2.1.1 Introduction

The human urge to automate our world seems unstoppable. Kassler (2001) claims the transfer of human intelligence into computer-controlled machines such as robots is analogous to the fundamental scientific aim to devise theories in a form that makes them reproducible. In essence, we transfer factual data and procedural theories into a computer so that the machine can carry out humanlike tasks.

The emphasis in the development of autonomous field robots is currently on speed, energy efficiency, sensors for guidance, guidance accuracy, and technologies such as wireless communication and Global Positioning System (GPS) (Grift, 2015).

2.1.2 Autonomous Robots

Both animals and robots manipulate objects in their environment to achieve certain goals. Animals use their senses (e.g. vision, touch, smell) to probe the environment. The resulting information, in many cases also enhanced by the information available from internal states (based on short-term or long-term memory), is processed in the brain, often resulting in an action carried out by the animal, with the use of its limbs.

Similarly, robots gain information on their surroundings using their sensors. The information is processed in the robot's brain, consisting of one or several processors, resulting in motor signals being sent to the actuators (e.g., motors) of the robot.

A robotic brain cannot operate in isolation. It needs sensory inputs, and it must produce motor output to influence objects in the environment. Thus, the main challenge in contemporary robotics is the development of robotic brains. However, the actual hardware, i.e., sensors, processors, motors etc., is certainly very important as well (Wahde, 2016).

2.1.2.1 History and Development

The Seekur robot was the first commercially available robot with mobile detection assessment and response system (MDARS) capabilities and was used by airports, utility plants, corrections facilities, and Homeland Security.

The Mars rovers MER-A and MER-B (now known as Spirit Rover and Opportunity Rover) can find the position of the sun and navigate their own routes to destinations by:

- Mapping the surface with 3D vision
- Computing safe and unsafe areas on the surface within the field of vision
- Computing optimal paths across the safe area towards the desired destination
- Driving along the calculated route
- Repeating this cycle until either the destination is reached or there is no known path to the destination.

The planned ESA Rover, ExoMars Rover, is capable of vision-based relative localization and absolute localization to autonomously navigate safe and efficient trajectories to targets by:

- Reconstructing 3D models of the terrain surrounding the Rover using a pair of stereo cameras
- Determining safe and unsafe areas of the terrain and the general "difficulty" for the Rover to navigate the terrain
- Computing efficient paths across the safe area towards the desired destination
- Driving the Rover along the planned path
- Building up a navigation map of all previous navigation data.

During the final NASA Sample Return Robot Centennial Challenge in 2016, a rover named Cataglyphis demonstrated fully autonomous navigation, decision-making, and sample detection, retrieval, and return capabilities (Hall, 2016). The rover relied on a fusion of measurements from inertial sensors, wheel encoders, light detection and ranging (LiDAR), and cameras for navigation and mapping, instead of using GPS or magnetometers. During the 2h challenge, Cataglyphis traveled over 2.6 km and returned five different samples to its starting position.

The Defense Advanced Research Projects Agency (DARPA) Grand Challenge and the DARPA Urban Challenge have encouraged development of even more autonomous capabilities for ground vehicles, also the goal for aerial robots since 1990 as part of the AUVSI International Aerial Robotics Competition.

Between 2013 and 2017, Total S.A. held the ARGOS Challenge to develop the first autonomous robot for oil and gas production sites. In the Challenge, the robots faced adverse outdoor conditions, such as rain, wind, and extreme temperatures (Total, 2015).

2.1.2.1.1 Delivery Robot

A delivery robot is an autonomous robot used to deliver goods. As of February 2017, the following companies were developing delivery robots (some with pilot deliveries in progress):

- Starship Technologies
- Dispatch
- Marble.

2.1.3 What Are Autonomous Robots?

An autonomous robot is designed and engineered to deal with its environment on its own and work for extended periods of time without human intervention. Autonomous robots often have sophisticated features that help them understand their physical environment and automate parts of their maintenance and direction that used to be done by human hands.

Autonomous robots typically go about their work without any human interaction unless that human interaction is necessary as part of their task. Many of these robots have sensors and other functional gear that helps them see any obstacles in their way or navigate rooms, hallways, or some environments. Complex delivery robots can even be programmed to use elevators and move throughout a multistory building with complete autonomy. However, autonomous robots still need to be maintained physically (Techopedia, 2019).

The working definition of an autonomous robot is the following: a robot is autonomous if it has the computational resources—in terms of both hardware and software—other than real-time interference from a human agent, to estimate how it is physically embedded in the environment to compute the best possible actions bounded by some constraints to perceive and move, if needed, to achieve a set of goals. According to this working definition, a robot's ability to estimate its current state (how it is physically embedded in the environment) is an essential component of autonomy. It has to have adequate computational resources at its disposal to take an action within bounds, to perceive the environment, and move, if needed, to achieve a given goal.

An autonomous agent should be able to act with the environment, so that with its embedded passive and active sensors, it can perceive and operate effectively.

The computational resources required are the following:

1. Hardware: Every embedded system needs a microprocessor. There is a variety of options for autonomous robots, from microcontrollers like programmable interrupt controllers (PICs) and automatic voltage regulators (AVRs) to microprocessors like Advanced RISC Machines (ARMs). Their selection may depend on the complexity of the automotive algorithms required.
2. Software: A well-performing robot needs to perform many functions simultaneously. A customized operation system (OS) might handle this. Most are versions of Linux, like ROS and RoBIOS. They can handle common tasks in robotics, such as machine vision and sensor interpretation.

2.1.4 Types of Autonomous Vehicles

Various kinds of autonomous vehicles (AVs) can operate with varying levels of autonomy. This section is concerned with underwater, ground, and aerial vehicles operating in a fully autonomous (non-tele-operated) mode. It also deals with AVs as a special kind of device, not with full-scale manned vehicles operating unmanned. The AV in question is likely to be designed for autonomous operation rather than being adapted for it, as would be the case for manned vehicles.

It should be noted that issues of control are pervasive regardless of the kind of AV being considered, but there are special considerations in the design and operation of AVs depending on whether the focus is on vehicles underwater, on the ground, or in the air (Meyrowitz, Blidberg, & Michelson, 1996).

2.1.4.1 Autonomous Underwater Vehicles

The development of autonomous underwater vehicle (AUV) systems began many years ago. Most of the original vehicles were extremely simple because of technological limitations, but their potential to serve both military and scientific purposes was apparent. For example, the Navy had requirements for drones, especially devices to meet search and survey needs, support mine countermeasures, and assist in understanding control and hydrodynamics questions. The need to understand the physics of the ocean enticed some of the ocean science community to begin work to develop sophisticated vehicle systems for data gathering; the oil and gas industry also had special interests in the use of AUV technology for the inspection of underwater structures and pipelines.

During the past two decades, the pace of AUV development has increased substantially. A number of recent efforts have been undertaken to determine, in a comprehensive way, the appropriate role for this technology in different application areas. In the United States, the academic groups addressing AUV technology include the University of Hawaii, Scripps, Stanford, University of California at Santa Barbara, FAU, Texas A&M, Massachusetts Institute of Technology, Woods Hole Oceanographic Institute (HOI), US National Parks Service (NPS), and a number of other universities. Two key issues have emerged: energy systems and high-level control. Without sufficient energy, nothing can be accomplished, and with minimal or possible suspension of communications, the demand for onboard decision-making becomes critical.

Industry plays an active role in planning the use of AUV technology, with applications being considered for deep ocean exploration, polar ocean exploration (the ice cover makes the use of standard technology impossible), and for the military or hazardous environments (e.g., areas of high-level chemical or radiation hazards). Furthermore, industry and government agencies alike may find AUVs are an attractive option for understanding and monitoring the health of the environment. Remote sensing from space-based satellites and from airplanes goes only so far in understanding the impact of the ocean and its estuaries on our environment.

As evidenced by the increasing attendance at the annual IEEE AUV Symposia and the biannual International Symposia on Unmanned Untethered Submersible Technology (ISUUST), there is a tremendous and growing international interest in AUV technology. Further, it is clear from the presented papers that the maturity of AUV technology is mostly in the hardware domain. It is now a nearly routine undertaking to design a hull structure to some set of operational parameters, construct the hull, populate it with sensors, effectors, computers, and batteries, derive a control algorithm, implement and install that algorithm on board, and effect a self-controlling vehicle. The most constraining physical parameter is the amount of energy that can be carried on board. New energy systems are being developed, however, which will shortly reduce the effect of this constraint.

One example of an experimental underwater vehicle is EAVE 111, developed by the Marine Systems Engineering Laboratory (MSEL) of the Autonomous Underwater Systems Institute (AUSI). EAVE I11 is a third-generation AUV characterized as an open-space frame test bed with excellent maneuverability, precise control, and an acoustic long baseline navigation system. Experiments include searches for underwater objects, navigation below an oil spill boom, acquisition of video images of undersea objects, and acoustic communication between two EAVE 111s for the purpose of demonstrating cooperative behavior and control between two autonomous systems.

R-One is an autonomous underwater free swimming robot equipped with a closed cycle diesel engine for long-term survey of mid-ocean ridges. An inertial navigation system (INS) cooperates with a Doppler sonar system to support accurate navigation when the robot swims in the vicinity of the seabed (see Table 2.1) (Meyrowitz, Blidberg, & Michelson, 1996).

Over a dozen different sensor payloads are being designed. These include a variety of water quality, optic, and acoustic sensors. The goal is modularity and rapid reconfiguration. An interface specification facilitates collaboration with other institutions with interests in building payloads for deployment.

The state of AUV software development is less mature. Architectures for vehicle software abound, but a significant factor in the acceptance of autonomous systems technology is the lack of real in-water experience to validate the potential utility of the alternatives. As discussed below, much work remains to be done in a number of enabling technologies, especially understanding the design and implementation of software to support navigation, communication, and the AUV's response to changing conditions in its internal state and in its environment (Meyrowitz, Blidberg, & Michelson, 1996).

TABLE 2.1

Data from the R-One Underwater Robot (Meyrowitz, Blidberg, & Michelson, 1996)

R-One Underwater Robot (Tamaki Ura)	
Species	**Cruising Type**
Mission	Long time diving survey
Launching date	11/1995
Dimension (L × diameter)	8.27 × 1.15 mØ
Mass	4,530 kg
Design depth	400 m
CPUs	2 × PEP-9000 VM40 (MC68040, 25 MHz)
Main thrusters	1.5 kW, 280 DVC
Vertical thrusters	2 × 0.75 kW, 280 DVC
Actuators	3 × 0.17 kW DC motor
Energy source	Closed cycle diesel engine system 5 kW, DC 280 V, 60 kWh 1,900 kg
Navigation	INS with Doppler sonar
Sensors	Depth gauge
	2 × bottom profiler
	Transponder link
	Radio link
	Radio link
	CTDO
	TV camera
Wet payload space	600 L

2.1.4.1.1 Control

The purpose of an AUV's control software is the same as that of any other AV: to allow the vehicle to sense, move about in, and interact with its world, to survive, and to carry out useful missions for its users. The controller can be thought of as a resource manager for the AUV, managing its sensor and effector use, its power consumption, its location, and its time to carry out missions for its users.

A basic function of an AUV's control software is managing its sensors. This includes determining what to "look" at, as well as deciding which sensors to have active. It also involves detecting and, if possible, correcting sensor errors, possibly by using redundant sensors or taking faulty sensors off-line. Sensor fusion is often necessary to provide "virtual sensors." For example, bottom topology may be important to a mission, yet there is no "bottom topology" sensor; rather, this virtual sensor is created by fusing information from (possibly) down-looking sonar, location sensors, motion sensors, etc.

Control also involves managing the AUV's effectors, including its thrusters or other means of movement. At a low level, the controller must ensure the effectors are operating properly and, if not, that steps are taken to correct or compensate for the problem. The controller may need to create "virtual effectors" by coordinating the activities of several real effectors.

Survival of the vehicle is an obvious responsibility of the vehicle controller. This includes not only internal homeostasis (e.g., ensuring a constant internal temperature or power consumption rate) and recovery from faults but also taking actions to maintain the current status in the world. For example, the AUV may need to maintain its station near a mine or hydrothermal vent in the presence of currents; failure to do so could result in mission failure and loss of the vehicle.

Finally, control software is responsible for providing a usable vehicle with which users can conduct missions. The level of intelligence aboard the vehicle can vary, depending on the vehicle and the users' needs. For some missions, for example, simple missions or those taking place in relatively static, well-known environments, it may be sufficient to have a relatively "dumb" AUV; the user specifies the mission in detail, and the AUV carries out the instructions to the letter. For other missions, for example, complicated missions, missions taking place in dynamic or uncertain environments, or missions that cannot be preplanned completely, it is advantageous to migrate some or all of the responsibility of planning from the user to the AUV's controller (Meyrowitz, Blidberg, & Michelson, 1996).

2.1.4.2 Autonomous Ground Vehicles

There are special challenges in building mobile vehicles capable of navigating through unknown potentially hazardous terrain. This is highly nontrivial even under the assumption of control with a human in the loop. Tele-operated vehicles, i.e., under fiber optic tether, have been demonstrated to transit over unmapped natural terrain, but this accomplishment has required extensive research addressing such issues as creating a sense of telepresence through a human–machine interface, high-speed mobility, long-range non-line-of-sight operation, ruggedness, and reliability (Aviles, 1990). If we wish to have a vehicle performing autonomously, a host of new research issues must be addressed to allow perceptual and problem-solving capabilities to reside entirely within the machine (Meyrowitz, Blidberg, & Michelson, 1996).

The autonomous ground vehicle (AGV) must be a highly competent problem solver to operate in natural and remote environments. We may not be able to provide it with accurate models of terrain. If maps are available, they may lack local detail and, in any case, will not represent the changes that can occur in dynamic situations where transient obstacles (such as other mobile vehicles and humans) will be encountered or where activities of the AGV itself might alter its environment during accomplishment of tasks. On the basis of information obtained from its sensors, the AGV must be capable of building its own maps to represent the environment and flexibly support its own reasoning for navigational planning and, if necessary, replanning. Moreover, as the world around the vehicle can change quickly, we want the software which implements control to be computationally efficient and supportive of real-time responses to events (Meyrowitz, Blidberg, & Michelson, 1996).

2.1.4.2.1 Progress Towards Autonomy

The wheeled robot SHAKEY (Meystel, 1991) is often cited as an early demonstration of automating perception and path planning. Through a combination of onboard computers and a radio link to larger computers elsewhere, SHAKEY used a scanning camera to obtain a wide field of view and attempted to keep track of its wheel rotation to support the calculation of its position on its internal map. The experiments with SHAKEY highlighted the need for research into more powerful computer vision, the integration of information from multiple sensors (such as visual and mechanical), and automated planning. SHAKEY also provided an early example of hierarchical control, an important architectural principle in the design of robotic software (Meyrowitz, Blidberg, & Michelson, 1996).

By the early 1980s, the Stanford Cart and the CMU Rover represented state-of the-art robotic mobility. The Stanford Cart was basically a camera platform on wheels; images broadcast from its on-board TV system provided knowledge of its surroundings as it moved through cluttered spaces. That movement was slow (1 m every 10–15 min) and not continuous. The Cart would stop after each meter to take and review new pictures and to plan a new path before moving on.

The CMU Rover was a more capable device designed to support a broader range of experiments in perception and control. Rapid processing of sensed data was facilitated by a dozen onboard processors and by a connection to a large remote computer; an omnidirectional steering system provided maximum mechanical flexibility. The cylindrical Rover, 1 m tall and a half-meter in diameter, was designed to achieve motion at ten times the speed of the Stanford Cart. Hierarchical control was also used, with three processing levels focused on planning, plan execution, and direction of actuators and sensors (Meyrowitz, Blidberg, & Michelson, 1996).

Even lacking planning capabilities, a robot fitted with a variety of sensors and the ability to wander about can serve as a sentry. This has been shown in the Robart-I and Robart-II mobile robots built at the Naval Postgraduate School and at the University of Pennsylvania's Moore School of Engineering. In particular, Robart-I could alter the direction of its motion if an obstacle was sensed; it relied on a combination of ultrasonic waves (to measure forward range), near-infrared proximity detectors, tactile feelers, and bumper switches. Robart-II, in keeping with its sentry mission, employed a microwave motion detector, a smoke detector, infrared sensors, an ambient temperature sensor, and physical contact sensors. Its sentry duties were supported by an ability to move at a fixed sensed distance from nearby walls and to recognize intersections by the disappearance of those walls (Meyrowitz, Blidberg, & Michelson, 1996).

2.1.4.3 Autonomous Air Vehicles

To date, there are few fully autonomous unmanned aerial vehicles (UAVs), much less ones slated for service. A notable exception is the cruise missile which can navigate between points using environmental cues with augmentation from other sources such as the GPS system. Cruise missiles exhibit moderate intelligence; however, a greater degree of machine intelligence (MI) can be envisioned in which the air vehicle interacts with its environment to modify its tactics to achieve a goal. In Japan, Sugeno has demonstrated limited use of fuzzy logic in tele-operated UAVs (Sugeno, Griffin, & Bastian, 1993). On a larger scale, but in simulation only, Gilmore, Roth, and du Fossat have demonstrated self-actuating behaviors and postulated methods for interactions among several fully autonomous UAVs operating in concert (Gilmore, Roth, & du Fossat, 1990).

The Association for Unmanned Vehicle Systems (AUVS) has sponsored a unique competition for universities to demonstrate a fully autonomous flying robot capable of navigating in a semi-structured environment, maintaining stability, searching for objects on the ground, and manipulating them once found. Conceived in 1990, the International Aerial Robotics Competition has grown in size and stature annually, attracting student teams from Europe, Canada, Asia, and the United States. The "aerial robotics" event brings academia, industry, and government together under the common goal of creating, on a small scale, some of the world's most advanced and robust autonomous air vehicles.

During the first year of the competition, most teams were challenged by stable autonomous flight. By the spring of 1993, teams had progressed to the point where autonomous takeoff, navigation-driven flight, hover, and landing were possible. Vehicles could also locate and manipulate (capture) specific objects on the ground. The goal of the competition is to demonstrate a higher level of reasoning in the autonomous behavior of a UAV than is currently being pursued by the governments of the world. For this reason, the requirements of the competition are rigorous and nontrivial (Meyrowitz, Blidberg, & Michelson, 1996).

2.1.5 Unmanned Aerial Vehicle (UAV) Technology

An unmanned aerial vehicle (UAV) is an aircraft without a human pilot on board. The vehicle is controlled either autonomously by attached microprocessors or telemetrically by an operator on the ground. UAVs can be used to execute observation or detection missions through automatic or remote control. They are mainly used in mapping applications, environmental change monitoring, disaster prevention response, resource exploration etc. Compared to other flying vehicles and satellite remote sensing technology, UAVs have two advantages when capturing aerial photographs: low cost and high mobility. However, they have many environmental restrictions on their use due to low flight stability. Therefore, how to use UAVs in different scenarios so that spatial information for qualitative and quantitative analysis can be reliably processed and produced is an important issue impacting their application (Liu et al., 2014).

2.1.5.1 A Typical UAV

There is a wide variety of UAV shapes, mechanisms, configurations, and characteristics. Since UAVs are usually developed for specific purposes, their hardware and software design can vary depending on task requirements. The following sections summarize the system design, implementation, and software of a typical present-day UAV (Liu et al., 2014).

2.1.5.1.1 System Design

The system design of a typical UAV includes the following:

1. Frame structure
2. Electromechanics
3. Flight controller
4. Telemetry control system.

1. Frame structure

    The frame structure is the shape of the aircraft. It is usually designed according to an aircraft's dynamic lifting method. For instance, fixed-wing aircraft (e.g., gliders) are able to fly using wings that generate lift via forward airspeed and wing shape. Another example is a rotary-wing aircraft (e.g., helicopter, quadcopter), which uses spinning rotors with aerofoil section blades to provide lift. The International Civil Aviation Organization (ICAO) defines a rotary-wing aircraft as "supported in flight by the reactions of the air on one or more rotors" (2009). Rotary-wing aircraft generally require one or more rotors to provide lift throughout the entire flight.
2. Electromechanics

    Electromechanical components of a typical UAV include the following: flight controller with multiple sensors (including GPS, gyroscope, barometer, and accelerometer), motors, propellers, speed controllers, and batteries (see Figure 2.1 for an example of a hexacopter). Different motor speeds and propellers provide different performance. For example, the combination of high-speed motors and short propellers brings more agility and mobility for the aircraft but lower efficiency and shorter battery life.
3. Flight controller

    A flight controller is a microprocessor on the aircraft that manipulates the power output of each motor to stabilize flight and respond to operator orders. There are many control algorithms, including variable pitch and servo thrust vectoring models. Variable pitch models

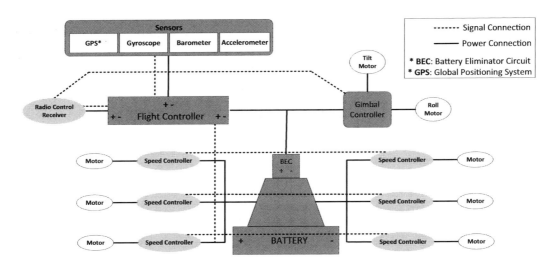

FIGURE 2.1 Hardware assembly of a hexacopter (Liu et al., 2014).

usually apply the cyclic differentially to non-coaxial propellers, allowing agile control and providing the potential to replace individual electric motors with belt-driven props hooked to one central motor (Cutler & How, 2012). Servo thrust vectoring uses differential thrust, as well as at least one motor mounted on a servo, which is free to change its orientation. This kind of algorithm is often used in bicopters and tricopters.

4. Telemetry control system

Common telemetry control systems use radio frequencies in various bands, such as FM, Wi-Fi, and microwave. The first general-use radio control systems in UAVs used single-channel analog equipment, and this allowed simple on-and-off switch control. More recent systems use pulse-code modulation (PCM) features to provide a computerized digital stream signal to the receiver, instead of analog-type pulse modulation.

2.1.5.1.2 System Implementation

When implementing a typical UAV (again taking a hexacopter as the example), we can divide the process into four main steps: frame assembly, electronics assembly, flight controller tuning, and optional equipment mounting.

The frame includes the body (to mount the flight controller and other electronics), arms (to mount the motors and speed controllers), and landing gear. Common materials for the frame assembly are aluminum and carbon fiber, which are light but have sufficient strength.

Electronic components include dynamic systems (propellers and motors), power connections (batteries and wiring), the flight controller, and telecommunication devices (e.g., radio system) (see Figure 2.2) (Liu et al., 2014).

Once the flight controller is well mounted, the variables must be tuned to adapt the controller to the frame and electronics assembly. PID (proportional-integral-derivative) is a generic control loop feedback mechanism widely used in UAV control systems. The PID controller attempts to minimize error by adjusting the process control inputs.

Finally, when the aircraft is ready to fly, various kinds of equipment can be mounted, depending on the task requirements. For example, a laser range finder and GPS unit integrated with a UAV affords the possibility of fetching 3D terrain information for geodesy inspection. UAVs with a digital camera and image telecommunication system allow practitioners to observe objects from high viewpoints and to explore unreachable or dangerous areas (Liu et al., 2014).

Development of Autonomous Vehicles   45

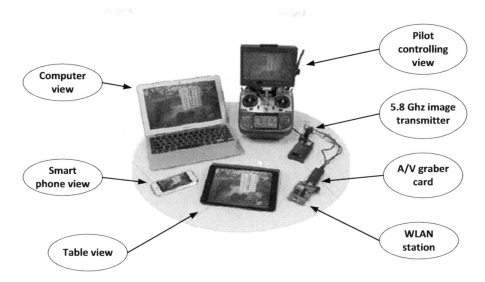

FIGURE 2.2 An example of a telecommunication control system (Liu et al., 2014).

2.1.5.2 UAV Control

Four control problems need to be considered when implementing UAV applications:

1. How many UAVs are required to achieve a task; this can be single or multiple (two or more).
2. Whether the application is model-based or model-free: i.e., whether a mathematical dynamic model or control law for the UAV should be derived.
3. Which of various control goals to pursue (e.g., stabilization/estimation of position or attitude, planning/tracking of path or target, obstacle/collision avoidance, cooperative formation flight, air/ground coordination, surveillance, or combinations thereof).
4. Whether the device will be fossil fuel- or electric-powered.

As an example, we illustrate a general attitude control architecture for UAVs in Figure 2.3. In the figure, the overall control system is a combination of: reference position $r_T(t)$; the position controller; desired dynamics $\phi^{des}(t)$, $\theta^{des}(t)$, $\psi^{des}(t)$; rotor speed differences between the nominal values $\Delta\omega_F$, $\Delta\omega_\phi$, $\Delta\omega_\theta$, $\Delta\omega_\psi$; motor dynamics; vertical force Fi and moment Mi generated by the i-th rotor; rigid body dynamics; components of angular acceleration $\ddot{\gamma}$, $\dot{p}(t)$, $\dot{q}(t)$, $\dot{r}(t)$ of the UAV in the body frame; and the actual position feedback $r(t)$ (Liu et al., 2014).

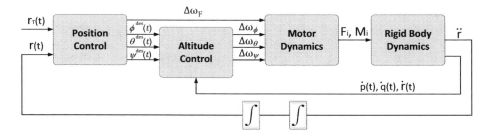

FIGURE 2.3 A general attitude control architecture for UAVs (Liu et al., 2014).

2.1.5.3 Software

Software is important in controlling a UAV to acquire information from an aerial perspective; such software is also known as a ground control station (GCS). A GCS is typically a software application running on a computer on the ground that communicates with a UAV via wireless telemetry. It displays real-time data on the UAV's performance and position and can serve as a remote cockpit. A GCS can also be used to control a UAV in flight, uploading new task commands and parameter configurations. Monitoring the live video stream is another common function of GCSs (Liu et al., 2014).

Stroumtsos, Gilbreath, and Przybylski (2013) developed GCS software for military use to eliminate the risk of disorientation and the misreading of numerical data by designing a graphical user interface. In addition, GCSs have been designed by applying other technologies, such as helmet-mounted displays (HMDs). Morphew, Shively and Casey (2004) showed the application of HMD to a target search task was more advantageous than using conventional computer monitors and joysticks. For some critical cases, simulation functions are required for GCS software. Reductions of schedule time, risk, and number of required test flights for complex aerospace tasks is a well-recognized benefit of utilizing prior simulation (Johnson & Mishra, 2002). Commercial UAV software is usually used to route craft through waypoints and provide functions such as fail-safes, return-to-home, and route editing. Recently, such software has been released on mobile device platforms, such as tablets and smart phones (Liu et al., 2014).

2.1.5.4 UAV Fields of Application

The market for drones will explode in the next 30 years in the fields of agriculture, energy, public safety/security, e-commerce/delivery, and mobility and transport. It is easy to find examples of the exploitation of drones for these different purposes in the literature. For instance, Lum and colleagues proposed using a drone to acquire multispectral images and the ground difference for precision agriculture (Lum, Mackenzie, Shaw-Feather, Luker, & Dunbabin, 2016). In the field of public safety and security, apart from the countless military applications, new sensors are being adapted for drones, including for X-ray cameras, IR cameras, and metal detectors. Applications for e-commerce and delivery are in the early stage of development. The impact of the weight on the battery duration and consequently on the distance remains problematic. However, the delivery of small objects is already a reality. Zipline (2017) describes the transportation of small medicines and blood in Africa using a fixed-wing UAV. In e-commerce, the big technology companies have made some initial proposals, such as the Prime Air service of Amazon (Amazon, 2016). The field of mobility and transport is evolving more slowly because of safety constraints and technology limitations, but it is also moving towards possible applications such as air taxis (Stimpson, Cummings, Nneji, & Goodrich, 2017).

UAVs are already being explored as useful tools in multiple civil scenarios. In many cases, UAV-based systems are required for surveillance and reconnaissance, monitoring, mapping and photogrammetry, automatic fault detection, or inventory tasks. This section focuses on inspection missions, for for example, photovoltaic plants, the environment, roads, cell towers, railway lines, mines, and buildings. Drones are starting to be applied in many other inspection scenarios, such as power lines, levees and embankments, confined spaces, ecology, wind turbines, cranes, and real estate. Traditional inspection procedures in most of these cases are costly, time consuming, repetitive, labor-intensive, and technically difficult. The use of UAV alleviates and improves maintenance and risk prevention processes (Besada et al., 2018).

Matsuoka and colleagues showed that using UAV to detect different failures of photovoltaic modules is much faster and more effective than traditional methods. The authors measured the deformation of a large-scale solar power plant using images acquired by a nonmetric digital camera on board a micro-UAS (Matsuoka et al., 2012). Arenella and colleagues documented a technique to detect hot spots in photovoltaic panels (defects causing destructive effects) by analyzing the sequence of thermal images (Arenella, Greco, Saggese, & Vento, 2017). Some researchers have employed UAVs to inspect a photovoltaic array field, using diverse thermal imaging cameras and a visual camera.

Monitoring environmental gases for risk assessment both indoors (gas leaks, fires, mining applications, etc.) and outdoors (agriculture biomass burning emissions, chemical and biological agent detection studies, etc.) may require long periods of observation and a large number of sensors. UAV may substantially complement existing ground sensor networks. For this purpose, a UAV has to be equipped with sensors capable of determining volatile chemical concentrations and detecting gas leakages. Kersnovski, Gonzalez, and Morton proposed a UAV with an onboard camera and a carbon dioxide gas sensor capable of performing autonomous gas sensing while simultaneously visually detecting predefined targets placed at locations inside a room. The system transmits the collected data in real time to a GCS for visualization and analysis through a Web interface (Kersnovski, Gonzalez, & Morton, 2017).

Soil pollution monitoring is another application of UAV technology. For example, Capolupo and colleagues proposed a multi-sensor approach to copper detection proposed system was able to predict copper accumulation points, using a combination of aerial photos taken by drones, micro-rill network modeling, and wetland prediction indices (Capolupo, Pindozzi, Okello, Fiorentino, & Boccia, 2015). UAVs may also be used to inspect contaminated areas, such as in fission reactors for leakage detection, in storage areas of nuclear sources, or even in hazardous scenarios of nuclear disasters. Tang and Shao focused on delivering a system that surveys forests, maps canopy gaps, measures forest canopy height, tracks forest wildfires, and supports intensive forest management (Tang & Shao, 2015). In marine ecology, UAVs may be used, for example, to produce very fine scale maps of fish nursery areas. Some authors have detailed the procedure of aerial photo acquisition (drone and camera settings) and post-processing workflow (i.e., 3D model generation using the motion algorithm and photo-stitching) (Besada et al., 2018).

Road inspection UAV-supported procedures may help to detect early signs of erosion and pavement distress. Branco and Segantine proposed a methodology to automatically obtain information about the conditions of highway asphalt from data collected through remote sensing using a UAV and specific image processing and pattern recognition techniques (Branco & Segantine, 2015).

Railway infrastructure inspection includes camera-based sensing and control methods. In one scenario, the UAV performs infrastructure inspection in close but difficult-to-access areas (such as long bridges or tracks separated from the road by, e.g., a river). A second scenario is oriented to the railway track and records the infrastructure, including tracks, sleepers, points, or cabling. Target detection is carried out using different image descriptors (Speeded-Up Robust Features (SURF), Scale Invariant Feature Transform (SIFT), Features from Accelerated Segment Test (FAST), and Shi-Tomasi), and edge detectors are used for line detection.

Power line detection is another area of research. Santos and colleagues presented a vision-based power line detection algorithm and tested it in multiple backgrounds and weather conditions (Santos et al., 2017). In the energy sector, the use of drones for the maintenance of power lines and transmission towers is already widespread. New diagnosis techniques based on drone use are emerging to improve the detection of problems. Electric tower detection, localization, and tracking were studied by Martinez and colleagues; these researchers proposed a combination of classic computer vision and machine learning (ML) techniques (Martinez, Sampedro, Chauhan, & Campoy, 2014). Priest described a cell tower inspection procedure in which an operator using a UAV and a processing device creates a model of the cell site and compares it to models created in subsequent inspections to determine significant differences (Priest, 2017). Finally, 5G advances, secure Internet of Things (IoT), and swarms of UAVs have been combined into an architecture to guarantee service in critical infrastructures (distributed generation plants, energy transmission and distribution networks, such as electricity cables and electrical isolators and natural gas/liquefied natural gas, tanks, pumps, and pipelines) (Zahariadis, Voulkidis, Karkazis, & Trakadas, 2017).

UAVs are also applied to construction management. In one application, the safety inspection of construction sites was addressed by De Melo and colleagues; they used a drone to provide visual assets to verify the safety checklists in two different projects (De Melo, Costa, Alvares, & Irizarry, 2017). Building inspection is another common application; for example, inspecting a rooftop using a UAV to extract information from a damaged area (Besada et al., 2018).

The project iDeepMon (Benecke, 2018) aims at enhancing shaft surveying technologies to create a fully UAV-based automated process, integrated into the overall control process of an autonomous mine. The mining environment poses defined constraints (variable lighting, etc.). Besada and colleagues described hybrid equipment using a helium gas-filled balloon, with remote-controlled quadcopter propellers, powerful LED lighting, rechargeable batteries, remote-controlled cameras, image stabilizers, and radio frequency transmitters for control and image visualization (Besada et al., 2018).

Many drone applications require photogrammetric data capture of complex 3D objects (buildings, bridges, monuments, antennas, etc.). Saleri and colleagues discussed an operational pipeline tested for photogrammetry in an archaeological site. Algorithms such as structure from motion and multi-view stereo image matching facilitate the generation of dense meshed point clouds (Saleri et al., 2013). Cefalu and colleagues detailed an automatic flight mission planning tool; it generates flight lines while aiming at camera configurations, which maintain a roughly constant object distance, provide sufficient image overlap, and avoid unnecessary stations, based on a coarse digital surface model and an approximate building outline (Cefalu, Haala, Schmohl, Neumann, & Genz, 2017).

The mission type obviously determines the measurement procedure to be completed (in terms of flight type, sensing payload, and required measurements) (González-Jorge, Martinez-Sánchez, Bueno, & Arias, 2017). All these applications need tools to accelerate and partially automate the creation of missions, calculate optimal trajectories, and automatically execute parts of the mission with the least human intervention to achieve cost-effectiveness. Although it is possible to automate the procedure for very specific cases, it is difficult to build a generalizable automated system (Besada et al., 2018).

2.2 Dynamics and Machine Architectures

2.2.1 Dynamic Vehicle Architectures

Various external forces and torques are caused by hydrodynamic forces, thruster forces, gravitational forces, buoyant forces, etc. Figure 2.4 shows the local (vehicle) coordinate system (oxyz) and the global coordinate system (OXYZ) (Yuh, 1994).

2.2.1.1 Rigid Body Motion

The equations of motion for a rigid body of mass m with an arbitrary origin are summarized below:

- Translational motion

$$F = m\left[\dot{U} + \Omega \times R_C + \Omega \times (\Omega \times R_C)\right] \quad (2.1)$$

- Rotational motion

$$G = \frac{d}{dt}([I]\Omega) + m(R_C \times \dot{U}) \quad (2.2)$$

where $F = [X, Y, Z]^T$ is the resultant external force, $G = [K, M, N]^T$ is the resultant external moment, $U = [u, v, w]^T$ is the velocity of the origin, $\Omega = [p, q, r]^T$ is the angular velocity about the origin, $R_C = [x_c, y_c, z_c]^T$ is the position of the center of mass in local coordinates, and $[I]$ is the inertia tensor with respect to the origin (Yuh, 1994).

2.2.1.2 Hydrodynamic Forces and Moments

The hydrodynamic forces and moments acting on a vehicle are described below, assuming the fluid rotation is negligible, and there is a current with a velocity U_f (Yuh, 1994).

- Added Mass: Since the density of water is similar to the density of an remotely operated vehicle (ROV), additional inertia terms must be introduced to account for the effective mass of surrounding fluid that must be accelerated with the vehicle. These added mass

Development of Autonomous Vehicles 49

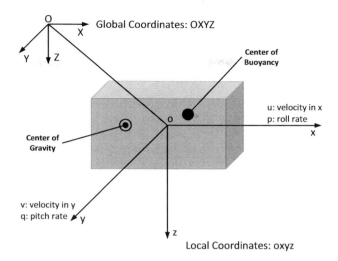

FIGURE 2.4 Vehicle coordinate systems (Yuh, 1994).

coefficients are defined as the proportionality constants which relate each of the linear and angular accelerations with each of the hydrodynamic forces and moments they generate. For example, the hydrodynamic force along the *x*-axis due to acceleration in the *x*-direction is expressed as:

$$X_A = -X_{\dot{u}}\dot{u} \text{ where } X_{\dot{u}} = \frac{\partial X}{\partial \dot{u}} \tag{2.3}$$

In a similar manner, all other added mass coefficients can be defined and assembled into an added mass matrix [*A*]. The force due to added mass can be obtained from:

$$\begin{bmatrix} F_A \\ G_A \end{bmatrix} = -\frac{d}{dt}\left([A] \begin{bmatrix} U_r \\ \Omega \end{bmatrix} \right) \tag{2.4}$$

where U_r is the vehicle velocity relative to the fluid (Yuh, 1994).
- Fluid Motion: For a vehicle traveling at low speeds in the ocean, the forces and moments exerted on the vehicle by fluid motion are significant and must be included in the dynamic model:

$$F_F = m_f U_f \text{ and } G_F = m_f \left(R_B \times U_f \right) \tag{2.5}$$

where m_f is the mass of the fluid displaced by the vehicle, and $R_B = \begin{bmatrix} x_b, & y_b, & z_b \end{bmatrix}^T$ is the position of the center of buoyancy in local coordinates. It should be noted that except for the special case of a neutrally buoyant vehicle, the mass of the fluid displaced by the vehicle will not be equal to the mass of the vehicle. Therefore, the forces and moments produced by vehicle motion and fluid motion cannot be conveniently combined into functions of relative motion only. However, in Equation (2.5), relative velocity can be used because the *added mass* coefficients are dependent only on the body geometry and not on m_f (Yuh, 1994).
- Drag: Drag is usually described as a force proportional to the square of the corresponding relative motion of the vehicle. For example, the drag force along the x-axis due to relative velocity in the *x*-direction is expressed as:

$$-X_{uu}|U_r|U_r,$$

where $X_{uu} = \dfrac{\partial^2 X}{\partial u_r^2}$, and the drag moment along the z-axis due to the angular velocity r is expressed as:

$$-N_{rr}r|r|,$$

where $N_{rr} = \dfrac{\partial^2 N}{\partial r^2}$. The drag force and moment are then denoted by F_H and G_H, respectively (Yuh, 1994).

2.2.1.3 Weight and Buoyancy

The gravitational force F_W and buoyant force F_B are defined in terms of the global coordinate system, so they must be transformed to the local coordinate system. The moments generated by these forces can be expressed in terms of the positions of the center of mass, C, and the center of buoyancy, B:

$$G_W = R_C \times F_W \text{ and } G_B = R_B \times F_B \qquad (2.6)$$

where R_C and R_B are the respective positions of the center of mass and the center of buoyancy in the local coordinate system (Yuh, 1994).

2.2.1.4 Thruster Force

The resultant force and moment of a thruster configuration consisting of N thrusters can be expressed as the vector sum of the force and moment from each individual thruster:

$$F_T = \sum_{}^{N} F_{Ti} \text{ and } G_T = \sum_{}^{N} G_{Ti} + \sum_{}^{N} R_{Ti} \times F_{Ti} \qquad (2.7)$$

where R_{Ti} is the position of the i-th thruster in local coordinates. All external forces and torques can be consolidated into the rigid body equations of motion, and the dynamic vehicle model can be described by:

$$M\dot{V} + C(V, U_f)V + G(X_p) = \Gamma \qquad (2.8)$$

$$\dot{X}_p = J(X_p)V \qquad (2.9)$$

$$\Gamma = RF_T \qquad (2.10)$$

where $V = [u, v, w, p, q, r]^T$; the vector X_p represents the vehicle position and orientation in global coordinates, the 6×6 inertia matrix $M = M_r$ (rigid body inertia) + M_a(added mass); C is a 6×6 matrix including all the nonlinear dynamic terms with inertia velocity terms, terms associated with the forces, and torques exerted on the vehicle by fluid motion, drag forces, and torques; G is a vector of gravity and buoyancy; $J(X_p)$ is a 6×6 Euler angle transformation matrix between the vehicle and global coordinate systems; the resulting six-dimensional vector represents the forces and torques generated by the thruster forces F_T with a matrix R that is given by the thrusters and the control surface configuration (Yuh, 1994).

2.2.2 Machine Architectures

The organization of a robotic system determines its capacity to perform tasks and react to events. The control structure of an autonomous robot must have both decision-making and reactive capabilities. Situations must be anticipated, and the correct actions decided by the robot. Tasks must be instantiated

Development of Autonomous Vehicles

and refined at execution time according to the actual context. The robot must react in a timely fashion to events. To meet these requirements, a robot control structure should have the following properties:

- Programmability: A useful robot cannot be designed for a single environment or task and programmed in detail. It should be able to achieve multiple tasks described at an abstract level. Its functions should be easily combined according to the task to be executed.
- Autonomy and adaptability: The robot should be able to carry out its actions and to refine or modify the task and its own behavior according to the current goal and execution context as perceived.
- Reactivity: The robot has to take into account events with time bounds compatible with the correct and efficient achievement of its goals (including its own safety).
- Consistent behavior: The reactions of the robot to events must be guided by the objectives of its task.
- Robustness: The control architecture should be able to exploit the redundancy of the processing functions. Robustness will require the control to be decentralized to some extent.
- Learning capabilities: The extensibility integration of new functions and definition of new tasks should be easy. The architecture should make learning possible.

Robot control architectures are at the core of the design of autonomous agents, and many authors have addressed them. The approaches differ, sometimes based on the philosophical standpoint: for example, some projects aim at imitating living beings, while others are more artificial intelligence (AI) oriented. One trend is to draw on biology and ethology; this has mainly yielded stimuli-response based systems. A second trend is to use symbolic representations and some reasoning capacities.

Together with action and perception capacities, a robot architecture has to be a suitable framework for the interaction between deliberation and action. Deliberation is a goal-oriented process, wherein the robot anticipates its actions and the evolution of the world, and also a time-bounded context-dependent decision-making process to ensure a timely response to events.

The robot will face high emergency situations requiring a first and immediate reflex reaction. But such situations often require second step actions to correct the robot's behavior more globally. This will need more information gathering or more searches in some representation space of the environment and of the robot state.

Hence, the architecture should include several task-oriented and event-oriented closed loops to achieve both anticipation capacities and real-time behavior.

The robot architecture can be decomposed into three levels (Figure 2.5) with different temporal constraints and manipulating different data representations (Alami, Chatila, Fleury, Ghallab, & Ingrand, 1998).

From bottom up, the levels are:

- A functional level: This includes all the robot's basic built-in action and perception capacities. These processing functions and control loops (image processing, obstacle avoidance, motion control, etc.) are encapsulated in controllable communicating modules. To make this level as hardware independent as possible and, hence, portable from one robot to another, it is interfaced with the sensors and effectors for a logical robot level. To accomplish a task, the modules are activated by the next level.
- A decision level: This level includes the capacities of producing the task plan and supervising its execution, while at the same time, being reactive to events from the previous level. This level may be decomposed into two or more layers, based on the same conceptual design, but using different representation abstractions or different algorithmic tools, and having different temporal properties. This choice is mainly application dependent. In the architecture of an autonomous agent, there are typically two such layers.

The architecture has been implemented in several mobile robots. The implementations differ in the number of layers of the decision level or in the tools used to realize each level (Alami, Chatila, Fleury, Ghallab, & Ingrand, 1998).

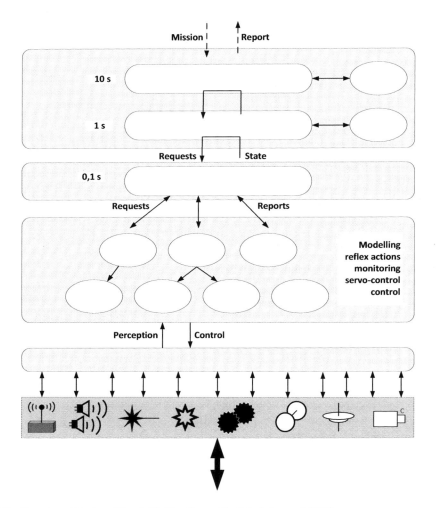

FIGURE 2.5 Generic architecture (Alami, Chatila, Fleury, Ghallab, & Ingrand, 1998).

2.3 Robots and Machine Intelligence

2.3.1 Machine Intelligence (MI)

Machine Intelligence (MI) has been defined as a unifying term for what others call Machine Learning (ML) and Artificial Intelligence (AI)" (Doorn, Bloem, Duivestein, & Ommeren, 2015).

2.3.1.1 What Is Machine Intelligence?

Existing at the intersection of ML and AI, MI is advanced computing that enables a machine to interact with its environment in an intelligent way.

2.3.1.2 What Is Artificial Intelligence?

AI is a set of algorithms, processes, and methodologies that allow a computer system to perform tasks that would normally require human-level intelligence. AI can appear as a component in a larger system or in the form of a computer application, digital agent, or autonomous machine.

2.3.1.3 What Is Machine Learning?

ML is a field of computing science focused on developing algorithms that enable a computer system to independently learn from and continuously adapt to data without being explicitly programmed for those data. ML is a crucial component in many AI systems.

2.3.1.3.1 Where Is Machine Learning Used?

- Recommender systems (e.g., Netflix or Amazon)
- Contextual Web searches (e.g., Google)
- Intelligent digital assistants (e.g., Cortana or Siri)
- Game-playing AI (e.g., AlphaGo or Cepheus)
- AVs
- Email spam filters.

2.3.1.4 Why Is Machine Intelligence Important?

MI technologies have experienced a global resurgence due to growing volumes and varieties of data, the utility of these data in training smart systems, and an increased awareness of the value of data in providing a competitive edge in business.

MI is expected to form the basis for most technological and business advancements for years to come. A report by McKinsey & Company says technologies that employ MI will have an economic impact of over $50 trillion by 2025.

MI allows organizations to operate more efficiently and effectively, using data to predict the future and manage the present. Computer systems with MI can perform a variety of tasks, including the following:

- Optimize and automate processes
- Extract and classify data
- Detect, analyze, and predict trends/patterns
- Enhance interactions with humans/the environment (Alberta Machine Intelligence Institute (AMII)).

2.3.2 State-of-the-Art Machine Intelligence

Today, scalability is even more important than intelligence. Investors are interested in plug-and play solutions and are looking for speed. Wit.ai (bought by Facebook at the beginning of 2015) and HyperScience are good examples. The former has developed an API for IoT products, and the latter has automated the data scientist, arguably helping to solve the shortage of data scientists (Doorn, Bloem, Duivestein, & Ommeren, 2015).

2.3.2.1 Robot Apps for Faster Growth

The cloud can enable robots to make the human world better. It can become an overall brain in a "World Wide Web for robots," in which robots upload their experiences. Other robots will be able to download these experiences, learn from them, and apply them in their own surroundings. The robotic vacuum cleaner is a good example. As soon as these devices encounter an unknown object, they don't know what to do. However, if they could send a picture to the cloud asking what it might be, other robots or humans could supply them with an answer.

Robot platforms featuring app stores for robot tasks are a logical continuation of what is happening with smartphones. There is general consensus that this will succeed. For the time being, the app stores for robots are filled with gadgets. Robotappstore.com has, for example, applications for popular robots such as the Aibo (dog) and the Roomba (vacuum cleaner) to let them create music or allow them to dance. This is interesting but not the kind of intelligence that is expected.

A recent estimate cited 2,948 investors in 423 new MI organizations. An updated list of names of these investors and the companies in which they invest can be found at: https://angel.co/artificial-intelligence. The companies are divided into the following categories: algorithms, computer vision, image recognition, ML, natural language processing, semantic Web, speech recognition, and intelligent systems (Doorn, Bloem, Duivestein, & Ommeren, 2015).

2.3.3 The Machine Intelligence Landscape

The first edition of the standard textbook *Artificial Intelligence: A Modern Approach* was published 20 years ago, following four decades of AI research. It brought a dozen or so barely related subfields together in a coherent framework. Authors Stuart Russell and Peter Norvig explain what this modern approach means in the preface:

"We have tried to synthesize what is now known into a common framework, rather than trying to explain each subfield of AI in its own historical context" (Russell & Norvig, 2009).

Shivon Zilis, venture capital investor at Bloomberg Beta, noticed the same thing in 2015 when she was planning to map the state of affairs in areas, companies, and products. She prefers the term machine intelligence (MI) rather than AI. Her overview contains an extensive list of approximately 240 companies and labels under the 35 labels that together form the converging field of MI (www.shivonzilis.com/). Ultimately, the success of MI is determined by human–machine interactions (HMIs).

2.3.3.1 Rethinking

One company that is rethinking how it does business is Tachyus, an oil and gas company. This company makes it clear which way things are heading: supporting the knowledge worker and directly contributing to the top line growth. Tachyus offers the industry an ML solution that can increase the production capacity of oil sources by 20%–30%.

Tachyus does not just focus on how the production of existing sources may increase; it also identifies the best locations for new drilling. This ML tool offers "Rethinking as-a-Service," basically as a practical accelerator of the process.

Another example is MetaMind. It was founded with money from Salesforce.com and Khosla Ventures. CEO David Socher created a user-friendly drag-and-drop application which makes it possible to start immediately. Image and text patterns are recognized and converted into useful analyses. MetaMind may be listed as an AI company, but it is a partner of the company Caresharing and is now capable of diagnosing diabetic retinopathy. MetaMind also collaborates with vRad to analyze strokes. This implies that easily applicable AI tools can quickly reach a certain vertical (Doorn, Bloem, Duivestein, & Ommeren, 2015).

2.3.3.2 Machine Intelligence Open Source Landscape

MI offers an open source toolbox. It has countless possibilities and requires more personal development. GitHub is an excellent source for this. Figure 2.6 shows a layered model of the open source landscape that can be read upside down or in circles.

The new drivers of the MI field are the many new supporting technologies. The core technologies are proven supporters of AI. Many fundamental and practical developments are taking place, such as adding pattern recognizing and ML. These devices traditionally focus on human–computer interactions and HMIs.

In short, ground-breaking developments in the field of AI and MI have led to many concrete rethinking applications (Doorn, Bloem, Duivestein, & Ommeren, 2015).

Machine Intelligence	
Supporting Technologies	Harward, Data Prep, Data Collection
Core Technologies	Artificial Intelligence, Deep Learning, Machine Learning, NPL Platforms, Predictive APIs, Image Recognition, Speech Recognition
Rethinking Humans/HCI/HMI	Augmented Reality, Gestural Computing, Robotics, Emotional Recognition
Rethinking Enterprise	Sales, Security/Authentication, Fraud Detection, HR/Recruiting, Marketing, Personal Assistant, Intelligence Tools
Rethinking Industries	Adtech, Agriculture, Education, Finance, Legal, Manufacturing, Medical, Oil and Gas, Media/Content, Consumer Finance, Philanthropies, Automotive, Diagnostics, Retail

FIGURE 2.6 Framework for the state-of-the-art open source landscape (Doorn, Bloem, Duivestein, & Ommeren, 2015).

2.3.4 Machine Intelligence versus Artificial Intelligence

2.3.4.1 Artificial Intelligence (AI) Definitions According to Different Authors

- The study of how to make computers do things at which people are presently better.
- Intelligent behavior in artifacts, i.e., perception, reasoning, learning, communicating, and acting in complex environments.
- The study of the computations that make it possible to perceive, reason, and act (Arroyo, 2015).

2.3.4.2 Machine Intelligence (MI) Definition

MI represents the ability to build machines that exhibit a high degree of sophistication and can operate autonomously in "their" environment. For example, a roach is a highly intelligent insect, so a "roach-like" machine would also be considered "intelligent." Put otherwise, it is computer intelligence grounded in reality—and it can be realized.

2.3.4.3 Two Approaches to AI/MI

- Symbol processing approaches:
  - Are based on the physical-symbol hypotheses
  - Apply logical reasoning and deduction to declarative knowledge bases
  - Follow the adage "in knowledge lies power"
  - Deduce consequences based on a first-order predicate calculus
  - Represent domain and problem knowledge
  - Are knowledge-based systems
  - Use top-down design methodologies
  - Have three levels: a knowledge level, a symbol level (usually represented using a LISt Processor (LISP) and other special-purpose AI languages), and a symbol processing implementation level.

- Sub-symbolic approaches: animat and subsumption:
  - Are based on the physical grounding hypotheses: signals are more appropriate units than symbols
  - Follow the adage "in the sensors lies the power"
  - Deduce consequences based on perception
  - Document emergent behavior
  - Use bottom-up design methodologies
  - Include neural networks, situated automata, control theory, dynamic systems.

MIL emphasizes the animat and subsumption approach to autonomous agent design and construction (Arroyo, 2015).

2.3.4.4 Application Areas of Classical Artificial Intelligence

- Natural language processing
- Intelligent database query
- Expert systems
- Robot plans
- Automatic programming
- Scheduling problems
- Perception
- Game playing
- Knowledge representation
- Theorem proving
- Logic programming (Arroyo, 2015).

2.3.4.5 Application Areas of Machine Intelligence

- Mathematics:
  - Dynamic programming
  - Nonlinear dynamics
  - Gradient search techniques
  - Fuzzy logic—Markov processes
  - Chaos
  - Opinion-guided reactions.
- Psychology:
  - Conditioned responses
  - Stimulus reinforcement.
- Ethology
- Reactive behavior:
  - No memory.
- Behavior with memory
- Reinforcement learning:
  - Supervised
  - Unsupervised.
- Evolutionary learning:
  - Genetic algorithms.

- Gradient search techniques:
  - Temporal differences
  - Reinforcement & Q-learning
  - Neural networks (Arroyo, 2015).

2.4 Programming Autonomous Robots

2.4.1 Autonomous Robots' Requirements

Many different robots have been developed to handle various situations of linear assets, buildings, ship hulls, or other human-made structures. However, most are limited to special situations or applications. To execute the desired tasks, autonomous robots, as well as all other technical systems, have to fulfil certain requirements. The requirements and their importance and focus depend on the individual application or tasks. However, a general set of requirements is the following:

1. Velocity and mobility: Vehicle speed and dynamics (ability to move) are two main aspects of robot design. Depending on the dimension of the linear asset, it may have to reach a relatively high velocity for sufficiently fast navigation between inspection areas or similar points of action. Another requirement is related to the desired manipulation and positioning capabilities of the system. This includes the precision of locomotion and its trajectory, since some inspection sensors need to be moved in a smooth and continuous way over the surface. The robot may also need to move sideways or to turn 360° to position sensors or tools. The system dynamics should be able to handle the various terrains and reach all positions of the asset.
2. Payload: Depending on the application, the system must be able to carry payloads of different weights. For example, in the case of steel piping, a payload of 5 kg or more is mandatory to carry ultrasonic inspection sensors. This requires a much bigger robot than a system which just needs a simple camera with a weight of several hundred grams. In other words, the dimension, adhesion, and motion components of the robot need to be adapted for the application.
3. Reliability and safety: An important nonfunctional aspect is the robustness of the system. If the autonomous robot fails frequently during one inspection task, it is not usable in practice. The requirements of reliability and safety include robust hardware, optimal controllers, and methods to detect and handle hazardous situations and to recover from them.
4. Usability: Velocity, maneuverability, and the ability to carry a certain payload are important, but they are only the basis of the general operability of the system. To bring a robotic system into application, it has to be more powerful, more efficient, and less dangerous than common approaches, for example, in terms of inspection devices. This includes aspects of maintainability and a broad range of other tasks. Therefore, it must be able to carry different payloads (e.g., inspection sensors or tools) depending on the desired task, parts need to be easily replaceable, and the operation must be faster and less complicated than existing approaches. Aspects like energy consumption, weight, or dimension of the system can be important as well. Based on the individual task, a robot developer has to decide which requirements have to be fulfilled and select a suitable locomotion and attraction principle (Seneviratne, Ciani, Catelani, & Galar, 2018).

2.4.2 Active Object Programming

This section reviews the main features of active object programming (A2OP), particularly the features that increase the expressivity of the object model (OM) and those required for autonomous robot software prototyping.

2.4.2.1 Main Features

Active object-oriented programming (A2OP) is not like object-oriented programming (OOP). Rather, A2OP is fully contained in the OM. This kind of programming has not been used on a large scale, perhaps because OOP has been powerful enough. With the increase in the needs of autonomous and reactive systems, as well as distributed systems through networks, A2OP may become more attractive.

In OOP, we think in terms of code structure: the sharing of data and functions/methods in classes/ objects. In A2OP, we think in terms of evolution and transition from a state to another. A2OP underlines the dynamic nature of active objects; a regular object code is basically static, even if we can add dynamic mechanisms.

To do so and to code active objects, we must provide any active object instance (AOI) with its own execution thread. These threads are like Unix threads, but this is oRis which creates and manages all the active object execution threads. Each execution thread is responsible for a set of instructions which are executed each time the thread is designated by oRis's scheduler. When the set of instructions arrives at its logical end, oRis's scheduler designates another thread. Designation can be made following two different modes. The first is fully sequential: from one oRis simulation cycle to the other, all the threads are designated in the same order. We hardly ever use this mode because it provokes an artificial side effect that does not exist in the real phenomena we want to simulate. Actually, it artificially introduces a priority between the threads and the instances. The second mode is random. More precisely, in the bounds of a simulation cycle, once a thread has executed its instructions one time, another is randomly designated. Each thread, or instance, can only be designated one single time inside a simulation cycle.

As we are using several threads, oRis gives the possibility of two kinds of multitasking. The first is cooperative. At the logical end of every main{}, there is preemption of the active thread, and the oRis's scheduler designates another one. The second multitasking mode requires to set, in milliseconds, the length of the execution duration. Once a thread is active, it will remain so until the preemption order by the end of the execution lap. Once more, oRis's scheduler designates another thread after each preemption. When using the cooperative mode, writing a main{} or any set of instructions that may be lying on an execution thread must follow precise rules. The most important concerns the logical size of the main{}. It must be short. We want to simulate the activities of entities which are supposed to live or to be active in the same time, or temporality.

Thus, if one particular instance stays active too much longer, to the detriment of the other instances, it may have a logical lead over the others. A temporal drift may occur, and it will be as if the time is not running at the same speed for all objects. From a formal point of view, the preemptive mode is the only one able to guarantee the trans-object's temporal uniformity. However, the cooperative mode, with a logically short main{}, could be another choice.

Any AOI can have as many execution threads as the programmer wants. This makes the AOP efficient for software prototyping. We are trying to design the behaviors of our autonomous robots. Thus, for any instance, we will be able to simulate any part, or function, of the robot using as many threads as we need: one execution thread for a specific part or function.

Parallel code gives the instance what we may call an execution autonomy. However, this autonomy leads any instance to another: the functional autonomy. This concept is the most important one when programming with the active OM. In object programming, when data or methods are set as public in a class and in its instances, any other instance may use these public data or call upon any methods. When an object is using other data or methods, it is doing so because it has been ordered to proceed this way by the main{} of the program. With active objects, the things could be quite different. In an active object, we may divide what could be private from what could be public. However, this makes no sense. Any active object is able to know who wants to read its data or use its methods. Thus, it may deny access. The object could act this way because it does not want a particular instance to use its methods, or because it is not, at this moment, able to execute the method. This functional autonomy, or ability to refuse to do something for another instance, opens a new way to conceive autonomous systems.

This autonomy could help us, not to solve conflicts that may occur inside the robot or with its environment, but to prevent conflicts from occurring. We must understand that OOP stands on the principle of code deconcentration. This means the logical function of the program is going to be divided into the

Development of Autonomous Vehicles

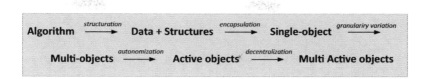

FIGURE 2.7 Programming scheme linking (Cozien, 2001).

classes. When using the A2OP, we try to maximize this principle, and we talk about code decentralization. Figure 2.7 shows the whole process from algorithm to multi-active object scheme. In procedural programming, conflicts occur because any part of the program handles the overall logical function and also because it is difficult for the programmer to anticipate all the situations that may provoke a conflict. In OOP, the programmer is able to check locally, class after class, if the objects do not include internal logical aberrations. However, in the overall program's main{}, some conflicts may remain.

In A2OP, we are still able to check if any object's logical function is self-coherent, but as there is no more overall main{}, we have to locally focus on each main{} of each object. This may seem like one more constraint. However, if our active object class is logically self-coherent, the main{}, as the autonomous execution capacity of the logical active object's function, is necessarily self-coherent. The functional autonomy should prevent conflicts from occurring because the programmer, in the active object main{}, can explicitly and exhaustively write the situations or set of conditions when the instance should refuse to be at another instance's disposal. This feature changes the programming style.

There is much more to say about this programming model, but we do not have enough space in this book. However, the three next sections give more information (Cozien, 2001).

2.4.2.2 oRis Programming Language

oRis programming language is a tool used for software prototyping. It is object oriented and allows A2OP. It can be twinned to a simulator and give full 3D visualization of the simulations. It is interpreted, not compiled. This last point is important. An interpreted language gives the programmer much more control of the code and its execution. Furthermore, it is much easier, and it requires far less time to go back to the code during the prototyping (and debugging) phase.

The syntax of oRis is quite close to C++, but the object concepts are used and are available to the programmer, as in Java. oRis is Prolog and Java friendly: it can execute both Prolog and Java instructions. It includes a number of packages, from the handling of 3D shapes to TCP/IP communications. It also includes a fuzzy logic package to design fuzzy inference engines. Many other packages are available.

oRis is fully interpreted and the concept of active object is explicitly achieved. Thus, objects can pilot and fully control the simulator when they are executed. This feature allows the programmer to design his or her own simulators and simulations, creating different simulators for different situations. Furthermore, it is easy to design the active objects that make the simulated environment of autonomous robots. As we use active objects for the environment, we can give them almost all the behaviors we want; this is convenient for mobile robot prototyping.

In oRis, all the active object instructions lie on one or several threads. oRis's scheduler designates them at random, and we may choose either a cooperative or a preemptive multitasking mode. In general, we use the second mode which guarantees no temporal drift.

One of the most important features of oRis is its ability to handle dynamic code. That is to say, that any instance or the programmer is able to change the inside code of any other instance or class during the execution of the program with no need to interrupt it. This leads to a very important concept in the active object scheme, *instance granularity*. Even if we can change the code of any instance, partly or totally, this instance will keep the name of its class. However, this particular instance will be quite different from the other instances of the same class. Let's assume we change, in a particular instance, the code of the *reaction* method which is executed each time a broadcasted message of a particular type is received. Thus, when the message is broadcasted, all the sensitive instances will execute the reaction method, but one will produce different effects, because the instructions of its reaction method have been changed.

```
class Mobile : Object2d
{ ...
 void new(void);
 void delete(void);
 void randomWalk(float angle,float step);
 void avoid(float direction,float angle,float step);
 void main(void);}

void Mobile::new(void)
{...}

void Mobile::delete(void)
{...}

void Mobile::randomWalk(float angle,float step)
{...}

void Mobile::avoid(float direction,float angle,float step)
{...}

void Mobile::main(void)
{Mobile obs;
 float distance,angle;
 obs=(Mobile)viewFirst(angle,distance,"Mobile",0,3*.size,0);
 if(obs)
    {if(distance<_threshold)
        {delete obs;
          delete this;}
      else avoid(angle,.pi./8,.size/2);}
 else randomWalk(.pi./8,.size/2);}
```

FIGURE 2.8 oRis code example (Cozien, 2001).

So the same cause (a broadcasted message) can produce different effects, depending on which instance is *responding*. This is the same concept as the dynamic link, itself a consequence of the polymorphic class link.

In A2OP, we are getting closer to the instance. So we translate the object concept from the static class point of view to the *dynamic* instance point of view.

Figure 2.8 gives a sample of the kind of code we may write with oRis. Obviously, the syntax is very similar to C++. oRis has many other features, some directly connected to active object issues, others to the ability to make simulations with 2D or 3D objects. As in C++ or Java, we write classes and their methods. When we want to write active object classes, the first and easiest way is to write a particular class named main{}. In this class, we put the *behavior* of the object. The instructions contained in the main{} are executed on a thread (Cozien, 2001).

2.4.2.3 Giving Autonomy to Objects

One way to explore the OM is to give more autonomy to the objects in the four following ways:

1. Autonomy of execution
2. Autonomy of perception
3. Autonomy of decision ≡ functional autonomy
4. Autonomy of code.

Note that we use the words "perception" and "decision," but there is no human connotation in these words.

The autonomy of execution lies in the use of parallel threads as previously described. The autonomy of perception underlines the local limitation of each object. Here, "perception" means the ability to read variables and, more generally, to access data in other objects. This local aspect means the objects cannot read all the variables or access all the objects. They have a perception/reading field. This field is logically defined by the third point: the autonomy of decision. As each active object is loading its own main{}, it loads its own instructions. These instructions consist mainly of how the object will act according to the information it gets in its perception/reading field.

However, there is a much more original way to give autonomy of decision to active objects. First, we must consider that the OM is basically a master ↔ servant model. When any object is requested to do something by the main{} or by any other object, it has no choice but to execute the request: this a synchronous method call. When using active objects, we do not change much. Any object can make a synchronous method call to another object on its own. In that case, the second object has to answer, but because of the multi-thread structure, requests and answers may be delayed. Thus, we are in a client ↔ server configuration. This means the server must answer as soon as it is available. If we go one step further, we may imagine that any active object can send what we may call an invitation to tender to all or at least a group of active objects. At this point, any other AO can make a bid. We are in a real negotiation process. However, this is a human point of view, and negotiation should imply intelligence.

Nevertheless, we could consider using the underlying principle of negotiation: the autonomy of proposal. If AO A makes a request to AO B, B may know who is making the call, and it may answer or act differently depending who is making the call. B may not answer or answer "no" to the request. This is unthinkable in the usual way of using (active) objects.

In oRis, we use a command to identify the origin of the request, as shown in Figure 2.9 (Cozien, 2001).

In void B :: dancing(void), we use the command isA() to let B know the type (class name) of who is making the request and to get this information by the command that(). We use quotation marks to give B to the right type: isA("A"). We do so to express the dynamic aspect of the process. The expression "A" may change during the running of the program, and any object belonging to "A" class or any class inheriting from "A" (polymorphic links) could match the isA("A").

The code in Figure 2.9 shows how the programmer can give to any object the ability to compute and to act differently depending on the origin of the request. We can extend this principle to all the means of communication between objects. From a human point of view, behaving this way seems obvious, but this is not a usual way to make object programs. We must then ask ourselves if introducing this kind of autonomy is helpful and if it increases the expressivity of the programming model (Cozien, 2001).

```
class A
{ ...
  B _b;
  void main(void); }

void A::main(void)
{ _b->dancing(void);
  ... }

class B
{ ...
  void dancing(void) }

void B::dancing(void)
{ if that isA(''A'') then moveMyBody()
  else println(''I am not dancing for you''); }
```

FIGURE 2.9 Autonomy of decision (Cozien, 2001).

	Dependent Behaviour	Autonomous Behaviour
Synchronous	Method call	Call for bid
Asynchronous	Client ←→ Server	Stigmergy

FIGURE 2.10 From synchronous dependence to asynchronous autonomy (Cozien, 2001).

2.4.2.4 Synchonous versus Asynchronous Communications

In A2OP and in any distributed programming model like a *client ↔ server* one, the synchronous call method is the easiest and the most efficient. However, as soon as we leave a *client ↔ server* structure and give autonomy to the objects, the object communication could be unappropriated in some cases. Furthermore, if the distributed aspect becomes geographical, as in Web applications, a fully synchronous communication may not work. In this latter case, and if we wish to maximize the autonomy, the objects should not communicate directly, one to one. Instead, we should work in terms of "call for bid."

If a call for bid is coherent with the functional autonomy given to active objects, it is not a fully asynchronous means of communication. If the call for bid can be made in an asynchronous way, however, when the requested object answers the bid, it does so in a synchronous way. Nevertheless, the requested object answering the bid can do so when and as it wants.

Figure 2.10 shows the cross-links between dependence and autonomy on the one hand, and synchronous and asynchronous means of communication on the other hand (Cozien, 2001).

The concept of "stymergy" is new in computer science. It is more commonly used in the field of social insects to denote the possibility of communicating through a common environment. It was first applied to insect nest construction. Entomologists argue that no one insect has the plan of the nest; every insect is guided by every other insect.

When we apply this to programming, any object can modify the environment (it writes in any variable) and any other object may read this new value which may, in turn, influence the action of this second object in the next cycle. Thus, the state of the whole system is the result of all the objects' actions and is completely recorded in the variables' values. When one object reads a (set of) variable(s) to make a choice, this choice is then the consequence of the actions of other objects.

Figure 2.10 is simple, but there is no easy way in practice to separate what is autonomous from what is dependent, what is synchronous from what is not. If we need reference marks when programming, we particularly need expressivity in our programming means, including in programming languages and any method for application conception and design. Using OOP and any of its extensions, "agent programming" gives great expressivity and the ability to explicitly work in the same (distributed) application on several scales (Cozien, 2001).

2.5 Adaptive Algorithms and Their Use

There is great interest in using mobile robots on land, under water, and in air, as sensor-carrying platforms to perform sampling missions, such as searching for harmful biological and chemical agents, carrying out search and rescue operations in disaster areas, or performing environmental mapping and monitoring.

Even though mobility introduces additional degrees of complexity in managing an untethered collection of vehicles, it allows the repositioning of the onboard sensors. This, in turn, can greatly expand the coverage and survivability of the sensor network. In the context of AUVs, many important issues of the deployment architecture have not been fully addressed, including the AUV size, cost, and coverage trade-off, the selection of appropriate information measures to guide and evaluate the mission and the distribution of computation and communication among the AVs.

Development of Autonomous Vehicles 63

Sampling is a broad methodology for gathering statistics about physical and social phenomena; it provides a data source for predictive modeling in oceanography and meteorology. Adaptive sampling denotes sampling strategies that can change depending on prior measurements or analysis and thus allow adaptation to dynamic or unknown scenarios. One such scenario involves the deployment of multiple underwater vehicles for the environmental monitoring of large bodies of water, such as oceans, harbors, lakes, rivers, and estuaries. Predictive models and maps can be created by repeated measurements of physical characteristics, such as water temperature, dissolved oxygen, current strength and direction, and bathymetry. However, because the sampling volume could be quite large, only a limited number of measurements are usually available. Intuitively, a deliberate sampling strategy based on models will be more efficient than a random sampling strategy (Popa et al., 2004).

2.5.1 Adaptive Sampling Algorithms

2.5.1.1 Problem Formulation

We define the generic adaptive sampling problem as follows:

Given N underwater vehicles sampling a space-time field distribution, the vehicles should be directed to sample such that:

- Uncertainty in the knowledge of the field distribution is minimized.
- Network utilization is optimized, given bandwidth constraints between vehicles that vary with the distance between them.
- Additional secondary objectives, such as minimizing energy consumption, obstacle avoidance, and sampling duration, are minimized.

Addressing the AS problem involves four different aspects, depicted in Figure 2.11 and listed below:

- Field variable estimation using adaptive sampling
- Sensor/robot localization using SLAM-type algorithms

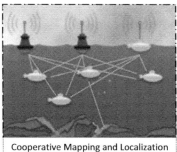
Cooperative Mapping and Localization

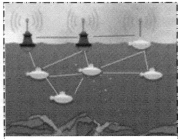
Distributed Network Architecture

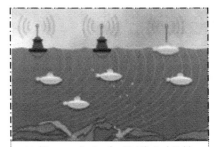
Multi-sensor Fusion for Distributed Fields

FIGURE 2.11 Schematic diagram of fundamental problems related to adaptive sampling (Popa et al., 2004).

- Robot repositioning in the presence of physical and energy constraints
- Optimization of communication protocol during reconfiguration.

Uncertainties in localization and in sensor measurements are especially relevant for underwater vehicles, since position estimates are often inaccurate due to navigational errors from dead reckoning. In what follows, we describe an approach based on model parameter estimation for the field variable, which we then use as an additional constraint to reduce the uncertainty in AUV localization. For example, if the field distribution is a linear function of the sample location, and assuming two of the sample coordinates can be measured accurately while the third one is inaccurate, we can certainly determine the third coordinate from the additional constraint imposed by the field distribution (Popa et al., 2004).

2.5.1.2 Information Measures and the Kalman Filter

Let $X_i[k]$ denote the 3D position of the i-th AUV at sample number k. Assuming that a kinematic model for the AUV is a sufficiently accurate representation, the i-th vehicle kinematics is usually nonlinear and non-holonomic and describes the state evolution as:

$$X_i[k+1] = X_i[k] + h\big(X_i[k], u_i[k]\big) + G w_i[k] \tag{2.11}$$

where U_i is the control input to the vehicle, and W_i is state measurement noise, assumed to be white, with zero mean and covariance matrix $E[w_i[k]w_i[k]] = Q_i$. Without loss of generality, we require that the N vehicles be synchronized in time at each sample point k. At each sample point, the i-th vehicle measurement model is written as:

$$g\big(X_i[k], A_i[k], Y_i[k]+V_i[k]\big) = 0 \tag{2.12}$$

where g is a nonlinear function of the vehicle position, $X_i[k]$, measurements of a noisy field variable $Y_i[k]$, with a white noise component with zero mean and covariance matrix V_i. The vector A is a set of known or unknown coefficients describing the field variable dependence with the sample location. If the set of coefficients is unknown but constant, the vector A will be added to the overall system state, and its evolution will be governed simply by $A[k+1] = A[k]$.

The simultaneous sampling and navigation estimation problem reduces to estimating the overall state vector $X[k] = \big(X_1[k]\ldots X_N[k], A\big)$, along with its covariance matrix $P[k]$.

A solution is available with the extended Kalman filter (EKF) by using Taylor expansion approximations of the nonlinear functions h, and g, and reducing the problem to a linear Kalman filter estimation problem. By the Taylor series expansion,

$$X[k] = \frac{\partial h(X_i[[k|k+1]], u_i[k+1])}{\partial X}(X_i[K+1] - X_i[k|k+1]) +$$

$$+ X_i[k|k+1] + G w_i[k],$$

$$0 = g(X_i[k], Y_i[k]+V_k) = g(X_i[[k|k+1]], Y_i[k]) + \frac{\partial g(X_i[[k|k+1]], Y_i[k])}{\partial Y} V_i[k]$$

$$+ \frac{\partial g(X_i[[k|k+1]], Y_i[k])}{\partial X}(X_i[k] - X_i[k|k+1]),$$

Development of Autonomous Vehicles

$$X_i[k|k+1] = (X_i[k+1] + [h]([k+1], u_1[k+1])]$$

If we denote

$$M_i[k] = \frac{\partial g(X_i[[k|k+1]][[k|kV]1], Y_i[k])}{\partial X}, V_i[k] = \frac{\partial g(X_i[[k|k+1]], Y_i[k])}{\partial Y} V_i[k],$$

$$Z_i[k] = g(X_i[[k|k+1]], Y_i[k]) - \frac{\partial g(X_i[[k|kV]1], Y_i[k])}{\partial X}[X_i[k|k+1]]$$

then, the linearized measurement equation becomes:

$$Z_i[k] = M_i[k]X_i[k] + V_i[k],$$

where the noise covariance is now:

$$E[\dot{V}_i[k]\dot{V}_i[k]] = \frac{\partial g(X_i[[k|k+1]], Y_i[k])}{\partial Y} V_i \cdot \frac{\partial g(X_i[[k|k+1]], Y_i[k])^T}{\partial Y} V V_i$$

and the overall Kalman filter formulation becomes the following:

- Prediction of state:

$$X_i[k|k+1] = (X_i[k+1] + [h([k+1], u_i[k+1])$$

- Prediction of covariance matrix:

$$P_i[k|k+1] = \frac{\partial h_{k+1}}{\partial X} P_i[k|k+1] \frac{\partial h_{k+1}}{\partial X}^T + GQ_iG^T$$

- Kalman gain:

$$K_i[k] = P_i[k|k+1]M_i[k]^T (M_i[k]P_i[k|k+1]M_i[k]^T + V_i[k])^{-1}$$

- State estimation equation:

$$X_i[k] = X_i[k|k+1]K_ig(X_i[k|k+1], M_i[k]^T + V_i[k])^{-1}$$

- Covariance matrix equation:

$$P_i[k] = (I - K_i[k]M_i[k])P_i[k|k+1]$$

A common measure of uncertainty is provided by the entropy of the probability density function for the estimated state:

$$H(k) = -\int p(x)\log(p(x))dx, \qquad (2.13)$$

which, for a Gaussian pdf, is related to the state covariance through:

$$H(k) = \frac{1}{2}\ln((2\pi e)^d |P[k]|), \qquad (2.14)$$

The effectiveness of the adaptive sampling algorithm (ASA) can be determined by comparing the entropy before and after a next sampling step. This could be expressed by the Kullback–Leibler divergence expressing the dissimilarity between two pdf's $p(x)$ and $q(x)$. The measure of difficulty in discriminating in favor of p against q can be expressed as:

$$D(p:q) = \int p(x)\log(p(x)/q(x))dx,$$

and the overall measure of dissimilarity is:

$$J(p,q) = D(p:q) + D(p:q),$$

which, for the behavior of the i-th AUV using the EFK approach outlined below, becomes:

$$J(k+1,k) = D(P[k+1] : P[k]) + D(P[k]:P[k+1])$$

The ASA will then seek to sample at a new location to minimize the divergence function. A more intuitive measure of effectiveness for the ASA algorithm is to require that at the new sampling point for the i-th AUV, the infinity norm of $P_i[k+1]$ is minimized (Popa et al., 2004).

2.5.1.3 Localization Uncertainty

Most AUVs are non-holonomic robots, and in this section, we consider an approximate kinematic model for AUSI's Solar-Powered Autonomous Underwater Vehicle (SAUV)-II. We use this model to illustrate the vehicle localization uncertainty by using dead reckoning. The solar AUV is driven by a steerable

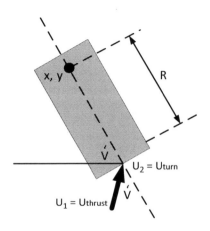

FIGURE 2.12 A simple kinematic model for SAUV-II (Popa et al., 2004).

rear thrust vector as shown in Figure 2.12 and is equipped with an axial velocity sensor measuring flow along the center line of the vehicle ($u_\|$), as well as steering thrust angle measurements and a compass measuring the vehicle orientation ($\dot{V}$). If we consider the 2D yaw ($\dot{V}$) model of the vehicle (e.g., the vehicle maneuvers at a constant depth), the kinematics can be written as:

$$x \dot{V} u_1 Cos \dot{V} Cos \dot{V} \dot{V} k u_1 Sin \dot{V} Sin \dot{V}$$

$$x \dot{V} u_1 Cos \dot{V} Sin \dot{V} \dot{V} k u_1 Sin \dot{V} Cos \dot{V} \qquad (2.15)$$

$$\dot{V} \dot{V} u_2$$

$$\dot{V} \dot{V} \frac{1k \dot{V}}{R} u_1 Sin \dot{V}$$

The vehicle state variable is $X = [x, y, \dot{V}, \dot{V}]$, and the measured outputs are $\dot{V}$, $\dot{V}$, and $u_\| \dot{V} u_1 \cos(\dot{V})$. The constant k is a lateral slip coefficient for the vehicle. SAUV-II is also equipped with a GPS unit that can only be used when the vehicle is surfaced. If the vehicle sample outputs occur at time intervals dt, the dead-reckoned estimate of position can be written as:

$$\hat{X}[k \dot{V} 1] \dot{V} \hat{X}[k] \dot{V} dt(f_1(\hat{X}[k])u_1[k] \dot{V} f_2(\hat{X}[k])u_2[k]),$$

and the state covariance matrix becomes:

$$P[k \dot{V} 1] \dot{V} E[(X[k \dot{V} 1] \dot{V} \hat{X}[k \dot{V} 1])(X[k \dot{V} 1] \dot{V} \hat{X}[k \dot{V} 1])^T]$$

where X[k+1] is the updated dead-reckoned position of the vehicle, including incremental measurement noise:

$$X[k \dot{V} 1] \dot{V} X[k] \dot{V} dt(f_1(X[k] \dot{V} dX[k])u_1[k] \dot{V}$$

$$\dot{V} f_2(X[k] \dot{V} dX[k])u_2[k] \dot{V} X[k] \dot{V} dt u_1[k] f_1(X[k]) \dot{V}$$

$$\dot{V} dt u_2[k] f_2(X[k] \dot{V} dt(u_1 \frac{\dot{V} f_1}{\dot{V} x} \dot{V} u_2 \frac{\dot{V} f_2}{\dot{V} x} \dot{V})d X[k].$$

Therefore, $P[k \dot{V} 1] \dot{V} (I \dot{V} J[k]) P[k] (I \dot{V} J[k]^T)$, where

$$J[k] \dot{V} dt(u_1[k] \frac{\dot{V} f_1 \hat{X}[k]}{\dot{V} X} u_2[k] \frac{\dot{V} f_2 \hat{X}[k]}{\dot{V} X}). \qquad (2.16)$$

When the vehicle is on the surface, the estimate of position can be set to the approximate steady-state solution of the Kalman filter, which fuses the GPS data with the dead-reckoning information by weighting the data commensurate to its uncertainty. If the dead-reckoning estimate is very imprecise, its weight in the equation below is almost zero:

$$\hat{X} \dot{V} w_1 \hat{X}_{k\,dead\,\dot{V}\,reckoning} \dot{V} w_2 \hat{X}_{GPS},$$

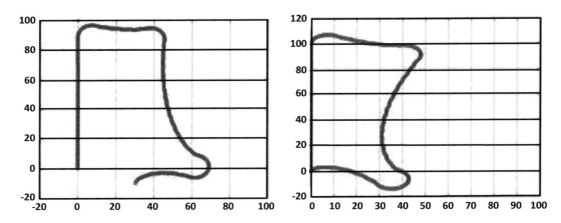

FIGURE 2.13 Actual course (top) and dead-reckoned course (bottom) of the non-holonomic AUV model (15), with a 5 m turn radius, using the closed-loop feedback scheme (18) (Popa et al., 2004).

$$w_1 \dot{V} \frac{\left\|P\left(\hat{X}_{GPS}\right)\right\|_{\dot{V}}}{\left\|P\left(\hat{X}_{GPS}\right)\right\|_{\dot{V}} \dot{V} \left\|P\left(\hat{X}_{k\,dead\,\dot{V}\,reckoning}\right)\right\|_{\dot{V}}}, \quad (2.17)$$

$$w_2 \dot{V} \frac{\left\|P\left(\hat{X}_{k\,dead\,\dot{V}\,reckoning}\right)\right\|_{\dot{V}}}{\left\|P\left(\hat{X}_{GPS}\right)\right\|_{\dot{V}} \dot{V} \left\|P\left(\hat{X}_{k\,dead\,\dot{V}\,reckoning}\right)\right\|_{\dot{V}}},$$

Assuming the AUV moves between waypoints at maximum thrust speed, the onboard controller will chart a course that tries to follow a straight line to the destination. The following example uses a PID gain between the estimate of the AUV heading and the desired heading:

$$u_1 \dot{V} \max\_speed, u_2 \dot{V} PID(\dot{V}_d \dot{V} \hat{\dot{V}}), \quad (2.18)$$

$$\dot{V}_d \dot{V} a\tan 2(x_d \dot{V} \hat{x}, y_d \dot{V} \hat{y})$$

To illustrate that dead reckoning by itself can introduce a very large error if GPS measurements are not used frequently enough, we simulated a "U" maneuver for the SAUV. The dead-reckoning estimate of the final AUV position was obtained by integrating the nominal kinematics, while the "truth" model was obtained by using a 10% vehicle thrust error and a 1° compass uncertainty. The results shown in Figure 2.13 indicate that the vehicle will be approximately 30 m off course, if it does not surface at various waypoint locations during the maneuver. This model forms the basis for adaptive sampling analysis under true mission conditions (Popa et al., 2004).

2.5.2 Other Algorithms

The following flowchart demonstrates the simple operation of a robot roaming around a room in a clockwise direction and avoiding obstacles (Permana, 2013) (Figure 2.14).

Development of Autonomous Vehicles

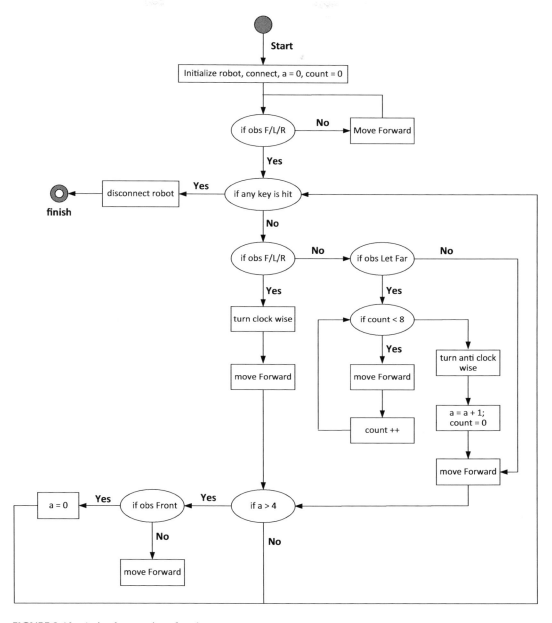

FIGURE 2.14 A simple operation of a robot.

At the beginning of the program, the robot will move in a straight line until it finds the wall or any obstacle. Once the wall has been found, the robot will then move along the wall in a clockwise direction. If it encounters any obstacle along the wall, it will move around the obstacle and continue moving in clockwise direction along the wall.

The process continues until any keyboard key is pressed, terminating the program. Figure 2.15 shows the approximate path inside the room (Permana, 2013).

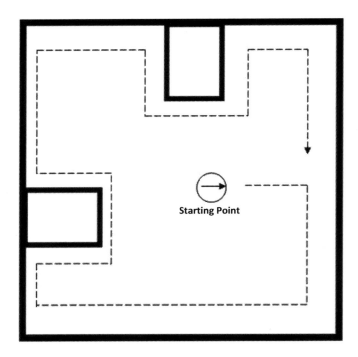

FIGURE 2.15 Approximate path of a robot inside a room (Permana, 2013).

REFERENCES

Alami R., Chatila R., Fleury S., Ghallab M., Ingrand F., 1998. An architecture for autonomy. *International Journal of Robotics Research*, 17, 315–337.

Alberta Machine Intelligence Institute (Amii). Machine intelligence, artificial intelligence & machine learning. www.amii.ca/machine-intelligence-artificial-intelligence-machine-learning/. Viewed: February 17, 2019.

Amazon, 2016. Amazon prime air. www.amazon.com/Amazon-Prime-Air/b?node=8037720011. Viewed: April 30, 2019.

Arenella A., Greco A., Saggese A., Vento M., 2017. Real time fault detection in photovoltaic cells by cameras on drones. In Image Analysis and Recognition, *Proceedings of the 14th International Conference, ICIAR 2017*, Montreal, Canada, 5–7 July 2017; F. Karray, A. Campilho, & F. Cheriet (Eds.), Cham, Switzerland: Springer International Publishing, 2017, pp. 617–625.

Arroyo A. A., 2015. EEL-5840 elements of (artificial) machine intelligence. University of Florida. EEL 5840 Class 1- Fall 2015.

Aviles W. A., 1990. Issues in mobile robotics: The unmanned ground vehicle program teleoperated vehicle. Mobile Robots V, Vol. 1388, SPIE, 1990.

Benecke N., 2018. iDeepMon: Intelligent deep mine shaft inspection and monitoring. Available online: https://eitrawmaterials.eu/project/ideepmon/. Accessed: 11 April 2018.

Besada J. A., Bergesio L., Campaña L., Vaquero-Melchor D., López-Araquistain J., Bernardos A. M., Casar J. R., 2018. Drone mission definition and implementation for automated infrastructure inspection using airborne sensors. *Sensors*, 18, 1170. doi:10.3390/s18041170.

Branco L. H. C., Segantine P. C. L., 2015. MaNIAC-UAV—A methodology for automatic pavement defects detection using images obtained by Unmanned Aerial Vehicles. *Journal of Physics: Conference Series*, 633, 012122.

Capolupo A., Pindozzi S., Okello C., Fiorentino N., Boccia L., 2015. Photogrammetry for environmental monitoring: The use of drones and hydrological models for detection of soil contaminated by copper. *Science Total Environment*, 514, 298–306.

Cefalu A., Haala N., Schmohl S., Neumann I., Genz T., 2017. A mobile multi-sensor platform for building reconstruction integrating terrestrial and autonomous UAV-based close range data acquisition. *International Archives of the Photogrammetry, Remote Sensing and Spatial Information Sciences-ISPRS Archives*, XLII(2W6), 63–70.

Cozien L. R., 2001. Active objects programming for military autonomous mobile robots software prototyping. French Department of Defense. Military Officers Academy of Saint Cyr - DGER -56381 Guer cedex - France.

Cutler M., How J. P., 2012. Actuator constrained trajectory generation and control for variable-pitch quadrotors. In *Proceeding of the AIAA Guidance, Navigation, and Control Conference and Collocated Conferences*, Minneapolis, MI, USA, August 2012.

De Melo R. R. S., Costa D. B., Alvares J. S., Irizarry J., 2017. Applicability of unmanned aerial system (UAS) for safety inspection on construction sites. *Safety Science*, 98, 174–185.

Doorn M. V., Bloem J., Duivestein S., Ommeren E. V., 2015. Machine intelligence. Executive introduction. Attribution-NonCommercial-ShareAlike 3.0 (CC BY-NC-SA 3.0). SOGETILabs.

Gilmore J., Roth S., Soniat du Fossat E., 1990. Unmanned aerial; vehicle command and control expert system. In *Association for Unmanned Vehicle Symposium*, Dayton, OH, USA, 1990.

González-Jorge H., Martínez-Sánchez J., Bueno M., Arias P., 2017. Unmanned aerial systems for civil applications: A review. *Drones*, 1, 2. doi:10.3390/drones1010002.

Grift T. E., 2015. Development of autonomous robots for agricultural applications. December 18, 2015.

Hall L., 2016. NASA awards $750K in sample return robot challenge for autonomous technology. September 8, 2016. www.nasa.gov/directorates/spacetech/centennial_challenges/feature/2016_sample_return_robot_challenge_award.html. Viewed: February 14, 2019.

Johnson E. N., Mishra S., 2002. Flight simulation for the development of an experimental UAV. In *Proceedings of the AIAA modeling and simulation technologies conference*, Monterey, CA, USA, August 2002.

Kassler M., 2001. Agricultural automation in the new millennium. *Computers and Electronics in Agriculture*, 30, 237–240.

Kersnovski T., Gonzalez F., Morton K., 2017. A UAV system for autonomous target detection and gas sensing. In *Proceedings of the 2017 IEEE Aerospace Conference*, Big Sky, MT, USA, 4–11 March 2017, pp. 1–12.

Liu P., Chen A. Y., Huang Y. N., Han J. Y., Lai J. S., Kang S. C., Wu T. H., Wen M. C., Tsain M. H., 2014. A review of rotorcraft Unmanned Aerial Vehicle (UAV) developments and applications in civil engineering. Copyright© 2014 Techno-Press, Ltd. ISSN: 1738-1584 (Print), 1738-1991 (Online).

Lum C., Mackenzie M., Shaw-Feather C., Luker E., Dunbabin M., 2016. Multispectral imaging and elevation mapping from an unmanned aerial system for precision agriculture applications. In *Proceedings of the 13th International Conference on Precision Agriculture*, St. Louis, MO, USA, 31 July–4 August 2016.

Martinez C., Sampedro C., Chauhan A., Campoy P., 2014. Towards autonomous detection and tracking of electric towers for aerial power line inspection. In *Proceedings of the 2014 International Conference on Unmanned Aircraft Systems (ICUAS)*, Orlando, FL, USA, 27–30 May 2014, pp. 284–295.

Matsuoka R., Nagusa I., Yasuhara H., Mori M., Katayama T., Yachi N., Hasui A., Katakuse M., Atagi T., 2012. Measurement of large-scale solar power plant by using images acquired by non-metric digital camera on board UAV. *International Archives of the Photogrammetry, Remote Sensing and Spatial Information Sciences-ISPRS Archives*, 39, 435–440.

Meyrowitz A. L., Blidberg D. R., Michelson R. C., 1996. Autonomous vehicles. *Proceedings of the IEEE*, 84(8), 1147–1164.

Meystel A., 1991. *Autonomous Mobile Robots*. Singapore: World Scientific.

Morphew M. E., Shively J. R., Casey D., 2004. Helmet mounted displays for unmanned aerial vehicle control. In *Proceedings of the Helmet- and Head-Mounted Displays IX: Technologies and Applications*, Orlando, FL, USA, September 2004.

Permana H., 2013. Autonomous robot – basic principles. April 4, 2013. www.codingepiphany.com/2013/04/04/autonomous-robot-basic-principle/. Viewed: May 10, 2019.

Popa D. O., Sanderson A. C., Komerska R., Mupparapu S., Blidberg R., Steven Chappel S., 2004. Adaptive sampling algorithms for multiple autonomous underwater vehicles. *Conference Paper*. July 2004.

Priest L., 2017. Detecting changes at cell sites and surrounding areas using unmanned aerial vehicles. U.S. Patent 20170318477A1, November 2, 2017.

Russell S., Norvig P., 2009. *Artificial Intelligence: A Modern Approach*, Third edition. Prentice Hall, 1152.

Saleri R., Pierrot-Deseilligny M., Bardiere E., Cappellini V., Nony N., De Luca L., Campi M., 2013. UAV photogrammetry for archaeological survey: The theaters area of Pompeii. In *Proceedings of the Digital Heritage International Congress (DigitalHeritage)*, Marseille, France, 28 October–1 November 2013; Vol. 2, pp. 497–502.

Santos T., Moreira M., Almeida J., Dias A., Martins A., Dinis J., Formiga J., Silva E., 2017. PLineD: Vision-based power lines detection for Unmanned Aerial Vehicles. In *Proceedings of the 2017 IEEE International Conference on Autonomous Robot Systems and Competitions (ICARSC)*, Coimbra, Portugal, 26–28 April 2017, pp. 253–259.

Seneviratne D., Ciani L., Catelani M., Galar D., 2018. Smart maintenance and inspection of linear assets: An Industry 4.0 approach. Acta IMEKO, Vol. 7, No. 1, article 9, March 2018, identifier: IMEKO-ACTA-07 (2018)-01-09.

Stimpson A., Cummings M., Nneji V. C., Goodrich K. H., 2017. Exploring concepts of operations for on-demand passenger air transportation. In *Proceedings of the 17th AIAA Aviation Technology, Integration, and Operations Conference*, Denver, CO, USA, 5–9 June 2017.

Stroumtsos N., Gilbreath G., Przybylski S., 2013. An intuitive graphical user interface for small UAS. In *Proceedings of Unmanned Systems Technology XV Conference*, Baltimore, MD, USA, April 2013.

Sugeno M., Griffin M., Bastian A., 1993. Fuzzy hierarchical control of an unmanned helicopter. In *Proceedings of IFSA Congress*, 1993.

Tang L., Shao G., 2015. Drone remote sensing for forestry research and practices. *Journal Forest Research*, 26, 791–797.

Techopedia, 2019. Autonomous robot. Copyright© 2019 Techopedia Inc. www.techopedia.com/definition/32694/autonomous-robot. Viewed: February 15, 2019.

Total, 2015. Enhanced safety thanks to the ARGOS challenge. July 1, 2015. www.total.com/en/media/news/news/enhanced-safety-thanks-argos-challenge?folder=7692. Viewed: February 14, 2019.

Yuh J., 1994. Learning control for underwater robotic vehicles. 0272-1708/94/$04.00©1 1994IEEE. April 1994.

Wahde M., 2016. *Introduction to Autonomous Robots*. Goteborg, Sweden: Department of Applied Mechanics. Chalmers University of Technology.

Zahariadis T., Voulkidis A., Karkazis P., Trakadas P., 2017. Preventive maintenance of critical infrastructures using 5G networks drones. In *Proceedings of the 2017 14th IEEE International Conference on Advanced Video and Signal Based Surveillance (AVSS)*, Lecce, Italy, 29 August–1 September 2017, pp. 1–4.

Zipline, 2017. www.flyzipline.com. Viewed: April 30, 2019.

Zivis, S., n.d. The current state of machine intelligence 3.0. www.shivonzilis.com/.

3
Autonomous Inspection for Industrial Assets

3.1 Autonomous Vehicle Inspection Platform

3.1.1 Autonomous Inspection and Maintenance

Remotely controlled and autonomous inspection and maintenance devices are used in different sectors for different purposes. For instance, the military uses unmanned aerial vehicles (UAVs) for inspection, and offshore oil and gas industries use underwater robots for maintenance.

The following presents autonomous or remotely controlled devices used in the inspection and maintenance of linear assets and explains their purpose (Seneviratne, Ciani, Catelani, & Galar, 2018):

1. **Railways**
   - Identification of obstacles and track irregularities using drones.
   - Inspection of rail profile, cracks, irregularities, and missing components using an autonomous robot vehicle.
   - Replacement of missing components, crack welding, etc. using an autonomous maintenance robot vehicle.
2. **Roads**
   - Identification of obstacles and damage using drones.
   - Inspection of roadway, road alignment, road profile etc. using an autonomous robot vehicle.
   - Repair of roadway (placement of asphalt/concrete), repair of pavement, maintenance of embankments, maintenance and cleaning of ditches, etc. using autonomous maintenance robot vehicle.
3. **Canals and Waterways**
   - Identification of debris, obstacles, and damages for the infrastructure through drones.
   - Inspection of waterway, sidewalls, berm, gates, etc. using an autonomous robot vehicle, both land and water.
   - Removal of debris and obstacles, repair of sidewalls, berm, etc. using an autonomous maintenance robot vehicle (both land and water).
4. **Power Lines**
   - Identification/inspection of power line damage, insulator defects, and tower damage using drones.
   - Cleaning of insulators and repair of line damage using an autonomous drone robot vehicle.

3.1.1.1 Smarter Drones

UAVs have attractive features, such as flexibility, adaptability, and a range of payloads. Sensors include high-resolution digital and infrared cameras, light detection and ranging (LiDAR), geographic information systems (GIS), sonar sensors, and ultrasonic sensors; most can be adapted to a UAV platform. A close-up photograph of a structure on an offshore platform, difficult for inspectors to reach, will show maintenance personnel how much corrosion/erosion has built up and suggest the situation of welds and other structural elements.

Drones equipped with forward-looking infrared (FLIR) or ultraviolet sensors can detect hot spots or corona discharge on conductors and insulators, signaling a potential defect or weakness in the component. LiDAR can be integrated with drones to survey a proposed right-of-way, show the infrastructure situation when seismic conditions are changing, or monitor the encroachment of vegetation. There are many more potential uses, and the examples are only a small fraction of the possible applications.

At present, UAVs are remotely operated; the next phase of UAV technology will be to deploy "smarter" machines that can fly autonomously. This technology will allow UAVs to sense and avoid other objects in their path, recognize features or components through various sensors (including cameras) using complex software algorithms such as image processing algorithms, and achieve situational awareness. This advanced technology will foster calculated decision-making, such as initiating focused inspections, issuing work orders for repairs, and starting maintenance work with the same robot or another autonomous robot integrated in the system.

In any industry, safety and cost are two of the most significant drivers of operation and maintenance and, thus, are always of high importance. Many industrial work areas are hazardous, so measures must be taken to secure the safety of users. Health safety and environmental (HSE) indicators can mitigate the risks, but the situation remains challenging when new technologies are introduced.

For instance, working on energized high-voltage transmission lines, sometimes hundreds of feet up in the air, can make the consequences of a mistake deadly. According to the US Bureau of Labor Statistics, 15 linemen were fatally injured in 2013 as a result of "exposure to harmful substances or environments."

Unmanned systems have the potential to greatly reduce the amount of risk exposure of the operational workforce. The safety of personnel involved in risky operational tasks can be ensured with this new technology (Seneviratne, Ciani, Catelani, & Galar, 2018).

3.1.1.2 Autonomous Robots

Many different robots have been developed to handle various situations on linear assets, buildings, ship hulls, or other human-made structures. However, most are limited to special situations or applications. To execute the desired tasks, autonomous robots, as well as all other technical systems, have to fulfil certain requirements. The requirements and their importance and focus depend on the individual application or tasks, but we can formulate a general set of requirements as follows (Seneviratne, Ciani, Catelani, & Galar, 2018):

1. Velocity and mobility: Vehicle speed and dynamics (ability to move) are two main aspects of robot design. Depending on the dimension of the linear asset, it may have to reach a relatively high velocity for sufficiently fast navigation between inspection areas or similar points of action. Another requirement is related to the desired manipulation and positioning capabilities of the system. This includes the precision of its locomotion and trajectory, since some inspection sensors need to be moved in a smooth and continuous way over the surface. The robot may also need to move sideways or turn 360° to position sensors or tools. The system dynamics should be able to handle the various terrains and reach all positions of the asset.
2. Payload: Depending on the application, the system must be able to carry payloads of different weights. For example, in the case of steel piping, a payload of 5 kg or more is mandatory to carry ultrasonic inspection sensors. This requires a much bigger robot than a system which just needs a simple camera with a weight of several hundred grams. In other words, the dimension, adhesion, and motion components of the robot need to be adapted for the application.
3. Reliability and safety: A further important nonfunctional aspect is the robustness of the system. If the autonomous robot fails frequently during one inspection task, it is not usable in practice.

The requirements of reliability and safety include robust hardware, optimal controllers, and methods to detect and handle hazardous situations and to recover from them.

4. Usability: Velocity, maneuverability, and the capability of carrying a certain payload are important, but they are only the basis of the general operability of the system. To bring a robotic system into application, it has to be more powerful, more efficient, and less dangerous than common approaches, for example, in terms of inspection devices. This includes aspects of maintainability and a broad range of other tasks. Therefore, it must be able to carry different payloads (e.g., inspection sensors or tools) depending on the desired task, parts need to be easily replaceable, and the operation must be faster and less complicated than existing approaches. Aspects such as energy consumption, weight, or dimension of the system can be important as well. Based on the individual task, a robot developer has to decide which requirements have to be fulfilled and select a suitable locomotion and attraction principle.

3.1.1.3 Autonomous Inspections

Traditionally, electric power suppliers have inspected power lines for encroaching trees, damage to structures, and deterioration of insulators by having employees traverse the lines on foot and climb the poles. This is time consuming and arduous, with a considerable element of risk. Now the task is often carried out by crews in manned helicopters using binoculars and thermal imagers to detect the breakdown of insulators. This too is not without hazard.

Recently, trials have tested the use of UAVs to inspect power lines, with considerable success. UAVs offer lower costs, do not create a hazard for aircrews, can operate in more adverse weather conditions, and are less obtrusive to neighboring communities or animals. Hover flight is essential for the inspection task. The UAV carries an electro-optic and thermal imaging payload, the data from which are available in real time to the operator and recorded. The UAV is automatically guided along the power line within a limited volume of airspace close to the lines using a distance measuring device. An important requirement of UAVs deployed in this role is that they must be flown close to high-voltage power lines, i.e., within their electromagnetic fields, without adverse effects on their control system or payload performance.

Oil and gas supplying companies are interested in UAVs for inspection and exploration purposes. UAVs offer a less expensive means of surveying the land where pipelines are installed. They also offer a means of patrolling the pipes to look for disruptions or leaks caused by accidents such as landslides or lightning strikes or for damage caused by vehicles or falling trees. In certain areas of the world, sabotage is not uncommon, so they look for this as well.

UAVs could be used in road and railway inspections and for certain maintenance purposes by traffic infrastructure agencies. In addition to being less expensive to operate than manned aircraft, they are more covert and will avoid distracting drivers.

Irrigation projects, river authorities, and water boards could use UAVs to monitor canals, waterways, pipelines, and rivers. UAVs could be used to monitor reservoirs for pollution or damage or to monitor pipelines for security purposes.

However, the use of UAVs in many of these cases will depend on the approval of the relevant regulatory authorities (Seneviratne, Ciani, Catelani, & Galar, 2018).

3.1.1.4 Autonomous Maintenance

Autonomous maintenance activities are mainly associated with robotic applications. Various industries, especially those dealing with high risk activities, are already using remotely operated robots for maintenance activities, for instance, marine repairs (repairs of ships offshore, offshore oil and gas platform maintenance, deep sea pipeline, and cable maintenance), oil refinery repairs, nuclear power plant repairs, etc. At the moment, because of the limited development of robots for maintenance purposes, complete maintenance cannot be performed in the abovementioned industries (Seneviratne, Ciani, Catelani, & Galar, 2018).

3.1.1.5 Conceptual Framework

With autonomous inspection devices and autonomous maintenance robots, a dynamic asset maintenance and management plan can be deployed with the help of big data technologies and available analytics. Right now, industries are using the devices separately for inspection and maintenance; the two have not yet been integrated. By integrating the two operations with the available Information and Communication Technologies (ICT), the asset maintenance and management process can be automated. The possible architecture for the ICT framework is shown in Figure 3.1. Moreover, the incorporation of artificial intelligence (AI) tools can make the whole process dynamic and autonomous (Seneviratne, Ciani, Catelani, & Galar, 2018).

Since linear assets have a common behavior and architecture across their length, the implementation of the concept may reduce costs, ensuring more effective operation and maintenance. The proposed framework is shown in Figure 3.2 (Seneviratne, Ciani, Catelani, & Galar, 2018).

3.1.2 Maintenance as a Combination of Intelligent IT Systems and Strategies

3.1.2.1 Introduction

Industries are searching for technological solutions to improve their performance in business. The growth of ICT has helped organizations in using advanced solutions, such as eMaintenance, to manage their processes effectively, in this case their maintenance activities. eMaintenance can be seen as a tool for integrating companies' production and maintenance operations through information technological solutions. Due to the rapid technological development, the research topic of eMaintenance is changing and redirecting its focal point constantly. This section identifies and describes the key components of eMaintenance.

A number of academic reviews on eMaintenance describing the development and implementation of eMaintenance systems have been published over the past decade. For example, the basic ideas of eMaintenance have been presented, and Maintenance Management (MM) has been described as composed of the pillar of IT, Maintenance Engineering (ME), and relationship management. Another classification is that MM consists of optimization, models, maintenance techniques, scheduling, and IT. eMaintenance

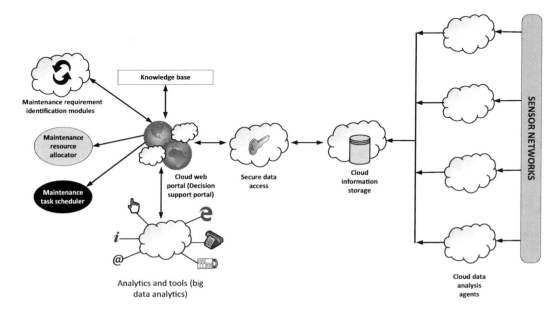

FIGURE 3.1 Proposed ICT infrastructure for the autonomous inspection and maintenance of linear assets (Seneviratne et al., 2018).

Autonomous Inspection for Industrial Assets 77

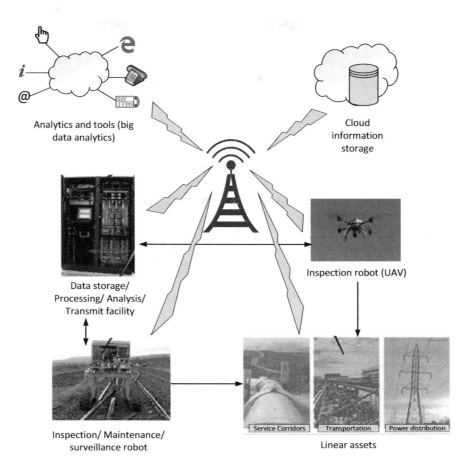

FIGURE 3.2 Proposed conceptual framework for autonomous inspection and maintenance of linear assets (Seneviratne et al., 2018).

defines the strategic vision, organization, service and data architecture, and IT infrastructure. However, a number of joint academic and industrial papers providing an updated view on eMaintenance systems or parts of them within the context of industrial applications have been published recently (Metso, Baglee, & Marttonen-Arola, 2018).

3.1.2.2 eMaintenance Concept

Karim et al., 2010, one of the first articles to explain eMaintenance examined a distributed intelligent and integrated system. This system integrated the control, maintenance, and technical management activities of a shop-floor organization with an intelligent system, now known as eMaintenance. eMaintenance allows the integration of production and maintenance operation systems. The importance of the maintenance function should be acknowledged, as it can impact the production operations and business process by ensuring system safety and by decreasing the costs of operations during the lifetime of systems.

eMaintenance as a system should be connected to other systems to assist in the collection, analysis, and definition of maintenance tasks which take advantage of the idea of "right data to right person at right time." The "strength" of eMaintenance is based on various data sources, and it utilizes different tools and techniques (Lee, Ni, Djurdjanovic, Qiu, & Liao, 2006).

eMaintenance can be described as an "intelligent maintenance center" because it collects data from a number of different sources and provides relevant data to be used in the development of maintenance tasks. It supports the use of data collection and transfer to remote use through a number of Internet-enabled technologies. eMaintenance technologies can be used with other maintenance strategies to share and exchange information, such as eIntelligence. eIntelligence is a term that covers eMaintenance but also other data and ICT-related aspects of business (Metso, Baglee, & Marttonen-Arola, 2018).

3.1.2.2.1 Data

The aim of eMaintenance is to connect maintenance and production data with an intelligent system to analyze a number of manufacturing parameters. In effect, it creates a "smart factory" environment. However, to be useful, eMaintenance solutions need to be integrated with other information systems which allow transferring data between different environments. It is important that all systems in the eMaintenance network can exchange information in an efficient and usable way. This is depicted in Figure 3.3 (Metso, Baglee, & Marttonen-Arola, 2018).

Figure 3.3 shows that eMaintenance includes monitoring, collection, recording, and distribution of real-time data and decision/performance support information. eMaintenance improves the performance of the maintenance process through effective data collection and distribution. Data are converted into information and generated to knowledge that is valuable in the decision-making process. Computerized maintenance management systems (CMMS) allow users to use maintenance data. These systems often contain work order control, labor management, equipment management, material control and purchase, and performance report modules. Enterprise resource planning (ERP) systems have been used in MM to collect, store, and analyze manufacturing data. However, eMaintenance is a much wider concept than the relatively narrow modules suggested in ERPs. eMaintenance solutions must have access to different data sources. The integration of different systems can be challenging. Data quality must be taken into consideration, and interconnectivity is also important for eMaintenance solutions, because data are transferred between heterogeneous environments. All systems must be able to interact in an efficient and usable way in eMaintenance solutions.

Traditional "fail and fix" maintenance practices are changing into a perspective of "predict and prevent." eMaintenance methodology has emerged due to the increase in Internet technologies, faster data transfer, and specific data analytics to collect and analyze large amounts of data quickly and effectively. Sensors enable the collection and delivery of data about the status of machines. These data are rarely used to support a continuous information flow throughout the entire maintenance process, because the infrastructures do not support data delivery, management, and analysis. A possible answer is to include smart machines in a remotely monitored network, where data are modeled and analyzed with embedded systems; this should allow a shift from predictive maintenance to intelligent prognostics, a systematic approach to monitor and predict potential machine failures continuously and to synchronize maintenance actions with production operation, maintenance resources, and spare parts.

However, data quality from different sources can be inadequate to support maintenance actions. Another issue is technical problems in transferring data from one system to another system. Relevant data are needed, and real-time data and data analysis can improve maintenance actions, but the existing body of knowledge has not yet succeeded in finding optimal solutions to these challenges (Metso, Baglee, & Marttonen-Arola, 2018).

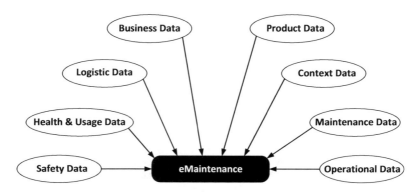

FIGURE 3.3 eMaintenance data access (Wandt et al., 2012).

3.1.2.3 Advantages and Challenges of eMaintenance

The advantages and challenges of eMaintenance are presented in Table 3.1. It should be noted that the same main items can be seen as both advantages and challenges for eMaintenance; for example, in theme 3, inspection and monitoring are difficult if real-time, remote, and distributed monitoring and analysis devices have not been developed (Metso, Baglee, & Marttonen-Arola, 2018).

eMaintenance offers possibilities to improve the development of new ways of implementing maintenance. Maintenance expert centers can be organized easily because data can be shared easily over the Internet. Experts can log into the systems remotely and give their instructions to maintenance employees. Maintenance support can be improved, and real-time data should always be available. Maintenance documentation should always be updated and available because the users could log in to the systems anywhere with any equipment. This is naturally an issue of security because eMaintenance is used over the Internet, and the same security risks exist as in other Internet-based solutions. Transparency and access to maintenance information and services across the maintenance operation chain have also been seen as a benefit in eMaintenance.

The positive impact of eMaintenance can be divided into two levels. The first is the "maintenance micro-level," where eMaintenance serves technicians, mechanics, and support engineers as a support to execute maintenance tasks. eMaintenance reduces the number of interfaces to information sources, improves fault diagnosis and knowledge sharing, and automates the procedures of technical administration. The second level is the "maintenance macro-level," where eMaintenance supports managerial maintenance planning, preparation, and assessment. It enables information-driven maintenance and

TABLE 3.1

Advantages and Challenges of eMaintenance

Potential Improvements	Challenges
Theme 1: Developing New Maintenance Types and Strategies	
Remote maintenance operations and decision-making; logging in anywhere and on any devices, manpower on the customer's site can be reduced, the use of expertise is easy, new features can be added	Remote maintenance; security and reliability over the Internet, human resource training, maintenance agreements
Business process integration and cooperative/collaborative maintenance; easy to design and synchronize maintenance with production, maximizing process throughput	Business process integration and cooperative/collaborative maintenance; data transform mechanism, communication, data protocols, safe network; the maintenance processes must be stable and capable
Fast online maintenance; real-time remote monitoring, alerts, high-rate communication to experts, maintenance support system	
Predictive maintenance; prognostics and health management systems	Predictive maintenance; difficult to integrate different techniques
Theme 2: Improving Maintenance Support and Tools	
Fault/failure analysis; development in sensors, signal processing, ICT, etc.	
Maintenance documentation; easy to fill out forms, remove bottlenecks between the plant floor and business system	Maintenance documentation; need to collect, record, and store information from different systems, multitasking, and multiuser operating environment
Theme 3: Improving the Maintenance Activities	
Fault diagnosis/localization; e-diagnosis offers online fault diagnosis for experts	Inspection/monitoring; problems with distributed monitoring
Repair/rebuilding; reducing downtimes by direct interaction	Modification/improvement—Knowledge capitalization and management; how to realize the knowledge-based operation and maintenance of plants

Source: Modified from Crespo-Marquez, and Iung (2008), Muller, Crespo-Marquez, and Iung (2008).

support processes. In, for example, aviation maintenance, eMaintenance implementation enables a more efficient use of digital product information and design data over the whole life cycle (Metso, Baglee, & Marttonen-Arola, 2018).

3.2 Inspection Communications and Transport Security

3.2.1 Security Issues of Autonomous Vehicle

3.2.1.1 CPS Security

An autonomous vehicle can be considered a specific type of cyber-physical system (CPS) and also a kind of Internet of Things (IoT) system. CPSs are complex, heterogeneous distributed systems, typically consisting of a large number of sensors and actuators connected to a pool of computing nodes. With the fusion of sensors, computing nodes, and actuators, which are connected through various means of communications, CPSs aim to perceive and understand changes in the physical environment analyze the impacts of such changes to the operation of the CPS, and make intelligent decisions to respond to the changes by issuing commands to control physical objects in the system, thereby influencing the physical environment in an autonomous way. As illustrated in Figure 3.4, the connections between actuation and sensing through the physical environment, and between sensors and actuators through one or multiple (distributed) computing or intelligent control node(s) form a feedback loop which aims at achieving a desired objective or steady state. As such, a CPS acts either with full autonomy or at least provides support for a human-in-the-loop mechanism as part of some semiautonomous control functions. This distributed closed-loop process allows a CPS to remotely influence, manage, automate, and control many industrial operations (Chattopadhyay & Lam, 2018).

Due to the operational nature of CPSs in most industrial control processes, CPS are also known as operational technology systems (OT systems).

The massive adoption of Internet-enabled devices (i.e., IPenabled sensors and actuators) in CPS systems has blurred the boundary between CPS and IoT. The concept of IoT stems from connected smart devices, which may or may not be interacting with physical objects. Hence, there are application scenarios in the classical OT domain that can be conveniently classified both as an IoT system and a CPS system, for example, a distributed set of sensor nodes (SNs) to monitor and control the energy usage of a

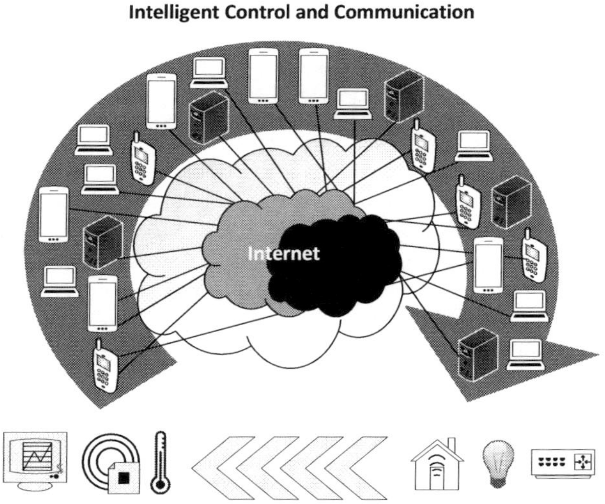

FIGURE 3.4 Interactions between sensor layer and actuator layer (Chattopadhyay & Lam, 2018).

Autonomous Inspection for Industrial Assets

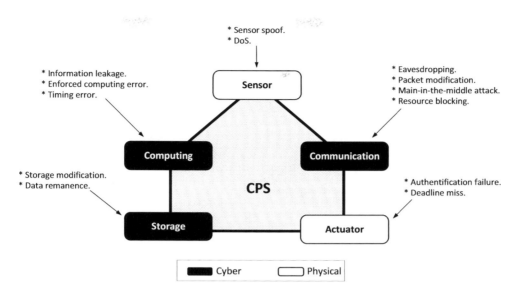

FIGURE 3.5 CPS attacks: generic model (Chattopadhyay & Lam, 2018).

manufacturing plant. Prominent examples of CPS and IoT systems and their corresponding applications include, among others, autonomous vehicles.

Attack techniques against a diversity of OT systems have similarities and can be classified as attacks on different CPS components, for example, communication, storage, actuator, sensor, and computing nodes. A few such attacks are shown in Figure 3.5 (Chattopadhyay & Lam, 2018).

However, this naive and isolated analysis of attacks, in the context of a specific CPS and the adoption of the corresponding countermeasures is grossly inadequate and misleading for several reasons.

- These generic attack studies tend to ignore the security objectives of the CPS, which aims to strike a balance between risks, cost, and convenience through the adoption of a hybrid of security control measures. Thus, a seemingly insecure mechanism may be operationally acceptable because it is operating within a controlled environment created by other security mechanisms of the system.
- Depending on the prevailing OT security practices, as well as the assumed adversarial model, it might be unnecessary to account for certain vulnerabilities.
- The generalization of attacks across all CPSs typically ignores the roles of Roots of Trust (RoT) and security perimeter modeling, which are the basis of many security by design approaches.

In principle, the security design of a CPS is a holistic process and viewed as a systems engineering problem. Addressing specific attacks in an isolated and ad hoc manner helps very little in security design practices. The literature mostly compromise generic attack studies in that they tend to study localized attacks in a generic setting of CPS. This trend of generic attack studies is exacerbated by the fact that no well-defined security standard aligns with the road safety standards for the AV.

In contrast, we emphasize the security design of AV as a system, indeed, a CPS. In the context of the specific vulnerabilities of AV with different degrees of autonomy, the technological challenges to safeguard AV are derived directly from the underlying safety objectives (Chattopadhyay & Lam, 2018).

3.2.1.2 AV Security Issues

The networking capabilities of a typical AV are depicted in Figure 3.6. The car is connected to the outside world with vehicle-to-infrastructure (V2I) and vehicle-to-vehicle (V2V) communication links. The V2I and V2V links provide the AV with, for example, traffic status information from traffic management

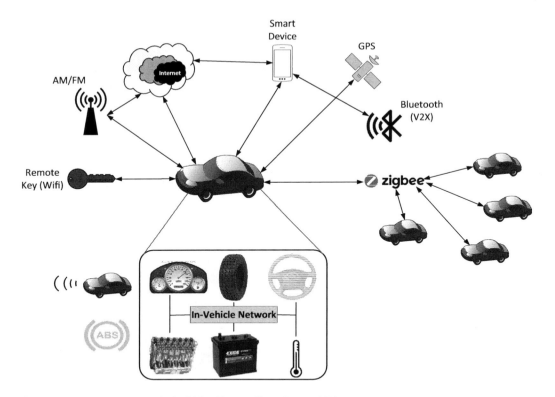

FIGURE 3.6 Exemplary networked vehicle (Chattopadhyay & Lam, 2018).

infrastructure or navigation-related information received from other AVs on the road. These connection interfaces represent attack surfaces that the adversaries will aim to exploit in order to obtain unauthorized access to the AV. Thus, it is of paramount importance that these V2I and V2V connections are mutually authenticated and the payload suitably protected from unauthorized disclosure and unauthorized modification (Chattopadhyay & Lam, 2018).

However, the AV is installed with a multitude of independent and overlapping CPS systems, such as adaptive cruise control, antilock braking systems, and assistive parking systems. Depending on the level of autonomy, more functionalities are assumed by the internal CPS controllers, with varying degree of human intervention. These CPS systems are supported by actuators and sensors, such as tire pressure sensor, crankshaft position sensor, light sensor, and obstacle sensors. Today, intra-vehicle communications for supporting such CPS controls happen mostly through wired connections known as a Control Area Network (CAN). The intra-vehicle communication protocols typically follow the CAN bus standard or the more recent FlexRay standard.

The security design of an AV requires that intra-vehicle networks are rigorously protected and access from outside of the vehicle, if applicable (e.g., for maintenance and services of the AV), be strictly controlled. It is evident from previous generic attack studies of AV security that, without proper security designs, even communications through these intra-vehicle wireline standards are potentially susceptible to security breaches. Specific studies of the intra-vehicle network security have been undertaken, but these are decoupled from the holistic view of the security-by-design principles.

In addition, connections to the Internet enable the AV to transmit operational data to the car manufacturer. As a critical system from the perspective of autonomy functions, it is crucial that the communication link between the AV and its manufacturer be authenticated and the payload be appropriately protected in accordance with the nature of the operational data being transmitted from the AV to the manufacturer (Chattopadhyay & Lam, 2018).

3.3 Obstacle Avoidance

3.3.1 Introduction

In robotics, obstacle avoidance is the task of satisfying some control objective subject to non-intersection or non-collision position constraints. In unmanned air vehicles, it is a hot topic. What is critical about obstacle avoidance in this area is the growing need of the usage of UAVs in urban areas for especially military applications where it can be very useful in city wars. Normally, obstacle avoidance is considered to be distinct from path planning in that one is usually implemented as a reactive control law while the other involves the pre-computation of an obstacle-free path along which a controller will then guide a robot. With recent advances in the autonomous vehicle sector, a good and dependable obstacle avoidance feature of a driverless platform is also required to have a robust obstacle detection module (Wikipedia, 2019).

The obstacle avoidance algorithm uses a series of rules based on (Hitchings, Engwirdal, Kajitani, & Vlacic, 1998):

- Distance to the obstacle as determined by the ultrasonic transceivers
- Type of sensor that detected the obstacle (infrared or ultrasonic)
- Position of the sensor(s) that detected the obstacle
- Speed and direction of the vehicle
- Number of sensors detecting the obstacle.

In the algorithm, the ultrasonic transducers perform two main functions (Hitchings, Engwirdal, Kajitani, & Vlacic, 1998):

a. To detect obstacles not detected by the infrared sensors. In this role, they act as a backup to the infrared sensors. First, the distance to the obstacle is calculated using the time that sound emitted from the ultrasonic transceiver takes to return after reflection from an obstacle. This distance is checked regularly, and if the vehicle has moved within a preset "safe" distance to the obstacle, the assumption is that the infrared sensors have not detected its presence. The vehicle then performs a corresponding avoidance maneuvre.

b. To act as a decision tool. The infrared sensors have a limited range of about 20 cm and, hence, cannot be expected to be used to obtain data about the larger environment in which the vehicle operates. If the vehicle reaches an obstacle in its forward path, it must make a decision on which direction it must turn in order to avoid the obstacle. The data obtained from the ultrasonic sensor assist this decision by indicating the presence of obstacles situated at longer distances from the vehicle (i.e., up to 4.5 m).

3.3.1.1 Two Sensor-Based Obstacle Avoidance Options for Autonomous Vehicles

3.3.1.1.1 Infrared Transceiver(s)

Infrared photodiodes are used to emit light at a peak wavelength of 935 nm. Phototransistors are used to detect infrared light reflected from obstacles around the vehicle. Eight infrared emitter/receiver pairs are situated around the horizontal periphery of the vehicle, forming an octagonal pattern. The emitters project infrared light at a beam width of 400 and this light is modulated to allow the receiving circuit to differentiate the received light from that of any background infrared light. This configuration gives the vehicle a detection range of about 20 cm with very few "blind spots" or areas where a small obstacle cannot be detected. The main limitation of using infrared light as a detection medium is that some object surfaces absorb infrared light which cannot be detected. This includes black, nonglossy surfaces. Another limitation is that the range of detection is only around 20 cm (Hitchings, Engwirdal, Kajitani, & Vlacic, 1998).

3.3.1.1.2 Ultrasonic Transceiver

An ultrasonic transceiver is used on the vehicle to overcome these infrared transceiver limitations. A transducer emits sound waves at a frequency of 40 kHz, and the same transducer receives the echoes reflected from obstacles in the vehicle's path (Jaffe, 1985; Everett, 1995). Unlike infrared sensors, the ultrasonic transceiver is not affected by the color of a material. However, it is affected by the density and rigidity of a material. An acoustically absorbent object may only be detected when it is in close range or may not be detected at all. Furthermore, small objects may not be detected because they do not reflect sufficient sound energy to trigger the sensor. The shape and orientation of an object can have a large effect on the reliability of the sensor. At a large angle, no energy is reflected back to the sensor (Everett, 1995). According to experiments performed on a slow moving robot, this effect is not significant at short distances. Namely, the small amount of energy that is scattered is detected as an echo. It is also important to point out that reflections from the floor are quite weak and, consequently, the floor is very seldom detected as a nearby object (Kolodko, 1997).

The main purpose of complementing the infrared transceivers with ultrasonic transceivers is that sound waves are reflected from a hard surface irrespective of color. This enables the vehicle to detect a wider range of obstacles (Hitchings, Engwirdal, Kajitani, & Vlacic, 1998).

One proposal is to use a Polaroid 6500 Series Sonar Ranging Module (Polaroid, 1991) driving a 600 Series Instrument Grade Electrostatic Transducer. The transceiver is mounted on the shaft of the four-phase hybrid stepper motor and, together with related circuitry, mounted on the robot. Scanning steps of 15° reliably detect obstacles as small as 8 mm diameter vertical poles (Borenstein & Koren, 1995). For omnidirectional robots, this means 24 ultrasonic range sensors should be mounted on a ring around the robot to cover all possible obstacles.

In contrast to this traditional construction, others propose using only one ultrasonic sensor mounted on a stepper motor turning around in steps of 15°. This configuration allows the forward 180° of the vehicle to be "scanned" by positioning the sensor in 15° steps, firing the sensor and waiting a preset maximum time for a return echo. The positioning of a single sensor in steps around the vehicle, while not the optimum solution in terms of the time to scan the vehicle's environment, alleviates the cross-talk problem in slow moving vehicles. The effect of cross talk between separate ultrasonic sensors occurs when they are mounted around the periphery of the vehicle (Borenstein & Koren, 1988; Masek, Kajitani, Ming, & Kanamori, 1997). With this method, the absolute reading accuracy of each sonar measurement is within 2% of the real distance. The measurement results do not drift significantly from one measurement to the next of the same distance. The maximum range error is up to ±1% of the measured value indicating good repeatability of distance to the obstacle within the range of 18 cm to 4.5 m (Kolodko, 1997). The performance of the sonar ranging module for distances to the obstacle greater than 4.5 m remains to be addressed and experimentally tested (Hitchings, Engwirdal, Kajitani, & Vlacic, 1998).

3.3.1.2 Description of Obstacle Avoidance Maneuver

3.3.1.2.1 Analysis of Obstacle Avoidance Process

Obstacle avoidance maneuvers of autonomous driving vehicles consist of two parts: vehicle collision risk assessment and obstacle avoidance path planning. To ensure the safety of the generated path, the path planning process of human drivers is first analyzed. Obstacle avoidance operations performed by human drivers are based on the understanding of dynamic traffic scenarios. However, there are many obstacle avoidance judgment criteria for drivers. The main basis of risk assessment is different for different drivers, even in the same scene. As shown in Figure 3.7, the whole behavior of obstacle avoidance can be divided into five sub-behaviors (Wang, Gao, Li, Sun, & Cheng, 2019).

- Stage 1: Scene recognition for obstacle avoidance: identify whether current scenario meets the criteria of obstacle avoidance, such as obstacles on the driving route, vehicle speed is greater than the obstacle, and there is enough space to complete an obstacle maneuver.

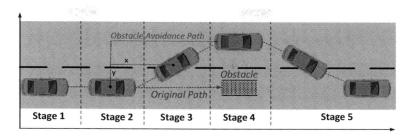

FIGURE 3.7 Obstacle avoidance process (Wang et al., 2019).

- Stage 2: Computation and path planning: a collision-free path is planned based on environmental information.
- Stage 3: Changing lane to bypass obstacles: in order to track the generated obstacle avoidance path, a lane changing maneuver is carried out.
- Stage 4: Overtaking: the obstacle is overtaken after the lane change.
- Stage 5: Back to original lane: when the obstacle is left behind over a certain distance, the subject vehicle goes back to the original lane (Wang, Gao, Li, Sun, & Cheng, 2019).

3.3.1.2.2 Safety Model of Obstacle Avoidance

According to the National Highway Traffic Safety Administration (NHTSA), nearly 30% of traffic accidents happen during lane changes. To reduce the loss and injury caused by accidents, scholars have proposed a lane change hazard perception model based on the behavior of drivers. Results show that the speed, movement state, and the properties of obstacles determine the severity of traffic accidents. To ensure safety, a time to predict risk advance of 0.3–1 s would be in accordance with the psychological and operation characteristics of drivers. Based on the abovementioned facts, a simplified obstacle avoidance model can be constructed by combining the behavior characteristics. Analysis of the operation characteristics of human drivers shows that the parameters of obstacles have a great influence on Stage 1. For example, when the mass and volume of obstacles ahead are large, drivers tend to perform obstacle avoidance maneuvers earlier. In Stages 3 and 4, drivers tend to increase the steering angle and reserve more lateral distance to bypass an obstacle when the obstacle has a larger volume or shows a trend of lateral movement. During the actual driving, the most difficult issue is dealing with moving other, such as other vehicles, pedestrians, and other traffic participants. Obstacle avoidance strategies aimed at static and moving obstacles are different in realistic scenarios.

For autonomous driving vehicles, the calculation results of Stage 2 affect the follow-up operation of obstacle avoidance directly. Further operations are more dependent on the control precision of actuators. In Stage 3, the probability of collision is greater than other stages. If no collision occurs in this stage, it means the subject vehicle avoids the obstacle successfully. Therefore, a safety model of obstacle avoidance needs to be built to ensure the safety of the obstacle avoidance path (Wang, Gao, Li, Sun, & Cheng, 2019).

As shown in Figure 3.8, the condition for collision avoidance between vehicle and obstacle is as follows (Wang, Gao, Li, Sun, & Cheng, 2019):

$$X_{ego} + w\sin\theta < X_{obs} + S, \tag{3.1}$$

where X_{ego} is the longitudinal displacement of the subject vehicle during the time of obstacle avoidance operation; X_{obs} is the longitudinal displacement of the obstacle vehicle during the time of the subject vehicle obstacle avoidance operation; S is the initial distance between the subject vehicle and the obstacle vehicle; w is the width of the subject vehicle; and θ is the angle between the x-axis of the vehicle coordinates and the lane. The value of θ is related to the lateral velocity (Wang, Gao, Li, Sun, & Cheng, 2019).

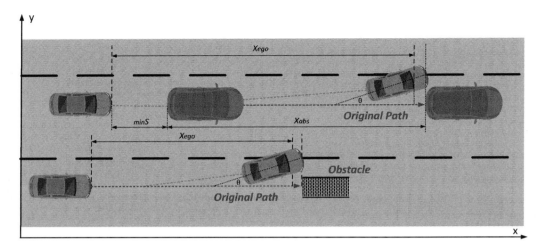

FIGURE 3.8 Longitudinal safety distance between ego vehicle and obstacle (Wang et al., 2019).

The longitudinal displacement of the subject vehicle during the time of obstacle avoidance operation can also be expressed by the equation (Wang, Gao, Li, Sun, & Cheng, 2019):

$$X_{ego} = \int_0^{t_c} \dot{X}_{ego}\, dt, \qquad (3.2)$$

The longitudinal displacement of the obstacle vehicle during the time of the subject vehicle obstacle avoidance operation could also be expressed by the equation:

$$X_{obs} = \int_0^{t_c} \dot{X}_{obs}\, dt, \qquad (3.3)$$

In order to ensure no collision occurs in the lane change progress, according to Equation (3.1), it can be deduced that the initial distance S should satisfy:

$$S = X_{ego} + w\sin\theta - X_{obs} > 0 \qquad (3.4)$$

Combining the above equations and the kinematic relationship of vehicles in the obstacle avoidance process, the minimum longitudinal safety distance required to complete obstacle avoidance without collision can be expressed by the equation:

$$\min S = \left\{ \left(\ddot{X}_{ego} - \ddot{X}_{obs}\right) + \int_0^{t_c} \left(\ddot{X}_{ego}(t)\, dt - \ddot{X}_{obs}(t)\, dt\right) + w\sin\theta, 0 \right\} \qquad (3.5)$$

In real traffic scenarios, the relative velocity of overtaking is relatively low, resulting in larger errors in the calculation results of Equation (3.5). Beyond this, to ensure driving safety, Equation (3.5) is improved with the consideration of time headway and driver behavior characteristics:

$$\min D = \min S + c\left(\dot{X}_{obs} - \dot{X}_{obs}\right) + d_0 + \sigma\left(\dot{X}_{obs} - \dot{X}_{obs}\right) \qquad (3.6)$$

where c represents headway, d_0 is safety breaking distance, σ is the risk factor of the obstacle, which can be expressed as:

$$\sigma = \frac{G_1 m_{obs} w_{obs}}{G_2 S} \exp G_3 \max\left\{\left(\dot{X}_{ego} - \dot{X}_{obs}\right), 0\right\} \tag{3.7}$$

where m_{obs} and w_{obs} are the mass and width of the obstacle, respectively. G_1, G_2, and G_3 are adjustment coefficients.

In general, the principle of human driver obstacle avoidance is similar to that of the artificial potential field. The basic idea is providing a collision-free path and setting apart an appropriate safety margin based on the types and the movement state of obstacles. The artificial potential field is ideal for describing the scenarios of obstacle avoidance. How to improve the artificial potential field and apply it to obstacle avoidance scenarios is an urgent problem (Wang, Gao, Li, Sun, & Cheng, 2019).

3.4 Inspection Modes and Content

3.4.1 A Simple Pathway to Drone-Based Asset Inspection

This leaves one key question: how can asset operators quickly exploit the potential of drones? We suggest a network-based approach that leverages existing market knowledge via a pilot project.

The first step is to set up a small project team with a dedicated, empowered leader. This person needs to have the mindset to perform in an agile environment and the clout to push the project through. A top-level sponsor is essential.

Once in place, the project leader should begin building a network of drone experts. This needs to encompass several areas. First, regulators and industry associations to help secure necessary licenses, as well as any publicly available funds. Second, established hardware, software, and drone-based service providers to determine the current state of the market, available technologies (off-the-shelf or customizable), and whether any shortcuts are possible, for example, via test runs with an existing service provider. Finally, startups and academia—this is where important developments happen and where top talent will most likely be found.

Next, while continuing to build the network, the team should use intermediate findings to design (and continuously adapt) the pilot project. This would include selecting a drone model and payload (ideally off-the-shelf to save time), identifying suitable assets for a trial run, and defining a timeline and KPIs. The KPIs should include UAV-specific indicators, as well as inspection-specific markers to monitor drone performance. It's important not to overengineer the trial—the goal should be proof of concept rather than a finished product.

After that, it's time to run the pilot. A key part of this should be agile monitoring of progress, with lessons learned compiled and fed back into the process. For example, has the UAV model proven itself, and does the process need refinement?

The final step should be focused on expanding the project. The best course of action will depend on the pilot's results, but a set of core questions will underpin all scenarios. These include whether the company should build all inspection capacity in house and develop knowledge independently or whether it should partner with or invest in a startup, university program, or drone service provider (Weichenhain, 2019).

Figure 3.9 shows the steps to implement drone-based asset inspection (Weichenhain, 2019):

3.4.2 The Drones Just Get Better

Looking forward, first movers will have the best chance of unleashing the technology's full potential. They will quickly be able to become more data driven in an era when data are king and leverage their know-how into a new business model, perhaps even providing drone services to others.

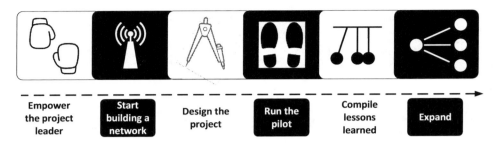

FIGURE 3.9 Steps to implement drone-based asset inspection (Weichenhain, 2019).

The prospect of future developments in UAV asset inspection also provides compelling reasons to act now. Advances in battery technology, for example, will enable greater and greater ranges, while new fuel cell, solar, tethering, and gas power systems could enable almost continuous operations.

And with drones able to stay aloft nearly all day, improved computing capacity and AI will enable them to perform real-time analysis. Growing computing muscle will also aid the development of so-called swarms of drones, which will call on AI to work together on increasingly complex tasks.

And how about this all being done autonomously? That too could soon be a reality. Several companies are already pushing the technological boundaries here. These include SkyX, a Canadian firm that is developing an autonomous, self-charging drone for the inspection of North America's vast oil and gas pipeline network (Weichenhain, 2019).

3.4.3 Robotics in Onshore Conditions

Most of the onshore oil and gas facilities, such as El Merk oil and gas development in Algeria, lie in inhospitable deserts where the logistical constraints of deploying equipment, materials, and labor across multiple and labor on multiple work fronts are critical challenges. Oil and gas facilities have extensive usage of all kinds of pipes and storage tanks during different stages of business, including exploration, extraction, transportation, processing, and distribution. Pipes and storage tanks need regular inspection and maintenance, especially those continuously used for long-distance transportation and long-term storage. Humanly inspecting these components is expensive and hazardous, so automated inspection and manipulation for these components are very much desired. Most of the robotic research for onshore facilities has been dedicated to development of in-pipe inspection robots (IPIRs) and tank inspection robots (Shukla & Karki, 2016).

3.4.3.1 Pipe Inspection

At onshore oil and gas facilities, pipes are used as a tool for transportation of oil, gas, and other fluids, from production sites to distribution sites. These pipes are mostly laid down under water or in an underground environment. In such environments, pipes are subjected to extreme weather conditions, such as extreme temperature and pressure, humidity, dust, and vibration. These unfavorable conditions lead to many troubles in pipes, such as corrosion, erosion, deposition, cracks, thermal cycling, pitting, shock loading, and joint failure. Any kind of leakage of petroleum products from pipes not only causes loss of revenue but also invites ecological disaster. Therefore, regular inspection and maintenance of transportation pipes are required for safe operation. The traditional way of digging and manually detecting the temporal position of these flaws in underground pipes is not only inconvenient but also expensive. Nondestructive testing (NDT) holds the key to any future development in this field. In-pipe robots are proposed as one solution to this problem. These IPIRs are inserted in the pipe from an inlet point and travel inside the pipe under external supervision. There are various causes and kinds of corrosion, and for this, there are appropriate inspection techniques, such as visual inspection, X-ray, eddy currents, acoustics, and ultrasonics. These robots are equipped with a sonar and acoustics-based leakage detector, HD pan–tilt–zoom camera, and LED light panel for the purpose of inspection and sending critical data back to the control center. Most IPIRs are tele-operated and connected by tethered cable to the operator at a remote location. There are five essential parameters to categorize IPIRs (Shukla & Karki, 2016):

1. Shape and size of robots.
2. Steering mechanism.
3. Propelling mechanism.
4. Detection technology.
5. Control mechanism.

3.4.3.2 Tank Inspection

Huge metallic tanks are used for storing the crude oil and gas at both offshore and onshore production plants. These tanks have many welded seams along welded plates, and these seams are prone to leakage due to corrosion and wear. Continuous storage of crude oil and other such products inside the metallic tanks generates many corrosive by-products such as iron sulfide and hydrogen sulfide. Bubbling H_2S does more damage to the roof than the bottom of the tank. The bottom of the tank is mainly damaged by the collection of large quantities of sludge material containing heterotrophic microorganisms, although it also contains many corrosion pits created by the internal reactions of these products. Human inspection requires completely emptying the tank and stopping all production for few weeks; this process is lengthy, expensive, and hazardous from a safety point of view. Automated inspection, while tanks are full, with continued operation of the plant is the motivation for research on mobile robots for in-tank inspections.

The main criterion for categorizing tank inspection robots is based on the principle of climbing the tank under inspection. There are two broad categories of climbing techniques, first, based on adhesion mechanisms and, second, on the locomotion principle. The most common adhesion mechanisms are magnetism, vacuum suction, and specific attachment devices such as rails or legs and grippers/clamps. The locomotion category can be further divided in four subgroups: wheels, tracks, legs, and actuator or arms based devices. Figure 3.10 shows the general functioning of permanent magnet adhesion-based locomotion principle during an out-tank inspection of an above ground tank (Shukla & Karki, 2016).

FIGURE 3.10 Metallic oil tank climbing the inspection robot (Kalra, Gu, & Meng, 2006a; Kalra, Shen, & Gu, 2006b).

There are various ways in which control system for navigation of the inspection robots can be designed. Wang, Huang, Hong, and Fang et al. (2008) have presented a fuzzy CMAC algorithm along with neural networks to establish the tracking control system for improving the performance of robot navigation. Fernandez, González, Feliu, and Rodríguez (2010) proposed a novel client/server architecture for autonomous operation of the inspection robot. Here, a client program is run on a local inspection route related to safe climbing and navigation of the robot, while a server program is run on the operator side in the control room. On the server side, programs are concerned with capturing visual data, detecting leakage, and performing some manipulations. Fuel tanks installed on a ship have thin metallic walls, and heavy climbing robots cannot be used because of the possible of deformation of this surface. A novel architecture called mother/child was proposed by Fischer, Tache, and Siegwart (2007) to deal with such a situation; the mother robot is a normal heavy climbing robot with high mobility used to climb through a track on the tank, and the child robot, which is very light in weight, carries only required detectors for inspection. All the locomotive and adhesive devices are installed on the mother robot; the child robot is only an extended inspecting device in this system. Most of these robots are semiautonomous, with an operator at a remote location supervising the whole mission. Kalra, Gu, and Meng (2006a) and Kalra, Shen, and Gu (2006b) presented a novel control mechanism for an inspection robot which can be operated in fully autonomous mode via wireless control. After analyzing all these models, it can safely be said that most are broadly based on the principle of teleoperation (Shukla & Karki, 2016).

3.4.3.3 Automated Gas Sampling

To determine the composition and quality of hydrocarbon, essential knowledge for chemical processing and knowing the price of fuel in international market, there is a routine process of collecting gas from the production site. Normally this job is done manually with an operator collecting the desired amount of the gas in a container by connecting it to supply valve. This process involves certain sets of protocol to collect a fully representative sample, but this often leads to sample variations depending on the individual operator's responses. Sample variation may be cause of error, but HSE is a far greater concern. Therefore, the Norwegian company Statoil has developed an automated gas sampling station using a standard six degrees of freedom (6-DOF) serial manipulator for this repetitive process (Shukla & Karki, 2016).

3.4.3.4 Aviation Robotics for Inspection

Normally, industry standard intelligent pigs are used for inspection of corrosion and other such integrity issues in transportation pipes, but such internal inspection always interferes with overall routine operation of the plant. Hence, external inspection is preferable. At present, external inspection of transportation pipes involves a process of manual inspection, where a group of workers drive a vehicle along the transportation pipes and perform a visual inspection to detect leakage or any other kind of damage. Such external manual inspection processes are highly inefficient, expensive, and hazardous. Therefore, using UAVs with appropriate sensors is considered a suitable option for the external automated inspection of the pipelines. The first UAV was tested in 1950 by Ryan Aeronautical for military reconnaissance; since then, this technology has come a long way in the service of humanity. UAVs are remotely controlled devices operating in a semiautonomous or autonomous architecture via a command center on the ground. A human pilot can be engaged in flying for not more than 5 h at a stretch, but a UAV can remain in the air up to 30 h making it suitable for long inspection operations. The absence of human pilots in UAVs not only makes flight safer but also reduces the cost of dangerous operations with higher performance efficiency in most cases.

UAVs are now extensively used for surveillance and reconnaissance in military missions, rescue missions, remote sensing and exploration, domestic policing, disaster relief, scientific research, archeology, film or photo shooting, forest fire detection and firefighting, pesticide spraying and geophysical surveys,

Autonomous Inspection for Industrial Assets

FIGURE 3.11 Quadrotor UAV used for surveillance of pipelines (Anon, 2012).

logistics and payload transport, and ad hoc communication gateways, to name just a few. Such inspection mechanisms are not only cheaper but also more efficient and robust, supplying inspection data round the clock without any interruption due to fatigue. In the oil and gas industry, apart from inspection of the pipelines, UAVs equipped with relevant sensors and data transmission systems can be widely used for surveillance of production plants, flare stacks, refineries, and transportation systems of petroleum products. Exploration of oil and gas in inhospitable and difficult remote areas is another challenging task where aviation robotics can be of great use. Tiffin et al. (2014) argued for the use of UAVs in Arctic zones to deploy tools for ice advisory systems. The first fully functional UAV to inspect plant assets, such as flare stacks, was at UK onshore oil refineries in 2010. With this technique, for the first time, operators got the chance to understand the condition of their equipment before any kind of shutdown and without exposing work personnel to a risky environment. This system has enabled engineers to continuously monitor critical components located at difficult heights in plant structures, such as chimney stacks, ducting, pipe racks, and vents, while allowing them to prioritize maintenance, relocate budgets, defer shutdowns, reduce the time for turnaround, and order replacement parts before turnaround. This success has motivated many operators around the world to use UAV technology for cost-effective and time-efficient inspection, replacing more expensive and dangerous conventional techniques. For an example, in 2012, British Petroleum (BP) set up a research team to develop a technology suitable for using UAVs to inspect their oil pipelines in Prudhoe, Alaska as shown in Figure 3.11. In the last 4 years, this technology has become standard operational practice, used extensively in routine remote surveillance of both onshore and offshore platforms. Beyond surveillance, UAV technologies can be helpful for oil spills and other such disasters where direct human intervention is not only hazardous but also inefficient. Belgium has deployed UAVs of B-Hunter class to monitor a marine oil spill in its portion of the North Sea. These big UAVs fly at high altitudes and carry various kinds of sensors, including visual and IR cameras for monitoring. When there is a gusher in an oil field, poisonous H_2S gas poses an acute danger to human life and causes a loss of revenue. The speed of spread and the density of gases are two immediately required parameters to prevent the situation from further deteriorating; in this case, the use of UAVs equipped with suitable sensors to collect data is one of the safest and reliable options (Shukla & Karki, 2016).

Like other robotic technologies in the oil and gas industry, UAVs work on the principle of teleoperation, where a flying machine as a remote system remains in autopilot mode, with a programmed navigation system equipped with Global Positioning Systems (GPS) and inertial sensors, while transmitting and receiving information signals from a ground-level operator (Shukla & Karki, 2016).

3.4.3.5 Wireless Sensor Network

Pipelines are the most inexpensive way to transport all kinds of fluids especially oil and gas, and these pipelines need to be continuously monitored for leakage and damage by a regular patrolling group of technicians. As mentioned, these types of inspection are not only inefficient and expensive but also hazardous. Therefore, as an alternative, sensors can be installed at regular intervals along the pipeline network itself. These installed sensors can be wired for their power supply and data transmission, but this technique suffers from unreliability in many ways, especially because any damage to the wire makes the whole inspection network completely dysfunctional. Therefore, Mohamed and Jawhar (2008), proposed an integrated network; multiple battery-powered wireless sensors are fit in to a wired sensor network to improve overall reliability. In this case, even if the wired sensor network does not function, the inspection process for detection of leakage and damage remains undisturbed supported by wireless sensors. BenSaleh, Qasim, Obeid, and Garcia-Ortiz (2013), reviewed many important wireless sensor network (WSN) techniques currently used for the inspection of overground, underground, and underwater pipelines. Petersen et al. (2007), Akhondi, Talevski, Carlsen, and Petersen (2010), and Adejo, Onumanyi, Anyanya, and Oyewobi (2013) have described in detail all the technical requirements and possible applications of WSNs in oil and gas facilities. Though WSN techniques are more reliable than a wired sensors network, they suffer from the "energy hotspot" problem and have a short life span because of unequal energy dissipation at different nodes. To overcome this problem, Yu and Guo (2012) proposed a novel data collection algorithm depending on the WSN data fusion strategy which eventually improves the network performance for both delay and energy. This strategy not only ensures the effective propagation of urgent data but also prolongs the life span of the overall network. To mitigate the hot-spot problem and optimally utilize energy of the wireless sensors, Anupama et al. (2014) proposed a sleep–wake cycle which enables each sensor in the network to wake up for only 10 s in a 10-min time interval, thus saving significant energy (Shukla & Karki, 2016).

Sand production in oil wells is a nother major problem for the petroleum industry because excessive sand production can cause severe erosion of transportation pipelines, valves, and separators. The most common solution for sand production is stopping or reducing production; therefore, Abdelgawad and Bayoumi (2011), proposed a sand monitoring system using WSN. The aim was to develop a mechanism which can be used for sand detection and sand production measurements, thus allowing operators to place sand removal machinery well in advance before any extreme situation occurs (Abdelgawad & Bayoumi, 2011). Apart from detection of sand, leakage, and damage, WSN techniques can be used to design anti-theft mechanisms for oil and gas pipes, as proposed by Zhang, Zhang, Wang, and Zhang (2011). Jawhar, Mohamed, and Shuaib (2007), Wu, Chatzigeorgiou, Youcef-Toumi, Mekid, and Ben-Mansour (2014), and Jawhar, Mohamed, Al-Jaroodi, and Zhang (2013) proposed cost-effective, efficient, and reliable ways of monitoring oil and gas pipelines by combining the standard WSN technique with UAVs, an in-pipe inspection legged robot, and an autonomous underwater vehicle. Leakage in pipes is mainly due to corrosion; therefore, rather than performing leakage detection as a reactive measure, Rahman and Hasbullah (2010) proposed using a WSN to detect early stages of corrosion as a proactive measure to stop final damage (Shukla & Karki, 2016).

Because of the linear nature of structure to be monitored (e.g., oil and gas pipelines), sensors are generally placed linearly along the structure; therefore, such networks are called linear sensor networks (LSNs). There are four kinds of LSNs depending on the node hierarchy and its increasing capacity to monitor areas from small to medium to large to very large. In the first small-range LSNs, only SNs are used for generating the sensory data and directly communicating them to the Network Control Center (NCC). Obviously, such networks cannot be used for long, complex, and uncertain environments, as the sensors generally have a limited range of communication. Sensing and communicating both drain huge amounts of energy, reducing the overall life cycle of these LSNs. The second kind of LSN has both an array of SNs and relay nodes (RNs). These RNs are dedicatedly used for aggregating sensory data sent by SNs and then relaying these data to the base station (BS) (Shukla & Karki, 2016).

3.5 Inspection Methods

3.5.1 Onboard Inspection Technologies

Onboard inspection sensors and technologies, such as nondestructive evaluation (NDE) systems, are critical for accurate and reliable inspection of civil structures. Payload capacity and fundamental sensor demands often dictate the implemented mechanisms, as is the case with lightweight climbing robots and small UAVs. The demands of the chosen sensor mechanism can also influence the speed of inspection, the choice of locomotive mechanism, and battery design (Lattanzi & Miller, 2017).

3.5.1.1 Imaging Methods

Structural distress often manifests itself as visually observable changes to the surface appearance of the structure, such as steel corrosion or concrete cracking. It may also manifest as excessive deformation of the structure itself. These sorts of damage phenomena can be captured through digital imaging; the most common sensors used by inspection robots are cameras, often in the form of video recording. This application is similar to human-led visual inspections and so is an intuitive approach for robots. However, there is typically a lack of back-end processing, which can result in highly trained human inspectors watching hours of inspection footage while looking for flaws. A growing number of studies address this issue. Using a ground-based robot, Torok, Fard, and Kochersberger (2012) developed a structural assessment system that performs 3D reconstructions of structural elements using Structure from Motion (SfM) methods (Hartley & Zisserman, 2000) and uses the resulting 3D model to detect anomalies in the structure, such as severe concrete cracking. The UAV-based roadway survey by Zhang and Elaksher (2012) used SfM methods as well. Lattanzi and Miller (2015) studied how 3D reconstruction algorithm demands affect the sensor payloads of UAVs. The dual use of images for both damage detection and flight planning was also evaluated (Ellenberg, Branco, Krick, Bartoli, and Kontsos, 2015). The UAV-based inspection approach developed by Eschmann, Kuo, Kuo, and Boller (2013) revealed the potential for using a combination of LiDAR, infrared, and digital imaging. Infrared imaging was also deployed onboard a UAV by Zink and Lovelace (2015) (Figure 3.12). In all of these studies, the core idea was to maximize the utility of image information through a variety of automated damage detection and analysis algorithms (Lattanzi & Miller, 2017).

Range-finding sensors, such as sonar, can yield a form of image in murky marine environments where imaging is difficult (DeVault, 2000). Infrared range finders have seen less use because of the

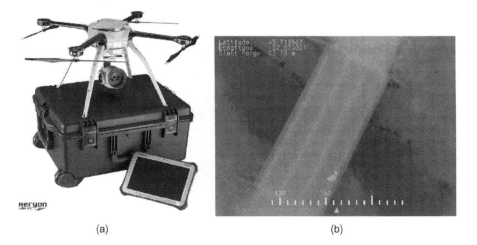

FIGURE 3.12 UAV bridge inspection robot and resulting infrared image: (a) Aeryon Skyranger used for inspection and (b) resulting infrared image of roadway surface, with detected delamination (Zink & Lovelace, 2015).

highly variable infrared responses of civil structure in changing environmental conditions. Laser range finders have been used to help guide the scanning and planning of inspection robots, for example, the AXBAM (Autonomous Exploration to Build a Map) system of Paul, Webb, Liu, and Dissanayake (2011). Notably, LiDAR has not been fully explored for robotic inspection, despite its widespread adoption as an inspection tool (Golparvar-Fard, Bohn, Teizer, Savarese, & Pea-Mora, 2011). This is largely because of the limited battery capacity and payload limitations of many robots (Eschmann, Kuo, Kuo, & Boller, 2013), though the potential exists. Fully 3D LiDAR systems also require a stable and static platform during scanning, which may not be possible with robots such as UAVs or Unmanned Maritime Vehicles (UMVs). Recent research into the use of UAV-mounted 2D LiDAR indicates one potential implementation strategy (Scherer et al., 2012).

3.5.1.2 Nondestructive Evaluation Methods

NDE methods are increasingly applied on robotic systems to capture the nonvisual damage phenomenon. Such technologies can also provide a level of quantitative structural assessment that is superior to the typically qualitative assessments of visual inspections.

Ultrasonic sensing has been the most commonly used NDE technology to assess steel structural integrity, with extensive application in the inspection of storage tanks. However, the application of ultrasound to metals normally requires direct contact between the transducer and the inspection surface, making it impractical for many applications. Magnetic flux leakage has seen use in robotic cable-stay inspections (Lattanzi & Miller, 2017).

Magnetic flux leakage has been adapted for the robotic inspection of restressed strands of reinforced concrete (Ghorbanpoor, Borchelt, Edwards, & Salam, 2000). Leibbrandt et al. (2012) implemented rebar corrosion assessment using half-cell potential methods. Other NDE methods that have been used for the inspection of concrete structures include ground-penetrating radar, thermography, and acoustic impact echo (Huston, Cui, Burns, & Hurley, 2011).

There have been several recent studies of robotic systems with a multisensor payload. The RABIT system, a collaborative project of Rutgers University and the Federal Highway Administration, is a ground-based robot with a variety of NDE technologies (Figure 3.13) (Gucunski et al., 2014).

One key advantage of the RABIT is that the various NDE scans are captured in tandem, allowing a composite analysis of the resulting data. The Betoscan similarly leverages a combination of NDE technologies, including ultrasonic, eddy current, and ground-penetrating radar (Figure 3.14) (Kurz, Boller, & Dobmann, 2013).

Several researchers have used robots to solve a long-standing problem in the sensor-based monitoring of structures, i.e., that of long-term sensor power supply and management. A prototype concept used crawling and ground-based robots to wirelessly power embedded sensors (Huston, Esser, Gaida, Arms, & Townsend, 2001). Mascareñas, Flynn, Farrar, Park, and Todd (2009) used a radio-controlled (RC) helicopter (Spectra G) to wirelessly power and interrogate embedded sensors on a bridge via an electromagnetic wave. In the study by Zhu, Guo, Cho, Wang, and Lee (2012), the robots themselves carried accelerometers as a payload, moving around the structure to sample vibration performance. The results were used to update a finite element model of a test bridge (Lattanzi & Miller, 2017).

The power requirements, weight, and logistical considerations of many NDE technologies have, historically governed overall robotic design. An important exception to this practice is the increasing use of general purpose UAVs and UMVs for inspections. These systems are chosen primarily because of their flexibility, compromising on the inspection payload by relying on digital imaging for sensing. Yet even these systems are seeing increasing customization to accommodate advanced NDE technologies (Mascareñas, Flynn, Farrar, Park, &Todd, 2009; Scherer et al., 2012).

3.5.2 Robotic Tunnel Inspection Systems

The use of robotic systems in the construction field is a common research area, and several studies review the advantages of the use of robotic platforms for construction and underground construction purposes.

Autonomous Inspection for Industrial Assets 95

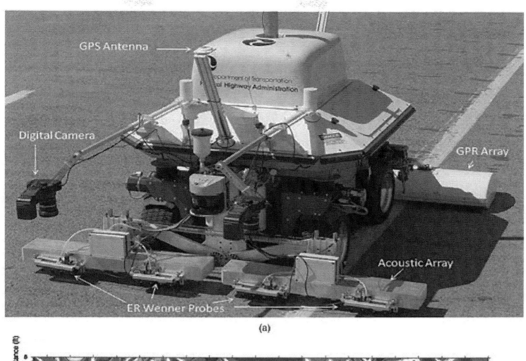

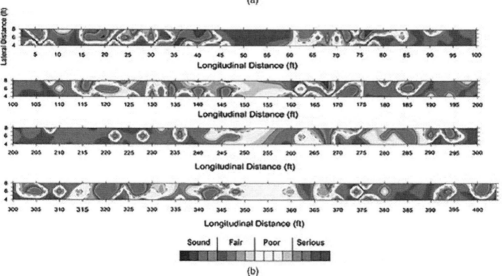

FIGURE 3.13 Federal Highway Administration (FHWA) Robotic Assisted Bridge Inspection Tool (RABIT) autonomous bridge deck assessment tool: (a) robot components and (b) resulting impact echo scan of roadway surface (Lattanzi & Miller, 2017).

Robotic systems can complete the inspection process with objective results and high efficiency. They also improve safety by performing inspection in dangerous environments, thus replacing humans.

Consequently, manual and (human) visual inspections are being replaced with more precise methods using mechanical, electronic, and robotic systems and processing data provided by cameras, laser, sonar, etc. The following review covers a variety of robotic systems using different kinds of sensors to detect defects on tunnels. Each subsection describes a different approach to inspect the tunnels (Montero, Victores, Martínez, Jardón, & Balaguer, 2015).

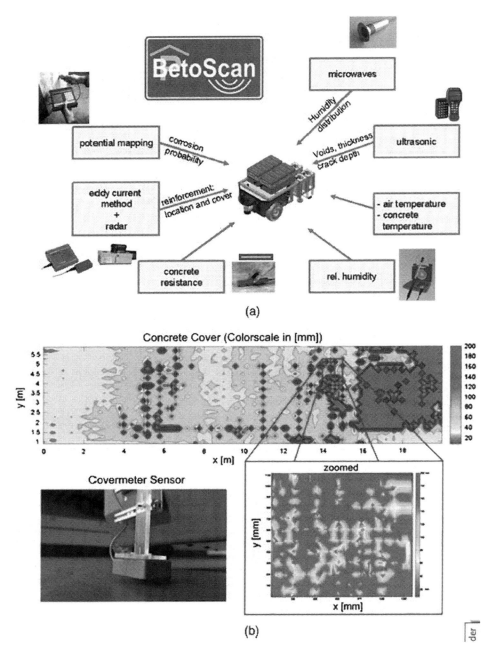

FIGURE 3.14 Betoscan multi-NDE system (Courtesy of BAM, Federal Institute for Materials Research and Testing, with permission): (a) schematic of the robot and (b) resulting concrete cover assessment from eddy current sensor (Lattanzi & Miller, 2017).

3.5.2.1 Visual Methods

In Figure 3.15, a small mobile robot is equipped with a CCD camera. The robot stays at a constant distance from the wall using a differential-drive wheel configuration, and a set of photos is taken. The camera is mounted on an anti-vibration device to stabilize the images. The robot goes through the tunnel performing the inspection; the data are processed after all the images are collected.

Autonomous Inspection for Industrial Assets 97

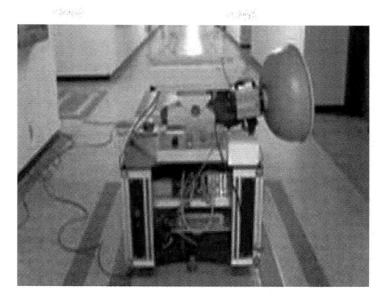

FIGURE 3.15 Robotic platform with camera used in tunnel inspections (Yu, 2007).

The inspection consists of the detection of cracks via computer vision algorithms (Montero, Victores, Martínez, Jardón, & Balaguer, 2015).

A similar robot can be found in Yao (2000, 2003). In this case, the mobile robot is equipped with 21 ultrasonic sensors and six video cameras. These sensors are mounted on the same plane and are in a semiring shape. The inspection consists of the scan of the tunnel lining to search for deformations. The experimental results show that this system can detect the deformed inner walls at divisions of 14 mm when the robot moves at 20 mm/s (Montero, Victores, Martínez, Jardón, & Balaguer, 2015).

3.5.2.2 Impact Methods

Figure 3.16 shows a system built with an industrial manipulator robot (Suda, 2004). The system consists of an 8-ton truck used as a base machine, tunnel cross section measuring systems, electronic distance measuring (EDM) instruments employed to measure impact locations, an impact unit with five hammers that generate impact sounds and is set on the robotic arm, a lifter that raises the robot up to ceiling level, and a computer unit that controls all these components (Montero, Victores, Martínez, Jardón, & Balaguer, 2015).

The system uses an impact acoustics method for the inspection procedure; it impacts the concrete wall with hydraulic hammers, converts the impact sounds into electric signals, and then analyzes them. The system is capable of finding exfoliation and cavities in a concrete lining. To maintain stability, the truck is equipped with outriggers on the nonmotorized wheels. Three people conduct the impact sound diagnosis: a supervisor, an operator, and a driver. The machine is operated from the touch panel of a computer situated at the operator console (Montero, Victores, Martínez, Jardón, & Balaguer, 2015).

Another example, seen in Figure 3.17, uses two lasers to perform a hammer-like inspection to detect inner defects in concrete structures like transportation tunnels (Fujita, 2012). The system is mounted on a motor vehicle, and the technique is based on the initiation and detection of standing Lamb waves (or natural vibration) in the concrete layer between surface and inner defects. One laser is used like a hammer to impact the surface and another is used to take the measurements. The system can detect various types of inner defects, like voids, cracks, and honeycombs. The accuracy of defect location is about 1–3 cm, and the detection depth is up to 5 cm (Montero, Victores, Martínez, Jardón, & Balaguer, 2015).

FIGURE 3.16 A robotic tunnel inspection system that uses the impact sound method (Suda, 2004).

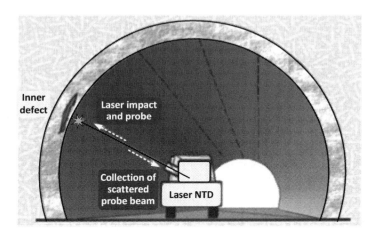

FIGURE 3.17 Schematic of the hammer-like laser remote inspection system (Fujita, 2012).

3.5.2.3 Laser Methods

The tunnel inspection system developed by Euroconsult and Pave metrics, shown in Figure 3.18, is based on cameras and laser sensors that allow scanning a tunnel's wall linings at speeds up to 30 km/h. The software of the system also allows the data from two different inspection runs to be rapidly compared, and structural changes and wall lining defects to be assessed (Montero, Victores, Martínez, Jardón, & Balaguer, 2015).

The measurement sensors for the condition survey are installed on a truck capable of running on rails or flat terrain. The vehicle comprises all the systems necessary for safe road and rail travel (lane occupation indicator, speed governor, electric power supply for all systems, signaling equipment etc.). It can hold up to six laser cameras. Each pair of laser-camera units inspects a 2 m wide section with an

FIGURE 3.18 The Tunneling system sensor structure (Montero et al., 2015).

accuracy of 1 mm. Using the six cameras, tunnels with a 9 m diameter can be inspected at the system's maximum resolution.

The system developed by Sano (2006) consists of a crack detecting vehicle equipped with laser sensors and CCD cameras. The vehicle is driven through the tunnel by an operator, and the cameras take pictures of the tunnel walls. The isolated images taken by the cameras are merged together into a surface map of the tunnel. After the map is obtained, a dedicated vision software detects cracks in it (Montero, Victores, Martínez, Jardón, & Balaguer, 2015).

3.5.2.4 Drilling Methods

The system shown in Figure 3.19 checks for voids behind lining by drilling holes with a mechanized crane (Oshima, 2005). It performs high-speed drills of 33 mm diameter by a combination of rotation and

FIGURE 3.19 Void detection rotary percussive drilling machine (Oshima, 2005).

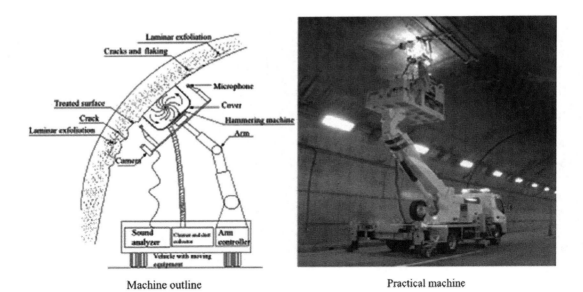

FIGURE 3.20 Mechanized hammer test (Mashimo & Ishimura, 2006).

striking the lining concrete of a tunnel surface to investigate the thickness of a lining and the height of a rear cavity with high accuracy (Montero, Victores, Martínez, Jardón, & Balaguer, 2015).

This example and others (such as methods based on a mechanized hammering tester installed on a crane, Figure 3.20) are described by Mashimo and Ishimura (2006). They define the status of road tunnel inspection and maintenance in Japan in 2006 (Montero, Victores, Martínez, Jardón, & Balaguer, 2015).

More examples of systems in Japan tunnels can be found in Asakura and Kojima (2003). These authors show the maintenance technology and typical deformation of Japanese railway tunnels, along with some methods of inspection and diagnosis and three case studies (Tsukayama Tunnel, Fukuoka Tunnel, and Rebunhama Tunnel). The examples of inspection methods include hammer testing on the lining (performed by an operator in this case), crack measurements in tunnel lining using line-sensor cameras mounted on a vehicle on rails, and the surface of the tunnel lining using an infrared camera and CCD cameras (Montero, Victores, Martínez, Jardón, & Balaguer, 2015).

3.5.2.5 Tunnel Cleaning

The mechanized cleaning truck shown in Figure 3.21 is an example of a tunnel maintenance system (Sanchez, 2013; Groso, 2013) used to prevent further damage. It was designed by engineers of Colas, Switzerland, in collaboration with operators of road networks. It was set up in 2012 (Montero, Victores, Martínez, Jardón, & Balaguer, 2015).

This tunnel cleaning system consists of a standard commercial truck equipped with eight mechanical arms with different types of brushes. The arms and the brushes have hydraulic actuators which provide movement and water flow for the cleaning process.

The mechanical arms can be positioned remotely to adapt to different tunnel geometries. This is achieved with a communication briefcase-like system controlled by an operator near the truck. A second operator is needed to drive the truck at a speed of 2 km/h while the tunnel walls are being cleaned with the brushes. The system can operate in tunnels with maximum height of 7.66 m. Only half of the tunnel section is covered each time, to avoid blocking the traffic on the free lanes (Montero, Victores, Martínez, Jardón, & Balaguer, 2015).

Autonomous Inspection for Industrial Assets 101

FIGURE 3.21 Tunnel cleaning system (Groso, 2013).

3.5.2.6 GPR Methods

Another commercial example is the IRIS Hyrail built by Penetradar (2013), shown in Figure 3.22. The system has a GPR sensor mounted on a telescopic piece in the front of a Hyrail vehicle (e.g., a vehicle able to go on road or rails). The GPR positioning device can be rotated to cover the sides and top of the tunnel walls, and the motorized boom can be retracted to avoid obstructions. Penetradar provides specialized software to manage data collection, data processing, and display of GPR data (Montero, Victores, Martínez, Jardón, & Balaguer, 2015).

FIGURE 3.22 IRIS Hyrail system inspecting a tunnel. Note the capability to be mounted on rails (Penetradar, 2013).

3.5.2.7 Small Tunnel Robots

When tunnels that need to be inspected have reduced dimensions, such as underground tunnels used to deploy power cables, the use of robotic platforms is appropriate. In this scenario, small tele-operated mobile robots can make inspections providing visual and concentration data of some poisonous gases, like the system by Zhuang (2008) shown in Figure 3.23 (Montero, Victores, Martínez, Jardón, & Balaguer, 2015).

This tele-operated robot (420 mm long, 320 mm wide, and 300 mm high) can operate in 1 m wide tunnels, move at a rate of 24 m/min, and has 2 h of autonomy. Its sensor system includes a pan–tilt–zoom camera, inclinometer, gyroscopes, gas sensors (CO, CH_4, CO_2, and O_2), thermometer, IR distance sensors, and ultrasonic sensors.

In other cases, cables are not inside small tunnels but along a greater one and are placed on the walls. Dian (2011) designed a robot based on a shrimp-rover vehicle (Estier, 2000) with six wheels able to go over the tunnel power cables while making the inspection. Unfortunately, this work is only theoretical, and the robot does not exist physically (Montero, Victores, Martínez, Jardón, & Balaguer, 2015).

Another type of small tunnel is a ventilation tunnel. Minichan (2011) designed three mobile robots to inspect the ventilation tunnels of the H-Canyon Facility in 2003, 2009, and 2011. Due to the toxic environment of the tunnels, only a robot can perform the inspection process. The control of the robots is remote, and the system is connected by a long tether to the control station. The inspection consists of a visual assessment with the images provided by the robot cameras. Figure 3.24 shows the three robot models (Montero, Victores, Martínez, Jardón, & Balaguer, 2015).

Not all tunnels are designed to carry vehicles, people, or cables. Water distribution is managed in tunnels too, and different solutions must be used to inspect this kind of structure. In this scenario, alternatives to mobile-wheeled robots include autonomous underwater vehicles (AUVs) (Kalwa, 2012) and remotely operated vehicles (ROVs) (Ageev, 2000; Loisy & François, 2010) which can exploit the use of sonar sensors for the mapping procedure (Montero, Victores, Martínez, Jardón, & Balaguer, 2015).

3.5.2.8 Embedded Sensors

All the systems mentioned in this section have their own sensors and publish the data obtained to perform the assessment of the structure. However, an alternative strategy involves the use of sensors embedded in the structures to be inspected, such as strain gauges; these are usually more precise and reliable. Esser (2000) implemented this method and developed a robot capable of remotely powering and

FIGURE 3.23 Cable tunnel inspection robot (Zhuang, 2008).

Autonomous Inspection for Industrial Assets

FIGURE 3.24 Three ventilation tunnel robotic inspection systems (Minichan, 2011).

collecting data from a network of embedded sensing nodes and providing remote data access via the Internet. The system uses Addressable Sensing Modules (i.e., ASMs) to sample data from a wide variety of sensors (e.g., peak displacement, peak strain, corrosion, temperature, inclination). This kind of system is useful in long tunnels where a wired sensor network is difficult to implement or in tunnels with complicated access (Montero, Victores, Martínez, Jardón, & Balaguer, 2015).

3.5.3 Unmanned Aircraft Systems (UAS) for Railway Applications

UAS technology is having a powerful and transformative impact on the rail industry. In railroad environments, UAS are particularly suitable for (Sherrock & Neubecker, 2018):

- Structural monitoring, especially for critical assets like bridges and tunnels, and for fault detection (i.e., diagnostics/prognostics).
- Environmental security monitoring, such as assessments of fire, explosions, earthquakes, floods, and landslides along the track.
- Physical security monitoring, including detection of intrusions, objects stolen or moved, graffiti, etc.
- Safety monitoring, for example, to early detect failures on track elements/devices or obstacles on the track.
- Situation assessment and emergency/crisis management to monitor accident scenarios and coordinate the intervention of first responders.

The use of UAS technology offers the following direct benefits for routine inspection activities (Sherrock & Neubecker, 2018):

- Reduced risk to staff and people and infrastructure in the project area.
- Reduced planning cycle.

- More efficient work processes.
- More flexible, affordable verification tools.
- Higher quality data available in larger quantities at lower costs.

When natural disasters strike, many railroad assets can be at risk. In such situations, it is critical to determine which part of the railroad needs repair prior to the movement of trains. UAS can gather information regarding the condition of the track or bridges, as well as the presence of debris on the right of way.

The aging of European rail infrastructure network causes many condition assessment challenges. Visual condition assessment of the rail system remains the predominate input to the decision-making process. Many railroads use machine-vision technology installed on rail-bound vehicles, but there are situations in which inspectors on foot or in hi-rail vehicles assess the track's surroundings. In the case of high or steep slope embankments, UAS can collect detailed information that could be missed by inspectors (Sherrock & Neubecker, 2018).

3.5.3.1 UAVs Suitable for Railway Applications

Two UAV types are available for railway operations: "rotary wing" shown in the top portion of Figure 3.25 and "fixed-wing" aircraft shown in the lower portion of Figure 3.25 (Sherrock & Neubecker, 2018).

Rotary-wing UAVs share many characteristics with manned helicopters. Rather than a continuous forward movement to generate airflow, these units rely on lift from the constant rotation of the rotor blades. There is no limit on how many blades an aircraft has, but the average is between four and eight. Unlike fixed-wing units, rotary wing units have the ability of vertical takeoff and landing, meaning they can be deployed virtually anywhere. This enables the aircraft to lift vertically and hover at a specific location. These UAVs can move in any direction, hovering over important areas, collecting the most intricate data. It is this ability that makes them so well suited for inspections where precision maneuvering is critical to the operation.

FIGURE 3.25 Rotary-wing and fixed-wing UAVs (Buchanan, 2016).

Autonomous Inspection for Industrial Assets 105

Fixed-wing UAVs are designed for higher speeds and longer flight distances. This type of UAV is ideal for coverage of large areas, such as aerial mapping and surveillance applications. This type of UAV can often carry heavier payloads than rotary UAVs. They glide efficiently and the single fixed wing drastically reduces the risk of mechanical failure. The maintenance and repair requirements for these units are often minimal, saving time and money. However, the current beyond visual line of sight (BVLOS) regulations limit the utility of fixed-wing UAVs. Several railways are predominantly using multi-rotor or hybrid vehicles that employ multiple rotors along with fixed wings to facilitate short takeoffs. Among the various types of UAVs, the one with the highest number of units worldwide is the rotary wing followed by fixed-wing UAVs.

The nano-type UAV is becoming prevalent in the UAS market space. The nano-type UAV is a palm-sized platform with a maximum takeoff mass of less than 30 g (approximately 1 oz). They utilize advanced navigation systems, full-authority autopilot technology, digital data links, and multi-sensor payloads. The operational radius for this type of platform is more than 1.5 km, and it can be flown safely in strong wind. Future development is anticipated to yield even smaller and more advanced nano-type UAVs with high levels of autonomy (Sherrock & Neubecker, 2018).

3.5.3.2 Sensors Employed for Railway Applications

Cameras are still the most common sensors used on a UAV. However, dynamic sensor technologies created for use with UAVs provide essential situational awareness and a level of detail often missed by the human eye and standard cameras. LiDAR sensors on UAVs, such as that shown in Figure 3.26, capture imagery which only a few years ago required an aircraft and a crew to collect. A LiDAR sensor mounted on a UAV, along with sophisticated software, can produce accurate three-dimensional images very quickly (Sherrock & Neubecker, 2018).

UAV payloads can integrate sensors of a different nature, such as temperature sensors or multispectral cameras, to provide diverse functionalities, depending on energy consumption and maximum allowed weight. Self-powered chemical sensors can be mounted on the aerial platform to provide quick and safe analyses of chemical or air samples near a derailment (Sherrock & Neubecker, 2018).

Current standard UAS technology allows the registration and tracking of position with GPS, or an Inertial Navigation System (INS), and orientation of the implemented sensors in a local or global coordinate system. UAS-based photogrammetry, or the practice of making measurements from imagery, now allows the collection of information from platforms that are remotely controlled or operated in a semiautonomous or autonomous manner, therefore eliminating the need for a pilot sitting in the vehicle

Thermographic camera   **Gas detector**

LiDAR laser scanner   **Multispectral camera**

FIGURE 3.26 Various UAV-mounted sensors (Jurić-Kaćunić, Librić, & Car, 2016).

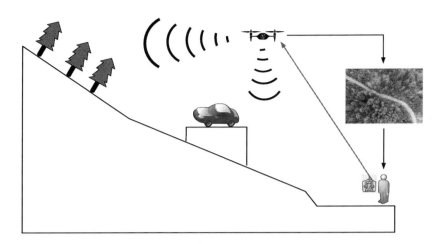

FIGURE 3.27 UAV mapping steep slopes and contours (Jurić-Kaćunić, Librić, & Car, 2016).

(Flammini, Pragliola, & marra, 2016). As described by the Croatian Association of Civil Engineers, UAS photogrammetry can be understood as a newer photogrammetric measurement tool with applications in the close-range domain that combines aerial and terrestrial photogrammetry to provide a real-time application and low-cost alternatives to the classical manned aerial photogrammetry. This approach can provide both an overview of a situation and detailed area documentation (Jurić-Kaćunić, Librić, & Car, 2016).

The collection of three-dimensional data by conventional surveying methods can be quite time consuming, expensive, and even dangerous for the field operator, especially on steep slopes and cuts where there are potential rock falls, landslides, or mudslides. Visual inspection of the terrain in such locations, just as geodetic data collection with classical methods, can result in incomplete and insufficiently detailed data, thus posing a risk to the company. The use of UAVs in such locations can complement, enhance, and even completely replace the classical methods of mapping, determining the volume, cross sections, contours, and other parameters that are necessary for remediation measures as illustrated in Figure 3.27 (Jurić-Kaćunić, Librić, & Car, 2016).

The challenge is to increase the level of automation to reduce the need for human interventions with the ongoing enhancement of UAV endurance and payloads, even in critical situations. The number of scenarios in which UAVs will be useful will be proportional to UAS performance growth (Sherrock & Neubecker, 2018).

REFERENCES

Abdelgawad A., Bayoumi M., 2011. Remote measuring for sand in pipelines using wireless sensor network. *IEEE Transactions on Instrumentation and Measurement*, 60(4), 1443–1452.

Adejo A. O., Onumanyi S. J., Anyanya J. M., Oyewobi S. O., 2013. Oil and gas process monitoring through wireless sensors networks: A survey. *Ozean Journal of Applied Sciences*, 6(2), 1077–1079.

Ageev M., 2000. Modernized TSL-underwater robot for tunnel and shallow-water inspection. In *Proceedings of the 2000 International Symposium on Underwater Technology, 2000, UT 00*, IEEE, 2000, Tokyo, Japan. pp. 90–95.

Akhondi M., Talevski A., Carlsen S., Petersen S., 2010. Applications of wireless sensor networks in the oil, gas and resources industries. In *Advanced Information Networking and Applications (AINA)*, IEEE, 2010, pp. 941–948.

Anon, 2012. BP Alaska: Unmanned aerial vehicle (UAV) pilot testing.

Anupama K., Kamdar N., Kamalampet S., Vyas D., Sahu S., Shah S., 2014. A wireless sensor network based pipeline monitoring system. In *International Conference on Signal Processing and Integrated Networks (SPIN)*, 2014, Noida, India. pp. 412–419.

Asakura T., Kojima Y., 2003. Tunnel maintenance in Japan. *Tunnelling and Underground Space Technology*, 18(2–3), 161–169.

BenSaleh M.S., Qasim S.M., Obeid A.M., Garcia-Ortiz A., 2013. A review on wireless sensor network for water pipeline monitoring applications. In *International Conference on Collaboration Technologies and Systems (CTS)*, 2013, San Diego, CA, USA. pp. 128–131.

Borenstein J., Koren Y., 1988. Obstacle a voidance with ultrasonic sensors. *IEEE Journal of Robotics and Automation*, RA-4(2), 213–218.

Borenstein J., Koren Y., 1995. Error eliminating rapid ultrasonic firing for mobile robot obstacle avoidance. *IEEE Transactions on Robotics and Automation*, 11(1), 132–138.

Buchanan K., 2016. Regulation of drones: New Zealand. April 2016.

Chattopadhyay A., Lam K.-Y., 2018. Autonomous vehicle: Security by design. October 01, 2018.

Crespo-Marquez A., Iung B., 2008. A review of e-Maintenance capabilities and challenges. *Journal on Systemics, Cybernetics and Informatics*, 6(1), 62–66.

DeVault J. E., 2000. Robotic system for underwater inspection of bridge piers. *IEEE Instrumentation & Measurement Magazine*, 3(3), 32–37.

Dian S., 2011. A novel shrimp rover-based mobile robot for monitoring tunnel power cables. In *2011 IEEE International Conference on Mechatronics and Automation*, Beijing, China: IEEE, 2011, pp. 887–892.

Ellenberg A., Branco L., Krick A., Bartoli I., Kontsos A., 2015. Use of unmanned aerial vehicle for quantitative infrastructure evaluation. *Journal of Infrastructure Systems*. doi:10.1061/(ASCE)IS.1943-555X.0000246, 04014054.

Eschmann C., Kuo C.-M., Kuo C.-H., Boller C., 2013. High resolution multi sensor infrastructure inspection with unmanned aircraft systems. *ISPRS—International Archives of the Photogrammetry, Remote Sensing and Spatial Information Sciences*, 2, 125–129.

Esser B., 2000. Wireless inductive robotic inspection of structures. In *Proceedings of IASTED International Conference Robotics and Applications*, Honolulu, HI, USA, 2000, pp. 14–16.

Estier T., 2000. Shrimp, a rover architecture for long range Martian mission. In *Proceedings of the Sixth ESA Workshop on Advanced Space Technologies for Robotics and Automation (ASTRA'2000)*, 2000, The Netherlands, December. pp. 5–7.

Everett H. R., 1995. *Sensors for Mobile Robots: Theory and Applications*. Natick, MA: A. K. Peters.

Fernández R., González E., Feliu V., Rodríguez A., 2010. A wall climbing robot for tank inspection. An autonomous prototype. In *IEEE Industrial Electronics*, 2010, pp. 1424–1429.

Fischer W., Tache F., Siegwart R., 2007. Inspection system for very thin and fragile surfaces, based on a pair of wall climbing robots with magnetic wheels. In *IEEE/RSJ International Conference on Intelligent Robots and System*, 2007, San Diego, CA, USA. pp. 1216–1221.

Flammini F., Pragliola C., Smarra G., 2016. Railway infrastructure monitoring by drones. In *International Conference on Electrical Systems for Aircraft, Railway, Ship Propulsion and Road Vehicles & International Transportation Electrification Conference (ESARS-ITEC)*, Naples, Italy.

Fujita M., 2012. Non-destructive remote inspection for heavy constructions. In *Conference on Lasers and Electro-Optics 2012Osa*, Washington, DC, USA, 2012.

Ghorbanpoor A., Borchelt R., Edwards M., Salam E. A., 2000. Magnetic-based NDE of prestressed and post-tensioned concrete members: The MFL system. Report No. FHWA-RD-00-026, Federal Highway Administration, Washington, DC, USA.

Golparvar-Fard M., Bohn J., Teizer J., Savarese S., Pea-Mora F., 2011. Evaluation of image-based modeling and laser scanning accuracy for emerging automated performance monitoring techniques. *Automation in Construction*, 20(8), 1143–1155.

Groso V., 2013. Camion de lavage de tunnels, 2013. 1.

Gucunski N., Boone S. D., Zobel R., Ghasemi H., Parvardeh H., Kee S.-H., 2014. Nondestructive evaluation inspection of the Arlington Memorial Bridge using a robotic assisted bridge inspection tool (RABIT). Nondestructive characterization for composite materials, aerospace engineering, civil infrastructure, and homeland security, Vol. 9063, SPIE, Bellingham, WA, USA.

Hartley R., Zisserman A., 2000. *Multiple View Geometry in Computer Vision*. Cambridge, UK: Cambridge University Press.

Hitchings M. R., Engwirdal A., Kajitani M., Vlacic Lj. B., 1998. Two sensor based obstacle avoidance for autonomous vehicles. Copyright© IFAC Intelligent Autonomous Vehicles, Madrid, Spain, 1998.

Huston D., Cui J., Burns D., Hurley D., 2011. Concrete bridge deck condition assessment with automated multisensor techniques. *Structure and Infrastructure Engineering*, 7(7–8), 613–623.

Huston D., Esser B., Gaida G., Arms S., Townsend C., 2001. Wireless inspection of structures aided by robots. Health monitoring and management of civil infrastructure systems, SPIE, Bellingham, WA, pp. 147–154.

Jaffe D. L., 1985. Polaroid ultrasonic ranging sensors in robotic applications. *Robotics Age*, 7(3), 23–30.

Jawhar I., Mohamed N., Al-Jaroodi J., Zhang S., 2013. An efficient framework for autonomous underwater vehicle extended sensor networks for pipeline monitoring. In *Robotic and Sensors Environments (ROSE)*, IEEE, 2013, Washington, DC, USA. pp. 124–129.

Jawhar I., Mohamed N., Shuaib K., 2007. A framework for pipeline infrastructure monitoring using wireless sensor networks. In *Wireless Telecommunications Symposium*, 2007, Pomona, CA, USA. pp. 1–7.

Jurić-Kaćunić D., Librić L., Car M., 2016. Application of unmanned aerial vehicles on transport infrastructure network. Građevinar, 68(4), 287–300.

Kalra L., Gu J., Meng M., 2006a. A wall climbing robot for oil tank inspection. In *2006 IEEE International Conference on Robotics and Biomimetics*, 2006, Kunming, China. pp. 1523–1528.

Kalra L., Shen W., Gu J., 2006b. A wall climbing robotic system for non-destructive inspection of above ground tanks. In *Canadian Conference on Electrical and Computer Engineering*, 2006, Ottawa, Canada. pp. 402–405.

Kalwa J., 2012. SeaCat inspects 24 km long water tunnel. *Ocean News Technology*, 18(4), 30.

Karim R., Kajko-Mattsson M., Mirijamdotter A., 2010. Fundamentals of the eMaintenance concept. In *The 1st International Workshop and Congress on eMaintenance*, 2010, 22-24 June, Luleå, Sweden.

Kolodko J., 1997. Technical report ultrasonic ranging sub-system. Technical Report TR-007/97, ICSL, University, Australia.

Kurz J. H., Boller C., Dobmann G., 2013. Condition assessment of civil infrastructure in Europe: Recent developments and what might be ahead. *Journal of Engineering Mechanics*, 702–711. doi:10.1061/(ASCE)EM.1943–7889.0000317.

Lattanzi D., Miller, G., 2015. 3D scene reconstruction for robotic bridge inspection. *Journal of Infrastructure Systems*. doi:10.1061/(ASCE)IS.1943-555X.0000229.

Lattanzi D., Miller G., 2017. Review of robotic infrastructure inspection systems. *Journal of Infrastructure Systems*. doi:10.1061/(ASCE)IS.1943-555X.0000353.

Lee J., Ni J., Djurdjanovic D., Qiu H., Liao H., 2006. Intelligent prognostics tools and eMaintenance. *Journal of Computers in Industry*, 57(6), Elsevier Science Publishers B.V., 476–489.

Leibbrandt A., Caprari G., Angst U., Siegwart R. Y., Flatt R. J., Elsener B., 2012. Climbing robot for corrosion monitoring of reinforced concrete structures. In *2nd International Conference on Applied Robotics for the Power Industry (CARPI)*, Zurich, Switzerland, New York, USA: IEEE, pp. 10–15.

Loisy F., François P., 2010. Underwater inspection experiment for a long tunnel of EDF's hydroelectric facilities. In *1st International Conference on Applied Robotics for the Power Industry*, Montréal, QC, Canada, 2010, pp. 1–4.

Mascareñas D., Flynn E., Farrar C., Park G., Todd M., 2009. A mobile host approach for wireless powering and interrogation of structural health monitoring sensor networks. *IEEE Sensors Journal*, 9(12), 1719–1726.

Masek V., Kajitani M., Ming A., Kanamori C., 1997. Mapping mobile robots' local environment by using multiple ultrasonic range sensor. *To Be Presented at Workshop on Intelligent Mechatronics*, Ritsumeikan University, Tokyo, Japan, 4–5 October 1997.

Mashimo H., Ishimura T., 2006. State of the art and future prospect of maintenance and operation of road tunnel. In *23th International Symposium on Automation and Robotics in Construction (ISARC)*, 2006, Tokyo, Japan. pp. 299–302.

Metso L., Baglee D., Marttonen-Arola S., 2018. Maintenance as a combination of intelligent it systems and strategies: A literature review. *Management and Production Engineering Review*, 9(1), 51–64. doi:10.24425/119400.

Minichan R., 2011. H-canyon air exhaust tunnel inspection vehicle development, ANS EPRRSD — 13th Robotics & Remote Systems for Hazardous Environments. In *11th Emergency Preparedness & Response American Nuclear Society*, Knoxville, TN, USA, 2011, p. 11.

Mohamed N., Jawhar I., 2008. A fault tolerant wired/wireless sensor network architecture for monitoring pipeline infrastructures. In *2008 Second International Conference on Sensor Technologies and Applications*, 2008, Cap Esterel, France. pp. 179–184.

Montero R., Victores J.G., Martínez S., Jardón A., Balaguer C., 2015. Past, present and future of robotic tunnel inspection. RoboticsLab, University Carlos III of Madrid, Spain. *Automation in Construction*, 59, 99–112.

Muller A., Crespo-Marquez A., Iung B., 2008. On the concept of e-maintenance: Review and current research. *Reliability Engineering & System Safety*, 93(8), 234–253.

Oshima K., 2005. Inspection of voids between lining and ground using rotary percussion system. *Tunnelling and Underground*, 36(4), 345–355.

Paul G., Webb S., Liu D., Dissanayake G., 2011. Autonomous robot manipulator-based exploration and mapping system for bridge maintenance. *Robotics and Autonomous systems*, 59(7–8), 543–554.

Penetradar, 2013. Penetradar GPR catalog 2013. p. 19.

Petersen S., Doyle P., Vatland S., Aasland C. S., Andersen T. M., Sjong D., 2007. Requirements, drivers and analysis of wireless sensor network solutions for the oil gas industry. In *IEEE Conference on Emerging Technologies and Factory Automation (ETFA)*, 2007, Patras, Greece. pp. 219–226.

Polaroid, 1991. Polaroid 6500 series sonar ranging module PID#6I5077. Polaroid Corporation, Cambridge, MA, USA.

Rahman M., Hasbullah H., 2010. Early detection method of corrosion on buried steel gas pipeline using wireless sensor network. In *International Conference on Computer and Automation Engineering (ICCAE)*, 2010, Singapore, Singapore; Vol. 3, pp. 553–556.

Sanchez S., 2013. "Octopus", le camion fribourgeois qui part à l'assaut des tunnels romands, La Liberté 2013. pp. 2–3.

Sano N., 2006. Investigation of cracks in lining concrete by using sequential image analysis. *Journal of Management in Engineering*, 62(3), 558–566.

Scherer S., Rehder J., Achar S., Cover H., Chambers A., Nuske S., Singh S., 2012. River mapping from a flying robot: State estimation, river detection, and obstacle mapping. *Robotics and Autonomous Systems*, 33(1–2), 189–214.

Seneviratne D., Ciani L., Catelani M., Galar D., 2018. Smart maintenance and inspection of linear assets: An Industry 4.0 approach. *ACTA IMEKO*, 7(1), 50–56. ISSN: 2221-870X.

Sherrock E., Neubecker K., 2018. Unmanned aircraft system applications in international railroads. Final Report DOT/FRA/ORD-18/04, February 2018, Office of Research, Development and Technology, Washington, DC, USA. U.S. Department of Transportation. Federal Railroad Administration.

Shukla A., Karki H., 2016. Application of robotics in onshore oil and gas industry—A review Part I. The Petroleum Institute, Abu Dhabi, United Arab Emirates. *Robotics and Autonomous Systems*, 75, 490–507.

Suda T., 2004. Development of an impact sound diagnosis system for tunnel concrete lining. *Tunnelling and Underground Space Technology*, 19(4), 328–329.

Tiffin S., Pilkington R., Hill C., Debicki M., McGonigal D., Jolles W., 2014. A decision support system for ice/iceberg surveillance, advisory and management activities in offshore petroleum operations. In *Offshore Technology Conference*, 2014, Houston, Texas.

Torok M., Fard M., Kochersberger K., 2012. Post-disaster robotic building assessment: Automated 3D crack detection from image-based reconstructions. *Computing in Civil Engineering*, 2012, 397–404.

Wandt K., Karim R., Galar D., 2012. Context adapted prognostics and diagnostics. In *International Conference on Condition Monitoring and Machinery Failure Prevention Technologies*, London, UK; Vol. 1, pp. 541–550.

Wang H., Huang X., Hong R., Fang C., 2008. A new inspection robot system for storage tank. In *7th World Congress on Intelligent Control and Automation*, 2008, Chongqing, China. pp. 7427–7431.

Wang P., Gao S., Li L., Sun B., Cheng S., 2019. Obstacle avoidance path planning design for autonomous driving vehicles based on an improved artificial potential field algorithm. *Energies*, 12, 2342. doi:10.3390/en12122342.

Weichenhain U., 2019. Drones: The future of asset inspection. Hamburg office, Central Europe. January 29, 2019. www.rolandberger.com/en/Publications/Drones-The-future-of-asset-inspection.html#!#&gid=1&pid=3. Viewed: August 15, 2019.

Wikipedia, 2019. Obstacle avoidance. January 25, 2019. https://en.wikipedia.org/wiki/Obstacle_avoidance. Viewed: August 06, 2019.

Wu D., Chatzigeorgiou D., Youcef-Toumi K., Mekid S., Ben-Mansour R., 2014. Channel-aware relay node placement in wireless sensor networks for pipeline inspection. *IEEE Transactions on Wireless Communications*, 13(7), 3510–3523.

Yao F., 2000. Development of an automatic concrete-tunnel inspection system by an autonomous mobile robot. In *Proceedings of the 9th IEEE International Workshop on Robot and Human Interactive Communication, RO-MAN 2000*, IEEE, 2000, Nashville, Tennessee, USA. pp. 74–79.

Yao F., 2003. Automatic concrete tunnel inspection robot system. *Advanced Robotics*, 17(4), 319–337.

Yu H., Guo M., 2012. An efficient oil and gas pipeline monitoring systems based on wireless sensor networks. In *2012 International Conference on Information Security and Intelligence Control (ISIC)*, 2012, Yunlin, Taiwan. pp. 178–181.

Yu S., 2007. Auto inspection system using a mobile robot for detecting concrete cracks in a tunnel. *Automation in Construction*, 16(3), 255–261.

Zhang C., Elaksher A., 2012. An unmanned aerial vehicle-based imaging system for 3D measurement of unpaved road surface distresses. *Computer-Aided Civil and Infrastructure Engineering*, 27(2), 118–129.

Zhang H., Zhang F., Wang Y., Zhang G., 2011. Wireless sensor network based antitheft system of monitoring on petroleum pipeline. In *Mechanic Automation and Control Engineering (MACE)*, 2011, Hohhot, China. pp. 4689–4692.

Zhu D., Guo J., Cho C., Wang Y., Lee K.-M., 2012. Wireless mobile sensor network for the system identification of a space frame bridge. *IEEE/ASME Transactions on Mechatronics*, 17(3), 499–507.

Zhuang F., 2008. A cable-tunnel inspecting robot for dangerous environment. *International Journal of Advanced Robotic Systems*, 5(3), 243–248.

Zink J., Lovelace B., 2015. Unmanned aerial vehicle bridge inspection demonstration project. Report No. 2015-40, Minnesota Department of Transportation, St. Paul, MN, USA.

4

Sensors for Autonomous Vehicles in Infrastructure Inspection Applications

4.1 Sensors and Sensing Strategies

Autonomous ground vehicles are important examples of mechatronic systems. Although they have clear promise, autonomous ground vehicles still have numerous challenges in sensing, control, and system integration. This section will look at these challenges in the context of off-road autonomous vehicles, automated highway systems, and urban autonomous driving.

4.1.1 Sensors

A sensor can detect the physical environment and forward the information for processing. Sensors commonly consist of two components: a sensitive element and a transducer. The sensitive element interacts with the input, and the transducer translates the input into an output signal that can be read by a data acquisition system (Matsson, 2018).

4.1.1.1 Sensor Error

The absolute sensor error is the difference between the sensor output and the true value. The relative sensor error is the difference divided by the true value.

4.1.1.2 Noise

Sensor noise is unwanted fluctuation in the sensor output signal when the true value is kept constant. The variance of the noise is an important parameter in sensor characteristics. White noise is a random signal where all frequencies contain equal intensity.

4.1.1.3 Drift

Sensor drift is an unwanted change in sensor output while the true value is kept constant.

4.1.1.4 Resolution

The resolution is the minimal change that the sensor can detect.

4.1.2 Inertial Sensors

4.1.2.1 Accelerometers

An accelerometer measures its own acceleration relative to an inertial reference frame. The function is comparable to a damped mass on a spring. When the sensor is exposed to an acceleration, the mass will be displaced. The displacement can be measured using the capacitive or piezoresistive effects.

A capacitive accelerometer uses the moving mass as a capacitor, changing the capacitance as it moves. A piezoresistive accelerometer uses the change in a material's electrical resistivity when it is deformed (Matsson, 2018).

4.1.2.2 Gyroscopes

A gyroscope measures angular rate relative to an inertial reference frame. Early gyroscopes used a spinning mass supported by gimbals. The conservation of angular momentum keeps the spinning mass leveled when the support is tilted, and the angular difference can be measured (Matsson, 2018).

4.1.2.2.1 Optical Gyroscopes

Optical gyroscopes use the Signac effect. If two pulses of light are sent in opposite directions around a stationary circular loop, they will travel the same inertial distance and arrive at the end simultaneously. But if the loop is rotating and two light pulses are again sent in opposite directions, the light pulse traveling in the same direction as the rotation will travel a longer inertial distance and arrive at the end later. Using interferometry, the differential phase shift can be measured and translated into angular velocity. This type of gyroscope is used in airplanes.

4.1.2.2.2 Vibrating Gyroscopes

Microelectromechanical system (MEMS) gyroscopes are commonly vibrating gyroscopes. This type of gyroscope consists of a vibrating mass mounted on a spring. If the mass is oscillating in the x-axis and a rotation about the z-axis is applied, an acceleration in the y-axis is induced. This acceleration is called Coriolis acceleration and is given by

$$a_{\text{cor}} = 2v * \Omega \tag{4.1}$$

where v is the velocity of the mass, and Ω is the angular rate of rotation.

The angular rate is, thus, given by the velocity of the oscillating mass and found by measuring the force which induces the Coriolis acceleration.

4.1.2.3 Rotary Encoders

Rotary encoders can be divided into absolute and incremental encoders. Absolute encoders can indicate the angular position of the shaft. The position is given by an encoded disc that rotates together with the shaft. Various different techniques are used to read the encoded disc, for example, mechanical or optical techniques. The incremental encoder cannot indicate the angular position, but it will indicate the incremental changes in angular rotation. Each increment of angular rotation produces an impulse in the sensor output (Matsson, 2018).

4.1.3 Absolute Measurements

4.1.3.1 Global Navigation Satellite System (GNSS)

Operational systems with global coverage are the United States' Global Positioning System (GPS) and Russia's GLONASS. Several other systems are scheduled to be operational by 2020, for example, Europe's Galileo and China's BeiDou-2. Other countries, such as India, Japan, and France, are also developing their own GNSS (Matsson, 2018).

4.1.3.1.1 GPS

GPS is divided into three segments: a space segment, a control segment, and a user segment.

4.1.3.1.2 Space Segment

The space segment originally consisted of 24 satellites divided into six circular orbits, with four satellites in each orbit. Today, there are a total of 31 operational satellites in the GPS constellation. The satellites are orbiting in the Medium Earth Orbit at an altitude of approximately 20,000.00 km. The orbits have a 55° inclination from the equator, and the orbital period is 12 h. The constellation ensures that at least four satellites are visible at any place on the earth at any given time.

4.1.3.1.3 Control Segment

The control segment is a global network of ground facilities. Its purpose is to control and maintain the system. The control segment consists of monitoring stations, ground antennas, and a master control station. There are 16 monitoring stations and 11 ground antennas spread around the world.

The monitoring stations track the satellites, collect GPS signals, and forward the information to the master control station.

The ground antennas communicate with the satellites via the S-band. They send commands and upload navigation data and program codes.

The master control station is located in Schriever Air Force Base in Colorado, United States. It commands and controls the satellites. It also collects data from the monitoring stations and computes the location of each satellite. The system is monitored to ensure system health and accuracy. Satellites can be repositioned to maintain an optimal constellation.

4.1.3.1.4 User Segment

The user segment consists of the receivers of the GPS signals. The receivers receive the coded signals and estimate position, velocity, and time.

4.1.3.2 Magnetometers

Magnetometers can measure the local magnetic field using the Hall effect. Magnetometers consist of a thin sheet of semiconducting material. In a magnetic-free environment, the electrons in the thin sheet are evenly distributed, and the potential difference is zero. When a magnetic field is present, the electrons will distribute unevenly, inducing a potential difference. The potential difference is measured and translated into magnetic flux density (Matsson, 2018).

4.1.4 Sensing

4.1.4.1 Sensing the Surroundings

In this chapter, we assume the "internal sensing" of a car, i.e., determination of vehicle speed, steering angle, engine, braking wheel torque, etc., is possible, and the process is similar for all three application domains of concern: off-road autonomous vehicles, automated highway systems, and urban autonomous driving. This section concentrates on the external sensing, as this will be different in each case.

The sensors used for the comparison and their coverage footprints are given in Figure 4.1a–c (Ozguner & Redmill, 2008).

The set of sensors used on roadway vehicles depends on the infrastructure available. Basic lane detection can be accomplished by vision, assuming clear detection opportunities (Redmill, 1997). However, aids installed on the roadway with different technologies are certainly useful. These include magnetic nails (Zhang, 1997; Tan, Rajamani, & Zhang, 1998; Shladover, 2007) and radar reflective stripes (Redmill & Ozguner, 1999; Farkas, Young, Baertlein, & Ozguner, 2007). Off-road vehicles don't need an infrastructure, but they do need more sensor capability, especially information on ground surface level detection, so as to compensate for the terrain, such as bumps and holes (Ozguner & Redmill., 2008).

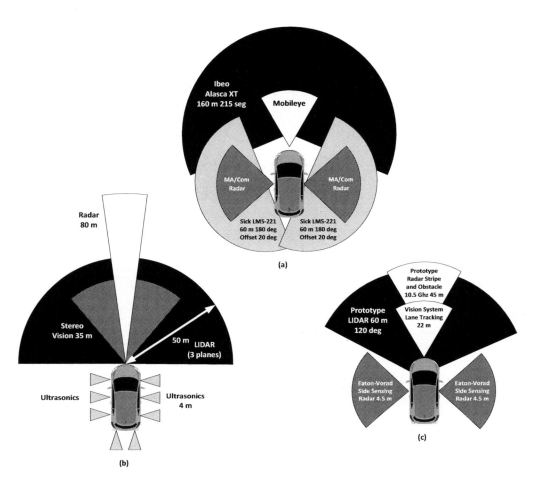

FIGURE 4.1 Sensor suite and footprints: (a) OSU-ACT, (b) OSU-ION, and (c) OSU Demo'97 cars (Ozguner & Redmill, 2008).

4.1.4.2 Sensor Fusion

There are some distinctions between sensor fusion for highway, urban, and off-road applications (Ozguner & Redmill, 2008):

- For pure highway applications, a very restricted approach may be appropriate, as there are a number of distinct concerns, for example, lane edge location and static or moving obstacles in relevant locations. If we simply fuse data related to specific tasks, however, we will not necessarily get a complete and integrated representation of the world.
- For off-road applications, the software or hardware needs to provide compensation for vibration and other vertical and rolling motions, for example, using the inertial measurement unit (IMU) and sensor data to specifically generate a "ground plane" that can be referenced for sensor validation and fusion. Sensor adjustments are also required to deal with dust, rain, and changing lighting conditions.
- For domains where there are many moving obstacles (i.e., urban applications), we may need to "track" individual obstacles all the time.
- Specific operations (parking, dealing with intersections, entering/exiting highways, etc.) may use totally separate sensing and sensor architectures. This may include information provided by the infrastructure or other vehicles through wireless communication.

In general, we can identify two approaches to sensor fusion: a grid, or occupancy map, and track identification.

In a grid map, the sensing architecture/sensor fusion is established by developing a discrete map of the vehicle's surroundings. All external sensors feed into this map, with obstacles sensed and related confidence levels recorded. The map is maintained internally in vehicle-centered world coordinates. The map doesn't rotate with the vehicle, but it does translate its movement. The sensor fusion algorithm implemented on OSU's 2005 Defense Advanced Research Project Agency (DARPA) Grand Challenge vehicle (see Chapter 2) uses such a grid occupancy approach (Redmill, Martin, & Ozguner, 2006).

Because traffic situations are highly dynamic, OSU-ACT has moved to an approach in which the sensor fusion algorithm is responsible for clustering and tracking all objects seen by the sensors.

The sensor fusion algorithm first uses information about the position and orientation of the sensors with respect to the vehicle to transform the returned information into a vehicle-centered coordinate system. The primary sensors, the LiDAR (light detection and ranging) suite, provide a cloud of points representing each reflection from some surface of all the targets in the world. Once the returns from the LiDARs are in vehicle-centered coordinates, the position and orientation of the vehicle with respect to the world are used to transform the LiDAR returns into world coordinates. After the LiDAR returns have been transformed into world coordinates, they are clustered into groups of points. The clustering algorithm places the laser returns into a disjoint set data structure using a union find algorithm. Ultimately, clusters of laser returns are found whose members are not farther than some maximum distance from each other.

Once the LiDAR returns have been clustered, the clusters are analyzed and those that can be identified as vehicles, based on shape and motion, are classified as such, and their centroids are estimated. All resulting clusters must be tracked using dynamic filters. Vehicle detections that are returned by the vision system or the radar sensors are matched to a LiDAR-generated cluster by looking for a LiDAR cluster within some distance threshold. If no suitable matching cluster is found, the detections may update or initialize a track without a corresponding LiDAR cluster. The output of the sensor fusion algorithm is a list of tracks. Each of the resulting tracks has a position and velocity, and the general size and shape of the point cluster supporting the track is abstracted as a set of linear features (Ozguner & Redmill, 2008).

4.2 Sensor Types: Introduction

The key components of autonomous vehicles are the sensors. These include cameras, LiDAR, radar, sonar, GPS, an IMU, and wheel odometry. Sensors in automotive vehicles are used to collect data that are analyzed by the computer in the autonomous vehicle and used to control the steering, braking, and speed of the vehicle. The information from environmental maps stored in the cloud and data uploaded from the other cars are also used to make decisions about vehicle control (Kocić, Jovičić, & Drndarević, 2018).

4.2.1 Autonomous Vehicle Sensor Categories

Automotive sensors fall into two categories: active and passive.

4.2.1.1 Active Sensors

Active sensors send out energy in the form of a wave and look for objects based on the information that comes back. One example is radar; it emits radio waves that are returned by reflective objects in the path of the beam.

4.2.1.2 Passive Sensors

Passive sensors simply take in information from the environment without emitting a wave; one example is a camera.

4.2.2 Automotive Radar

As with cameras, many ordinary cars already have radar sensors as part of their driver assistance systems—adaptive cruise control (ACC), for example.

There are two types of automotive radar: 77 and 24 GHz. A third type, 79 GHz radar, will be offered soon on passenger cars. A 24 GHz radar is used for short-range applications, while 77 GHz radar is used for long-range sensing.

Radar works best at detecting objects made of metal. It has a limited ability to classify objects, but it can accurately tell the distance to a detected object. However, unexpected metal objects at the side of the road, such as a dented guard rail, can provide unexpected returns for development engineers to deal with (Dawkins, 2019) (Figure 4.2).

Radar has been used in the automotive industry for decades and can determine the velocity, range, and angle of objects. It is computationally lighter than other sensors and can work in almost all environmental conditions.

Radar sensors can be classified according to their operating distance ranges: short-range radar (SRR), 0.2–30 m; medium-range radar (MRR), 30–80 m range; long-range radar (LRR), 80 m to more than 200 m.

LRR is the detector sensor used in ACC and highway automatic emergency braking systems (AEBS). Currently deployed systems using only LRR for ACC and AEBS have limitations and might not react correctly to certain conditions, such as a car cutting in front of another vehicle, detecting thin profile vehicles such as motorcycles being staggered in a lane, and setting distance based on the wrong vehicle because of the curvature of the road. To overcome the limitations in these examples, a radar sensor could be paired with a camera sensor in the vehicle to provide additional context to the detection (Ors, 2017).

4.2.2.1 Radar in Autonomous Vehicle Research Platforms (AVRPs)

Radar units on an AVRP are valuable resources for environment interactions, such as measuring distance, velocity, and angle with respect to an object, or tracking objects. Radar works by transmitting radio waves in a chosen direction and reading the reflected waves. When the time it takes for the signal to return is known, the distance can be calculated. If the object is moving with respect to the radar unit, a Doppler shift will occur. A Doppler shift is a change in the radio wave frequency. If an object is moving away, the reflected radio wave frequency will decrease, and if an object is moving closer, the frequency

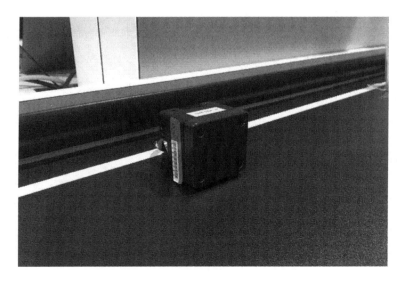

FIGURE 4.2 Automotive radar (Dawkins, 2019).

will increase. The velocity of the object can be determined by the change in frequency. When both objects are moving, the velocity of the object with the radar unit must be taken into account.

Having radar units on an AVRP allows the researcher to run studies that involve blind spot detection, ACC, lane changing assist, parallel and perpendicular parking, crash avoidance, and general safety during operation. Radar can have a longer detection range capability than LiDAR. For autonomous vehicles, it is typically broken down into three different ranges: long range, medium range, and short range. The longer the range, the narrower the typical detection window. LRR is narrower and used more for ACC and crash avoidance. MRR and SRR can be used for rear collision warning, blind spot detection, cross traffic warning, etc.

The operating frequency of the radar unit can affect its performance, specifically for distance and resolution. Two common frequencies for radar in autonomous vehicles are 24 and 77 GHz, with wavelengths of 12.5 and 3.89 mm, respectively. The higher the operating frequency, the better the resolution of the radar but the more attenuation caused by weather and atmospheric conditions. Higher frequencies offer better resolution because the wavelength is shorter, so it's easier for it to hit objects that the lower frequency, longer wavelength radio waves would miss. Radar can be more reliable than LiDAR in questionable conditions such as rain, fog, snow, or extremely dusty conditions because of these longer wavelengths. Having radar units in addition to LiDAR allows researchers to fully run tests without the AVRP's capabilities being limited because it only has one kind of sensor.

Forward-facing radar units can be mounted in several places, such as the front bumper, on the roof near the top of the windshield, and/or at the front corners of the vehicle. Rear-facing units can be mounted on the rear bumper or on the roof at the rear of the vehicle. Radar units can also be mounted on the roof facing out to the sides of the vehicle to expand areas of detection. Figure 4.3 shows mounting locations and detection zones. In many cases, the designer may choose to install the radar units behind plastic bumper covers to allow the AVRP to keep some of its stock appearance (Wicks, Steve, & Asbeck, 2017).

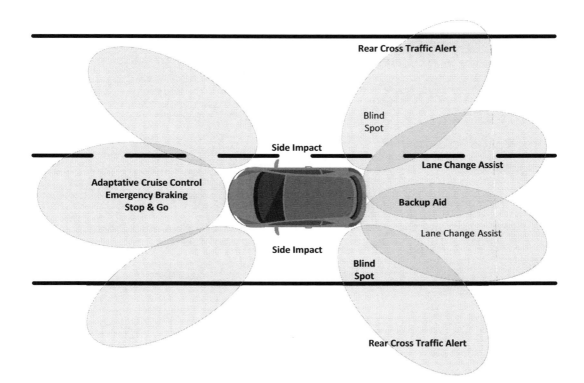

FIGURE 4.3 Radar mounting locations and their detection zones (Wicks, Steve, & Asbeck, 2017).

4.2.2.2 Radar Characteristics

- Accurate distance information.
- Relatively long range.
- Robust in most weather conditions.
- Can be hidden or protected behind body panels.
- Immune to effects of illumination or darkness.
- Fixed aim and field of view but able to employ multiple radar sensors as needed.
- Field of view (horizontal): ~15° (long range) to ~90° (short range).
- Range: ~250 m.
- Resolution: ~0.5° to ~5° (Schoettle, 2017).

4.2.2.3 Radar Sensing Fundamentals

In contrast to an ultrasonic sensor, an advanced driver assistance system (ADAS) radar sensor uses electromagnetic (EM) waves to determine an object's range, velocity, and angle in its field of view.

Various radar techniques are used to measure the range and velocity of an object. Texas Instruments applies the frequency-modulated continuous-wave (FMCW) method in its ADAS radar products (Pickering, 2017).

An FMCW radar transmit/receive (TX/RX) block works as follows (the numbers in Figure 4.4 correspond to those below):

1. A synthesizer generates a "chirp"—a sinusoid whose frequency increases linearly with time.
2. The chirp is transmitted by the TX antenna.
3. The chirp is then reflected by an object and received by the RX antenna.
4. A mixer combines the RX and TX signals to produce a composite IF signal.
5. A high-speed analog-to-digital converter (ADC) digitizes the IF signal.
6. A digital signal processor (DSP) performs a fast Fourier transform (FFT) to extract the frequency information, identify the objects, and determine their characteristics.

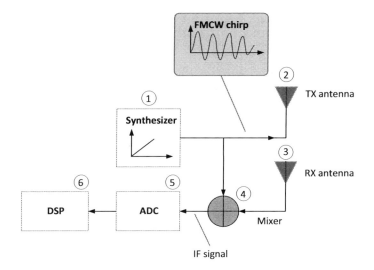

FIGURE 4.4 An FMCW radar transmits a synthesized chirp and combines it with the received signal to generate an IF signal that is analyzed to determine the object range (Pickering, 2017).

Sensors for Autonomous Vehicles   119

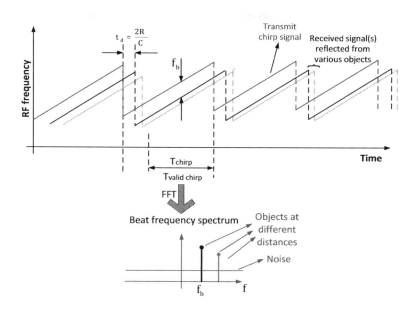

FIGURE 4.5 For each object, the IF signal contains a tone with frequency proportional to distance (Pickering, 2017).

The RX signal is delayed by the time needed for the TX signal to travel to the object and return. The IF signal, therefore, contains the information required to measure distance and distinguish between different objects (Pickering, 2017).

Figure 4.5 shows the received FMCW signals. These comprise different delayed and attenuated copies of the transmitted signal corresponding to various objects. Each object is a tone whose frequency (f_b) is proportional to the distance of the object from the radar (R).

The process of detecting multiple objects and their respective distances from the radar involves taking an FFT of the IF beat-frequency signal and identifying the tones that correspond to detected objects.

The resolution of the FMCW architecture is a function of the chirp bandwidth. The AWR1x family of devices (discussed later) supports a chirp bandwidth of up to 4 GHz in a single sweep, giving a range resolution of less than 5 cm.

The IF bandwidth determines unambiguous velocity, or the ability to separate objects with similar velocity. The radar front end in the AWR1x portfolio supports an IF bandwidth of 15 MHz, giving a maximum range of greater than 250 m (820 ft) and an unambiguous velocity of up to 300 kph (186 mph) (Pickering, 2017).

4.2.2.4 Automotive Applications of Radar Sensing

Automotive radar applications can be divided (see Figure 4.6) into SRR, MRR, and LRR, based on the range of objects to be detected:

- Ultra-short-range radar (USRR) is an emerging ADAS application for park-assist systems. These systems, part of SAE Level 1, typically use light-emitting diodes (LEDs) or steering-wheel vibrations to warn drivers of obstacles.
- Driver-assist features such as lane departure and blind spot warnings employ SRR. Current SRR systems utilize the 24–29 GHz frequency band but are expected to move to a higher frequency band in the future.
- Automatic emergency braking and ACC use LRR systems. Current LRR systems operate in the 76–77 GHz frequency range. Higher levels of automated driving demand higher range and resolution, so future front radar systems will likely use both the 76–77 GHz and 77–81 GHz frequencies and a combination of LRR and MRR systems.

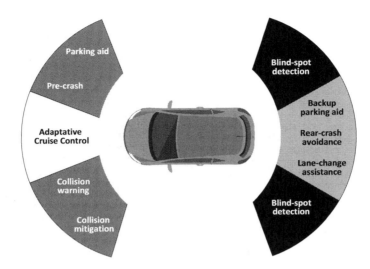

FIGURE 4.6 Automotive uses for radar include short-range (blind spot detection, backup parking) and long-range (ACC, pre-crash detection) applications (Pickering, 2017).

Satisfying higher SAE automation levels will require radar sensors to analyze complex scenarios: detect hazards, measure their properties, and categorize them into groups with distinct properties (distance, velocity, angle, height, etc.). Finally, the sensors must be able to coordinate with other systems so that the vehicle can take appropriate corrective action (Pickering, 2017).

4.2.3 Light Detection and Ranging (LiDAR)

LiDAR is one of the most hyped sensor technologies in autonomous vehicles and has been used since the early days of self-driving car development.

LiDAR systems emit laser beams at eye-safe levels. The beams hit objects in the environment and bounce back to a photodetector. The returned beams are brought together as a point cloud, creating a three-dimensional image of the environment (Dawkins, 2019) (Figure 4.7).

This is highly valuable information, as it allows the vehicle to sense everything in its environment, be it vehicles, buildings, pedestrians, or animals. This explains why so many development vehicles feature a large 360° rotating LiDAR sensor on the roof.

FIGURE 4.7 LiDAR (light detection and ranging) (Dawkins, 2019).

Sensors for Autonomous Vehicles 121

While LiDAR is powerful, it's also the most expensive sensor in use. Since their introduction, LiDAR sensors have had great size and cost reductions, but some of the more widely used and recognized models still cost a lot more than radar or camera sensors, and some even cost more than the vehicle they are mounted on. Some of the high-end sensors run into thousands of dollars per unit. However, there are many researchers and startups working on new LiDAR technologies, including solid-state sensors, which are considerably less expensive (Dawkins, 2019).

4.2.3.1 How LiDAR Sensors Work

LiDAR sensors measure the distance to an object by calculating the time taken by a pulse of light to travel to an object and back to the sensor.

LiDAR in automotive systems typically uses a 905 nm wavelength that can provide up to 200 m range in restricted fields of view (FOVs); some companies are now marketing 1,550 nm LiDAR with longer range and higher accuracy.

It is important to note that LiDAR requires optical filters to remove sensitivity to ambient light and to prevent spoofing from other LiDARs. It is also important to note that the laser technology used has to be "eye-safe." More recently, the move has been to replace mechanical scanning LiDAR that physically rotates the laser and receiver assembly to collect data over an area spanning up to 360° with solid-state LiDAR (SSL). SSLs have no moving parts and are therefore more reliable, especially in an automotive environment. SSLs currently have lower FOV coverage, but their lower cost provides the possibility of using multiple sensors to cover a larger area (Ors, 2017).

4.2.3.2 LiDAR in Autonomous Vehicle Research Platforms (AVRPs)

LiDAR plays a large part in autonomy by helping with digital mapping, ranging, object detection, and object tracking. It works by emitting pulses of laser light at a surface and measuring the time of return of the reflected light. Knowing the speed of light and the time it takes for the light to be reflected allows the distance to be calculated. Using several beams of light, all firing up to several thousand pulses a second, allows the creation of a point cloud in the observed direction(s). Rotating the unit allows the creation of a 3D visualization of the surrounding environment. A 2D LiDAR is also available and can be used for object detection and ranging, if the objects are within a detectable plane.

LiDAR units used in autonomy can have ranges up to a few hundred meters and an accuracy up to 10–12 cm. LiDAR units can have a vertical FOV that is either symmetrical at horizontal or greater below horizontal than above, i.e., +2.0° to −15.0°. Units with a greater FOV below horizontal are better utilized when mounted higher so their beams can reach an effective range. Horizontal FOVs have a variety of ranges, for example, 360°, 270°, 180°, 110°, and 85° (Wicks, Steve, & Asbeck, 2017).

With some LiDAR units, the user has control over some of its operational properties, such as rotational rate and the number of returns. Depending on the LiDAR unit, there can be several different numbers of laser returns. The more laser returns there are, the larger the point cloud generated and the more the LiDAR can detect. Each return represents a part of a single laser beam pulse that did not fully hit an object. For example, if half a beam hits the edge of a street sign, and the other half hits a wall behind it, two points will be returned, one for the edge of the sign and one for the wall. Utilizing multiple returns can help with resolution, but it can also create a problem if the computing system cannot handle the amount of data being collected through the LiDAR unit. It is up to the user to decide the rotation rate, if applicable, and the return mode to maximize the use of the LiDAR system.

Another useful aspect is the intensity of return. Every object that reflects light back to the unit will have a different return intensity. By analyzing the relative intensity data, we can try to classify the object detected, ground, water, vegetation, building, etc. This will contribute to the platform's ability to know whether a detected object is an obstacle or not.

A disadvantage of LiDAR is that it may not operate reliably in all weather conditions, such as rain, fog, snow, or extremely dusty conditions. This can be a result of the wavelength of light that most LiDAR units use and its interaction with small particles in the air. Velodyne uses wavelengths of 903 nm; SICK and Ibeo both use 905 nm wavelengths (Velodyne LiDAR Products; SICK Products; Ibeo Feature Fusion).

4.2.3.3 LiDAR Characteristics

- Accurate distance and size information.
- Ability to discern high level of detail (shape, size, etc.), especially for nearby objects and lane markings.
- Useful for both object detection and roadway mapping.
- Immune to effects of illumination or darkness.
- Fixed aim and FOV, but multiple LiDAR sensors can be employed as needed (some LiDAR systems are capable of 360° within a single piece of equipment).
- FOV (horizontal) is 360° (maximum).
- Range is ~200 m.
- Resolution is ~0.1° (Schoettle, 2017).

4.2.4 Cameras

Cameras are already commonplace on modern cars. Since 2018, all new vehicles in the United States are required to fit reversing cameras as standard. Any car with a lane departure warning (LDW) system will use a front-facing camera to detect painted markings on the road.

Autonomous vehicles are no different. Almost all development vehicles today feature some sort of visible light camera for detecting road markings—many feature multiple cameras to build a 360° view of the vehicle's environment. Cameras are very good at detecting and recognizing objects, so the image data they produce can be fed to artificial intelligence (AI)-based algorithms for object classification.

Some companies, such as Mobileye, rely on cameras for almost all of their sensing. However, they have drawbacks. For example, visible light cameras have limited capabilities in conditions of low visibility. In addition, using multiple cameras generates a lot of video data to process, which requires substantial computing hardware. Infrared (IR) cameras offer superior performance in darkness and additional sensing capabilities (Dawkins, 2019) (Figure 4.8).

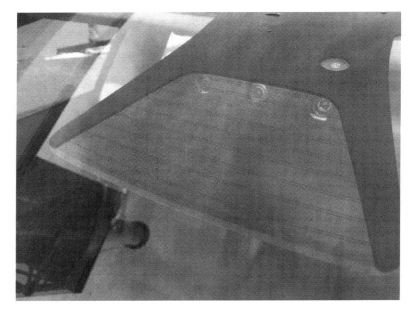

FIGURE 4.8 Cameras are already commonplace on modern cars (Dawkins, 2019).

4.2.4.1 Camera Technology

The camera sensor technology and resolution play a very large role in the camera's capabilities. Cameras, similar to the human eye, are susceptible to adverse weather conditions and variations in lighting. But cameras are the only sensor technology that can capture texture, color, and contrast information, and the high level of detail captured by cameras allows them to be the leading technology for classification. These features, combined with the ever-increasing pixel resolution and the low price point, make camera sensors indispensable for ADAS and autonomous systems.

Some examples of ADAS applications enabled by cameras are:

- ACC: These systems currently detect full-width vehicles like cars and trucks; they need to be able to classify motorcycles and keep a constant distance.
- Automatic high beam control (AHBC): These systems currently do high–low beam switching; they need to evolve to be able to detect oncoming vehicles and contour the ray of light accordingly.
- Traffic sign recognition (TSR): These systems currently recognize speed limits and a limited subset of signs; future systems need to understand supplemental signs and context ("Speed limit in effect 10 am to 8 pm"), detect traffic signals to adapt ACC, stop, slow down, etc.
- Lane keep systems (LKS): These systems currently detect lane markings; future systems need to detect drivable surfaces, adapt to construction signs and multiple lane markings etc.

Inside the cabins of autonomous vehicles, driver monitoring and occupancy tracking for safety are being joined by gesture recognition and touchless controls, including adding context to the gesture recognition based on gaze tracking. In autonomous vehicle (AV) systems, driver monitoring takes on the added use of checking if the driver is prepared to retake control if needed (Ors, 2017).

4.2.4.2 Cameras in Autonomous Vehicle Research Platforms (AVRPs)

Cameras are typically much cheaper than LiDAR and radar units. They can support the other three technologies in terms of object detection, tracking, blind spot detection, etc., through data fusion. With the capability of capturing color and light contrasts, cameras add the ability for an AVRP to perform lane detection, capture text and information on signs, and detect objects. These abilities extend the AVRP's detection and navigation capabilities beyond what LiDAR, radar, and ultrasonic are able to provide.

Common types of cameras used on AVRPs include regular color cameras, monochrome cameras, and stereo cameras. Monochrome cameras produce images in a single hue. This allows them to capture the actual light intensity values for each pixel, rather than using a color filter array (CFA) like a color camera does. In turn, this provides a better gradient, which can allow monochrome images to be better at tasks such as edge detection. Stereo cameras emulate human vision by using multiple cameras. This gives the platform the ability to create 3D images that can be processed to determine depth.

Some important camera specifications for consideration when selecting camera hardware include resolution, f-stop number, image sensor, frames per second (FPS), saturation capacity, and dynamic range.

- Resolution is the number of pixels an image is composed of. Better resolution is great, but it affects processing time.
- F-stop is the ratio of the lens focal length to the aperture diameter, expressed as *f/N*. A higher f-stop will give a greater field depth.
- Image sensors can be either charge-coupled devices (CCDs) or complementary metal-oxide semiconductors (CMOSs). CCD has a high image quality with low noise, while CMOS has a relatively lower image quality and lower tolerance to noise (Mehta, Patel, & Mehta, 2015).
- FPS is the number of consecutive images that can be taken in 1 s.

- Saturation capacity or well depth is the number of electrons, after being converted from intensity, which can be stored in a pixel before it becomes saturated and begins to affect surrounding pixels.
- Dynamic range is the range of luminosity of an image.

Since cameras rely on ambient light, the quality of images can be greatly affected by the time of day and weather. High dynamic range (HDR) imaging can help with image quality caused by under- or over-lit conditions, glare, etc. HDR works by taking multiple photos with different light exposures and fusing the images together. This broadens the range of luminosity of a usable picture. This greater range of luminosity allows better images, which, in turn, can improve perception of the environment and strengthen fusion of LiDAR, radar, and camera data.

Typical camera-mounting locations include front- and rear-facing cameras located behind the front windshield and rear window, respectively, to protect them from the elements. Forward-facing cameras are essential for lane detection. Mounting cameras facing outward from the sides of the AVRP assists with object detection and tracking. Integrating cameras into the side mirrors can assist with blind spot detection. Figure 4.9 shows some typical camera locations and the capabilities they bring to an AVRP (Wicks, Steve, & Asbeck, 2017).

4.2.4.3 Camera System Characteristics

- Color vision possible (important for sign and traffic signal recognition).
- Stereo vision when using a stereo, 3D, or time-of-flight (TOF) camera system.
- Fixed aim and FOV but able to employ multiple cameras as needed.
- FOV (horizontal): ~45° to ~90°.
- Range: No specific distance limit (mainly limited by an object's contrast, projected size on the camera sensor, and camera focal length), but realistic operating ranges of ~150 m for monocular systems and ~100 m (or less) for stereo systems are reasonable approximations.
- Resolution: Large differences across different camera types and applications (Schoettle, 2017).

4.2.5 Ultrasonic Sensors

Ultrasonic sensors have been commonplace in cars since the 1990s; they are used as parking sensors and are very inexpensive. Their range is limited to just a few meters in most applications, but they are ideal for providing additional sensing capabilities to support low-speed use cases (Dawkins, 2019).

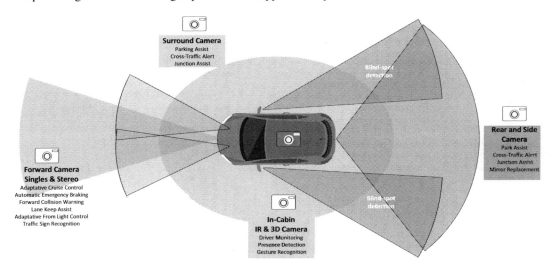

FIGURE 4.9 Common camera locations and their uses (Ors, 2017).

FIGURE 4.10 Ultrasonic sensors (Dawkins, 2019).

Figure 4.10 shows a type of short-range sensor using high frequency sound to measure distance to an object. These sensors are very low cost, often bumper mounted at the rear or (less often) at the front and side of vehicles at low levels. Main applications include low-speed maneuvering, parking, and proximity alerts; typical range is below 10 m (Dawkins, 2019).

4.2.5.1 Ultrasonic Sensors in Autonomous Vehicle Research Platforms (AVRPs)

To improve an AVRP's ability to detect objects in the immediate area, ultrasonic sensors may be required. They work by emitting high-frequency acoustic waves and waiting for a return of the waves. The time it takes for the acoustic waves to return determines the distance of the object. Sensors can detect obstacles anywhere from a few centimeters away to 15 m. However, in autonomous vehicles they are typically used for short-range object detection, parking assist, and blind spot detection.

Fusing ultrasonic data with LiDAR and radar data can greatly reduce the undetected areas surrounding the vehicle.

Ultrasonic sensors can be inhibited by dirt and particle buildup on the sensor face, causing false positives. Another source of error is the air through which the acoustic waves travel. Properties of the air, such as temperature and humidity, can affect the speed of sound, which, in turn, affects the accuracy of the sensor. Some sensors are equipped with a built-in temperature compensation to mitigate a temperature-induced source of error.

Ultrasonic sensors are typically integrated into the front and rear bumpers of the vehicle, as shown in Figure 4.11. The viewing angle of ultrasonic sensors is relatively limited, so they work best with straight-on object detection. Since they have a limited viewing angle, placing many of them along the length of a bumper or vehicle side can mitigate false distance readings and missed objects (Wicks, Steve, & Asbeck, 2017).

4.2.5.2 Characteristics of Ultrasound

Ultrasound is an acoustic wave with a very high frequency, beyond human hearing. Since the audible frequency range is said to be between 20 Hz and 20 kHz, ultrasound generally means acoustic waves above 20 kHz. Because of their echolocation (biological ultrasonic radar), bats can hear sounds up to 200 kHz, way beyond the capabilities of the human ear.

Ultrasound has several characteristics which make it useful for electronics applications. First, it is inaudible to humans and, therefore, undetectable by the user. Second, ultrasound waves can be produced

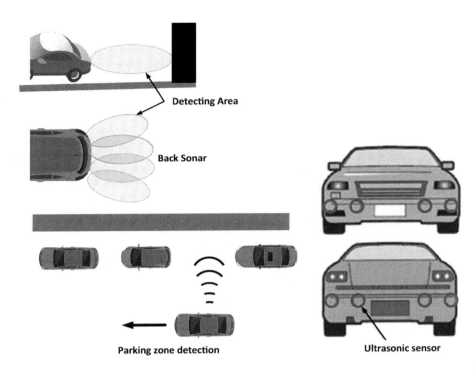

FIGURE 4.11 Ultrasonic mounting locations (Hikita, 2010).

with high directivity. Third, they are a compressional vibration of matter (usually air). Fourth, they have a lower propagation speed than light or radio waves.

The fact that ultrasound is inaudible to human ears is an important factor in ultrasound applications. For example, a car parking sensor system generates sound pressure of more than 100 dB to ensure clear reception. This is the equivalent of the audible sound pressure experienced when standing close to a jet engine.

Ultrasound's high frequency (short wavelength) enables narrow directivity, similar to its radio wave equivalent, microwaves. This characteristic is used in kidney stone treatments, where ultrasound emitted from outside the body is focused on the stone to break it down. Since the energy level is low, it does not harm the body.

Because ultrasound is a vibration of matter, it can also be used to examine the characteristics of that matter. Ultrasonic diagnosis uses this feature to detect and visualize the variance in reflectance and transmittance corresponding to the water content and density of the matter in the medium, for example, an organ in your body.

Ultrasound travels in the air at around 340 m/s, like other sounds. The time it takes for an ultrasound wave to travel 10 cm is approximately 3 ms, as opposed to 3.3 ns for light and radio waves. This allows measurement using low-speed signal processing (Hikita, 2010).

4.2.5.3 Ultrasonic Sensors for Parking

Parking sensors use a type of sonar. The term sonar is an acronym for sound navigation and radar; it is used to calculate the distance and/or direction of an object from the time it takes a sound wave to travel to the target and back. An ultrasonic sensor is a speaker or microphone that emits or receives ultrasound. There is also a type that can handle both emission and reception. Vehicle parking sensors are equipped with this type of sensor.

Ultrasound sensors were initially used in vehicles to detect obstacles when parking (see Figure 4.12) but are now evolving into automatic parking systems (Hikita, 2010).

Sensors for Autonomous Vehicles 127

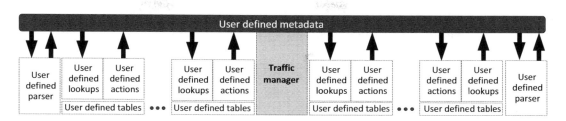

FIGURE 4.12 Examples of parking sensor systems in vehicles (Hikita, 2010).

This system controls steering, acceleration, and braking automatically, based on the parking zone and location information gained from the ultrasonic sensor, to achieve parallel parking and garage parking. For rear sonar, two to four ultrasonic sensors are mounted on the rear bumper to detect an obstacle up to 2–2.5 m away. The distance is communicated to the driver in real time using varying buzzer sounds. Even a wire fence can be detected if it is close enough. The main characteristics of ultrasonic sensors for rear sonar are directivity, ringing time, sensitivity, and sound pressure.

Directivity of an ultrasonic sensor corresponds to the size and shape of the vibrating surface (that is emitting the ultrasound) and the frequency at which it vibrates. Figure 4.13 indicates the directivity of a disc-type transducer (Hikita, 2010).

Narrower directivity can be achieved at higher frequency when the size remains the same or at larger size when the frequency remains the same. Differences in detection distance with varying frequency at the same size are indicated in Figure 4.14 (Hikita, 2010).

By using higher frequency and selecting an appropriate amplifier (gain), we can increase the influence of ground objects, such as wheel stoppers. While narrower vertical directivity improves sensor usability, wider horizontal directivity can provide wider coverage with fewer sensors. "Asymmetric" ultrasonic sensors (see Figure 4.15) are used in such situations (Hikita, 2010).

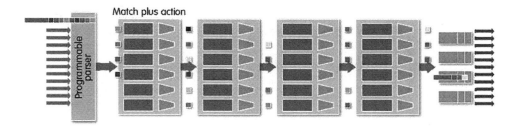

FIGURE 4.13 Emitting surface diameter and directivity (calculated value) for disc-type transducer (Hikita, 2010).

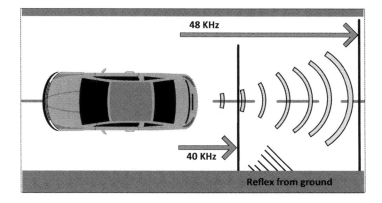

FIGURE 4.14 Frequency (directivity) and influence of the ground (Hikita, 2010).

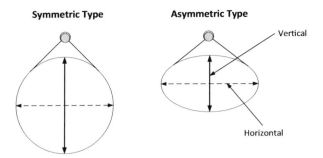

FIGURE 4.15 Directivity types (Hikita, 2010).

Since rear sonar ultrasonic sensors both send and receive ultrasound, unless the emitting sound wave dissipates without ringing quickly, it cannot start receiving its reflection. The shorter the ringing time, the closer the detection range can be.

The casing for Murata's ultrasound sensor, which also acts as an ultrasound emitter, is precision-made to resonate with the driving frequency. This is the reason for its high sensitivity and sound pressure. Figure 4.16 shows ultrasonic sensor MA40MF14-5B, most frequently used for vehicle sonar, and its directivity (Hikita, 2010).

The system is characterized by asymmetric directivity (110°×50°), short ringing time, high sound pressure, sensitivity, and reliability. There are 40, 48, 58, and 68 kHz models (Hikita, 2010).

4.2.5.4 Ultrasonic Sensing Fundamentals

As the name implies, an ultrasonic sensing system operates by transmitting short bursts of sound waves and measuring the time taken for the sound to travel to a target object, be reflected, and return to the receiver (Figure 4.17). The distance to the object is a function of the travel time and the speed of sound in air, approximately 346 m/s (Pickering, 2017).

Objects closer to the transmitter produce a stronger echo than those more distant. To avoid false positives, the system ignores all inputs below the noise value (i.e., the response when no object is present) plus a margin. This threshold determines the maximum range.

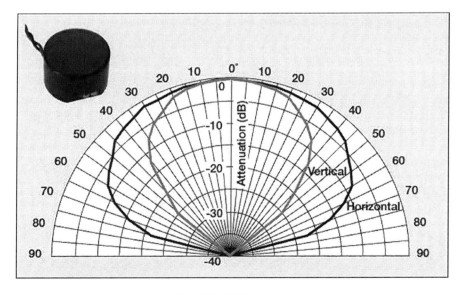

FIGURE 4.16 MA40MF14-5B and its directivity (Hikita, 2010).

Sensors for Autonomous Vehicles 129

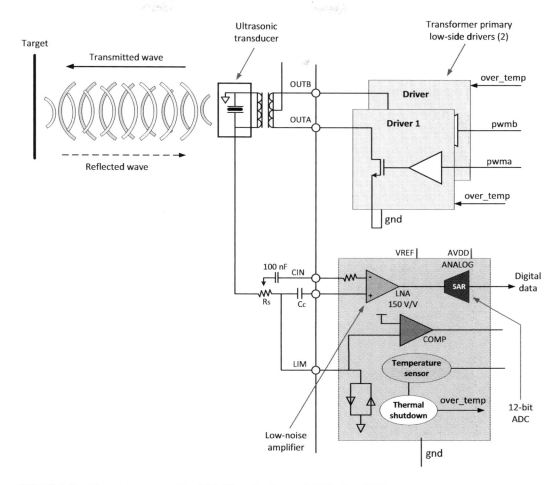

FIGURE 4.17 Ultrasonic system with a PGA450 analog front end (Pickering, 2017).

Key ultrasonic sensor specifications include frequency, sensitivity, and directivity. The system also includes a tunable transformer used to excite the transducer. The transformer is center-tapped to double the voltage. A tuning capacitor typically matches the resonant frequency between the transducer and transformer.

Air temperature, humidity, and wind all affect the speed of sound in air. If multiple sensors are used, they must be spaced so that the sensor signals don't interfere with each other (Pickering, 2017).

4.2.5.5 Automotive Applications of Ultrasonic Sensing

Ultrasonic sensing is usually used for short-distance applications at low speeds, such as park assist, self-parking, and blind spot detection. However, convenience features like kick-to-open lift gates are starting to appear. Also known as a smart trunk opener, this feature lets the vehicle owner open the trunk hands-free by making a kicking motion with his or her foot under the rear bumper.

For maximum coverage, an automotive ultrasonic system typically uses multiple sensors located in the wing mirror and front and rear bumpers. Their responses are repeatable and linear, which translates well into visual representations of target distance. Furthermore, the response does not depend on surface color.

Ultrasonic sensing is a more cost-effective choice than cameras, as these have poor close-distance detection. Although IR sensing is cheaper than ultrasonic sensing, it's less accurate and cannot function properly in direct sunlight (Pickering, 2017).

4.3 Sensors for Military Missions

There are two main categories of payloads onboard an unmanned aerial vehicle (UAV): one allows the vehicle to navigate and the other allows it to perform its main task. The distinction between them is sometimes very subtle. In some cases, a payload aimed to perform a task can be used for navigation purposes (e.g., cameras), and navigation payloads may, in some instances be integrated with other sensors to perform a specific task (e.g., GPS/inertial navigation system (INS) with LiDAR). This section describes common sensors for missions that are used in many typical payloads onboard UAVs (Valavanis & Vachtsevanos, 2015).

4.3.1 Navigation Sensors

The GNSS (including the GPS) and INS are at the core of most UAV guidance and navigation systems. Their complementary nature has been recognized, and as a result, GPS and INS sensors are the preferred sensor couple for the majority of autopilot systems (Valavanis & Vachtsevanos, 2015). Researchers have spent considerable effort proposing approaches to the integration of GPS and INS, such as uncoupled integration, loosely coupled integration, tightly coupled integration, and deeply coupled integration (Grewal, Weill, & Andrews, 2007).

GPS and INS are not the only two sensors used for navigation. They can be complemented with altimeters (laser-based, barometric, etc.) to enhance the estimation of the vehicle state. IR attitude sensors are also typically found in micro UAVs. Cao and Chao (2010) recently surveyed UAV autopilot alternatives and typical sensor combinations (Valavanis & Vachtsevanos, 2015).

An estimator is at the heart of integration (usually a form of a Kalman filter). It estimates position, velocity, attitude, GPS errors, and inertial sensor errors. Because of their complementary nature, GPS and INS are often the preferred core sensor suite. However, researchers have also investigated the integration of other sensor combinations, such as GPS with computer vision (Dusha & Mejias, 2011, 2012) and INS with computer vision (Merz & Duranti, 2006). Factors such as the trade-off between cost and accuracy of INS sensors and the susceptibility of GPS to spoofing and jamming have contributed to increased interest in alternative sensor combinations (Valavanis & Vachtsevanos, 2015).

While alternative integration schemes using other sensors are an attractive option, the low cost and future navigation availability and integrity that space-based augmentation systems (SBAS) such as Wide Area Augmentation System (WAAS), European Geostationary Navigation Overlay Service (EGNOS), Geo Augmented Navigation system (GAGAN), and Multi-functional Satellite Augmentation System (MSAS) will provide cannot be ignored. Submeter accuracy for civilian users will also be possible with the commissioning of Galileo and Compass, and the modernization of GLONASS. This, in turn, will encourage interoperability and the possibility of a triple-frequency civilian-band GPS over the next decades (Valavanis & Vachtsevanos, 2015).

4.3.1.1 Electro-Optical (EO) Sensors

It is difficult today to find a UAV without an EO sensor. They are standard onboard systems on aerial vehicles. However, a remaining challenge is the processing and interpretation of the information acquired by EO sensors. Perception through EO sensors can be seen as one of the most important tasks of a UAV, whether the EOs are used for navigation or surveillance (as an end application). This section introduces some of the most common EO sensors in UAVs (Valavanis & Vachtsevanos, 2015).

4.3.1.1.1 Visible Spectrum
A visible spectrum camera operates on about a 390nm (3.9 µm) to 750nm (7.5 µm) wavelength. There are two main categories: digital still cameras and machine vision cameras (including surveillance and webcam).

Digital still cameras offer very high resolution but cannot provide a continuous stream of images. The number of images they can provide usually depends on the amount of internal memory. This type of camera is applied in remote sensing and aerial photography.

Machine vision cameras have relatively lower resolution but can provide a continuous stream of images up to a few hundred FPS. The speed is related to the resolution and output format used (digital or analog). They are suitable for processes or tasks that require very fast perception of the environment. Common output protocols and interfaces for machine vision cameras include IEEE 1394, Camera Link, SD/HD Analog, USB, GigE Vision, and, in coming years, Thunderbolt cameras. They can provide color or gray-level (or both) images. The data representation or color space usually varies from one manufacturer to another. Typical color spaces or models used in most machine vision cameras are RGB, YUV, YPbPr, and YCbCr, etc. (Valavanis & Vachtsevanos, 2015).

Regardless of the end use and camera type, geometric models are necessary. They provide parameters to correct lens distortions, perform the mapping or representation of 3D objects onto the 2D surface called image, and allow manipulation of the data acquired. These parameters are normally estimated through a regular calibration process. For more details on camera models and calibration theory, refer to Forsyth and Ponce (2002) and Szeliski (2011).

4.3.1.1.2 Infrared

An infrared (IR) camera is a device that detects and converts light in the same way as common visible spectrum cameras but is sensitive to light at longer wavelengths. These cameras form an image using IR radiation in the spectrum at wavelengths from 5,000 nm (5 µm) to 14,000 nm (14 µm). IR cameras are used to convey a measure of thermal radiation of bodies. The intensity of each pixel can be converted for use in temperature measurement, with the brightest parts of the image (colored white) representing the warmest temperatures. Intermediate temperatures are shown as reds and yellows, and the coolest parts are shown in blue. IR images are often accompanied by a scale next to a false color image to relate colors to temperatures. The resolution of IR cameras (up to a maximum of 640×480 pixels) is considerably lower than the resolution of optical cameras. Furthermore, the price of IR cameras is considerably higher than their visible spectrum counterparts. IR cameras can be categorized in two main groups (Valavanis & Vachtsevanos, 2015):

- Cooled IR detectors.
- Uncooled IR detectors.

4.3.1.1.3 Hyperspectral Imaging

Sensors in this category acquire image data simultaneously in multiple adjacent spectral bands. This gives a wealth of data, but processing and interpretation of these data require good knowledge of what specific properties are to be measured and how they relate to the actual measurement done by the sensor. For example, a single cell position in an image will have a set of brightness (or intensity) levels for each wavelength (spectral band). Different materials under examination by a hyperspectral sensor will often exhibit different intensity versus wavelength relationships. Hence, with some prior knowledge, hyperspectral image data can be useful for identifying the type or composition of materials.

Cost and complexity are the two major drawbacks of this technology. Storage is another limiting factor given the multidimensional nature of hyperspectral datasets. The availability of graphics processing units (GPU) as powerful parallel processing hardware could bring this technology a step closer to widespread adoption. However, further research efforts are needed to create analytical techniques and algorithms to unleash the full potential of hyperspectral imaging (Valavanis & Vachtsevanos, 2015).

4.3.1.2 Radio Wave Sensors

4.3.1.2.1 Airborne Radio Detection and Ranging (Radar)

Radar is a radio system used to determine range, altitude, direction, or speed of objects. The system transmits controlled radio pulses which are reflected back by objects. The distance to objects is estimated by measuring the signal return time. The received power declines as the fourth power of the range, assuming transmitting and receiving antennas are in the same location, hence the need for high transmitting power in most cases. Speed can be estimated by tracking the change in distance over time or by exploiting the Doppler effect (Stimson, 1998).

Airborne radar has been in operation since WWII and can be considered an integral part of systems such as ground proximity warning systems (GPWSs) or traffic collision avoidance systems (TCASs). In the automotive industry, radar is starting to appear in the form of collision warning systems (Valavanis & Vachtsevanos, 2015).

In a UAV context, the main drawback of radar is the high power consumption. The size, weight, and power (SWaP) challenges that are faced by UAVs are well known. However, new systems such as synthetic aperture radar (SAR) (Soumekh, 1999) are beginning to make radar technology a feasible option onboard UAVs (Hanlon, 2008).

4.3.1.2.2 Light Detection and Ranging (LiDAR)

LiDAR operates in a manner similar to radar, in that it can estimate distance by measuring the time of return of a signal reflected from an object. LiDAR systems have been widely used in atmospheric and meteorology research (Couch & Rowland, 1991; Kiemle & Ehret, 1997) and remote sensing (Dubayah & Drake, 2000; Lefsky & Cohen, 2002).

Given its ability to provide very high definition (under 2.5 cm), LiDAR is a common sensor for mapping and infrastructure inspection (Yuee & Zhengrong, 2009).

However, LiDAR faces many of the same SWaP obstacles as radar technology. Hence, LiDARs are not often found in micro/small size UAVs but have been flown in light general aviation aircraft (ARA, 2011). Recent advances in technology include the Riegl LMS-Q160 LiDAR in <15 kg and 60 W range (Riegl, 2011), which is paving the path for the use of LiDARs in smaller UAVs in the future (Valavanis & Vachtsevanos, 2015).

4.3.2 Applications of Sensors in UAV Platforms

4.3.2.1 Collision Avoidance

A sense-and-avoid capability equivalent to the human pilot see-and-avoid is one of the key prerequisites for UAVs to gain routine access to civil airspace. Sense and avoid begins with the detection of potential conflicting air traffic, and as a result, considerable research has been dedicated to addressing the "sensing" aspect of sense and avoid.

An important design choice in the development of sensing and detection systems is the type of sensor used to collect information about the surrounding environment; it must account for the physical and resource limitations of the UAV platform and has implications for the target detection algorithms and other data processing techniques that are employed.

The cooperative or uncooperative nature of the system will also have an impact on the choice of sensors. Cooperative implies that all vehicles in the air share information through common communication links. Uncooperative denotes that vehicles do not communicate with each other; this implies there is no other way to detect other vehicles than a self-contained passive detection system (Valavanis & Vachtsevanos, 2015).

4.3.2.2 Remote Sensing: Power Line Inspection and Vegetation Management

Power line infrastructure is a large and expensive asset to manage, costing many millions of dollars per year. Inspecting the power lines for failures (or worn, broken parts) and vegetation encroachment is a significant task, particularly for large electricity companies (Li & Bruggemann, 2012).

Traditionally, power line inspection is carried out by manned crews visiting the power lines by ground vehicle. However, this is time consuming, and due to terrain or proximity limitations, it can be difficult. Unpiloted aircraft is an ideal technology for streamlining the inspection task carried out by ground crews. UAVs can achieve similar results to ground inspection but with greater efficiency. In many cases,

UAVs can gain visibility of power lines in difficult locations inaccessible by ground vehicle (Valavanis & Vachtsevanos, 2015).

Alternatively, inspecting power lines with a manned aerial inspection aircraft involves precise flying at low altitude above the infrastructure to be inspected. This is a potentially hazardous task for a human pilot due to fatigue from sitting in aircraft for long periods of time and the intense concentration and effort required to ensure that the infrastructure to inspect is within the sensor swath or FOV (usually LiDAR and high-resolution camera) (Li & Bruggemann, 2012). A LiDAR and camera combination provides excellent sources of data with complementary characteristics. However, LiDAR data files are large because of the amount of data LiDAR can capture, often in the order of hundreds of gigabytes or more.

The processing of LiDAR data captured by a UAV is an important task which currently must be done after flight missions (Valavanis & Vachtsevanos, 2015). A future avenue is to perform LiDAR processing onboard the aerial vehicle. One of the processing tasks is to distinguish the wanted features from unwanted features in the data. For example, trees, buildings, and other obstacles which are not near the power line infrastructure are unwanted. Data such as the precise height, location of power lines, and their displacement from potentially hazardous objects such as vegetation (which may start bushfires) are required (Valavanis & Vachtsevanos, 2015).

4.4 Sensor-Based Localization and Mapping

Vehicle self-localization is an important but challenging issue in current driving assistance and autonomous driving research activities. This section investigates two kinds of methods for vehicle self-localization: active sensor based and passive sensor based. Active sensor-based localization was initially proposed for robot localization and recently introduced into autonomous driving. Simultaneous localization and mapping (SLAM) techniques are examples of active sensor-based localization. Passive sensor-based localization technologies are categorized based on the type of sensors, the GNSS, inertial sensors, and cameras (Kamijo & Yanlei, 2015).

4.4.1 Introduction

Autonomous driving technologies are expected to significantly improve driving safety and convenience by alleviating the burden of a driver. They are presently implemented as a form of an ADAS to partially aid drivers. It is also anticipated that fully autonomous cars will soon emerge as the key component of intelligent transportation systems, replacing human drivers. The announcement of Google's self-driving cars in May 2014 was an encouraging step towards commercializing autonomous vehicles in the near future (Gibbs, 2014). Major automotive manufacturers have announced plans to market autonomous vehicles in the next decade (Zolfagharifard, 2013; Bora, 2015).

Autonomous driving technology made obvious progress when the DARPA Grand Challenge, held in 2007 (see Chapter 2), evaluated autonomous navigation technologies for urban environments. Most of the successful competitors concentrated on environment perception, precision localization, and navigation to execute various urban driving skills, including lane changes, U-turns, parking, and merging into moving traffic (Montemerlo et al., 2008; Leonard et al., 2008; Patz, Papelis, Pillat, Stein, & Harper, 2008; Urmson et al., 2008). Figure 4.18 shows Stanford University's autonomous research vehicle used in DARPA 2007; it was equipped with multiple active sensors to conduct sensing and localization (Kamijo & Yanlei, 2015).

Obviously, the perception of the environment is an essential function in autonomous driving to prevent collisions. The precision of localization is also significant, especially in an urban environment. Because it will be operated on a constructed urban road, the autonomous vehicle has to obey the traffic rules. For example, when turning, as demonstrated in Figure 4.19, the vehicle has to change to the right lane to achieve the right turning action. This decision is made based on the knowledge about the positioning, which needs to be lane level. The lane changing also needs information from the digital map, including the semantic description of traffic rules (Kamijo & Yanlei, 2015).

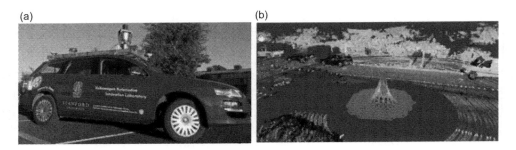

FIGURE 4.18 (a) Stanford University's autonomous research vehicle. (b) A 2-dimensional color histogram showing the vehicle's location (Levinson et al., 2011).

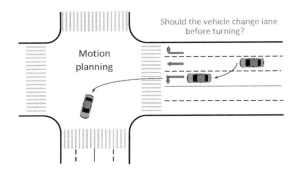

FIGURE 4.19 Motion planning for the right-turning case (Kamijo & Yanlei, 2015).

Based on the type of sensors used, the localization methods can be categorized into active sensor based and passive sensor based. Passive sensors, such as cameras and GNSS receivers, collect data, including light, radiation, heat, or signals in the surrounding environment. Active sensors include transmitters that send out a signal, a light wavelength, or electrons to be bounced off the target, with data gathered by the sensor upon their reflection. The most popular active sensor for localization is the LiDAR sensor.

Many approaches make use of active sensors, such as 2D LiDAR and Velodyne. Good examples of these are in Montemerlo et al. (2008), Leonard et al. (2008), Patz, Papelis, Pillat, Stein, and Harper (2008), and Urmson et al. (2008). These authors describe practical approaches used by DARPA in tests emulating real conditions. Active sensors have also been discussed at the IEEE International Conference on Intelligent Transportation Systems (ITSC) and IEEE Intelligent Vehicles Symposium (IV). The active sensor is popular because it can simplifies the underlying distance estimation while producing remarkably good results. Such simplification is achieved by acquiring dense clouds of 3D points with a laser (Kamijo & Yanlei, 2015).

Active sensors are very expensive, sometimes costing more than the vehicle. Even assuming a drastic decrease in the cost of these sensors, they still present a critical problem because of their excessively high energy consumption. It is necessary to consider lower-cost alternatives, such as passive sensors, for example, cameras (Ros, Sappa, Ponsa, & Lopez, 2012).

However, the collision avoidance function definitely needs an active sensor. Moreover, the more accurate the drivability map is, the less error is introduced into localization and motion planning. Active sensors are also preferred for map generation (Kamijo & Yanlei, 2015).

4.4.2 Localization and Mapping with Active Sensors

SLAM techniques are able to construct or update a map of an unknown environment while simultaneously keeping track of an agent's location within it. SLAM has become a key component in robotic navigation, which has seen significant progress in the last two decades. SLAM is used in autonomous driving for vehicle self-localization and mapping. Most applications of active sensors use SLAM techniques. This section focuses on active sensor-based SLAM and explains the application of active sensors to localization and mapping (Kamijo & Yanlei, 2015).

4.4.2.1 Localization in SLAM

4.4.2.1.1 SLAM Algorithm

The SLAM algorithm is found in Durrant-Whyte and Bailey (2006), Bailey and Durrant-Whyte (2006), and Paull, Saeedi, Seto, and Li (2014).

A large variety of different estimation techniques have been proposed to address the SLAM problem, i.e., to compute an estimate of the agent's location and a map of the environment. For example, Smith and colleagues (Smith & Cheeseman, 1987; Smith, Self, & Cheeseman, 1990) proposed using the extended Kalman filter (EKF). As shown in Figure 4.20, a mobile robot moves through an unknown environment and makes relative observations of landmarks. The estimates of these landmarks, m_j, are all necessarily correlated because of the common error in estimated vehicle location, x_k. A consistent full solution to the combined localization and mapping problem requires a joint state of EKF. The joint state is composed of the vehicle pose and every landmark position $\{x_k \cdot m_j\}$, and these must be updated following each landmark observation, $z_{k,j}$ (Kamijo & Yanlei, 2015).

Unfortunately, EKF covariance matrices lead to a quadratic number of landmarks, and updating them requires quadratic time.

FastSLAM, introduced by Montemerlo, Thrun, Koller, and Wegbreit (2002, 2003), has its basis in recursive Monte Carlo sampling or particle filtering. Here, the probability distribution is on the trajectory, $x_{0:k}$, rather than the single pose, x_k, because, when conditioned on the trajectory, the map landmarks become independent (Durrant-Whyte & Bailey, 2006). This is a key property of FastSLAM and the reason for its speed. Thus, the map is represented as a set of independent landmarks, with linear complexity, rather than a joint map covariance with quadratic complexity (Kamijo & Yanlei, 2015).

4.4.2.1.2 Curb-Based Localization

Curb detection is an important capability for autonomous ground vehicles in urban environments. It is particularly useful for path planning and safe navigation. Another important task that can benefit from curb detection is localization (Kamijo & Yanlei, 2015).

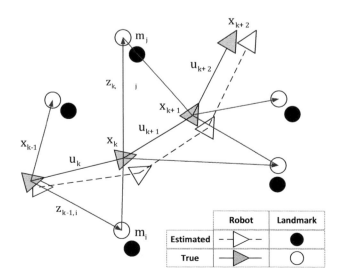

FIGURE 4.20 Demonstration of SLAM problem (Durrant-Whyte & Bailey, 2006).

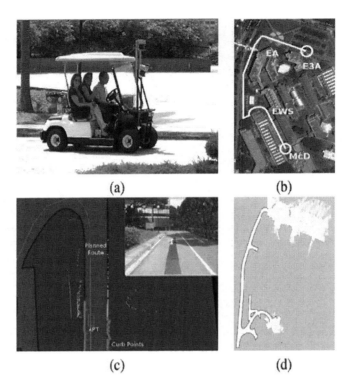

FIGURE 4.21 (a) Autonomous vehicle in operation. (b) Pickup–drop-off points. (c) Snapshot of curb localization estimate. (d) Curb map augmented by planar patches (Qin et al., 2012).

There are several approaches to identifying curbs using 2D LiDAR (see Zhang, 2010; Qin et al., 2012). Qin et al. (2012) mounted a single tilted 2D LiDAR in the front of a vehicle to detect the curb. Because of the height difference between the road shoulder and the road surface, the jump can be observed in the ranging data of LiDAR. Based on this idea, the direction and distance of the curb relative to the vehicle can be decided; the curb is used as a static reference for vehicle localization. This research group conducted an autonomous system demonstration on July 2011 using curb-only road feature-based localization; this is shown in Figure 4.21. The results showed that the lateral position of vehicle was accurate. However, the longitudinal variance changed remarkably along the drive. It is very interesting to note that when the vehicle was approaching the intersections and turns, the longitudinal positioning error decreased significantly based on the longitudinal information of the curb features at the intersection (Kamijo & Yanlei, 2015).

As 2D LiDAR does not obtain dense point clouds at once, a limited number of points can be detected as curbs in each frame. Three-dimensional LiDAR sensors (e.g., multilayer LiDAR) can overcome the lack of data. In Montemerlo, Becker, Bhat, Dahlkamp, Dolgov, and Ettinger (2008) and in Hata, Osorio, and Wolf (2014a), curb-like obstacles were detected by analyzing the ring compression of a multilayer LiDAR, as shown in Figure 4.22. However, when obstacles such as pedestrians and cars were present in the street, the object could be detected as a curb. The regression filter was introduced to estimate the curb shape and to remove points that did not follow the road model. The experiment showed that the longitudinal error was responsible for the localization error (mean 1.49 m), despite a relatively low lateral error of approximately 0.52 m (Hata, Osorio, & Wolf, 2014a).

4.4.2.1.3 Road Mark-Based Localization
Some LiDAR sensors return IR reflective intensity information. Hata and Wolf (2014b) extended the curb detection-based localization and proposed to extract all road marks on the road surface based on the LiDAR intensity for localization. The road mark detector was developed based on the intensity

Sensors for Autonomous Vehicles

FIGURE 4.22 Laser ring compression level. Obstacles can be roughly distinguished as walls, grass, road, and curbs using ring compression data (Hata, Osorio, & Wolf, 2014a).

FIGURE 4.23 Road marking detection through histogram thresholding. The lines parallel to the vehicle are the road bounds estimated by curb detector; the white points correspond to the detected road marking (top: crosswalk; bottom: double line) (Hata & Wolf, 2014b).

histogram thresholding. The thresholding result is shown in Figure 4.23. The localization results showed that both lateral and longitudinal errors are reduced by the integration of road marks and curb compared to the curb only. But the mean of longitudinal error is still larger than 1 m (Hata & Wolf, 2014b).

4.4.2.1.4 Landmark and Building-Based Localization

Besides the designed traffic features on the road area, the objects along the roadside can be localized. Choi proposed a hybrid map-based SLAM (Choi, 2014). Choi described the environment by using a grid map and a feature map together. The feature model selected thin and tall objects like street lamps or trees as landmarks. The grid map included the geometric information of surrounding buildings, as shown in Figure 4.24.

The feature-based SLAM approach generally had the worst performance. It produced huge errors, especially where no landmark measurement was found. The grid-based SLAM approach showed better results, and the hybrid map-based SLAM achieved the best performance. Error accumulation was observed in the results, however, and this can be considered the inherent weakness of SLAM (Kamijo & Yanlei, 2015).

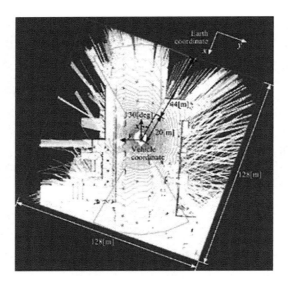

FIGURE 4.24 A local map (square boundary: a grid map area, two circle sector contours: feature map areas) (Choi, 2014).

4.4.2.2 Map Construction Using SLAM

EKF-SLAM and particle filter-based SLAM are more popular for online localization. A graph-based formulation of the SLAM problem was proposed by Lu and Milios (1997). Since then, graph-based SLAM methods have undergone a renaissance and now are state-of-the-art techniques in their speed and accuracy (Grisetti, Kummerle, Stachniss, & Burgard, 2010).

GraphSLAM extracts a set of soft constraints from the data and represents these in a sparse graph. Motion constraints link any two consecutive robot poses, and measurement constraints link poses to landmarks. GraphSLAM obtains the map and the robot path by resolving these constraints into a globally consistent estimate. The solution of GraphSLAM can be considered a least squares problem. In large-scale mapping problems, researchers have found GraphSLAM can handle large number of features and even incorporate GPS information into the mapping process (Thrun & Montemerlo, 2006).

Levinson integrated GPS, IMU, wheel odometry, and LiDAR data acquired by an instrumented vehicle to generate high-resolution 2D road surface maps using GraphSLAM. To generate a pure road surface map, dynamic objects should be excluded, as they cause a hole effect. To overcome this, data should be collected at multiple times. However, if the location of multiple measurements is set by GPS alone, ghosting occurs. After using GraphSLAM, the hole is filled, and the ghost image is removed. Levinson, Montemerlo, and Thrun (2007) conducted an online localization using the map created by this process. In a variety of urban roads, the vehicle was better able to localize in real time than previously created maps, with errors of less than 0.1 m, far exceeding the accuracy possible with GPS alone. Moreover, the proposed map-based localization succeeded in GPS-denied environments, such as in tunnels and under bridges (Levinson, Montemerlo, & Thrun, 2007).

Three-dimensional environmental models have also been proposed; these could improve reliability and accuracy, especially on unusually featureless roads. The Stanford University group built a 3D point cloud map for a real urban environment (Levinson & Thrun, 2010). With this 3D map, the vehicle was able to drive autonomously in several urban environments that were previously too challenging. As one example, the vehicle participated in an autonomous vehicle demonstration in downtown Manhattan in which several blocks of 11th Avenue were closed to regular traffic. The vehicle operated fully autonomously and stayed in the center of its lane, never hitting a curb or other obstacles.

Obviously, localization with an accurate pre-prepared map is preferable, because it can reduce the error accumulation in SLAM. Many groups have already developed their own 3D maps for research into autonomous driving. Figure 4.25 shows several examples: Stanford University, University of Freiburg, and Toyota Technological Institute (Kamijo & Yanlei, 2015).

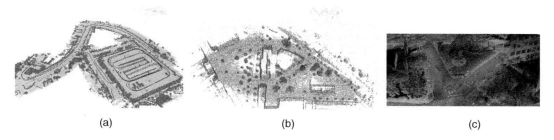

(a) (b) (c)

FIGURE 4.25 (a) Three-dimensional map of Stanford parking garage acquired with an instrumented car. This map was subsequently used to realize autonomous parking behavior (Grisetti, Kummerle, Stachniss, & Burgard, 2010). (b) Point cloud map acquired at the University of Freiburg. (c) Mobile Mapping System (MMS) point cloud map at Toyota Technological Institute (Yoneda, Tehrani, Ogawa, Hukuyama, & Mita, 2014).

4.4.3 Localization with Passive Sensors

4.4.3.1 Global Navigation Satellite System (GNSS)

GNSS is a space-based satellite navigation system that freely provides location and time information to users. GPS, operated by the United States, was the representative positioning system. In the open sky field, the accuracy of GPS positioning is less than 0.1 m (Takasu & Yasuda, 2009.). But the land vehicle navigation system typically has to operate in the areas where the GNSS signal is either blocked or reflected, such as urban canyons. Other satellite navigation systems, such as GLObal NAvigation Satellite System (GLONASS) of Russia, Quasi-Zenith Satellite System (QZSS) of Japan, Galileo of Europe, and BeiDou of China, are now in operation or are about to start operation. The multiple GNSSs reduce the probability of outage and improve the positioning accuracy for the urban environment (Kubo, Hou, & Suzuki, 2014).

The most challenging problem in the urban environment is that GNSS suffers from the non-line of sight (NLOS) and multipath effect. Various technologies have been developed to mitigate these effects. These include antenna-based (Mcgraw, Young, Reichenauer, Stevens, & Ventrone, 2004; Braasch & Spilker, 1996), receiver-based (Van Dierendonck, Fenton, & Ford, 2001; Braasch, 2001), and navigation-processor-based (Kubo, Suzuki, Yasuda, & Shibazaki., 2005; Viandier, Nahimana, Marais, & Duflos, 2008; Hsu, 2013) techniques.

Most multipath mitigation algorithms do not consider the effect of signal reflection as an aid to the position estimation. With the development of ranging technologies, 3D building information became available to estimate the multipath and NLOS effects. Meguro, Murat, Takiguchi, Aman and Hashizume (2009) used an omnidirectional IR camera, installed on the roof of the vehicle, to estimate the area of the sky and identify NLOS signals. As an extended idea of NLOS exclusion, Bauer, Obst, Streiter, and Wanielik (2013) built a shadow map to represent the satellite reception conditions real time. The NLOS measurement can be detected and excluded from the positioning solution. Finally, Obst, Bauer, Reisdorf, and Wanielik (2012) used a dynamic 3D map to exclude the potential multipath signal from the observation set for a vehicle-based loosely coupled GNSS/INS integration system.

4.4.3.2 Inertial Navigation System (INS)

An INS uses multiple onboard sensors, such as speedometers and gyro sensors, to continuously calculate via dead reckoning the position, orientation, and velocity of a moving object. The INS system can accurately provide relative vehicle position in a short time, but its accuracy degrades with time (Glaser, Burkle, & Niewels, 2013).

To overcome the disadvantages associated with the standalone operation of GNSS and INS, the two systems are often integrated so that their drawbacks are minimized. In early studies of this integration, the research focused on evaluating the integration system performance under the open sky (Cao et al., 2002; Hide & Moore, 2004). More recently, the integration system has been studied in different environments. To overcome the multipath interference when the GPS signal is reflected by external agent, Milanes, Naranjo, Gonzalez, Alonso, and Pedro (2008) proposed a dynamic integration system using a decision unit able to choose the correct signal from GPS and INS. Godha and Cannon employed constraints to describe the behavior of a typical land vehicle in the GPS/INS integrated system when a GPS outage occurs in urban areas (Godha & Cannon, 2007). Noureldin, Karamat, Eberts, and El-Shafie (2009) improved the microelectromechanical system (MEMS)-based inertial sensor errors to enhance the positioning accuracy during a GPS outage.

4.4.3.3 Vision-Based Object Detection

Another passive sensor, the camera, is widely used for autonomous driving. Most vision-based technologies aim to detect objects in front of the vehicle. Here we focus on the technologies for static object detection which can potentially be used for localization (Kamijo & Yanlei, 2015).

The most important information on the road surface is lane marking. Vision-based lane detection technology has received considerable attention since the mid-1980s (Dickmanns & Zapp, 1987; Kluge, 1994; Yenikaya, Yenikaya, & Düven, 2013). Techniques vary from monocular (Bertozzi & Broggi, 1996; Aly, 2008; Gopalan, Hong, Shneier, & Chellappa, 2012) to stereo vision (Bertozzi & Broggi, 2012). General lane detection first performs inverse perspective mapping (IPM) and then conducts the Hough transformation or random sample consensus (RANSAC)-based line detection (Aly, 2008). Lane detection has achieved quite high accuracy under good light conditions. But developments have mainly focused on keeping the vehicle on the lane and avoiding collisions (Kamijo & Yanlei, 2015).

The distance from the vehicle to stop lines or crossroads is important for autonomous driving, because the vehicle needs to make a smooth deceleration and stop before the stop line. Seo and Rajkumar proposed detecting the stop line based on the assumption that it is perpendicular to the lane; they tracked the stop line using a Kalman filter to reduce false alarm detection (Seo & Rajkumar, 2014). Marita, Negru, Danescu, and Nedevschi (2011) detected stop lines and crossroads using a stereo camera and used the depth information to do localization at intersections. Arrow mark recognition has also been proposed for localization at intersections (Wu & Ranganathan, 2013).

Curb detection is usually conducted using a stereo camera. Because of the height difference between the road surface and road shoulder, the curb is represented as an edge in the depth image. The most direct method of curb detection is to find the curb line from the depth map (Oniga, Nedevschi, & Meinecke, 2007). A more sophisticated algorithm for curb detection was developed by combining the texture and depth information, as shown in Figure 4.26 (Enzweiler, Greiner, Knoppel, & Franke, 2013). Traffic light detection and traffic sign detection (Gu, Tehrani, Yendo, Fujii, & Tanimoto, 2012; De Charette & Nashashibi, 2009) could be an aid in localization, because these objects are static, and their positions can be added to the map (Kamijo & Yanlei, 2015).

Even though the technologies of vision-based detection are quite mature, some points need to be discussed when we apply those technologies to localization. For example, the lane detection cannot determine the absolute position of the vehicle. In addition, using the multiple lane detection method for

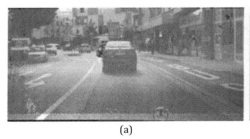
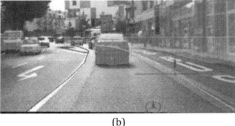

(a)   (b)

FIGURE 4.26 Integrated urban curb recognition system. (a) Input image and corresponding Semi-Global Matching disparity image. (b) Stixel World and freespace, curb classifier search area (short horizontal line), curb classifier region of interest (ROI) with the highest posterior probability, recovered curb location (Enzweiler, Greiner, Knoppel, & Franke, 2013).

Sensors for Autonomous Vehicles 141

absolute positioning is difficult, because of the occlusion of surrounding vehicles in an urban environment. But stop lines, curbs, and traffic signs can provide absolute positioning information for localization. We need distance information from the camera to the object of interest for localization; therefore, the stereo camera is preferable for localization (Kamijo & Yanlei, 2015).

4.5 Sensor Fusion, Sensor Platforms, and Global Positioning System

4.5.1 Sensor Fusion

Sensor fusion is an approach used to combine data delivered from disparate sources such that coherent information is created. The resulting information is more certain than it would be if the sources were used individually. This is especially important when different kinds of information are combined. For example, it is important to have a camera on the autonomous vehicle to replicate human vision, but the information on an obstacle's distance is best gained through sensors like LiDAR or radar. The fusion of camera data with LiDAR or radar data is very important, as they are complementary. Combining information from LiDAR and radar gives more accurate information about the distance of obstacles ahead of the vehicle or the general distance of the objects in the environment (Kocić, Jovičić, & Drndarević, 2018).

When the autonomous vehicle moves in the map, it tries to localize itself by observing landmarks and sensing its own motion. These observations, also called measurements, can be done not only with one sensor but also with more sensors of the same kind or with different types of sensors.

Sensor fusion means combining data from a set of heterogeneous or homogeneous sensors into a coherent and enhanced description of the surrounding environment (Tirindelli, 2016).

4.5.1.1 Types of Sensor Fusion

A. Sensor fusion for 3D object detection

Current trends in autonomous vehicle development include the increased use of LiDAR. Fusion of camera and LiDAR data gives an optimal solution in terms of hardware complexity of the system; only two types of sensors are integrated. Such fusion is also optimal for system coverage, as the camera provides vision data and LiDAR provides obstacle detection data. The image data are fused with 3D point cloud data; as a result, a 3D box and its confidence are predicted. One of the novel solutions for this problem is the PointFusion network (Xu, Jain, & Anguelov, 2018). This method has been applied in 3D object detection (Kocić, Jovičić, & Drndarević, 2018).

Novel approaches in achieving sensor fusion using neural networks tend to treat each signal as a different neural network, then to integrate the resulting representations into a new neural network through high-level fusion, as shown in Figure 4.27. The first benefit of this solution is avoiding lossy input predictions by having low-level processing of each signal individually. The second benefit is a conceptually simple solution, a future trend in neural network development under the motto "small neural nets are beautiful" (Iandola & Keutzer, 2017).

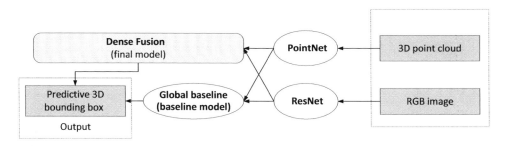

FIGURE 4.27 Block diagram of 3D object detection and 3D point cloud data (Xu, Jain, & Anguelov, 2018).

B. Sensor fusion for occupancy grid mapping

Occupancy grid mapping is used for navigation and localization of autonomous vehicles in dynamic environments (see Figure 4.28). Sensor data from cameras and LiDAR are fused. The camera provides high-level 2D information, such as color, intensity, density, and edge information, and LiDAR provides 3D point cloud data. The usual approach to occupancy grid mapping is to independently filter all grid cells. However, the new trend is to use super pixels for the grid map; the grid cells occupied by an obstacle are included (Oh & Kang, 2016).

C. Sensor fusion for moving object detection and tracking

Moving object detection and tracking is one of the most challenging aspects of the autonomous vehicle domain. Since solving this problem is crucial for autonomous driving, the ratability and performance of the solution are very important. Hence, all existing sensors mounted on the vehicle are generally used. The most common is fusion of camera, radar, and LiDAR sensor data.

Early approaches to moving object detection and tracking focused on fusing sensor data; these included tracking, along with additional information from a SLAM module. Additional fusion has been done on the track level to have an overall perception of the environment (Kocić, Jovičić, & Drndarević, 2018).

A new approach in this field performs detection at the radar and LiDAR levels, sends regions of interest from the LiDAR point clouds into the camera-based classifier, and then fuses all this information together. The information from the fusion module feeds the tracking module for the list of moving objects. The perceived environment is improved by including the object classification from multiple sensors. A block diagram of a multiple sensor perception system is presented in Figure 4.29 (Chavez-Garcia & Aycard, 2016).

When camera, radar, and LiDAR sensor data are fused, it is usual to apply low-level fusion on radar and LiDAR data that have been pre-processed but not otherwise run through any type of model to extract features or object information. This fused information is part of a high-level fusion block that considers camera inputs as well. In this context, low-level fusion solves localization and mapping, while detection and classification are results of high-level fusion. Combining low-level fusion as an input to high-level fusion is a trend in perception in autonomous vehicles (Kocić, Jovičić, & Drndarević, 2018).

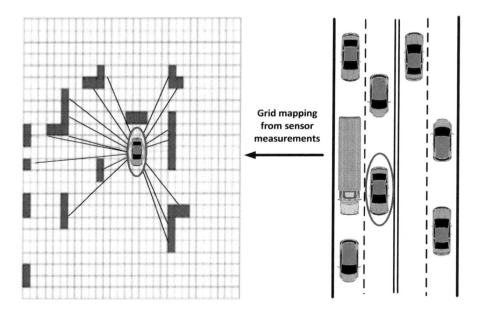

FIGURE 4.28 Occupancy grid mapping (Oh & Kang, 2016).

Sensors for Autonomous Vehicles

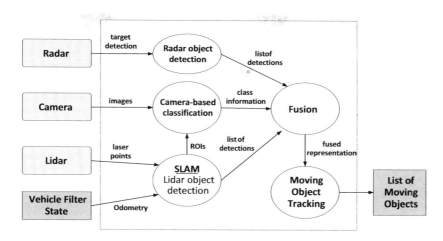

FIGURE 4.29 Multiple sensor perception system (Chavez-Garcia & Aycard, 2016).

The performance of data association and movement classification can be improved using vision object class and shape information to choose the method for object detection. Based on the distance of the object from the vehicle, a tracking system can be switched between a point and a 3D box model (Cho, Seo, Kumar, & Rajkumar, 2014). Information gained from a camera is critical in localization and tracking tasks, and research is exploring contextual information about urban traffic environments to improve the tracking system capability (Kocić, Jovičić, & Drndarević, 2018).

4.5.1.2 Sensor Fusion Advantages

A system using sensor fusion has the following advantages over a single sensor system:

- Improved accuracy: When multiple independent measurements of the same property are fused, the accuracy of the resulting value is better than the one achieved with a single sensor.
- Robustness and reliability: Multiple sensor suites have an inherent redundancy which enables the system to provide information even in case of partial failure.
- Robustness against measurement outliers: By increasing the dimensionality of the measurement space, the system becomes more robust to outliers (noncoherent measurements).
- Reduced ambiguity and uncertainty: Joint information reduces the set of ambiguous interpretations of the measured value.
- Extended spatial and temporal coverage: One sensor can look where others cannot or can perform a measurement while others cannot.
- Increased confidence: A measurement of one sensor is confirmed by measurements of other sensors covering the same domain.

Sensor fusion is not a new concept, but the advances in sensor technology and processing techniques, combined with improved hardware, make real-time fusion of data possible in complex applications such as SLAM (Tirindelli, 2016).

4.5.1.3 Need for Sensor Fusion

For self-driving cars to be accepted by society, they need to demonstrate significantly lower probability of collision than human drivers. A 2016 study by Virginia Tech Transportation Institute (Blanco, 2016) found self-driving cars would be comparable to or a little better than humans for severe crashes but would be significantly better at avoiding low severity level crashes (level 3).

The level 3 crash rate was calculated at 14.4 crashes per million miles driven for humans and 5.6 crashes for self-driving cars.

To keep things in perspective, we could estimate an average person in the United States will drive 900,000 miles in his or her lifetime (12,000 miles/year × 75 years). Also note that the above study used only Google self-driving car data. These cars are known to have a full suite of sensors (multiple LiDAR, radar, camera, ultrasonic, GPS, and other sensors).

Just like the human driver, the car has to integrate the information from multiple sensors and make the best decision possible in the circumstances. But it has to be much better than human drivers to get people to start adopting the technology. As each sensor technology has limitations, fusing multiple inputs reliably is daunting. Incorrect or poor implementation of the sensor fusion could end the possibility of a vehicle's use (Thakur, 2017).

4.5.1.4 Challenges to Sensor Fusion

The objective of sensor fusion is to determine the environment around the vehicle trajectory with enough resolution, confidence, and latency to navigate the vehicle safely.

Figure 4.30 illustrates the challenge of sensor fusion. Row 1 shows the ideal case when two sensors agree on an object, and the object is detected early enough to navigate the car. Row 2 shows a case where each of the sensors classifies the object differently. In this case, the best option may be to agree that it is a big enough object to avoid if possible. Row 3 is a similar situation where a person on a bicycle may be identified as a person or a bicycle. At any rate, it is an unidentified large moving object that needs to be avoided (Thakur, 2017).

The last two rows show smaller objects that pose difficult questions. Is it better to run over a small dog than to risk braking and getting rear-ended? Can the pothole be detected and classified early enough to navigate? Is the pothole or object small enough to run over?

These questions will take a longer time to resolve as they require improvements in sensing technology, computing, public acceptance, and legislation. The 80/20 Pareto principle would imply that the last 20% of the problems for self-driving cars will represent 80% of the time it takes to bring them to mass market (Thakur, 2017).

4.5.2 Sensor Platforms

Sensor platforms are a subset of smart sensors. Like smart sensors, they feature a microcontroller, a wired/wireless interface, and memory. However, sensor platforms are designed for nonspecific applications. They have the functionality to integrate with unspecified external sensors and an interface for programming the microcontroller to perform a specific function. These platforms are very useful for rapid prototyping—sensor hardware and actuators can be physically or wirelessly connected to the platform's sensor interface (digital or analog). Most sensor platforms are accompanied by an integrated

Object_list	RADAR	Camera	LIDAR	Sensor Fusion
Car@150m	car	Don't See it (Noise)	car	car
Not_Classified@100m & low light	car	bicycle	person	Evaluate TTC & brake if unresolved?
@50m	bicycle	cyclist	person	Person on bicycle
Not classified	Don't See it (Noise)	dog	dog	Brake or ignore?
Potholes & stuff	cone	pothole	pothole	What can be safely ignored?

FIGURE 4.30 The challenge of sensor fusion—illustrated (Thakur, 2017).

development environment (IDE) and sample application code that can be combined to program the sensors or actuators. If the prototype is successful, it can form the basis of a specific smart sensor design. If not, the sensors can be quickly reconfigured and retested or they can be discarded and the platform reused for a different application. However, this flexibility has some drawbacks: sensor platforms are physically larger and more expensive than smart sensors, and the user may not require all of the features available on the sensor platform.

Sensor platforms are very popular among hobbyists, designers, researchers, and educators. Many sensor platforms exist, including Parallax Basic Stamp, Netmedia's BX-24, Phidgets, and MIT's Handyboard. The most common sensor platforms in the health, wellness, and environmental domains are Arduino, Shimmer, and Smartphones (McGrath & Ní Scanaill, 2014).

4.5.2.1 Arduino I/O Board

The Arduino was developed in Italy in 2005 to allow nonexperts, including artists, designers, and hobbyists, to rapidly create interactive prototypes. The Arduino I/O board is commonly referred to as "an Arduino." However, the term "Arduino" describes not only the open source hardware sensor platform but also open source software and the active user community.

Traditionally, Arduino I/O boards were developed as a member of the Atmega MegaAVR MCU family, although there is nothing to prevent someone from creating an Arduino-compatible I/O board using a different MCU. For example, the Arduino UNO R3 board (shown in Figure 4.31) is based on an Atmel ATmega328, an 8-bit Atmega core microcontroller. The functionality of Arduino I/O boards can be extended by plugging an Arduino shield into the four expansion connectors found on most official Arduino I/O boards. There are currently hundreds of Arduino shields, providing radio and sensor functionality or prototyping functionality to the Arduino board.

A key feature of Arduino is its open source ethos. Arduino releases all of its hardware CAD design files under an open source license, and these designs can be used or adapted for personal and commercial use, provided Arduino is credited. As a result, several commercial vendors sell official Arduino and Arduino-compatible sensor platforms and shields. The newly introduced Galileo board from Intel is part of the Arduino Certified product line (McGrath & Ní Scanaill, 2014).

The Arduino I/O board MCU is programmed using the open source Arduino programming language (an implementation of the Wiring programming language) and the open source Arduino IDE (based on the Processing IDE). It can also be programmed using Atmel's AVR Studio or other software environments (McGrath & Ní Scanaill, 2014).

FIGURE 4.31 Arduino UNO R3 board (McGrath & Ní Scanaill, 2014).

4.5.2.2 Shimmer

Shimmer is a wireless sensor platform that can capture and transmit sensor data in real time or log the data to a microSD card. Shimmer is a mature sensor platform, targeted at researchers, clinicians, and scientists who want to capture data from sensors without worrying about the underlying sensor electronics, power, or casing. Its small size makes it ideal for body-worn applications, and it has also been used for ambient sensing applications (Burns, 2010).

Shimmer is the name given to the main board, as shown in Figure 4.32; it features an MSP430 MCU, two radios (802.15.4 and Bluetooth), a microSD connector, a 3-axis accelerometer, and a 20-pin Hirose connection. The Shimmer main board is a smart sensor itself, having the capability to perform basic motion sensing, data capture, processing, and wireless transmission of sensed data. The sensing capability of Shimmer can be extended by connecting a Shimmer daughterboard with the required sensing functionality or the Shimmer prototyping board to support the connection of other sensors. Shimmer has daughter boards for complex motion sensing, vital signs and biophysical sensing, and environmental and ambient sensing applications (McGrath & Ní Scanaill, 2014).

Shimmer can be programmed in two ways. First, Shimmer provides a number of precompiled firmware images for its various sensor modules that can be downloaded to Shimmer using the Shimmer Bootstrap Loader and a Universal Serial Bus (USB) docking station. Second, firmware developers can write their own firmware code using open source TinyOS modules, available from the TinyOS repository on code.google.com.

On the application side, Shimmer can interface with a PC or smartphone using a Matlab, Labview, C#, or Android module which are available on the Shimmer website or with the PAMSys physical activity monitoring platform (McGrath & Ní Scanaill, 2014).

4.5.2.3 Smartphones and Tablets

Smartphones and tablets contain many integrated sensors, which the operating system employs to improve the user experience. These may include motion and location sensors (accelerometers, gyroscopes, magnetometers, and pressure sensors), optical sensors (ambient light sensors, proximity sensors, image sensors, and display sensors), silicon microphones, and many environment sensors. The major mobile operating systems (iOS, Android, and Windows 8) provide sensor frameworks that allow application developers to easily access these real-time data streams within their applications. "Run Keeper" is an example of a popular mobile application that uses a smartphone's embedded sensors for an alternative use.

These mobile devices compare very favorably to other sensor platforms in many aspects. The sensors are integrated into an existing, frequently used technology, ensuring that user compliance will be high. The combination of a high-performance microcontroller and substantial memory storage can support

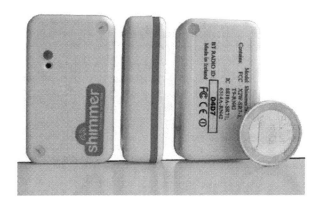

FIGURE 4.32 Shimmer Kinematic Sensors (main board, kinematic daughter board, and battery enclosed in a custom sensor casing) (McGrath & Ní Scanaill, 2014).

complex data analysis over long periods of time. And the integration of a sensor framework into the software development environment greatly simplifies the development process.

However, there are also disadvantages to sensing using mobile devices. The sensor framework abstracts the sensor details and control, making it difficult to achieve the granularity realized with a discrete sensor. Moreover, sensing is a secondary function of the device and can be paused or killed by the operating system when a higher priority task occurs. Finally, Android and Windows 8 run on various hardware configurations; this means the programmer has no control over the sensor specification or how its data are translated by the sensor framework. In fact, the only way to ensure the sensor data are accurate is to use a discrete sensor wirelessly connected to the phone. Mobile devices can also be used as aggregators for internal and external sensors (McGrath & Ní Scanaill, 2014).

4.5.3 Global Positioning System (GPS)

4.5.3.1 Introduction to Navigation

The five basic forms of navigation are as follows:

1. Pilotage, which essentially relies on recognizing landmarks to know where you are.
2. Dead reckoning, which relies on knowing where you started from, plus some form of heading information and some estimate of speed.
3. Celestial navigation, which uses time and the angles between local vertical and known celestial objects (e.g., sun, moon, or stars).
4. Radio navigation, which relies on radio frequency sources with known locations (including GPS satellites).
5. Inertial navigation, which relies on knowing your initial position, velocity, and attitude and thereafter measuring your attitude rates and accelerations. It is the only form of navigation that does not rely on external references.

These forms of navigation can be used in combination as well.

Kalman filtering exploits a powerful synergy between the GPS and an INS. This synergy is possible, in part, because INS and GPS have complementary error characteristics. Short-term position errors from the INS are relatively small, but they degrade without bound over time. GPS position errors are not as small over the short term, but they do not degrade with time. The Kalman filter is able to take advantage of these characteristics to provide a common, integrated navigation implementation with performance superior to that of either subsystem (GPS or INS). By using statistical information about the errors in both systems, it is able to combine a system with tens of meters' position uncertainty (GPS) with another system whose position uncertainty degrades at kilometers per hour (INS) and achieve bounded position uncertainties in the order of centimeters [with differential GPS (DGPS)] to meters.

A key function of the Kalman filter is the statistical combination of GPS and INS information to track drifting parameters of the sensors in the INS. As a result, the INS can provide enhanced inertial navigation accuracy during periods when GPS signals may be lost, and the improved position and velocity estimates from the INS can then be used to make GPS signal reacquisition happen much faster when the GPS signal becomes available again.

This level of integration necessarily penetrates deeply into each of these subsystems, in that it makes use of partial results that are not ordinarily accessible to users. To take full advantage of the offered integration potential, it is necessary to delve into technical details of the designs of both types of systems (Grewal, Weill, & Andrews, 2007).

4.5.3.2 GPS

The GPS is part of a satellite-based navigation system developed by the US Department of Defense under its NAVSTAR satellite program (Grewal, Weill, & Andrews, 2007).

4.5.3.2.1 GPS Orbits

The fully operational GPS includes 24 or more (March 28, 2000) active satellites approximately uniformly dispersed around six circular orbits with four or more satellites each. The orbits are inclined at an angle of 55° relative to the equator and are separated from each other by multiples of 60° right ascension. The orbits are non-geostationary and approximately circular, with radii of 26.560 km and orbital periods of one-half sidereal day (≈11,967 h). Theoretically, three or more GPS satellites will always be visible from most points on the earth's surface, and four or more GPS satellites can be used to determine an observer's position anywhere on the earth's surface 24 h/day (Grewal, Weill, & Andrews, 2007).

4.5.3.2.2 GPS Signals

Each GPS satellite carries a cesium and/or rubidium atomic clock to provide timing information for the signals transmitted by the satellites. Internal clock correction is provided for each satellite clock. Each GPS satellite transmits two spread spectrum, L-band carrier signals—an L_1 signal with carrier frequency $f_1 = 1,575.42$ MHz and an L_2 signal with carrier frequency $f_2 = 1,227.6$ MHz. These two frequencies are integral multiples $f_1 = 1,540 f_0$ and $f_2 = 1,200 f_0$ of a base frequency $f_0 = 1,023$ MHz. The L_1 signal from each satellite uses binary phase-shift keying (BPSK), modulated by two pseudorandom noise (PRN) codes in phase quadrature, designated as the C/A-code and P-code. The L_2 signal from each satellite is BPSK modulated by only the P-code (Grewal, Weill, & Andrews, 2007).

- Compensating for propagation delays

    This is one motivation for the use of two different carrier signals, L_1 and L_2. Because delay varies approximately as the inverse square of signal frequency f (delay α f^{-2}), the measurable differential delay between the two carrier frequencies can be used to compensate for the delay in each carrier (Grewal, Weill, & Andrews, 2007).

- Code division multiplexing

    Knowledge of the PRN codes allows users independent access to multiple GPS satellite signals on the same carrier frequency. The signal transmitted by a particular GPS signal can be selected by generating and matching or correlating the PRN code for that particular satellite. All PRN codes are known and are generated or stored in GPS satellite signal receivers carried by ground observers. A first PRN code for each GPS satellite, sometimes referred to as a precision code or P-code, is a relatively long, fine-grained code with an associated clock or chip rate of $10 f_0 = 1,023$ MHz. A second PRN code for each GPS satellite, sometimes referred to as a clear or coarse acquisition code or C/A-code, is intended to facilitate rapid satellite signal acquisition and hand over to the P-code. It is a relatively short, coarser grained code with an associated clock or chip rate $f_0 = 1,023$ MHz. The C/A-code for any GPS satellite has a length of 1,023 chips or time increments before it repeats. The full P-code has a length of 259 days, during which each satellite transmits a unique portion of the full P-code. The portion of P-code used for a given GPS satellite has a length of precisely one week (7.000 days) before this code portion repeats. Accepted methods for generating the C/A-code and P-code were established by the satellite developer in 1991 (Grewal, Weill, & Andrews, 2007).

- Navigation signal

    The GPS satellite bit stream includes navigational information on the ephemeris of the transmitting GPS satellite and an almanac for all GPS satellites, with parameters providing approximate corrections for ionospheric signal propagation delays suitable for single-frequency receivers and for an offset time between satellite clock time and true GPS time. The navigational information is transmitted at a rate of 50 Bd (Grewal, Weill, & Andrews, 2007).

4.5.3.2.3 Selective Availability

Selective availability (SA) is a combination of methods used by the US Department of Defense to deliberately derate the accuracy of GPS for "nonauthorized" (i.e., non-US military) users. The current satellite configurations use only pseudorandom dithering of the onboard time reference, but the full configuration can include truncation of the transmitted ephemerides. This results in three grades of service provided to GPS users. SA was removed on May 1, 2000 (Grewal, Weill, & Andrews, 2007).

- *Precise positioning service (PPS)* is the full-accuracy, single-receiver GPS positioning service provided to the United States and its allied military organizations and other selected agencies. This service includes access to the unencrypted P-code and the removal of any SA effects.
- *Standard positioning service (SPS)* provides GPS single-receiver (stand-alone) positioning service to any user on a continuous, worldwide basis. SPS is intended to provide access only to the C/A-code and the L_1 carrier. SPS also guarantees user-specified levels of coverage, availability, and reliability.
- *Standard positioning service with SA*: The horizontal-position accuracy is currently advertised as 100 m, the vertical-position accuracy as 156 m, and time accuracy as 334 ns—all at the 95% probability level.

4.5.3.3 Future Satellite Systems

In Europe, activities supported by the European TRIPARTITE Group [European Space Agency (ESA), European Commission (EC), and EUROCONTROL] are underway to specify, install, and operate a future civil GNSS: GNSS-2 or GALILEO.

Based on the expectation that GNSS-2 will be developed through an evolutionary process, as well as long-term augmentations [e.g., GNSS-1 or European GNSS Navigation Overlay Service (EGNOS)], short- to midterm augmentation systems (e.g., differential systems) are being targeted.

The first steps towards GNSS-2 will be made by the TRIPARTITE Group. The augmentations will be designed such that the individual elements will be suitable for inclusion in GNSS-2 at a later date. This design process will provide the user with maximum continuity in the upcoming transitions.

In Japan, the Japanese Commercial Aviation Board (JCAB) is developing the MSAS (Grewal, Weill, & Andrews, 2007).

4.5.3.4 Applications

Both GPS and GLONASS have evolved from dedicated military systems into true dual-use systems. Satellite navigation technology is utilized in numerous civil and military applications, ranging from golf and leisure hiking to spacecraft navigation (Grewal, Weill, & Andrews, 2007).

4.5.3.4.1 Aviation

The aviation community has propelled the use of GNSS and various augmentations (e.g., WAAS, EGNOS, and MSAS). These systems provide guidance *en route* through precision approach phases of flight. Incorporation of a data link with a GNSS receiver enables the transmission of an aircraft's location to other aircraft and/or to air traffic control (ATC). This function is called automatic dependent surveillance (ADS) and is used in the Pacific Ocean Region (POR). Key benefits are ATC monitoring for collision avoidance and optimized routing to reduce travel time and fuel consumption (Grewal, Weill, & Andrews, 2007).

4.5.3.4.2 Spacecraft Guidance

The space shuttle utilizes GPS for guidance in all phases of its operation (e.g., ground launch, orbit and reentry, and landing). NASA's small satellite programs use and plan to use GPS, as does the military on SBIRLEO (space-based infrared low earth orbit) and GBI (ground-based interceptor) kill vehicles (Grewal, Weill, & Andrews, 2007).

4.5.3.4.3 Maritime

GNSS has been used by both commercial and recreational maritime communities. Navigation is enhanced on all bodies of waters, from oceanic travel to river ways, especially in bad weather (Grewal, Weill, & Andrews, 2007).

4.5.3.4.4 Land

The surveying community depends on DGPS to achieve measurement accuracies in the millimeter range. Similar techniques are used in farming, surface mining, and grading for real-time control of vehicles and in the railroad community to obtain train locations on adjacent tracks. GPS is a key component in intelligent transport systems (ITS). In vehicle applications, GNSS is used for route guidance, tracking, and fleet management. Combining a cellular phone or data link function with this system enables vehicle tracing and/or emergency messaging (Grewal, Weill, & Andrews, 2007).

4.5.3.4.5 Geographic Information Systems (GIS), Mapping, and Agriculture

Applications include utility and asset mapping and automated airborne mapping, with remote sensing and photogrammetry. GIS, GPS, and remote sensing have now matured enough to be used in agriculture. GIS companies such as Environmental System Research Institute (Redlands, California) have developed software applications that allow growers to assess field conditions and their relationship to yield. Real-time kinematic and differential GNSS applications for precision farming are being developed, including soil sampling, yield monitoring, and chemical and fertilizer applications. Some GPS analysts are predicting that precision site-specific farming is "the wave of the future" (Grewal, Weill, & Andrews, 2007).

REFERENCES

Aly M., 2008. Real time detection of lane markers in urban streets. In *Proceedings of IEEE Intelligent Vehicles Symposium*, Eindhoven, Netherlands, June 2008, pp. 7–12.

ARA, 2011. Airborne research Australia. Flinders University.

Bailey T., Durrant-Whyte H., 2006. Simultaneous localization and mapping (SLAM): Part II. *IEEE Robotics & Automation Magazine*, 13(3), 108–117.

Bauer S., Obst M., Streiter R., Wanielik G., 2013. Evaluation of shadow maps for non-line-of-sight detection in urban GNSS vehicle localization with VANETs-the GAIN approach. In *Proceedings of IEEE International conference on Vehicular Technology 2013 (VTC Spring)*, Dresden, Germany, June 2013, pp. 1–5.

Bertozzi M., Broggi A., 1996. Real-time lane and obstacle detection on the GOLD system. In *Proceedings of IEEE Intelligent Vehicles Symposium*, Parma, Italy, September 1996, pp. 213–218.

Bertozzi M., Broggi A., 2012. GOLD: A parallel real-time stereo vision system for generic obstacle and lane detection. *IEEE Transactions on Image Processing*, 7(1), pp. 62–81.

Blanco M., 2016. Automated vehicle crash rate comparison using naturalistic data, Jan 2016. Virginia Tech Transportation Institute. Figure 1, p. IV.

Bora K., 2015. Nissan gets into self-driving mode, says its autonomous cars will be ready by 2020. May 18, 2015. www.ibtimes.com/nissan-gets-self-driving-mode-says-its-autonomous-cars-will-be-ready-2020-1926447. Viewed: March 29, 2019.

Braasch M. S., Spilker B. W., 1996. Multipath effects. In B. W. Parkinson & J. J. Spilker (Eds.), *Global Positioning System: Theory and Applications*, Washington, DC: AIAA.

Braasch M. S., 2001. Performance comparison of multipath mitigating receiver architectures. In *Proceedings of IEEE International Conference on Aerospace*, Big Sky, MT, USA; Vol. 3, pp. 1309–1315.

Burns A., 2010. SHIMMER™ – A wireless sensor platform for noninvasive biomedical research. *IEEE Sensors*, 10(9), 1527–1534.

Cao F. X., Yang D. K., Xu A. G., Ma J., Xiao W. D., Law C. L., Chua H. C., 2002. Low cost SINS/GPS integration for land vehicle navigation. In *Proceedings of IEEE 5th International Conference on Intelligent Transportation Systems*, Singapore, Singapore, September 2002, pp. 910–913.

Cao Y., Chao H., 2010. Autopilots for small unmanned aerial vehicles: A survey. *International Journal of Control, Automation, and Systems*, 8(1), 44.

Chavez-Garcia R. O., Aycard O., 2016. Multiple sensor fusion and classification for moving object detection and tracking. *IEEE Transactions on Intelligent Transportation Systems*, 17(2), 525–534.

Cho H., Seo Y., Kumar B. V. K. V., Rajkumar R. R., 2014. A multi-sensor fusion system for moving object detection and tracking in urban driving environments. In *2014 IEEE International Conference on Robotics and Automation (ICRA)*, Hong Kong, China, 2014, pp. 1836–1843.

Choi J., 2014. Hybrid map-based SLAM using a Velodyne laser scanner. In *Proceedings of IEEE 17th International Conference on Intelligent Transportation Systems*, Qingdao, China, October 2014, pp. 3082–3087.

Couch R. H., Rowland C. W., 1991. Lidar in-space technology experiment: NASA's first in-space lidar system for atmospheric research. *Optical Engineering*, 30(1), 88–95.

Dawkins T., 2019. Autonomous cars 101: What sensors are used in autonomous vehicles? January 14, 2019. https://levelfivesupplies.com/sensors-used-in-autonomous-vehicles/#. Viewed: February 26, 2019.

De Charette R., Nashashibi F., 2009. Real time visual traffic lights recognition based on spot light detection and adaptive Traffic lights templates. In *Proceedings of IEEE Intelligent Vehicles Symposium*, Xi'an, China, June 2009, pp. 358–363.

Dickmanns E. D., Zapp A., 1987. A curvature-based scheme for improving road vehicle guidance by computer vision. In *Cambridge Symposium Intelligent Robotics Systems, International Society for Optics and Photonics*, Cambridge, MA, USA, February 1987, pp. 161–168.

Dubayah R. O., Drake J. B., 2000. Lidar remote sensing for forestry. *Journal of Forestry*. 98(6), 44–46.

Durrant-Whyte H., Bailey T., 2006. Simultaneous localization and mapping: Part I. *IEEE Robotics & Automation Magazine*, 13(2), 99–110.

Dusha D., Mejias L., 2011. Fixed-wing attitude estimation using temporal tracking of the horizon and optical flow. *Journal of Field Robotics*, 28(2), 372.

Dusha D., Mejias L., 2012. Error analysis and attitude observability of a monocular GPS/visual odometry integrated navigation filter. *International Journal of Robotics Research*, 31(6), 714–737.

Enzweiler M., Greiner P., Knoppel C., Franke U., 2013. Towards multi-cue urban curb recognition. In *Proceedings of IEEE Intelligent Vehicles Symposium*, Gold Coast, QLD, Australia, June 2013, pp. 902–907.

Farkas D., Young J., Baertlein B., Ozguner U., 2007. Forward looking radar navigation system for 1997 AHS demonstration. In *IEEE Conference on Intelligent Transportation Systems*, Boston, MA, USA, November 2007, pp. 672–676.

Forsyth D. A., Ponce J., 2002. *Computer Vision: A Modern Approach*. Upper Saddle River, NJ: Prentice Hall.

Gibbs S., 2014. Google's self-driving car: How does it work and when can we drive one? May 29, 2014. www.theguardian.com/technology/2014/may/28/google-self-driving-car-how-does-it-work. Viewed: March 29, 2019.

Gu Y., Tehrani M. P., Yendo T., Fujii T., Tanimoto M., 2012. Traffic sign recognition with invariance to lighting in dual-focal active camera system. *IEICE Transactions on Information and Systems*, 95(7), 1775–1790.

Glaser C., Burkle L., Niewels F., 2013. An inertial navigation system for inner-city ADAS. In *Proceedings of IEEE 16th International Conference on Intelligent Transport Systems*, The Hague, Netherlands, October 2013, pp. 1503–1508.

Godha S., Cannon M. E., 2007. GPS/MEMS INS integrated system for navigation in urban areas. *GPS Solutions*, 11(3), 193–203.

Gopalan R., Hong T., Shneier M., Chellappa R., 2012. A learning approach towards detection and tracking of lane markings. *IEEE Transactions on Intelligent Transportation Systems*, 13(3), 1088–1098.

Grewal M. S., Weill L.R., Andrews A. P., 2007. *Global Positioning Systems, Inertial Navigation, and Integration*. Hoboken, NJ: Wiley-Interscience.

Grisetti G., Kummerle R., Stachniss C., Burgard W., 2010. A tutorial on graph-based SLAM. *IEEE Intelligent Transportation Systems Magazine*, 2(4), 31–43.

Hanlon M., 2008. ScanEagle UAV gets Synthetic Aperture Radar (SAR).

Hata A. Y., Osorio F. S., Wolf D. F., 2014a. Robust curb detection and vehicle localization in urban environments. In *Proceedings of IEEE Intelligent Vehicles Symposium*, Dearborn, MI, USA, June 2014, pp. 1257–1262.

Hata A., Wolf D., 2014b. Road marking detection using LIDAR Reflective intensity data and its application to vehicle localization. In *Proceedings of IEEE 17th International Conference on Intelligent Transportation Systems*, Qingdao, China, October 2014, pp. 584–589.

Hide C. D., Moore T., 2004. Multiple model Kalman filtering for GPS and low-cost INS integration. In *Proceedings of NAV/AIS04, Location and Timing Applications*, Long Beach, CA, USA, November 2004.

Hikita M., 2010. An introduction to ultrasonic sensors for vehicle parking. www.newelectronics.co.uk/electronics-technology/an-introduction-to-ultrasonic-sensors-for-vehicle-parking/24966/. Viewed: February 27, 2019.

Hsu L. T., 2013. Integration of vector tracking loop and multipath mitigation technique and its assessment. In *Proceedings of the 26th International Technical Meeting of the Satellite Division of the Institute of Navigation (ION GNSS 2013)*, Nashville, Tennessee, USA, September 2013, pp. 3263–3278.

Iandola F., Keutzer K., 2017. Keynote: Small neural nets are beautiful: Enabling embedded systems with small deep-neural-network architectures. In *2017 International Conference on Hardware/Software Code sign and System Synthesis (CODES+ISSS)*, Seoul, 2017, pp. 1–10.

Ibeo Feature Fusion. www.ibeo-as.com/had/ibeo-feature-fusion/. Viewed: February 27, 2019.

Kamijo S., Yanlei G., 2015. Autonomous vehicle technologies: Localization and mapping. *Fundamentals Review*, 9(2), 131–141.

Kiemle C., Ehret G., 1997. Estimation of boundary layer humidity fluxes and statistics from airborne Differential Absorption Lidar (DIAL). *Journal of Geophysical Research*, 102(D24), 29189–29203.

Kluge K., 1994. Extracting road curvature and orientation from image edge points without perceptual grouping into features. In *Proceedings of IEEE Intelligent Vehicles Symposium*, Paris, France, October 1994, pp. 109–114.

Kocić J., Jovičić N., Drndarević V., 2018. Sensors and sensor fusion in autonomous vehicles. School of Electrical Engineering, University of Belgrade, Belgrade, Serbia. November 2018.

Kubo N., Suzuki T., Yasuda A., Shibazaki R., 2005. An effective method for multipath mitigation under severe multipath environments. In *Proceedings of the 18th International Technical Meeting of the Satellite Division of the Institute of Navigation (ION GNSS)*, Long Beach, CA, USA, September 2005, pp. 2187–2194.

Kubo N., Hou R., Suzuki T., 2014. Decimetre level vehicle navigation combining multi-GNSS with existing sensors in dense urban areas. In *Proceedings of the 2014 International Technical Meeting of the Institute of Navigation*, San Diego, CA, USA, January 2014, pp. 450–459.

Lefsky M. A., Cohen W. B., 2002. Lidar remote sensing for ecosystem studies. *BioScience*, 52(1), 19–30.

Leonard J., How J., Teller S., Berger M., Campbell S., Fiore G., 2008. A perception-driven autonomous urban vehicle. *Journal of Field Robotics*, 25(10), 727–774.

Levinson J., Montemerlo M., Thrun S., 2007. Map-based precision vehicle localization in urban environments. *Robotics: Science and Systems*, 4, 1.

Levinson J., Thrun S., 2010. Robust vehicle localization in urban environments using probabilistic maps. In *Processing of IEEE International Conference on Robotics and Automation*, Anchorage, AK, USA, 2010, pp. 4372–4378.

Levinson J., Askeland J., Becker J., Dolson J., Held D., Kammel S., 2011. Towards fully autonomous driving: Systems and algorithms. In *Proceedings of IEEE Intelligent Vehicles Symposium*, Baden-Baden, Germany, June 2011, pp. 163–168.

Li Z., Bruggemann T., 2012. Towards automated power line corridor monitoring using advanced aircraft control and multi-source feature fusion. *Journal of Field Robotics*, 29(1), 4–24.

Lu F., Milios E., 1997. Globally consistent range scan alignment for environment mapping. *Autonomous Robots*, 4(4), 333–349.

Marita T., Negru M., Danescu R., Nedevschi S., 2011. Stop-line detection and localization method for intersection scenarios. In *Proceedings of IEEE International Conference on Intelligent Computer Communication and Processing*, Cluj-Napoca, Romania, August 2011, pp. 293–298.

Matsson F., 2018. Sensor fusion for positioning of an autonomous vehicle. Design and implementation of an unscented Kalman filter. Degree project in mathematics, Second cycle, 30 credits Stockholm, Sweden.

McGrath, Michael J., Ní Scanaill C., 2014. *Sensor Technology. Healthcare, Wellness and Environment Applications*. Springer Nature.

Mcgraw G. A., Young R. S. Y., Reichenauer K., Stevens J., Ventrone F., 2004. GPS multipath mitigation assessment of digital beam forming antenna technology in a JPALS dual frequency smoothing architecture. In *Proceedings of the 2004 National Technical Meeting of the Institute of Navigation*, San Diego, CA, USA, January 2004, pp. 561–572.

Megur J. I., Murat T., Takiguchi J. I., Aman Y., Hashizume T., 2009. GPS multipath mitigation for urban area using omnidirectional infrared camera. *IEEE Transaction on Intelligent Transportation Systems*, 10(1), 22–30.

Mehta S., Patel A., Mehta J., 2015. CCD or CMOS Image sensor for photography. In *2015 International Conference on Communications and Signal Processing (ICCSP)*, April 2015, pp. 0291–0294. doi:10.1109/ICCSP.2015.7322890.

Merz T., Duranti S., 2006. Autonomous landing of an unmanned helicopter based on vision and inertial sensing. In *Experimental Robotics IX* (p. 352), Vol. 21. Berlin/Heidelberg: Springer.

Milanes V., Naranjo J. E., Gonzalez C., Alonso J., Pedro T., 2008. Autonomous vehicle based in cooperative GPS and inertial systems. *Robotica*, 26(05), 627–633.

Montemerlo M., Thrun S., Koller D., Wegbreit B., 2002. Fast-SLAM: A factored solution to the simultaneous localization and mapping problem. In *Proceedings of Innovative Applications of Artificial Intelligence Conference*, Edmonton, Alberta, Canada, July 2002, pp. 593–598.

Montemerlo M., Thrun S., Koller D., Wegbreit B., 2003. Fast-SLAM 2.0: An improved particle filtering algorithm for simultaneous localization and mapping that provably converges. In *Proceedings of the International Joint Conference on Artificial Intelligence*, Acapulco, Mexico, August 2003, pp. 1151–1156.

Montemerlo M., Becker J., Bhat S., Dahlkamp H., Dolgov D., Ettinger S., 2008. Junior: The Stanford entry in the urban challenge. *Journal of field Robotics*, 25(9), pp. 569–597.

Noureldin A., Karamat T. B., Eberts M. D., El-Shafie A., 2009. Performance enhancement of MEMS-based INS/GPS integration for low-cost navigation applications. *IEEE Transaction on Vehicular Technology*, 58(3), 1077–1096.

Obst M., Bauer S., Reisdorf P., Wanielik G., 2012. Multipath detection with 3D digital maps for robust multi-constellation GNSS/INS vehicle localization in urban areas. In *Proceedings of IEEE Intelligent Vehicles Symposium*, Alcala de Henares, Spain, June 2012, pp. 184–190.

Oh S., Kang H., 2016. Fast occupancy grid filtering using grid cell clusters from LIDAR and stereo vision sensor data. *IEEE Sensors Journal*, 16(19), 7258–7266.

Oniga F., Nedevschi S., Meinecke M. M., 2007. Curb detection based on elevation maps from dense stereo. In *Proceedings of IEEE International Conference on Intelligent Computer Communication and Processing*, Cluj-Napoca, Romania, September 2007, pp. 119–125.

Ors O. A., 2017. RADAR, camera, LiDAR and V2X for autonomous cars. NPX, May 24, 2017. https://blog.nxp.com/automotive/radar-camera-and-lidar-for-autonomous-cars. Viewed: February 26, 2019.

Ozguner U., Redmill K., 2008. Sensing, control, and system integration for autonomous vehicles: A series of challenges. *SICE Journal of Control, Measurement, and System Integration*, 1(2), 129–136.

Patz B. J., Papelis Y., Pillat R., Stein G., Harper D., 2008. A practical approach to robotic design for the DARPA urban challenge. *Journal of Field Robotics*, 25(8), 528–566.

Paull L., Saeedi S., Seto M., Li H., 2014. AUV navigation and localization: A review. *IEEE Journal of Oceanic Engineering*, 39(1), 131–149.

Pickering P., 2017. Radar and ultrasonic sensors strengthen ADAS object detection. August 22, 2017. www.electronicdesign.com/automotive/radar-and-ultrasonic-sensors-strengthen-adas-object-detection. Viewed: February 27, 2019.

Qin B., Chong Z. J., Bandyopadhyay T., Ang M. H., Frazzoli E., Rus D., 2012. Curb-intersection feature based Monte Carlo localization on urban roads. In *Proceedings of IEEE International Conference on Robotics and Automation*, Saint Paul, MN, USA, May 2012, pp. 2640–2646.

Redmill K., 1997. A simple vision system for lane keeping. In *1997 IEEE Intelligent Transportation Systems Conference*, Boston, MA, USA, November 1997, pp. 212–217.

Redmill K., Ozguner U., 1999. The Ohio State University automated highway system demonstration vehicle, (SAE Transactions 1997). *Journal of Passenger Cars*, SP-1332, Society of Automotive Engineers, 1999.

Redmill K., Martin J. I., Ozguner U., 2006. Sensing and sensor fusion for the 2005 Desert Buckeyes DARPA grand challenge offroad autonomous vehicle. In *2006 IEEE Intelligent Vehicles Symposium*, Tokyo, Japan, June 2006, pp. 528–533.

Riegl, 2011. RIEGL Laser Measurement Systems GmbH.

Ros G., Sappa A., Ponsa D., Lopez A. M., 2012. Visual slam for driverless cars: A brief survey. In *Proceedings of IEEE Intelligent Vehicles Symposium 2012 Workshops*, Alcalá de Henares, Spain, June 2012.

Schoettle B., 2017. Sensor fusion: A comparison of sensing capabilities of human drivers and highly automated vehicles. Report No. SWT-2017-12, August 2017, Sustainable worldwide transportation. The University of Michigan, Ann Arbor, MI, USA.

Seo Y. W., Rajkumar R., 2014. A vision system for detecting and tracking of stop-lines. In *Proceedings of IEEE 17th International Conference on Intelligent Transportation Systems*, Qingdao, China, October 2014, pp. 1970–1975.

Shladover S. 2007. PATH at 20-history and major milestones. *IEEE Transactions on Intelligent Transportation Systems*, 8(4), 584–592.

SICK Products. www.sick.com/de/en/product-portfolio/c/PRODUCT_ROOT. Viewed: February 27, 2019.

Smith R., Cheeseman P., 1987. On the representation of spatial uncertainty. *The International Journal of Robotics Research*, 5(4), 56–68.

Smith R., Self M., Cheeseman P., 1990. *Estimating Uncertain Spatial Relationships in Robotics. Autonomous Robot Vehicles* (pp. 167–193). New York: Springer.

Soumekh M., 1999. *Synthetic Aperture Radar Signal Processing with Matlab Algorithms.* New York: Wiley.

Stimson G. W., 1998. Introduction to Airborne Radar. Mendham: SciTech Pub.

Szeliski R., 2011. *Computer Vision: Algorithms and Applications.* New York/London: Springer; Terranean: Terranean Mapping Technologies.

Takasu T., Yasuda A., 2009. Development of the low-cost RTK-GPS receiver with an open source program package RTKLIB. In *International Symposium on GPS/GNSS*, Jeju, Korea, November 2009.

Tan H. S., Rajamani R., Zhang W. B., 1998. Demonstration of an automated highway platoon system. In *1998 Proceedings of the American Control Conference*, Philadelphia, PA, USA, June 1998, pp. 1823–1827.

Thakur R., 2017. Infrared sensors for autonomous vehicles. We are IntechOpen, the world's leading publisher of Open Access books Built by scientists, for scientists. December 20, 2017. doi:10.5772/intechopen.70577.

Thrun S., Montemerlo M., 2006. The graph SLAM algorithm with applications to large-scale mapping of urban structures. *The International Journal of Robotics Research*, 25(5), 403–429.

Tirindelli P., 2016. Sensor fusion of raw GPS measurements for autonomous vehicle localization. Master in Artificial Intelligence. April 21, 2016.

Urmson C., Anhalt J., Bagnell D., Baker C., Bittner R., Clark M. N., 2008. Autonomous driving in urban environments: Boss and the urban challenge. *Journal of Field Robotics*, 25(8), 425–466.

Valavanis K. P., Vachtsevanos G. J., 2015. *Handbook of Unmanned Aerial Vehicles.* ISBN 978-90-481-9706-4. Dordrecht Heidelberg New York London: Springer.

Van Dierendonck A. J., Fenton P., Ford T., 2001. Theory and Performance of Narrow Correlator Spacing in a GPS Receiver. *Navigation*, 39(3), 265–283.

Velodyne LiDAR Products. http://velodynelidar.com/products.html. Viewed: February 27, 2019.

Viandier N., Nahimana D. F., Marais J., Duflos E., 2008. GNSS performance enhancement in urban environment based on pseudo-range error model. In *Proceedings of. IEEE/ION Symposium Position, Location and Navigation*, Monterey, CA, USA, pp. 377–382, May 2008.

Wicks A. L., Steve C., Asbeck A. T., 2017. The design of an autonomous vehicle research platform. Thesis submitted to the Faculty of the Virginia Polytechnic Institute and State University in partial fulfillment of the requirements for the degree of Master of Science in Mechanical Engineering. August 3, 2017. Blacksburg, Virginia. Copyright 2017, Denver Hill Walling.

Wu T., Ranganathan A., 2013. Vehicle localization using road markings. In *Proceedings of IEEE Intelligent Vehicles Symposium*, Gold Coast, QLD, Australia, June 2013, pp. 1185–1190.

Xu D., Jain A., Anguelov D., 2018. PointFusion: Deep sensor fusion for 3D bounding box estimation. In *Proceedings of CVPR*, Salt Lake City, UT, USA, 2018.

Yenikaya S., Yenikaya G., Düven E., 2013. Keeping the vehicle on the road: A survey on on-road lane detection systems. *ACM Computing Surveys (CSUR)*, 46(1), 2.

Yoneda K., Tehrani H., Ogawa T., Hukuyama N., Mita S., 2014. LIDAR scan feature for localization with highly precise 3-D map. In *Proceedings of IEEE Intelligent Vehicles Symposium*, Dearborn, MI, USA, June 2014, pp. 1345–1350.

Yuee L., Zhengrong L., 2009. Classification of airborne LIDAR intensity data using statistical analysis and Hough transform with application to power line corridors. In *Digital Image Computing: Techniques and Applications (DICTA '09)*, 2009, Melbourne, Australia: IEEE.

Zhang W. B., 1997. National automated highway system demonstration: A platoon system. In *1997 IEEE Intelligent Transportation Systems Conference*, Philadelphia, PA, USA, November 1997.

Zhang W., 2010. Lidar-based road and road-edge detection. In *Proceedings of IEEE Intelligent Vehicles Symposium*, San Diego, CA, USA, June 2010, pp. 845–848.

Zolfagharifard E., 2013. It really is hands free! Self-driving Mercedes-Benz is unveiled - and it should be available within seven years. September 12, 2013. www.dailymail.co.uk/sciencetech/article-2418526/Self-driving-Mercedes-Benz-sale-2020-unveiled.html. Viewed: March 29, 2019.

5

Data Acquisition and Intelligent Diagnosis

5.1 Data Acquisition Principle and Process for Laser Scanning, Visual Imaging, Infrared Imaging, UV Imaging

5.1.1 Laser Scanning

5.1.1.1 Introduction

Automated restitution methods for object acquisition have gained more and more importance in recent years. Automatic image matching and laser scanning, often called LiDAR (light detection and ranging), have revolutionized 3D data acquisition for both topographic and close-range objects. In contrast to the "classical" manual data acquisition techniques, like terrestrial surveying and analytical photogrammetry, which require manual interpretation to derive a representation of the sensed objects, these new automatic recording methods allow an automated dense sampling of an object's surface in a short time (Pfeifer & Briese, 2007).

Laser scanning is an emerging data acquisition technology that has remarkably broadened its application field to seriously compete with other surveying techniques. Scanning can be airborne or terrestrial. Terrestrial laser scanning is a reasonable alternative method in many kinds of applications that previously used ground-based surveying or close-range photogrammetry (Lovas, 2010).

Photographic images record passive solar or artificial radiation backscattered by objects in the camera's field of view (FOV). The backscatter strength is typically resolved in one of the following ways:

 i. Spatially, by pixels in the image plane;
 ii. Chromatically, by recording in different wavelength bands;
 iii. Radiometrically, by quantizing the photo current with typically 8–12 bits.

In contrast, laser scanning achieves spatial resolution by scanning the instantaneous FOV with the help of mechanical devices, for example, a moving mirror, over the entire FOV. The backscatter is recorded for one wavelength only, i.e., monochromatically. Nonpassive radiation is used for the measurement, but the backscatter of laser energy is emitted by the sensor system itself. The time lag between emission of a laser beam and detection of its backscattered echo is measured as well. With the group velocity of the light and the light speed in the atmosphere, the time difference can be transformed to the range between emitter and detector.

Both photographs and the recorded echoes of laser beams are impaired—to a small extent only—by ambient light, i.e., energy not originating from the location of the specific sensed objects but stray light. Photographs record texture and color; laser scanning primarily measures ranges but also monochromatic reflectance. In both cases, however, data are acquired area-wise in a systematic manner. Recording electromagnetic radiation by a map (ping) that generates an image of object space is the basis for both. In images, measurements can be performed automatically or manually (Pfeifer & Briese, 2007).

In recent years, with the development of sensor fusion techniques, navigation solutions, including inertial measurements, and urban modeling, mobile laser scanning has gained momentum, as can be seen in the sensor manufacturers' product lists. A significant paradigm change can be observed in geodesy,

155

for example, in direct orientation instead of indirect orientation, in surface detection instead of point measurements, and in complex 3D models instead of simple coordinates.

All scanners are based on the same principle: the scanner emits a laser beam through the ground/object and computes the distance by measuring the traveling time or the phase difference of the laser beam. The emission rate of cutting-edge sensors is in the 100–200 kHz range. The direction of the beam is determined by different types of rotating or oscillating mirrors that enable the scanning of the area of interest. In airborne and mobile laser scanning, the position of the sensor is given by high accuracy Global Navigation Satellite Systems (GNSSs) and inertial navigation systems (INSs).

Laser scanning is often referred to as LiDAR. ALM (airborne laser mapping) or ALS (airborne laser scanning) abbreviations are also widely used, while "terrestrial laser scanning" is used for ground-based laser scanning (Lovas, 2010).

5.1.1.2 Airborne Laser Scanning (ALS)

ALS is an active remote sensing technology that is able to rapidly collect data from huge areas. The resulting dataset can be the basis for digital surface and elevation models. ALS is often coupled with airborne imagery; in this coupling, the point clouds and images can be fused, resulting in enhanced quality 3D products.

The basic principle is as follows: the sensor emits a laser pulse through the terrain in a predefined direction and receives the reflected laser beam. If the speed of light is known, the distance of the object can be calculated; see Figure 5.1 (Lovas, 2010).

Airborne LiDAR systems are composed of the following subsystems:

- Laser sensor and computing, data storage unit
- INS/IMU (inertial measurement unit)
- GNSS
- GNSS ground station(s).

The components are shown in Figure 5.2:

A collection of point clouds from airborne platforms always requires that the path of the platform, i.e., its position and angular attitude, is observed continuously. With a pulse repetition rate (PRR) of 100 kHz, 100,000 range and scanner angle measurements are made per second, and for each one, the sensor coordinate system has its own exterior orientation. For increasing accuracy, some form of ground control is necessary, requiring methods of strip adjustment.

Data in ALS are acquired strip-wise. A strip length of 20 km is not uncommon, but the strip length cannot be made arbitrarily long because of drift errors in the IMU, which are corrected after flight

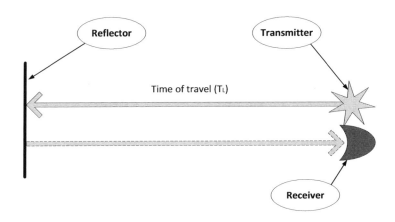

FIGURE 5.1 Time-of-flight laser range measurement (Lovas, 2010).

Data Acquisition and Intelligent Diagnosis 157

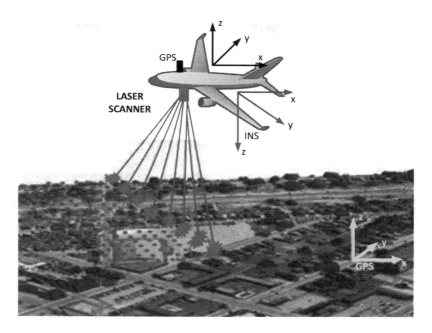

FIGURE 5.2 Principle of airborne LiDAR (Lovas, 2010).

maneuvers. Wider areas are acquired by placing strips next to each other with an overlap to avoid gaps. Larger overlaps can increase accuracy in strip adjustment (see below) or increase the point density (e.g., by a strip overlap of 50%).

Laser scanning from a ground-based or water-based moving platform is similar to ALS, except the laser scanning is not primarily vertical and data acquisition follows the allowed routes (e.g., streets) rather than a systematic scanning of the project area (Pfeifer & Briese, 2007).

5.1.1.2.1 LiDAR Work Phases and Data Processing

Based on the particular application or specific requirements of a project, different work flows are executed. However, the main steps are often the same for many applications. A typical LiDAR application has the following work phases:

- Planning (coverage, point density, flying parameters, etc.)
- Deployment of GNSS base stations (if needed)
- Calibration of equipment (e.g., to determine certain misalignments)
- LiDAR surveying.

LiDAR data processing can be grouped in many ways. Assuming a typical airborne LiDAR project to derive a digital surface model (DSM), for example, for mobile communication companies to support precisely deployed base stations and antennas, the following data processing steps are required:

- Georeferencing: Transformation of the point cloud (usually from WGS84) to the local coordinate system.
- Noise removal: Filtration of points not reflected from the surface (e.g., reflected from a bird or below the ground) because of multipath reflection or faulty time measurement.
- Coarse classification: Classification of ground points, above ground points, water etc.; also point density adjustment, interpolation.
- Modeling: DSM/digital elevation model (DEM) generation, feature extraction (segmentation and/or classification).

- Data fusion: Warping of airborne images onto DSM etc.
- Measurements on model, advanced feature extraction.

Many of these data processing steps can be automated; certain procedures built into processing software are capable of classifying features, recognizing building roofs, trees, etc.

LiDAR data are usually stored in LAS format, but many manufacturers apply self-developed file formats. The LAS file format is a public file format for the interchange of LiDAR data between vendors' customers. This binary file format is an alternative to proprietary systems or generic ASCII file interchange systems used by many companies (Lovas, 2010).

5.1.1.2.2 Intensity Data

Recent LiDAR systems are capable of measuring the signal strength (i.e., intensity) of the reflected laser pulse. Different objects have different reflectivity; therefore, the intensity values can support object recognition and identification.

Note that LiDAR intensity values vary based on light and weather circumstances; therefore, they are seldom used alone without additional data for classification (Lovas, 2010).

5.1.1.3 Terrestrial Laser Scanning

While ALS competes with photogrammetry, interferometric radar (Lovas, 2010) and so-called window scanners have a FOV similar to a conventional area camera or a panoramic scanner (Pfeifer & Briese, 2007). Traditional close-range photogrammetry does not have a wide application field and is not considered a competing technology in planning projects (Lovas, 2010).

In the early 2000s, terrestrial laser scanning was used only for specific tasks, for example, surveying and modeling complex systems (e.g., cooling/heating pipes) of a factory. Recently, terrestrial laser scanning has broadened its application area and is used in projects conventionally considered to be in the field of traditional surveying (e.g., small-scale topographic surveys, road construction, in-door surveying of buildings).

The principle of terrestrial laser scanning is very similar to that of airborne surveying. The sensor continuously emits a laser beam towards an object, receives it back, and computes the distance of the object. The beam is directed by rotating or oscillating mirrors (usually the same mirror can operate both in rotating and oscillating mode). The main components of a terrestrial laser scanner are shown in Figure 5.3 (Lovas, 2010).

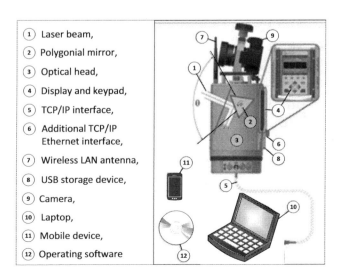

FIGURE 5.3 Components of terrestrial laser scanner (Lovas, 2010).

Data Acquisition and Intelligent Diagnosis

The main difference is that the sensor is not moving but mounted on a tripod or any kind of structure. Therefore, no positioning solution is needed; however, newer scanners are able to be directly connected to Global Positioning System (GPS) receivers to directly obtain the scanner station location (Lovas, 2010).

Normally, one scan is not enough to collect data covering the entire object of interest. If the scanner is placed inside, occlusions prevent all details being seen from one standpoint. Scanning the outside of an object requires more standpoints to scan the object from all sides. Three observations are made to measure one point on the object surface: the range r and two angles, α, the horizontal angle, and β, the vertical angle. In the sensor coordinate system, the coordinates (x, y, z) of the point are obtained by a conversion from the spherical to the Cartesian coordinate system:

$$(x, y, z) = r(\cos\alpha \cos\beta, \sin\alpha \cos\beta, \sin\beta)$$

If only the object itself is of interest, it is sufficient to determine the relative orientation of the scans. If the object also has to be placed in a superior coordinate system, absolute orientation becomes necessary. If the superior coordinate system is earth-fixed, this is called georeferencing (Pfeifer & Briese, 2007).

5.1.1.3.1 Data Processing

The workflow of the terrestrial laser scanning procedure can be summarized as follows (Lovas, 2010):

- Preparation (planning, preliminary geodetic measurements, etc.)
- Scanning
- Registration/georeferencing (if required and if preparation and scanning was done)
- Selecting area of interest (optional)
- Filtering and converting data
- Segmenting and classifying data
- Modeling (triangulation, rendering, fitting geometrical elements onto a point cloud)
- Measurements on model
- Visualization
- Application-dependent products (e.g., cross sections for architects).

As in most engineering projects, project planning and preparatory work must be emphasized, including creating the geodetic network (if needed), checking field of view, and planning scan station locations. Besides saving costs, optimal scan station locations achieve the required coverage, point density, and visibility of dedicated parts of the object and ensure the desired accuracy.

Scanning usually starts with panorama scanning, i.e., a low resolution surveying of the surrounding area. Then the area to be mapped can be selected on this point cloud or specified by corner coordinates or angular ranges. Specific points (e.g., control points or dedicated points of a structure whose displacement has to be measured) can be marked with specific targets. These are special stickers or objects (usually in a form of disc, cylinder, or sphere) with extremely high reflectivity. The scanner software is able to recognize the reflectors in range and provide the coordinates of the middle of the reflectors. Careful planning is crucial when deploying the reflectors. If repeatability is an issue (e.g., in monitoring and quality control projects), the reflector locations have to be marked in such a way that they will remain until the next survey. The protection of reflectors also has to be solved for scans in open areas.

In registration and georeferencing, the resulting point cloud(s) and images have to be transformed into a given coordinate system. In many cases, there is no need to transform the data into a local coordinate system; measurements can be executed in the scanner's own coordinate system. The images taken by the scanner camera or by a camera mounted on a scanner are usually warped onto the point cloud by the scanner software.

Scanners capture points reflected from all objects in the FOV and range; thus, selecting the area of interest (i.e., points reflected from the object(s) to be mapped) has to be done before modeling and any kinds of measurements.

Before modeling, further preprocessing steps are needed: converting the point cloud into the required format (depending on the processing software's requirements), filtering outliers, and interpolating points into a predefined grid (if needed). In other words, achieving a clean dataset that meets the requirements of further processing steps.

In some applications, segmentation and/or classification of certain areas in the point cloud are needed, but in most cases, the entire dataset has to be modeled. Modeling can cover wide ranges of procedures, for example, triangulating surfaces, fitting geometric elements on the point cloud, detecting edges, and creating a vector model.

Some applications, such as engineering surveys, need to derive particular values, for example, displacement between two or more dedicated points or the measurement of certain distances. In some cases, these values can be obtained without modeling, simply executing measurements on the point cloud. Note that there are specific point cloud processing software packages available on the market.

Visualization is more important in laser scanning than in other geodetic procedures; the results must be presented to the customers, users, and decision makers in an easily understandable form.

Specific applications may require specific products, such as cross and longitudinal sections for architectural design purposes, specific distance and volume calculations of artifacts for archaeological surveys, deformation measurements at special parts of structures, and surface material features. These kinds of specific products often require intense consultation with experts in the area.

Note that it is recommended to include a detailed description of the end product of laser scanning in the project contract. The raw point cloud or even a 3D model can be useless for the user if the required measurements, evaluations, or any kinds of assessments cannot be executed. For cross-checking purposes, consultation with independent experts or, at least, gathering information from other projects, is highly recommended (Lovas, 2010).

5.1.1.4 Mobile Laser Scanning

The most recent application of laser scanning is the mobile mapping application, where the scanner is mounted on a mobile platform, mostly on a passenger car or truck. Mobile laser scanning is used in projects where big areas (or long corridors) are covered on the ground and data are to be acquired for those areas (Lovas, 2010).

5.1.1.4.1 Data Processing

The main work phases of processing mobile laser scanned data are as follows:

- Calculating position and orientation of the sensor platform
- Georeferencing the point cloud and registering the images
- Coarsely classifying points (e.g., ground, vegetation, building, and other)
- Measuring, evaluating, and modeling according to the particular application.

Sensor position/trajectory and orientation are generally supported by Kalman filtering. Georeferencing and registration of point clouds are the main differences between terrestrial and mobile laser scanning. Since an urban environment has areas where no GNSS signal is available (or there is a less accurate signal), and INS provides sufficient accuracy for a limited range, careful planning of measurement is needed. These factors have to be considered during the accuracy assessment (Lovas, 2010).

5.1.2 Visual Imaging

5.1.2.1 What Is Visual Information?

Two kinds of information are associated with a visual object (image or video): information about the object, called its metadata, and information contained within the object, called visual features. Metadata are alphanumeric and generally expressible as a schema of a relational or object-oriented database.

Visual features are derived through computational processes—typically image processing, computer vision, and computational geometric routines—executed on the visual object (Gupta & Jain, 1997).

The simplest visual features that can be computed are based on pixel values of raw data, and several early image database systems used pixels as the basis of their data models. These systems can answer such queries as the following (Gupta & Jain, 1997):

- Find all images for which the 100th to 200th pixels are orange if orange is defined as having a mean value of (red=255, green=130, and blue=0).
- Find all images that have about the same color in the central region of the image as this particular one. The "central region" of the image can be specified by a coordinate system, and the expression "about the same color" is usually defined by computing a color distance. A variant of the Euclidean distance is often used to compare two color values.
- Find all images that are shifted versions of this particular image, in which the maximum allowable shift is D.

If the user's requirements are satisfied with this class of queries, data modeling for visual information is almost trivially simple. More realistically, however, a pixel-based model suffers from several drawbacks. First, it is very sensitive to noise; therefore, a couple of noise pixels may be sufficient to cause it to discard a candidate image for the first two queries. Second, translation and rotation invariance are often desirable properties for images. For example, for the third query, if the database contains a 15° rotated version of this image, the rotated version may not be reported by the system. Third, apart from noise, variations in illumination and other imaging conditions affect pixel values drastically, leading to incorrect query results (Gupta & Jain, 1997).

That is not to say pixel-oriented models are without merit. Significant video segmentation results can be obtained by measuring pixel differences over time. For example, an abrupt scene change can be modeled by finding major discontinuities in time plots of cumulative pixel difference over frames (Hampapur, 1995). However, information retrieval based only on pixel values is not very effective (Gupta & Jain, 1997).

Consider a database of aerial images in which the only objects of interest are buildings, ground vehicles, aircraft, roads, and general terrain. Also imagine that a human interpreter draws bounding rectangles for each region in an image in which one or more of these five kinds of objects appear and labels the regions accordingly. Now we have a fairly precise specification of the information contained in the images. That information can be directly modeled by a relational database schema that maintains the location (bounding box) of each object type and a time stamp for each image. With some additional geometric processing added to this relational model, we can answer very complex queries (Gupta & Jain, 1997), such as the following:

- Is there any location where more than five ground vehicles are close to a building located in the middle of the general terrain?
- Have there been any changes in the position of the aircraft at this location in the past couple of hours?
- Which approach roads have been used by ground vehicles over the past few days to come close to the aircraft?

While these queries are meaningful, the most crucial part of information retrieval—information extraction—is performed by a human using his or her knowledge and experience in aerial image interpretation. The reason this task requires a human is simple. Fully automatic interpretation of aerial images is an unsolved research problem. If the human extracts the useful information, we can then use a spatial database system to organize and retrieve it. In a real-life aerial surveillance situation, this approach is unrealistic. For a battlefield application, the territory under surveillance is large enough to need several camera-carrying aircraft. Images from every aircraft, each image several MB in size, stream in at the video rate of 30 frames per second. The high influx of images means error-free interpretation takes a long time; hence, the simple image database scenario we painted is not practical for any time-critical operation (Gupta & Jain, 1997).

5.1.2.2 What Is Image Processing?

Image processing is a method to convert an image into digital form and perform some operations on it to get an enhanced image or to extract some useful information. It is a type of signal dispensation in which input is an image, like a video frame or photograph, and output may be an image or characteristics associated with that image. An image processing system usually treats images as two-dimensional signals and applies preset signal processing methods.

It is a rapidly growing technology, with applications in various aspects of a business. Image processing also represents a core research area within engineering and computer science disciplines (Mary, 2011).

Image processing basically includes the following three steps (Mary, 2011):

- Importing the image with an optical scanner or by digital photography.
- Analyzing and manipulating the image, including data compression and image enhancement, and spotting patterns that are not visible to the human eye, as in satellite photographs.
- Outputting an altered image or report based on image analysis.

5.1.2.3 Purpose of Image Processing

The purpose of image processing can be divided into five groups (Mary, 2011):

1. Visualization: Observing objects that are not visible.
2. Image sharpening and restoration: Creating a better image.
3. Image retrieval: Seeking the image of interest.
4. Measurement of pattern: Measuring various objects in an image.
5. Image recognition: Distinguishing the objects in an image.

5.1.2.4 Methods Used for Image Processing

The two types of methods used for image processing are analog and digital image processing.

Analog techniques of image processing can be used for hard copies like printouts and photographs. Image analysts use various fundamentals of interpretation for these visual techniques. The image processing is not confined to an area that has to be studied but includes the knowledge of the analyst. Association is another important tool in image processing using visual techniques. In association, analysts apply a combination of personal knowledge and collateral data to image processing.

Digital processing techniques help the manipulation of the digital images by using computers. Raw data from imaging sensors on satellite platform contain deficiencies. To get over such flaws and to obtain original information, the raw data must undergo various phases of processing. These are preprocessing, enhancement and display, and information extraction (Mary, 2011) (Figure 5.4).

5.1.2.5 Image Processing Applications

1. Intelligent transportation systems: This technique can be used in automatic number plate recognition and traffic sign recognition.
2. Remote sensing: For this application, sensors capture pictures of the earth's surface in remote sensing satellites or multispectral scanners mounted on an aircraft. These pictures are processed by transmitting them to the earth station. They are used in flood control, city planning, resource mobilization, agricultural production monitoring, etc.

Data Acquisition and Intelligent Diagnosis 163

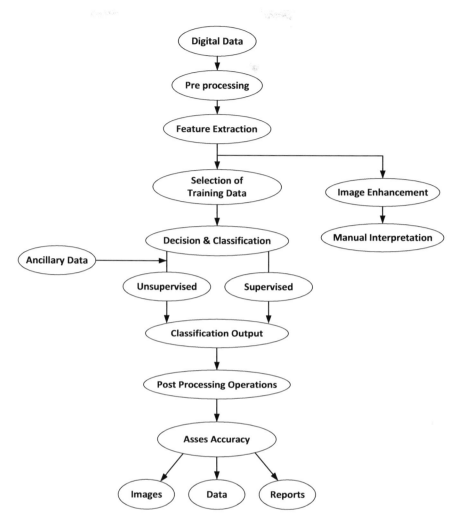

FIGURE 5.4 Flow chart showing phases in digital image processing (Mary, 2011).

3. Moving object tracking: This application enables the measurement of motion parameters and the acquisition of a visual record of the moving object. The different approaches to tracking an object are:
   - Motion-based tracking
   - Recognition-based tracking.
4. Defense surveillance: Aerial surveillance methods are used to continuously keep an eye on the land and oceans. This application is also used to locate the types and formation of naval vessels on the ocean surface. The important duty is to divide the various objects present in the water body part of the image. Different parameters, such as length, breadth, area, perimeter, and compactness, are established to classify each object. It is important to recognize the distribution of these objects in all directions to explain all possible formations of the vessels. We can interpret the entire oceanic scenario from the spatial distribution of these objects.
5. Biomedical imaging techniques: Various types of imaging tools, such as X-ray, ultrasound, and computer-aided tomography (CT), are used for medical diagnosis. X-ray, magnetic resonance imaging (MRI), and CT are shown in Figure 5.5 (Mary, 2011).

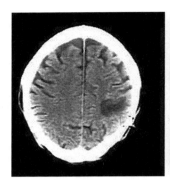 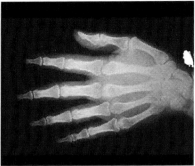

FIGURE 5.5 Representational image of X-ray, MRI, and CT (Mary, 2011).

Some biomedical imaging applications are the following (Mary, 2011):
- Heart disease identification: Important diagnostic features, such as size of the heart and its shape, are required to classify heart diseases. To improve the diagnosis of heart diseases, image analysis techniques are applied to radiographic images.
- Lung disease identification: In X-rays, the regions that appear dark contain air while regions that appear lighter are solid tissues. Bones are more radio opaque than tissues. The ribs, heart, thoracic spine, and diaphragm are clearly seen on an X-ray film.
- Digital mammograms: These are used to detect breast tumors. Mammograms can be analyzed using image processing techniques, such as segmentation, shape analysis, contrast enhancement, and feature extraction.

6. Automatic visual inspection system: This application improves the quality and productivity of products.
    - Automatic inspection of incandescent lamp filaments: This involves examination of the bulb manufacturing process. Because there is no uniformity in the pitch of the wiring in a lamp, the filament of the bulb becomes fused within a short period of time. In this application, a binary image slice of the filament is created and the silhouette of the filament is fabricated from it. Silhouettes are used to recognize the nonuniformity in the pitch of the lamp's wiring. This system is used by General Electric Corporation.
    - Automatic surface inspection systems: In metal industries, it is essential to detect surface flaws. For instance, it is essential to detect any kind of aberration on the rolled metal surface in hot or cold rolling mills in a steel plant. Image processing techniques, such as texture identification, edge detection, fractal analysis, etc., are used for detection.
    - Faulty component identification: This application identifies the faulty components in electronic or electromechanical systems. A higher amount of thermal energy is generated by faulty components. Infrared images are produced from the distribution of thermal energy in the assembly. The faulty components can be identified by analyzing the infrared images.

5.1.3 Infrared Imaging

5.1.3.1 Introduction

What we typically think of as "light" is really electromagnetic radiation that our eyes can see. We perceive the world in the colors of the rainbow, red through violet. But these colors of light are actually a very small portion of the electromagnetic spectrum.

Our eyes are capable of seeing only a very narrow region of the electromagnetic spectrum, and we need special instruments to extend our vision beyond the limitations of the unaided eye. As the energy of

Data Acquisition and Intelligent Diagnosis

light changes, so too does its interaction with matter. Materials that are opaque at one wavelength may be transparent at another. A familiar example of this phenomenon is the penetration of soft tissue by X-rays. What is opaque to visible light becomes transparent to reveal the bones within.

Extending human vision with electronic imaging is one of the most powerful techniques available to science and industry, particularly when it enables us to see light in the infrared (IR) portion of the spectrum. IR means "below red," as IR light has less energy than red light. We typically describe light energy in terms of wavelength, and as the energy of light decreases, its wavelength gets longer. IR light, having less energy than visible light, has a correspondingly longer wavelength. The IR portion of the spectrum has wavelengths ranging from 1 to 15 µm or about 2 to 30 times longer wavelengths (and 2–30 times less energy) than visible light.

IR light is invisible to the unaided eye but can be felt as heat on our skin. Warm objects emit IR light, and the hotter the object, the shorter the wavelength of IR light emitted. This IR "glow" enables rescue workers equipped with longwave IR sensors to locate a lost person in a deep forest in total darkness, for example. IR light can penetrate smoke and fog better than visible light, revealing objects that are normally obscured. It can also be used to detect the presence of excess heat or cold in a piece of machinery or a chemical reaction (Janos Technology, 2019).

5.1.3.2 What Is Infrared (IR) Imaging?

IR imaging is a technique of capturing the IR light from objects and converting it into visible images interpretable by a human eye.

The IR region is spread across the 10–100 µm wavelength in the electromagnetic region which can be distributed into three bands: the near-IR region from 0.7 to 1.3 µm; the mid-IR region from 1.3 to 3 µm; the thermal-IR region in the remaining part of the band. While the first two are used in general electronic applications like remote control and illumination IR photography, thermal IR is used in thermal imaging. The main difference is that the first two are used in reflective type of applications, while thermal IR is emanated from the object, not reflected by it.

At absolute zero, perfect order is believed to exist in the atomic structure, with no collisions and minimal entropy. Any object above the absolute zero temperature has atomic chaos and collisions, resulting in thermal energy being radiated, most of which falls in the IR band. If these radiations can be detected by some means, objects can be visualized through their radiation patterns without the need of an optical source. This forms the basis of IR imaging (Thakur, 2011).

5.1.3.3 Types of Infrared Imaging

Illumination-based imaging: single lens reflex (SLRs) and digital cameras featuring IR night vision often rely on the basic fact that the charge-coupled devices (CCDs) and complementary metal-oxide semiconductor (CMOS) sensors used in them are sensitive to the near-IR region which comprises the nonthermal part. Thus, a nearby source of IR illumination during the night time or the sun during the day time serves as the primary source of IR radiation which is then reflected in varying degrees by the object being photographed, producing an IR image.

IR filters block all other forms of visible light from reaching the film. Very interesting in-camera effects with dream-like or lurid coloring, called the Wood Effect after the IR photography pioneer Robert W. Wood, appear because of the reflection of IR rays from foliage. The sources of illumination can be incandescent lamps with an IR filter in front of them, LED illuminators, or laser-type illuminators based on laser diodes (Thakur, 2011) (Figure 5.6).

Thermal imaging: The wider part of the IR band comprises thermal IR, which is emitted from almost everything above absolute zero. Thus, it is important in IR imaging. This class of IR imaging includes the following steps (Thakur, 2011):

1. The IR light emanating from objects in the scene is focused by means of a special lens.
2. A phased array of IR detectors scans the light to create a detailed temperature pattern called a thermogram.

FIGURE 5.6 Illumination-based thermal imaging (Thakur, 2011).

3. This thermogram is converted into electrical impulses which are fed into a dedicated signal processing chip to convert the electrical data into a format suitable for viewing purposes.
4. This information is sent to the display unit where it appears as an image.

5.1.3.4 Data Processing of Infrared Thermography (IRT)

With the recent development of advanced excitation technologies, a new research line is gaining ground. This new research deals with data processing algorithms, which are used not only to improve the level of detection of the IR thermography (IRT) technology but also to characterize the detected defects to automate the inspection process (Ibarra-Castanedo et al., 2004). Important data processing techniques used in IRT include statistical moments, principal components analysis, dynamic thermal tomography (DTT), polynomial fit and derivatives, and pulsed phase thermography (PPT) (Usamentiaga et al., 2014).

5.1.3.4.1 Statistical Moments

Data obtained with IRT are a sequence of numerical values. These numerical values can be treated by statistical functions describing certain behaviors to detect significant changes between some values and others. Different statistical moments offer different results (Usamentiaga et al., 2014).

The term moment is used to represent the expected values of the different powers from a random variable (Madruga, Ibarra-Castanedo, Conde, Lopez-Higuera, & Maldague, 2010) and to determine the degree to which data fit a given type of distribution. Skewness is the third standardized statistical moment from a distribution. The mathematical formula used to calculate skewness can be seen in Equation (5.1), where μ is the mean value and σ is the standard deviation of the random variable x. E is the mathematical expectancy, defined as Equation (5.2), where n is the number of data points (Usamentiaga et al., 2014).

$$k_3 = \frac{E\left[(x-\mu)^3\right]}{\sigma^3}, \tag{5.1}$$

$$E[X] = \frac{1}{n-1}\sum_{i=1}^{n} x_i, \tag{5.2}$$

Using skewness, it is possible to measure the asymmetry of the probability distribution from a random variable of real parameters. This method is an appropriate processing technique for IR images; the application of this statistical method is only slightly affected by nonuniform heating or by the shape of the surface of the tested material (Madruga, Ibarra-Castanedo, Conde, Lopez-Higuera, & Maldague, 2008).

Kurtosis is the fourth standardized statistical moment from a distribution. It is generally defined as a measure that reflects the degree to which a distribution has a peak shape (Albendea, Madruga, Cobo, & Lopez-Higueral, 2010). In particular, kurtosis provides information about the height of the distribution in relation to the value of the standard deviation. Mathematically, it is defined as Equation (5.3) (Usamentiaga et al., 2014.):

$$k_4 = \frac{E\left[(x-\mu)^4\right]}{\sigma^4}, \tag{5.3}$$

The temperature distribution of a defect in an image has a kurtosis value that differs from an area without defects, depending on the thermal diffusivity of the defect area. Therefore, it is possible to estimate the kurtosis values for every pixel in the image sequence and to obtain a unique image showing these values: a kurtogram. The kurtogram gives an indication of the location of defects on the subsurface and their thermal diffusion (Madruga, Ibarra-Castanedo, Conde, Maldague, & Lopez-Higuera, 2009).

5.1.3.4.2 Principal Component Analysis

Principal component analysis (PCA) is a statistical technique to synthesize information. Its objectives are, first, to reduce the number of variables in a dataset, losing the least amount of information possible, and, second, to highlight the differences and similarities in data (Usamentiaga et al., 2014).

Smith (2002) describes the steps for applying this method to a set of data, and Cramer and Winfree (2005) show the effectiveness of this method for reducing thermographic data.

Processing based on PCA uses a set of statistic orthogonal functions, known as empirical orthogonal functions (EOFs), to decompose the thermal sequence of the surface temperature variation of a specimen obtained after a pulsed active thermography test of its principal components. In this way, data can be reduced without deleting useful information. These principal components are obtained from the singular value decomposition (SVD) of the thermic temporary data matrices. The method is called principal component thermography (PCT) (Usamentiaga et al., 2014).

DTT means "layering" a test specimen in different layers corresponding to different depths to observe the distribution of the thermal properties. This technique is based on the analysis of the surface temperature evolution after applying an external thermal excitation (Swiderski, 2008).

This algorithm is applied to a thermal image sequence where the evolution of the temperature is observed over time. To utilize this technique, the time evolution of each pixel in the image is fitted with different order polynomials. While the low-order polynomials describe the behavior of the areas without defects, higher-order polynomials describe the variations of the defects. The differential for each pixel is expressed as Equation (5.4) (Usamentiaga et al., 2014):

$$\Delta T(i,j,\tau) = Th(i,j,\tau) - Tl(i,j,\tau), \tag{5.4}$$

The DTT algorithm returns two different images: the maxigram, where the maximum values of ΔT are observed, and the timegram, which indicates the time at which these maximum values take place (Usamentiaga et al., 2014).

5.1.3.4.3 Pulsed Phase Thermography

PPT is based on the phase calculation of a sequence of images, in which the time history of each pixel describes the thermal propagation of an external energy excitation (Maldague, Galmiche, & Ziadi, 2002; Ibarra-Castanedo & Maldague, 2004). PPT transforms the image sequence into the frequency domain using the DFT, as seen in Equation (5.5), where i is the imaginary number, n is the frequency increment, and Re_n and Im_n are the real and imaginary parts of the DFT (Usamentiaga et al., 2014.):

$$F_n = \sum_{k=1}^{N-1} T(k)e^{\frac{2\pi i k n}{N}} = \text{Re}_n + \text{Im}_n, \quad (5.5)$$

Finally, the phase of each pixel is defined as Equation (5.6) (Usamentiaga et al., 2014):

$$\phi_n = a\tan\left(\frac{\text{Im}_n}{\text{Re}_n}\right) \quad (5.6)$$

This technique combines the advantages of modulated and pulse IRT (Maldague & Marinetti, 1996).

5.1.3.5 Applications

Originally developed to augment military capabilities, IR imaging has found use in numerous industrial and medical applications and is a seasoned tool in military, astronomy, and remote sensing. It offers an alternative to security officials who no longer have to frisk suspects manually for dangerous weapons. Meteorology departments have been able to develop eyes that can observe the earth's atmosphere 24-7 without depending on the sun for illumination. In addition, IR related to the temperatures of clouds offers insights into the height and moisture content of the cloud cover and into climatic changes (Thakur, 2011) (Figure 5.7).

The Spitzer Space Telescope launched a few years ago to take deep space images is based on the IR band. Near-IR cameras using active illumination from an IR source have been used in photography and are becoming a standard add-on feature in modern digital cameras. They also comprise the generation-0 of night vision systems. IR photography remains a popular form of photography, as it produces images with surreal colors and details otherwise oblivious to the human eye.

Thermal IR has been the driving technology behind the first-, second-, third-, and fourth-generation night vision devices. It helps us see when there is no source of light, even penetrating layers of fog or smoke. Firefighting departments have endorsed thermal vision gear as indispensable for any firefighter to locate survivors beneath rubble.

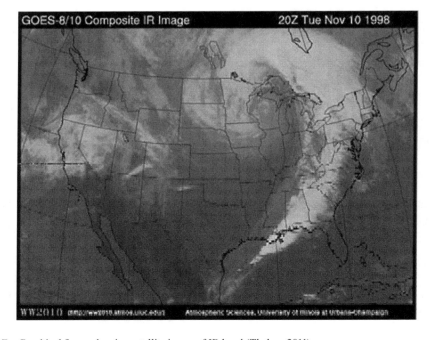

FIGURE 5.7 Graphical figure showing satellite image of IR band (Thakur, 2011).

In the medical field, IR imaging has been successfully deployed for oncology, respiratory problems, vascular disorders, skeletal diseases and tissue viability, cancer detection, etc., under the common name DITI (for digital IR thermal imaging). Forward-looking IR or FLIR helped detect the spread of H1N1 Swine Flu in airports in 2009.

IR imaging is an indispensable tool in industry for nondestructive testing of faults in materials, weld verification, fault detection, etc. A noncontact testing method is more desirable than any contact testing procedure (Thakur, 2011).

5.1.4 Ultraviolet (UV) Imaging

UV imaging has a wide variety of scientific, industrial, and medical applications, for instance, in forensics (Krauss & Warlen, 1985), industrial fault inspection (Chen, Wang, & Yu, 2008), astronomy, skin condition monitoring (Fulton, 1997), and remote sensing (McElhoe & Conner, 1986; Smekens, Burton, & Clarke, 2015). To date, scientific-grade UV cameras, which have elevated quantum efficiencies in this spectral region, have been applied in this context. However, these systems are relatively expensive (typical units cost thousands of dollars) and can be power intensive, since they may incorporate thermoelectric cooling. Although these units may provide high signal-to-noise ratios, a lower price point solution could expedite more widespread implementation of UV imaging (Wilkes et al., 2016).

It is important to distinguish between UV light and UV-fluorescence imaging. Although both use UV lighting, they are entirely different. UV imaging starts by passing the emission of a UV-emitting LED, lamp, or diode or looking at a subject illuminated with UV light that is reflected off the item being inspected. The reflected-UV light is then captured by the camera. The wavelength of the UV light is not converted or shifted in this process.

While UV-fluorescence imaging also requires illuminating a surface with UV light, the fluorescent material absorbs the UV light and electrons are released, causing the material to radiate light at a longer wavelength. The light emitted during this process is usually in the visible range, and in industrial applications, it will usually be blue light. In this type of reaction, light energy in will always exceed light energy out.

UV imaging inspection isn't used often in machine vision. However, as UV-sensitive cameras and UV-emitting light sources, particularly LED lighting, have become widely available and less costly, new applications are emerging. Monochromatic UV sources, such as lasers and LEDs, are desirable in machine vision applications because when paired with appropriate bandpass filters, camera optics don't need to be achromatic, significantly lowering cost.

Images formed with monochromatic illumination are always sharper than images made with broader UV sources, and capability naturally increases as the wavelength used to image the item being inspected is shortened. With UV illumination, smaller features can often be formed and detected more easily and accurately. This is why monochromatic UV (excimer) lasers and optical imaging are used in producing integrated circuits.

The UV band is broad, spanning a wavelength range from 10 (below this are X-ray wavelengths) to 400 nm (above this are visible wavelengths). A system's cameras, optics, filtering, and illumination must be carefully selected according to the UV range being imaged. Otherwise, because of internal camera filtering and the optical lenses being used, most visibly-optimized CCDs and CCD cameras and lens systems will block all of the deep-UV (DUV) and most of the near-UV spectrum.

The near-UV range, between 290 and 400 nm, is most commonly used in industrial imaging applications. This range is typically subdivided into UV-A (320–400 nm) and UV-B (290–320 nm) radiation. Standard optical glasses absorb light and cannot be used for imaging in the region below 290 nm, known as the UV-C or DUV portion of the spectrum. Instead, lenses incorporating fused silica, fused quartz, or calcium fluoride are designed for these applications. Below 180–190 nm, air absorbs UV light. This UV portion is often referred to as the vacuum UV (VUV), since imaging can only take place in a very high vacuum or nitrogen environment (MIDOPT) (Figure 5.8).

Because UV wavelengths are shorter and easily scattered, some of the most common applications for true UV imaging include detecting scratches and digs on polished or highly specular surfaces. When dark field illumination is used to enhance the scattering effect, scratches that aren't apparent in a visible

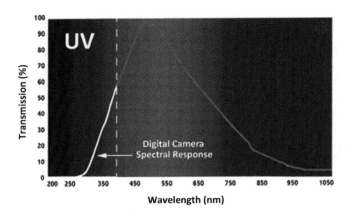

FIGURE 5.8 Example of UV imaging (MIDOPT).

image can become easier to image in UV. UV photolithography processes are used in the production of computer chips. Patterns are optically imaged onto a silicon wafer covered with a film of UV light-sensitive material (photoresist). The photoresist is then further processed to create the actual electronic circuits on the silicon.

Another application involving reflected-UV light is detecting surface contamination. Since UV light tends to be absorbed by organic materials, traces of oil or grease can sometimes be detected on surfaces, particularly in the DUV. Petroleum-based products can also appear differently in UV light, and this can be useful in identifying the nature and source of oil spills. It is sometimes possible under UV illumination to distinguish different paints or finishes if repairs have been made to antiques or other valuable objects (MIDOPT).

5.1.4.1 Ultraviolet (UV) Imaging Sensors

As interest in digital reflected-UV imaging has grown, a small number of camera manufacturers and lens designers have specifically addressed the demands of this market and have produced off-the-shelf UV cameras based on silicon CCDs. UV sensitivity is often overlooked by CCD and CMOS camera manufacturers, who usually publish spectral response curves for their sensors that stop at 400 nm, the edge of human vision and the beginning of the near-UV region of the spectrum. This omission has sometimes made it difficult for scientists and engineers to select a suitable camera for reflected-UV imaging applications. In fact, many commercial visible-light CCD and CMOS cameras have UV-blocking layers incorporated into the optical path to prevent undesirable chromatic aberrations in the image, making them virtually useless for UV imaging.

Most of the emerging breed of UV-specific cameras are based on thinned CCD arrays and are packaged for the machine vision and industrial inspection market. The thinning process removes silicon material that prevents UV radiation from reaching the active layer in the detectors. This thinning process shortens the cut-on wavelength of the sensor to as low as ~200 nm. Some newer CCDs are being built with ITO (indium tin oxide) instead of polysilicon gates. The ITO material is more transparent in the near-UV band and allows shorter wavelengths of light to be detected, as with thinning, but with a lower manufacturing cost.

As mentioned earlier, silicon CCD cameras are more sensitive in the visible and near-IR bands than in the UV band, even with thinning, and the UV imaging system designer must carefully control the spectrum of light that reaches the sensor. Camera filters which pass near-UV light while blocking visible and near-IR are always required, unless the illumination itself is purely UV.

Another method for enhancing a silicon sensor's UV response is a wave shifting coating such as metachrome. These fluorescent materials are applied directly onto the CCD surface during manufacture. Ultraviolet A (UVA) light that would normally be absorbed in the silicon before generating carriers is converted to visible light, which is then easily detected by the CCD.

Data Acquisition and Intelligent Diagnosis

UV image converters and image intensifiers are also commercially available and are especially useful when the UV signal level is very low. These devices convert UV photons into electrons using an electrically charged photocathode on the front of an evacuated image tube. The photoelectrons are converted to visible light via a green phosphor on the tube's anode and can then be directly viewed by the operator. Relay lenses can be used to reimage the viewing screen onto a CCD or CMOS camera for applications demanding a video signal (Richards, 2006a).

5.1.4.2 Ultraviolet (UV) Imaging Applications

As with IR imaging, the applications for reflected-UV imaging are diverse. As more UV cameras become available to commercial customers, the list of practical applications will certainly grow. One way to look at UV imaging is that it is all about absorption. Many common materials (especially those based on organic molecules) strongly absorb near-UV light due to electronic transitions. Changes or modifications to the surface of the material can affect this UV absorption, making the changes easier to detect. In contrast, near-IR imaging applications are often all about transmission. Many materials that are opaque in the visible band are actually quite transparent in the near-IR band. These materials include ink, paint, fabric dye, silicon wafers, thin paper, and plastic. Many practical near-IR applications, thus, require that something be rendered transparent. The near-IR and near-UV bands seem to be complementary in nature in terms of imaging applications.

Some of the most interesting applications for reflected-UV imaging are the following (Richards, 2006a):

- Imaging surface texture not apparent to visible-light imaging.
- Detecting changes in painted or coated surfaces due to variances in UV reflectance.
- Imaging UV lasers, LEDs, and other UV light sources.
- Detecting sun damage, bite marks, and bruises on skin.
- Evaluating the efficacy of sunblock and the uniformity of its application to skin.
- Detecting trace evidence not apparent to fluorescence imaging, IR, or visible light.
- Detecting both natural and manmade white camouflage in snowy conditions.
- Visualizing markings on flowers and butterflies that are only visible in the near-UV band.
- Visualizing repairs, cracks, and damage to teeth.

5.1.4.3 Ultraviolet (UV) Imaging Opens New Applications

Industrial machine vision has traditionally centered on visible-light imaging cameras and visible-light illumination. The simplest machine-vision applications are replacements for human workers, who see in the portion of the electromagnetic spectrum in wavelengths between 400 and 750 nm. A human-replacement machine-vision system might consist of a monochrome video camera combined with a software algorithm to detect the presence or absence of the cap on a tube of toothpaste and the degree to which it has been tightened. For a system like this, the lighting can be provided by simple tungsten lamps.

The machine-vision industry has done very well with this sort of application, using both color and monochrome video cameras that closely match the spectral response of the human eye. Yet there is a great deal of light in the electromagnetic spectrum that the human eye cannot see. This invisible light often carries significant amounts of interesting information (Richards, 2006b).

5.1.4.4 Reflected-UV Imaging

UV imaging has begun to emerge as an inspection modality for some industrial processes. Relative to X-ray or IR machine vision, UV machine vision is still in its infancy, but the field is growing, as commercial UV hardware drops in price and increases in diversity. New applications are emerging, as more users integrate off-the-shelf UV cameras into production environments and experiment with them.

UV light interacts with materials in a unique way, enabling features and characteristics to be observed that are difficult to detect by other methods. UV light tends to be strongly absorbed by many materials, making it possible to visualize the surface topology of an object without the light penetrating into the interior. Because of its short wavelength, it tends to be scattered by surface features that are not apparent at longer wavelengths. Thus, even smaller features can be resolved or detected via UV light scattered off them.

It is important to distinguish between reflected-UV imaging and UV-fluorescence imaging. They are different techniques with different characteristics, but because they both involve UV light, they are often confused. Reflected-UV imaging starts with the illumination of a surface with UV light. The UV light is reflected or scattered and is then imaged by a camera sensitive in the UV band. The wavelength of the UV light is not shifted during the process.

UV-fluorescence imaging also starts with active illumination of a surface with UV light, but the detected signal is in the visible or IR band. The fluorescent material absorbs the UV excitation, then reradiates at a longer wavelength. The emitted fluorescence is not reflected light; it tends to be a diffuse emission.

The UV band is broad, spanning the range of wavelengths, from the start of the X-ray band at 10 nm to the edge of human visual sensitivity at 400 nm. There are two main classes of industrial UV imaging applications, involving two different bands of the UV spectrum, and the cameras, optics, and illumination must be selected accordingly. As discussed above, the band of the spectrum between 300 and 400 nm is commonly known as the near-UV band. It is divided into the UV-A and UV-B sub-bands. Below 300 nm, standard optical glass becomes very absorbing. This region of the spectrum is known as the DUV band, or alternatively, the UV-C band. Machine-vision systems in the DUV band generally operate around 250–280 nm.

One of the most common applications for reflected-UV imaging is the detection of scratches in a surface. The shorter UV wavelengths tend to scatter more strongly off surface features than the visible or near-IR bands. So, for example, scratches not apparent in a visible image may be visible only to a person with excellent eyesight with great difficulty when visible light strikes at a very oblique angle. In contrast, in a UV image taken at 365 nm, the scratches can be seen quite easily (see Figure 5.9) (Richards, 2006b).

As a result, UV imaging enables automated systems to detect scratches and digs on optical surfaces such as lenses or windows. In the semiconductor industry, photolithography requires inspection of photomasks with very fine lines and features to find defects that may be submicron in size. Confocal microscopes operating in the DUV band at 248 or 266 nm (laser wavelengths that can be generated by krypton fluoride and frequency-quadrupled Nd:YAG lasers, respectively) can be used to image these features with much greater clarity than in the visible band and can be used to find tiny defects in the silicon wafer starting material. Detection of these defects early in the production process can greatly improve yields and reduce waste.

Other reflected-UV applications involve the detection of small amounts of surface contamination. Since UV light tends to be absorbed by organic materials, traces of oil or grease are sometimes detectable on many surfaces, particularly in the DUV band (see Figure 5.10). It is also possible to distinguish new paint from old in some situations, even when the two types of painted surfaces look identical in the visible band (see Figure 5.11) (Richards, 2006b).

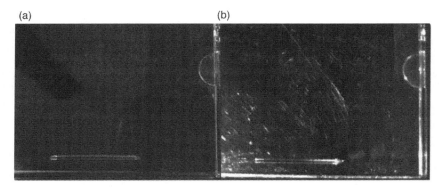

FIGURE 5.9 CD jewel case is imaged in both visible (a) and 365-nm UV lighting (b). Scratches are not apparent in the visible image but are clear in the UV image (Richards, 2006b).

Data Acquisition and Intelligent Diagnosis 173

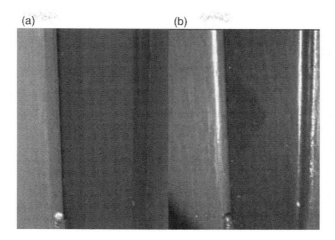

FIGURE 5.10 Images of a metal cabinet show brown painted trim containing an oil stain. The stain shows as a darkened area using 365-nm UV imaging (b) but is not apparent in the visible image (a) (Richards, 2006b).

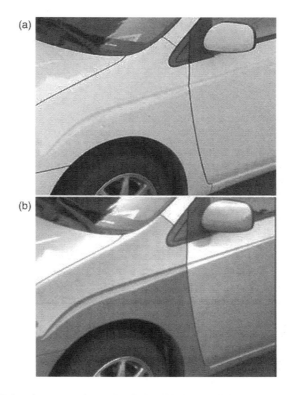

FIGURE 5.11 (a) With UV imaging, new paint can be distinguished from old in some situations, such as a white Toyota Prius that has had the driver's side fender replaced after an accident. (b) The new paint is relatively unoxidized and has a UV-inhibiting clearcoat, and so the fender looks darker in the 320–400-nm UV band relative to the older paint on the rest of the car. UV machine vision can ensure that clear coat is uniformly applied (Richards, 2006b).

5.1.4.5 Reflected-UV Imaging Applications in Forensics

It is well known that UV light has properties that make it a very powerful investigative tool for forensics, particularly because it makes many substances fluoresce. Less well known is the power of reflected-UV imaging to reveal hidden evidence. It does this for several reasons (Richards, 2010):

- Absorption: UV light is highly absorbed by many commonly encountered organic materials, yet is reflected by many inorganic materials like stone and metal. If these organic materials are on a surface with higher UV reflectance, the substances will often stand out more strongly than visible-light or near-IR images. The reverse is true as well—traces of inorganic materials like salt stand out on a dark organic surface like a wooden table.
- Lack of penetration: UV light does not penetrate even very thin layers of materials, making surface topology more apparent, since normally translucent surfaces appear opaque. The high energy of UV photons makes them interact strongly with the electrons in atoms and molecules. Many materials look very dark when imaged with UV light.
- Highly scattered UV waves: UV light waves have a short wavelength, so they are scattered much more readily by small surface imperfections on a smooth surface than either visible or near-IR light. Scratches and dust are much more apparent; therefore, the optics industry uses UV imaging to inspect lens surfaces, for example. Some of the texture imaging can be accomplished by raking-illuminated visible-light photography, though UV has advantages over raking light.

These three properties of the interaction between materials and UV light make reflected-UV imaging very useful for certain applications in forensic imaging. Three primary applications that are well documented in the literature are (Richards, 2010):

- Imaging of bite marks and other pattern injuries on skin.
- Imaging of shoeprints on surfaces where visible-light contrast is low.
- Imaging of latent fingerprints.

The latter application requires an imaging system that works in the shortwave UV band, unless the fingerprints are made while the fingers are coated with a substance that absorbs near-UV light, like sunscreen.

Two other forensic applications that are less well documented are the following (Richards, 2010):

- Imaging traces of certain substances on certain classes of surfaces.
- Imaging changes in surface texture on smooth surfaces caused by physical contact.

Forensic investigators tend to image what they know is already there, because of the difficulties inherent in reflected-UV imaging with film. Thus, these last two applications have historically received very little attention, because the presence of the anomaly may only be apparent in the UV band. Unless the photographer has a means of scanning the scene with a UV imaging scope or video camera, he or she might never know to photograph in a certain area of a crime scene with reflected UV to discover invisible forensic evidence.

In some cases, traces of materials and changes in surface texture can be imaged with raking light illumination or by imaging the surface at a highly oblique angle. This is not always possible due to geometric constraints. In some cases, UV imaging works better than raking light imaging, especially in situations where the surface anomaly is subtle (Richards, 2010).

5.2 Cloud Data Post-Processing Technology

5.2.1 Introduction

Organizations maintain historical data for future analysis. Huge volumes of data accumulate and can be used by organizations to improve performance. Organizations are interested in using historical data to gain valuable insights. The problem is finding the right data in an enormous dataset. Data may come from different databases, different locations, and in different formats. In addition, there can be problems with data quality: data may have inconsistent, blank or null, and noisy values. Therefore, data must be preprocessed before they can be mined.

Data preprocessing includes four stages: cleaning data, integrating data, reducing data, and transforming data.

- Cleaning data: Filling empty values, ignoring noisy data, and correcting inconsistent data.
- Integrating data: Merging the data from multiple data sources.
- Reducing data: Reducing the dataset while maintaining the integrity of the original.
- Transforming data: Transforming or consolidating data into forms appropriate for mining.

Data preprocessing is challenging, as it involves extensive manual effort and time. There are a number of different tools and methods for preprocessing: sampling selects a representative subset from a large population of data; transformation manipulates raw data to produce a single input; denoising removes noise from data; normalization organizes data for more efficient access; and feature extraction pulls out specified data that are significant in a particular context. Preprocessing techniques are also useful for association rule mining algorithms like Apriori, Partitioned, Pincer-Search algorithms and many more.

Data mining: Once data are preprocessed, they can be mined. Simply stated, data mining is a collection of techniques for efficient automated discovery of previously unknown, valid, novel, useful, and understandable patterns in large databases. The patterns must be actionable so that they may be used in an enterprise's decision-making process. Data mining is often complex and may require many steps before useful results are obtained.

Data mining can be directed or undirected. Directed data mining attempts to explain or categorize a particular target field such as income or response. Undirected data mining attempts to find patterns or similarities among groups of records without the use of a particular target field or collection of predefined classes.

Once data are mined, they can be post-processed to generate insights.

Data post-processing includes two stages: visualization and summation.

- Data visualization: Communicating data clearly and effectively through the graphical representation.
- Data summation: Collecting patterns to summarize data.

After data are extracted, it is important to visualize the extracted knowledge and summarize it in such a form that users can gain insight for better decision-making. More information on post-processing is given in the next section.

5.2.2 Post-Processing Techniques

As noted in the previous section, data post-processing methods are divided into two categories (Tamilselvi, Sivasakthi, & Kavitha, 2015):

- Data visualization
- Data summation.

5.2.2.1 Data Visualization

After data are extracted, it is important to visualize the extracted knowledge in such a form that the user can gain insight into data for better decision-making. Various techniques are available for information visualization.

Scatter plot matrix technique. Scatter plots are organized in matrix form and use Cartesian coordinates to plot data points. The relationship between two variables, also known as correlation, is represented by the scatter plots. The correlation between two variables may be positive or negative. If the data points are distributed uniformly in the scatter plot, the correlation between the variables is low or zero. High correlation may be positive or negative depending on the relationship between variables. If the value of one variable increases with the increment of the value of another variable, and if data points are represented by a straight line, the correlation is said to be a highly positive one. If the value of one variable decreases with the increment of the value of another variable, and data points make a straight line, the correlation is called a highly negative one (Tamilselvi, Sivasakthi, & Kavitha, 2015) (Figure 5.12).

Parallel coordinates: This technique maps a multidimensional point onto a number of parallel axes. Initially in this method, coordinates start mapping with one axis; more axes may gradually be lined up as per requirements. A line is used to connect the individual coordinate mappings. This method may be extended to n-dimensions, but there is a practical limit which depends on the screen display area (Figure 5.13).

Data visualization technologies are used in steering computation, in aiding directed analysis, in query interfaces to complex multimedia databases, and in information presentation and navigation (Tamilselvi, Sivasakthi, & Kavitha, 2015).

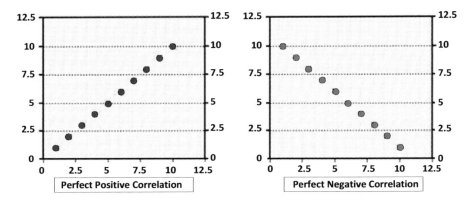

FIGURE 5.12 Perfect positive and negative correlation between two variables (Tamilselvi, Sivasakthi, & Kavitha, 2015).

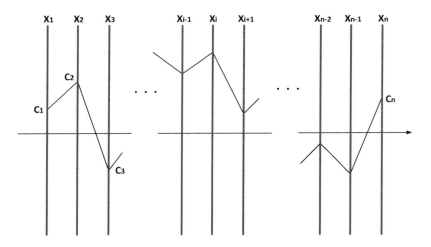

FIGURE 5.13 Parallel axes (Tamilselvi, Sivasakthi, & Kavitha, 2015).

5.2.2.2 Data Summation

Knowledge discovery in both structured and unstructured datasets stored in large repository database systems requires the data to be summarized. Data summation methods in the unstructured domain usually involve text. In summation, documents that share similar characteristics are grouped together. With the ever-growing number of text documents in large database systems, algorithms for text summation in the unstructured domain, such as document clustering, are often limited by the dimensionality of the data features. Summation is closely related to compression, machine learning, and data mining (Tamilselvi, Sivasakthi, & Kavitha, 2015).

5.2.3 Post-GPS Data Processing

One way to filter GPS data in post-processing is to use different curve-fitting techniques, such as moving average smoothing, digital smoothing, polynomial filter, and local regression smoothing. The use of span parameters to construct curves with respect to nearby points makes it possible to include previous and subsequent points for a better fit. Big players in the market of GPS devices use this fitting technique (e.g., uTrack from Garmin).

Examples include various types of machinery, as the smoothed GPS path might be used for further visualization and the creation of a less noisy machinery movement path. For example, pavers move along the road in one direction without major turns (ideally) and maintain a constant and preferred speed. Speed deviations are more likely than directional deviations. Therefore, their movements can be easily described, and simple smoothing techniques might be enough to filter out outliers and keep the machinery path consistent in case of GPS signal loss.

Compared to the pavers, roller movements are problematic—even in post-processing. The roller moves at higher speed and has more freedom in space, time, direction, and even configuration (a tandem roller has two articulation points). Simple smoothing techniques in this case might lead to the loss of valuable data, for example, the point where the curving or reverse movement starts and ends. Therefore, we should control the difference between the original coordinates and the smoothed coordinates, so we do not lose valuable data (Bijleveld, Vasenev, Hartmann, & Dorée, 2014).

5.2.3.1 Post-Processing

After GPS field data are collected, they can be transferred to a PC for post-processing. All collected GPS data need some form of post-processing whether differential correction or editing. After they are transferred, the data should be viewed on the PC for verification. While the data file is onscreen, unwanted positions and/or features, as well as attribute information, can be identified, and preliminary editing can begin. Depending on the GPS unit and data collection software, this step may or may not include converting the data to a GIS format at the time of download. Either way, it is important to have clear procedures for post-field data management and verification (National Park Service, 2003).

Suggested steps in post-processing GPS data are the following (National Park Service, 2003):

1. Transfer GPS files which may include GPS data files, base files, etc., into a single file.
2. Run differential correction.
3. Combine data files.
4. Verify data onscreen.
5. Perform preliminary editing.

    Editing tools are used to remove unwanted mistakes during collection, positions collected during uncooperative satellite times, and intersecting vertices between area or line features. Mistakes or errors not removed in this step can be edited during the finalizing GIS data step after the GPS data are exported to a GIS format. GIS software packages usually provide more editing capability and flexibility.

Editing suggestions:
- Points require almost no editing but may require close inspection. Does the point line up with other features in the GIS?
- Lines and polygons may require extensive editing. Due to errors in GPS signals, or obstructions like canopies or large buildings, there may be cases where a road feature appears like a zigzag, or the polygon feature collapses upon itself. Delete positions that deviate from the trend of a straight line, remove positions that cross over each other at intersections, and delete a series of positions collected by standing too long at the beginning of a trial or failing to pause during collection.
- Attribute validation can be done during editing to enforce attribute structure and ensure attribute values are filled in appropriately.

6. Clear out the GPS unit for the next job or person. Do not leave data files on the GPS unit if they have been properly downloaded and verified.
7. Charge or refresh internal or external batteries (National Park Service, 2003).

5.2.3.2 Watch Outs!

This section describes some of the things to watch out for when post-processing GPS data.

- File management: A GPS field project may generate tens or hundreds of files over a course of a field season. Take time to set up a coherent file management scheme that organizes data transferred from the GPS to the PC and a logical sequence of file naming conventions that enable file changes to be tracked over time.
- File transfer troubleshooting: One of the most common problems when transferring data is the transfer flow rate through the computer's RS-2332 Serial port (COM 1 or COM 2). If a program like ActiveSync is holding onto the COM port, the GPS unit will not be able to transfer. Steps to free up a COM port include:
  - Close Active Sync if open and try port again.
  - Access Control Panel | Ports and view the COM settings. If settings say "in use" or "reserved," try removing the COM 1 port using Control Panel | Remove Hardware and specify the COM 1 port. Once removed, putting the computer on reboot will re-establish contact with the COM port and shake lose any remnant program holding onto it.
- Timely action: The longer data sit unedited, the less accurate the editing, and the more suspect the information.
- Combining files: For multiple files to be combined, the data's file structure must be identical. This translates into using the same data dictionary file for every GPS unit and every data file collected.
- Reference features: By collecting point features like anchors or beginning and end points, zigzagging line features can be edited more easily. Since points are always more accurate than an instantaneous position along a line, a line can be straightened and a polygon closed more accurately using the reference point feature as a guide or snapping feature in the GIS (National Park Service, 2003).

5.2.4 Knowledge Discovery in Databases (KDD)

Knowledge discovery in database (KDD) refers to advanced data management processes that have become very attractive for both researchers and practitioners. The steps in KDD are the following: domain understanding, data collection, data preprocessing, data mining, and post-processing of the derived knowledge.

The data that are to be processed by a knowledge acquisition algorithm are usually noisy and incomplete and often inconsistent. Therefore, many steps must be performed before data analysis itself, and preprocessing procedures are continuously developing. Moreover, a result from a machine learning algorithm, such as a decision tree, a set of decision rules, or an artificial neural network, may not be appropriate from the viewpoint of customers or commercial applications. As a result, a conceptual description (model, knowledge base) produced by such an inductive process usually has to be post-processed. Post-processing procedures usually include varied pruning routines, rule filtering, or even knowledge integration (Bruha & Famili, 2000).

Zhang, Zhang, and Yang (2003) argue for the importance of data preparation for three reasons: real-world data are impure; high-performance mining systems require quality data; and quality data yield high-quality patterns. Real data can disguise useful patterns if data are incomplete, noisy, and inconsistent. Data preparation generates new datasets that are smaller than the original ones by selecting relevant data, recovering incomplete data, purifying data, reducing data, and resolving data conflicts (Diaz, Herrera, Izquierdo, & Pérez-Garcia, 2010).

5.2.4.1 Post-Processing

KDD post-processing components can be categorized into the following four groups (Bruha & Famili, 2000): knowledge filtering, interpretation and explanation, evaluation, and knowledge integration. In the case of machine learning algorithms, such as trees or decision rules trained with noisy data, the results cover few training data because the induction algorithms try to subdivide the training dataset. To overcome this problem, the decision trees or rules should be shrunk, by either post-pruning (decision trees) or truncation (decision rules). New knowledge can be implemented in an expert system or used by an end user. In the latter case, the knowledge results should be documented for the end user's interpretation. Another possibility is to display the knowledge and transform it into a form understandable to the end user. We can check the new knowledge for potential conflicts with previously induced knowledge, summarize the rules, and combine them with domain-specific knowledge provided for the given task. Once a learning system has induced concept hypotheses (models) from the training set, these hypotheses should be evaluated (or tested) for classification accuracy, understanding, computational complexity, and so on (Diaz, Herrera, Izquierdo, & Pérez-Garcia, 2010).

5.3 Cloud Data Intelligent Diagnosis

Data are valuable assets of today's businesses, and timely and accurate analysis of available data is essential to make the right decisions and compete in today's ever-changing business environment.

Most businesses face an analytics challenge: they build ever-growing data warehouses but lack analytical competence and resources. Data owners typically are domain experts who understand the processes generating the data and what the data represent. However, they typically are not expert analysts, so they struggle with the application of advanced analytics. Passing data on to an analyst results in a communication challenge that requires the domain expert to explain the data context and generate a problem statement that the analyst can use as the basis for analyzing the data. After analysis, the analyst has to present the results in such a way that the data owner can relate them to his or her context and derive valuable information for future decision-making.

The next challenge is application; IT support staff must turn analytical results into software that can be integrated into operational systems. With the introduction of data analysis functionality in databases and a standardized language for model descriptions like PMML (predictive model markup language) defined by the Data Mining Group, integration may become simpler in the future. Under the assumption that the analysis tool is able to create a PMML description for the model in question, and the database implements the underlying analysis algorithm, the PMML description can simply be included in a database script (e.g., PL/SQL for Oracle databases) that will be used to analyze data in the operational system. However, it will take many years before data analysis is standard in databases, and a large variety of models can be transferred in that way.

Commercial data analysis software aimed at a business context is either too simplistic because the manufacturer has decided to provide only limited functionality that nonexpert users can handle or too complex and provides advanced functionality aimed directly at expert analysts. To overcome both the analytics challenge and the communication challenge, software tools must empower domain experts and data owners to run advanced analytics themselves with as little help from analysts as possible. One approach is to hide complex analytics under a layer of automation that provides an interface allowing users to work goal-oriented instead of method-oriented (Wang, 2008).

5.3.1 What Is Data Intelligence?

Data intelligence refers to all the analytical tools and methods companies employ to form a better understanding of the information they collect to improve their services or investments. Data intelligence focuses on analysis and interaction with information in a meaningful way to promote better decision-making in the future.

When collecting data for intelligence purposes, businesses and organizations use a variety of sources, including business performance metrics, data mining of consumers and users, and other descriptive sources.

Unlike business intelligence, which focuses on organizing data and presenting it in a way that makes it easier to understand and to derive business intelligence insights, data intelligence is more concerned with the analysis of information itself. To achieve this, data intelligence experts divide data into five major types: descriptive, prescriptive, diagnostic, decisive, and predictive. They focus on understanding data, uncovering alternative explanations, resolving issues, and identifying future trends to improve decisions.

Today, data intelligence incorporates both artificial intelligence (AI) and machine learning tools, thus permitting organizations to analyze enormous amounts of data much faster and more reliably than if done manually. Moreover, data intelligence helps expedite analytics by arranging data neatly and establishing clearer models for warehousing and scrubbing large data sets (SISENSE, 2019).

Specific criteria for what makes data intelligent include the following (Hallen, 2012):

1. Data are clear and unambiguous: Data values can be defined and measured in a repeatable fashion.
2. Data are concise: The data represent the smallest number of data points that would lead to the same action. If we need 90% certainty to take action, it's the amount of data that will safely give us that.
3. Data are directly linked to action: Based on different values of those data, different decisions will be made and implemented.

In short, intelligent data are a direct input to analysis, very specifically, to the right analysis needed to decide between decision A or B (Hallen, 2012).

5.3.1.1 Application of Data Intelligence

Much like business intelligence, data intelligence is a vital part of any organization's efforts to improve its services and create forward-looking strategies.

One of the most common uses of data intelligence is to understand consumer preferences. By using data mining techniques to collect information on habits, shopping preferences, online trends, and other individual information, companies can tailor their services better and understand tendencies across their target demographics.

Another useful deployment of data intelligence is understanding a company's investments and their effectiveness. By collecting data that provide a greater context and using the descriptive analytics and prescriptive analytics tools in their data analytics software, companies can determine whether the dollars they spend are providing returns or would be better allocated elsewhere.

For organizations whose services are performing adequately but not improving or experiencing growth, data intelligence can provide an idea of areas where services may be optimized and possibly reveal different approaches that may be more effective over the long run (SISENSE, 2019).

5.3.2 Artificial Intelligence (AI) Technique in Diagnostics

AI can be defined as the ability of computer software and hardware to do those things that we, as humans, recognize as intelligent behavior. Traditionally those things include such activities as:

1. Searching: Finding the best solution in a large search space.
2. Recognizing patterns: Finding items with similar characteristics or identifying an entity when not all its characteristics are stated or available.
3. Making logical inferences: Drawing conclusions based upon a given hypothesis.

Approaches to diagnostic reasoning include: rule-based systems, model-based diagnostics, learning systems, case-based reasoning systems, and probabilistic reasoning systems. The first three use AI. This section explains how these AI-based techniques are used in system diagnosis (Poongodai & Bhuvaneswari, 2013).

5.3.2.1 System Diagnosis

System diagnosis is the process of inferring the cause of any abnormal or unexpected behavior. During system operation, indications of correct or incorrect functioning of a system may be available. In complex applications, the symptoms of incorrect (or correct) behavior may be observed directly or may need to be inferred from other variables that are observable during system operation. Monitoring is a term that denotes observing system behavior. The capability for monitoring a system is a key prerequisite for diagnosing problems in it.

Diagnostic applications make use of system information from the design phase, such as safety and mission assurance analysis, failure modes and effects analysis, hazards analysis, functional models, fault propagation models, and testability analysis.

In any complex system, given the complexity of the data and the variety of failure modes, a diagnostic system based on AI can be designed. A comprehensive analysis of the data can extract a set of uncorrelated features to detect various fault modes and specify the actionable knowledge to be applied to solve the problems.

The system will take sensor values as input and ideally perform the following:

- Fault detection: Detecting that something is wrong.
- Fault isolation: Determining the location of the fault.
- Fault identification: Determining what is wrong, i.e., determining the fault mode.
- Fault prognostics: Determining when a failure will occur based conditionally on anticipated future usage (Poongodai & Bhuvaneswari, 2013).

5.3.2.2 AI Approach

As mentioned above, three AI approaches to diagnostics and prognostics are the following (Poongodai & Bhuvaneswari, 2013):

1. Rule-based approach.
2. Model-based approach.
3. Data-driven approach.

The following subsections discuss each of these in more detail.

5.3.2.2.1 Rule-Based Approach

Rule-based expert systems have wide application for diagnostic tasks. Procedures can be broken down into multiple steps and encoded into "rules." A rule describes the action(s) that should be taken if a symptom is observed. A set of rules can be incorporated into a rule-based expert system and then used to generate diagnostic solutions. When this approach is selected, expertise and experience are available, but deep understanding of the physical properties of the system is either unavailable or too costly to obtain.

Two primary reasoning methods may be employed to generate the diagnosis results (Poongodai & Bhuvaneswari, 2013):

1. Forward chaining.
2. Backward chaining.

In forward chaining, the process examines rules to see which ones match the observed evidence. If only one rule matches, the process is simple. However, if more than one rule matches, a conflict set is established and is examined using a predefined strategy that assigns priority to the applicable rules. Rules with higher priority are applied first to obtain diagnostic conclusions.

If the starting point is a conclusion, a backward chaining algorithm collects or verifies evidence that supports the hypothesis. If the supporting evidence is verified, the hypothesis is reported as the diagnostic result (Poongodai & Bhuvaneswari, 2013).

The advantages of rule-based systems include (Poongodai & Bhuvaneswari, 2013):

- Increased availability and the reusability of expertise but at high cost.
- Increased safety, if the expertise must be used in hazardous environments.
- Increased reliability for decision-making.
- Fast and steady responses.
- Consistent performance.
- Built-in explanation facility.

The challenges are (Poongodai & Bhuvaneswari, 2013):

1. Determining the order in which the rules should be matched.
2. Determining the completeness, consistency, and correctness of derived rules.

5.3.2.2.2 Model-Based Techniques

Model-based reasoning systems refer to inference methods used in expert systems based on models of the physical world. Models might be quantitative, i.e., based on mathematical equations, or qualitative, i.e., based on cause/effect models. They may include representation of uncertainty, behavior over time, normal behavior, or abnormal behavior.

The main focus of application development is developing the model. At run time, an engine combines this model and observed data to derive conclusions, such as a diagnosis or a prediction.

Model-based algorithms encode human knowledge via a (more or less) hand-coded representation of the system (Poongodai & Bhuvaneswari, 2013). Hand-coded models use qualitative, rather than numerical, variables to describe the physics of the system. Model-based AI techniques include (Poongodai & Bhuvaneswari, 2013):

1. Rule-based expert systems
2. Finite-state machines
3. Qualitative reasoning.

5.3.2.2.3 Data-Driven Approaches

Data-driven approaches are derived directly from routinely monitored system operating data. They rely on the assumption that the statistical characteristics of the data are stable, unless a malfunctioning event occurs in the system.

Data-driven approaches automatically fit a model of system behavior to historical data, rather than hand-coding a model. They can use "conventional" numerical algorithms, such as linear regression or Kalman filters, or they can use algorithms from the machine learning and data mining AI communities, such as neural networks, decision trees, and support vector machines (Poongodai & Bhuvaneswari, 2013).

Data-driven techniques have the ability to (Poongodai & Bhuvaneswari, 2013):

- Transform high-dimensional noisy data into lower-dimensional information for detection and diagnostic decisions
- Handle highly collinear data of high dimensionality
- Substantially reduce the dimensionality of the monitoring problem
- Compress the data for archiving purposes
- Provide monitoring methods
- Facilitate model building via identification of dynamic relationships among data elements.

The main drawback of data-driven approaches is that their efficacy is highly dependent on the quantity and quality of system operational data (Poongodai & Bhuvaneswari, 2013).

The engineering processes needed to relate system malfunctioning events using a data-driven diagnosis approach typically involve the following steps (Poongodai & Bhuvaneswari, 2013):

- Determination of high-impact malfunctions
- Data selection, transformation, denoising, and preparation
- Data processing
- Testing and validation
- Fusion.

REFERENCES

Albendea P., Madruga F., Cobo A., López-Higueral J., 2010. Signal to Noise Ratio (SNR) comparison for pulsed thermographic data processing methods applied to welding defect detection. In *Proceedings of the QIRT 2010*, Quebec, Canada, 27–30 July 2010; Vol. 20, pp. 1–8.

Bijleveld F. R., Vasenev A., Hartmann T., Dorée A. G., 2014. *Real-Time and Post Processing of GPS Data in the Field of Visualizing Asphalt Paving Operations*. the Netherlands: University of Twente, Department of Construction Engineering & Management.

Bruha I., Famili A., 2000. Post-processing in machine learning and data mining. *SIGKDD Explorations Newsletter*, 2(2), 110–114. doi:10.1145/380995.381059.

Chen Z., Wang P., Yu B., 2008. Research of UV detection system based on embedded computer. In *World Automation Congress 2008*, Hawaii, HI, USA. IEEE.

Cramer K. E., Winfree W. P., 2005. The application of principal component analysis using fixed eigenvectors to the infrared thermographic inspection of the space shuttle thermal protection system. In *Proceedings of the Quantitative Infrared Thermography Conference (QIRT)*, Neuquen, 2–4 November 2005, pp. 28–30.

Díaz J. L., Herrera M., Izquierdo J., Pérez-García R., 2010. The tasks of pre and post-processing in Data Mining applied to a real world problem. In *International Environmental Modelling and Software Society (iEMSs) 2010 International Congress on Environmental Modelling and Software Modelling for Environment's Sake, Fifth Biennial Meeting*, Ottawa, Canada.

Fulton J. E., 1997. Utilizing the ultraviolet (UV detect) camera to enhance the appearance of photodamage and other skin conditions. *Dermatologic Surgery*, 23, 163–169.

Gupta A., Jain R., 1997. Visual information retrieval. *Communications of the ACM*, 40(5), 70–79.

Hallen E., 2012. Big data vs intelligent data (and what Startups can do with it). September 5, 2012. www.klaviyo.com/blog/big-data-vs-intelligent-data-startups. Viewed: May 20, 2019.

Hampapur A., 1995. Designing video data management systems. *Ph.D. Dissertation*, The University of Michigan, Ann Arbor, MI, USA, 1995.

Ibarra-Castanedo C., Gonzalez D., Klein M., Pilla M., Vallerand S., Maldague X., 2004. Infrared image processing and data analysis. *Infrared Physics Technology*, 46, 75–83.

Ibarra-Castanedo C., Maldague X., 2004. Pulsed phase thermography reviewed. *Quantitative InfraRed Thermography Journal*, 1, 47–70.

Janos Technology, 2019. What is IR imaging. Copyright© 2019 Janos Tech. All Rights Reserved. www.janostech.com/knowledge-center/optical-reference-guide/what-is-ir-imaging.html. Viewed: May 14, 2019.

Krauss T., Warlen S., 1985. The forensic science use of reflective ultraviolet photography BT—The forensic science use of reflective ultraviolet photography. *Journal of Forensic Science*, 30, 262–268.

Lovas T., 2010. Data acquisition and integration 4. Laser scanning. Copyright© 2010 University of West Hungary Faculty of Geoinformatics.

Madruga F., Ibarra-Castanedo C., Conde O., Lopez-Higuera J., Maldague X., 2008. Automatic data processing based on the skewness statistic parameter for subsurface defect detection by active infrared thermography. In *Proceedings of the QIRT 9–Quantitative Infrared Thermography*, Krakow, Poland, 2–5 July 2008, pp. 2–5.

Madruga F. J., Ibarra-Castanedo C., Conde O. M., Maldague X. P., Lopez-Higuera J. M., 2009. Enhanced contrast detection of subsurface defects by pulsed infrared thermography based on the fourth order statistic moment, kurtosis. In *Proceedings of the SPIE Defense, Security, and Sensing. International Society for Optics and Photonics*, Orlando, FL, USA, 13 April 2009, pp. 72990U–72997U.

Madruga F. J., Ibarra-Castanedo C., Conde O. M., Lopez-Higuera J. M., Maldague X., 2010. Infrared thermography processing based on higher-order statistics. *NDT & E International*, 43, 661–666.

Maldague X., Galmiche F., Ziadi A., 2002. Advances in pulsed phase thermography. Infrared Physics Technology, 43, 175–181.

Maldague X., Marinetti S., 1996. Pulse phase infrared thermography. *Journal of Applied Physics*, 79, 2694–2698.

Mary R., 2011. Introduction to image processing. Copyright© 2019 EngineersGarage. www.engineersgarage.com/articles/image-processing-tutorial-applications. Viewed: May 17, 2019.

McElhoe H. B., Conner W. D., 1986. Remote measurement of sulfur dioxide emissions using an ultraviolet light sensitive video system. *Journal of the Air Pollution Control Association*, 36, 42–47.

MIDOPT. Midwest Optical Systems, Inc., Palatine, IL, USA. https://midopt.com/solutions/monochrome-imaging/ultraviolet-imaging/. Viewed: May 17, 2019.

National Park Service (NPS). 2003. Post-processing. www.nps.gov/gis/gps/gps4gis/postprocess.html. Viewed: May 20, 2019.

Pfeifer N., Briese C., 2007. *Laser Scanning – Principles and Applications*. Austria: Vienna University of Technology. Institute of Photogrammetry and Remote Sensing. doi:681.7.055:621.375.826.

Poongodai A., Bhuvaneswari S., 2013. AI technique in diagnostics and prognostics. International Journal of Computer Applications (0975–8887). *2nd National Conference on Future Computing*, Puducherry, India, February 2013.

Richards A., 2006a. Digital reflected-ultraviolet imaging. April 2006.

Richards A., 2006b. UV imaging opens new applications. Chief technology officer at Oculus Photonics, Goleta, CA, USA; July 1, 2006. www.vision-systems.com/non-factory/life-sciences/article/16736672/uv-imaging-opens-new-applications. Viewed: May 17, 2019.

Richards A., 2010. Reflected ultraviolet imaging for forensics applications. Adjunct Professor, Brooks Institute of Photography. Partner, Oculus Photonics, Santa Barbara, CA, USA. Senior Research Scientist, FLIR Commercial Systems. March 28, 2010.

SISENSE, 2019. Data intelligence. Copyright© 2019 Sisense Inc. www.sisense.com/glossary/data-intelligence/. Viewed: May 20, 2019.

Smekens J. F., Burton M. R., Clarke A. B., 2015. Validation of the SO_2 camera for high temporal and spatial resolution monitoring of SO_2 emissions. *Journal of Volcanology and Geothermal Research*, 300, 37–47.

Smith L. I., 2002. A tutorial on principal components analysis. Cornell University, New York, USA, 51, 52.

Swiderski W., 2008. The characterization of defects in multi-layered composite materials by thermal tomography methods. In *Proceedings of the Tenth Annual Conference of the Materials Research Society of Serbia*, Herceg Novi, Montenegro, September 2008; Vol. 115, pp. 800–804.

Usamentiaga R., Venegas P., Guerediaga J., Vega L., Molleda J., Bulnes F. G., 2014. Infrared thermography for temperature measurement and non-destructive testing. *Sensors*, 14, 12305–12348. doi:10.3390/s140712305.

Tamilselvi R., Sivasakthi B., Kavitha R., 2015. An efficient preprocessing and post processing techniques in data mining. *International Journal of Research in Computer Applications and Robotics*, 3(4), 80–85.

Thakur A., 2011. Infrared imaging or thermal imaging. Copyright© 2019 EngineersGarage. All rights reserved. www.engineersgarage.com/articles/infrared-ir-imaging. Viewed: May 14, 2019.

Wang H. F., 2008. *Intelligent Data Analysis: Developing New Methodologies Through Pattern Discovery and Recovery.*, Taiwan, ROC: National Tsing Hua University.

Wilkes T. C., McGonigle A. J. S., Pering T. D., Taggart A. J., White B. S., Bryant R. G., Willmott J. R., 2016. Ultraviolet imaging with low cost smartphone sensors: Development and application of a raspberry Pi-based UV camera. Sensors (Basel), 16(10), 1649.

Zhang S., Zhang C., Yang Q., 2003. Data preparation for data mining. *Applied Artificial Intelligence*, 17, 375–381.

6
Three-Dimensional Visualization

6.1 Overview
6.1.1 Visualization
6.1.1.1 Introduction

The progress made in hardware technology allows today's computer systems to store very large amounts of data. Researchers from the University of Berkeley estimate that every year about 1.5 exabytes (= 1.5 million terabytes) of data are generated, of which a large portion is available in digital form. It is possible that in the next three years, more data will be generated than in all of human history to date.

The data are often automatically recorded via sensors and monitoring systems. Even simple transactions of everyday life, such as paying by credit card or using the telephone, are typically recorded by computers. Many variables are usually recorded, resulting in data with a high dimensionality. The data are collected because people believe they are a potential source of valuable information, providing new insights or a competitive advantage (at some point). Finding valuable information hidden in the data, however, is a difficult task. With today's data management systems, it is only possible to examine quite small portions of the data. If the data are presented textually, the amount of data that can be displayed is in the range of some 100 data items, but this is a drop in the ocean when dealing with datasets containing millions of data items. If there is no possibility to adequately explore the large amounts of data that have been collected because of their potential usefulness, the data become useless and the databases become "data dumps."

Information visualization and visual data analysis can help to deal with the flood of information. The advantage of visual data exploration is that the user is directly involved in the data analysis process. A large number of information visualization techniques have been developed over the last two decades to support the exploration of large datasets. In this chapter, we give an overview of information visualization and visual exploration using a classification based on the relations of the data type to the visualized, the visualization technique, and the interaction technique (Berthold & Hand, 2002).

6.1.1.2 Benefits of Visual Data Exploration

For data analysis to be effective, it is important to include the human in the data exploration process and combine the flexibility, creativity, and general knowledge of the human with the enormous storage capacity and the computational power of today's computers. Visual data mining aims at integrating the human in to the data analysis process, applying human perceptual abilities to the analysis of large datasets available in today's computer systems.

The basic idea of visual data mining is to present the data in some visual form, allowing the user to gain insight into the data, draw conclusions, and directly interact with the data. Visual data analysis techniques have proven to be of high value in exploratory data analysis. Visual data mining is especially useful when little is known about the data, and the exploration goals are vague. Since the user is directly involved in the exploration process, shifting and adjusting the exploration goals can be done in a continuous fashion as needed.

Visual data exploration can be seen as a hypothesis generation process; the visualizations of the data allow the user to gain insight into the data and come up with new hypotheses. The verification of the hypotheses can be done via data visualization but may also be accomplished by automatic techniques from statistics, pattern recognition, or machine learning. In addition to the direct involvement of the user, the main advantages of visual data exploration over automatic data analysis techniques are the following:

- Visual data exploration can easily deal with highly nonhomogeneous and noisy data.
- Visual data exploration is intuitive and requires no understanding of complex mathematical or statistical algorithms or parameters.
- Visualization can provide a qualitative overview of the data, allowing data phenomena to be isolated for further quantitative analysis.

As a result, visual data exploration usually allows faster data exploration and often provides more interesting results, especially in cases where automatic algorithms fail. In addition, visual data exploration techniques provide a much higher degree of confidence in the findings of the exploration. These facts lead to a high demand for visual exploration techniques and make them indispensable in conjunction with automatic exploration techniques (Berthold & Hand, 2002).

6.1.1.3 Visual Exploration Paradigm

Visual data exploration usually follows a three-step process: overview, zoom and filter, and details-on-demand (also called the information-seeking mantra; see Shneiderman, 1996). First, the user needs to get an overview of the data. In the overview, the user identifies interesting patterns or groups in the data and focuses on one or more of them (zoom and filter). Then, to analyze the patterns, the user must drill down and access details of the data.

Visualization technology may be used for all the three steps of the data exploration process. Visualization techniques are useful for showing an overview of the data, allowing the user to identify interesting subsets. In this step, it is important to keep the overview visualization while focusing on the subset using another visualization technique. An alternative is to distort the overview visualization in order to focus on the interesting subsets. This can be performed by dedicating a larger percentage of the display to the interesting subsets while decreasing screen utilization for uninteresting data. To further explore the interesting subsets, the user needs a drill-down capability to observe the details of the data. Note that visualization technology provides the base visualization techniques for all the three steps and also bridges the gaps between them (Keim & Ward, 2002).

6.1.2 Three-Dimensional Visualization

Briefly stated, three-dimensional or 3D approaches try to create visualizations that are closer to real-world metaphors or to improve space usage by adding an extra dimension. The user is able to rotate and move 3D objects and navigate inside a 3D world. The use of 3D software visualization has the potential to aid in the development process. Three-dimensional software visualization may transform the way that knowledge gathering activities take place during software engineering phases.

Some approaches propose using a 2D layout seen in a 3D perspective with interaction limited to 2D, i.e., a 2.5D approach (Teyseyre & Campo, 2008).

In fact, there is an ongoing debate on 2D vs. 3D in the information visualization area. To identify and analyze the strengths and weaknesses of 3D/2D, we first review a categorization of 3D visualizations (Stasko & Wehrli, 1993):

1. Augmented 2D views: These are typical 2D visualizations where the third dimension is added for aesthetic purposes. For example, Figure 6.1a shows a 3D presentation of a traditional 2D bar chart sorting algorithm (Carson, Parberry, & Jensen, 2007).
2. Adapted 2D views: These are 2D visualizations extended to 3D to encode additional information. For example, Figure 6.1b shows a 3D visual representation of a software release history that displays the structure of the system in 2D planes and uses the third dimension to display historical information (Gall, Jazayeri, & Riva, 1999).

3. Inherent 3D application domain views: This category includes computations involving inherent 3D entities. For instance, Figure 6.1c represents a software system and its relationships using a metaball metaphor, i.e., a 3D modeling technique commonly used to represent complex organic shapes and structural relationships in biology and chemistry (Rilling & Mudur, 2002).

In general, the use of the third dimension in category 3 is not discussed in the literature. However, recent research in specific domains shows that 2D and 3D presentations are useful for different task types, and, hence, combined 2D/3D displays are suggested. Yet the question of the benefits offered by 3D over 2D remains salient in the other categories.

Several authors state that when two dimensions are enough to show information, it is not desirable to add a third. This extra dimension should be only used to visualize a dataset semantically richer. However, other authors think 3D presentations facilitate perception of the human visual system. They believe the inclusion of aesthetically appealing elements, such as 3D graphics and animation, can greatly increase a design's appeal, intuitiveness, and memorability (Brath, Peters, & Senior, 2005). For example, when Irani and Ware compared 2D UML diagrams to geon diagrams (3D shaded solids), they found users could identify substructures and relationship types with much lower error rates for geon diagrams than UML diagrams (Irani & Ware, 2003). In addition, the use of 3D presentations provides greater information density than 2D ones (Robertson, Card, & Mackinlay, 1993). An experiment (Ware & Franck, 1996) suggests larger graphs can be interpreted if laid out in 3D and displayed with stereo and/or motion depth cues to support spatial perception. This extra dimension also helps to provide a clear perception of relations between objects by integrating local views with global views and by composing multiple 2D views in a single 3D view (Irani & Ware, 2003). Finally, 3D graphics' similarity with the real world enables us to represent the world in a more natural way. The representation of objects can be done according to their associated real concept; therefore, the interactions can be more powerful

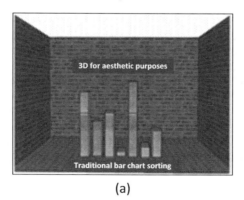

(a)

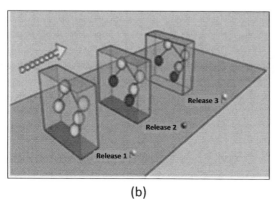

(b)

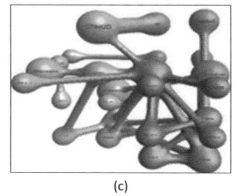

(c)

FIGURE 6.1 Three-dimensional view categorization: (a) 2D augmented view: a bubble sort visualization; (b) 2D adapted view: software release history; (c) Inherent 3D view: metaballs (Teyseyre & Campo, 2008).

(ranging from immersive navigation to different manipulation techniques), and the animations can be more realistic (Teyseyre & Campo, 2008).

There are some problems with 3D, however, such as intensive computation, more complex implementation, and user adaptation and disorientation. The first problem can be addressed using powerful and specialized hardware. Development complexity can be reduced using 3D toolkits and frameworks, 3D modeling languages, or 3D software visualization frameworks.

That leaves the problem of user adaptation. Most users just have experience with classical WIMP (Windows, Icons, Menus, Pointing) 2D desktop metaphors. Therefore, the interaction with 3D presentations and possibly the use of special devices demand considerable adaptation efforts (Teyseyre & Campo, 2008).

Furthermore, it is often difficult for users to understand 3D spaces and perform actions inside them. In particular, as a consequence of a richer set of interactions and more degrees of freedom, users may be disoriented. For example, Plaisant, Grosjean, and Bederson (2002) suggest 3D representations only marginally improve the screen space problem while increasing the complexity of interaction. When Cockburn and McKenzie (2002) evaluated the effectiveness of spatial memory in 2D and 3D, they found navigation in 3D spaces can be difficult, and even simple tasks can be problematic.

To overcome these limitations, 3D enhanced interfaces have been proposed. These interfaces might offer simpler navigation, more compelling functionality, safer movements, and less occlusion than 3D reality (Shneiderman, 2003). For instance, one way to reduce disorientation is to constrain user navigation with lateral or linear movements or to use physical laws such as gravity (Bowman & Hodges, 1995). Another possibility is to include automatic camera assistance during the transition phase from one focus object to the other. Several approaches have been proposed using landmarks to help users to orient in a 3D world (Teyseyre & Campo, 2008).

Finally, occlusion may distort the user's perception of the data space when the information space is dense (Chuah, Roth, Mattis, & Kolojejchick, 1995). Occlusion is a serious problem because when objects are occluded, they are invisible to the user (Teyseyre & Campo, 2008).

To sum up, there is a vast literature on the advantages and disadvantages of 3D versus 2D with conflicting results. Table 6.1 summarizes 3D visualization strengths and weaknesses (Teyseyre & Campo, 2008). Ultimately, if used in ways that exploit their strengths while avoiding their weaknesses (Mullet et al., 1995), 3D visualizations have the potential to aid and improve the development process.

6.1.2.1 Software Visualization

It is a well-known fact that developing software systems is complex and requires a number of cognitive tasks, such as search, comprehension, analysis, and design, among others. Software visualization can be a helpful tool to enhance the comprehension of computer programs. In fact, in a recent survey of 111 researchers from software maintenance, reengineering, and reverse engineering, 40% of the respondents said software visualization was very necessary for their work and another 42% found it important but not critical (Koschke, 2002).

The aim of software visualization is not to create impressive images but to create images that evoke the user's mental images for better software comprehension (Diehl, 2007). Through visualization, engineers can obtain an initial perception of how software is structured, understand the software logic, and

TABLE 6.1

Three-Dimensional Strengths and Weaknesses (Teyseyre & Campo, 2008)

Strengths	Weaknesses
• Greater information density.	• Intensive computation.
• Integration of local views with global views.	• More complex implementation.
• Composition of multiple 2D views in a single 3D view.	• User adaptation to 3D metaphors and special devices.
• Enhanced perception of the human visual system.	• More difficult for users to understand 3D spaces and perform actions in it.
• Familiarity, realism, and real-world representations.	• Occlusion.

explain and communicate the development. Software visualization combines techniques from different areas, including software engineering, data mining, computer graphics, information visualization, and human–computer interactions (Teyseyre & Campo, 2008). More precisely, software visualization is a specialized area of information visualization that can be defined as:

> a representation of computer programs, associated documentation and data, that enhances, simplifies and clarifies the mental representation the software engineer has of the operation of a computer system.
>
> **(Mili & Steiner, 2002)**

Software visualization in 2D has been extensively studied, and many techniques for representing software systems have been proposed. However, there is a demand for effective programs to understand techniques and methods.

Although the question of the benefits offered by 3D over 2D still remains to be answered, a growing area of research is investigating the application of 3D graphics to software visualization with good results. Researchers are trying to find new 3D visual representations to overcome some of the limitations of 2D and exploit 3D's richer expressiveness. Three-dimensional software visualization has been studied in such areas as algorithm animation for educational purposes, debugging, 3D programming, requirements engineering, software evolution, cyberattacks, ontology visualization and semantic Web, mobile objects, visualization for reverse engineering, software maintenance, and comprehension at different levels of abstraction (source code, object-oriented systems, and software architectures), among others (Teyseyre & Campo, 2008).

6.1.2.2 Definition of 3D Visualization Software

Three-dimensional visualization software is used to view and interrogate 3D models and other deliverables created using mechanical computer-aided design (MCAD) software (Lifecycle Insights).

6.1.2.2.1 Capabilities Provided

MCAD software provides some combination of the following capabilities (Lifecycle Insights):

- Three-dimensional model conversion to lightweight formats: 3D models in CAD applications can be exorbitantly large because they contain the full definition of how the geometry was created, often using parametric feature-based approaches. When applications in this technology category import such models, they convert them into lightweight models that are dramatically smaller and more responsive.
- Three-dimensional model interrogation: Some non-engineering stakeholders only need to be able to interrogate 3D models to obtain the information they need to do their jobs. This includes taking measurements, creating cross sections, and checking other geometric characteristics. This set of capabilities is common for the applications in this technology category.
- Markup and review: A critical activity in engineering is the design review. This involves engineering peers checking the validity of the design and looking for mistakes and errors. These applications enable engineers to do so by marking up the 3D model with highlights and annotations.
- Procedure development and validation: Organizations such as manufacturing and service often need to do more than simply interrogate a 3D model. They must develop procedures that represent how the product will be manufactured on the shop floor or serviced in the field. Software providers for these applications have added capabilities that allow users to develop such procedures and then validate that they can, in fact, be completed.
- Specialized visualization: Visualization technologies started out as generic tools that could be used in a wide variety of use cases. Since then, more and more organizations have voiced needs to produce specific types of deliverables, and software providers have enhanced their applications to accommodate them. Today, a wide range of specialized 3D visualization applications use organizational-specific terminology.

- Mobile 3D visualization: While some activities in 3D visualization applications require a user to sit at a desk, many do not. Software providers have been active in moving their applications to mobile platforms like tablets and smartphones. As a result, engineering and non-engineering stakeholders alike can work on the go.

6.1.3 Visualization Techniques

There are many techniques to visualize data. In addition to standard 2D/3D techniques such as *x-y* (*x-y-z*) plots, bar charts, line graphs, and maps, there are a number of more sophisticated techniques. These correspond to basic visualization principles that may be combined to implement a specific visualization system (Keim & Ward, 2002).

6.1.3.1 Geometrically Transformed Displays

Geometrically transformed display techniques aim at finding "interesting" transformations of multi-dimensional datasets. The class of geometric display methods includes techniques from exploratory statistics, such as scatterplot matrices (Andrews., 1972; Cleveland, 1993), and techniques that can be subsumed under the term "projection pursuit" (Huber, 1985). Other geometric projection techniques include Prosection Views (Furnas & Buja, 1994; Spence, Tweedie, Dawkes, & Su, 1995), Hyperslice (Van Wijk & Van Liere, 1993), and the well-known parallel coordinates visualization technique (Inselberg & Dimsdale, 1990). The parallel coordinates technique maps the *k*-dimensional space onto the two display dimensions by using *k* axes that are parallel to each other (either horizontally or vertically oriented), evenly spaced across the display. The axes correspond to the dimensions and are linearly scaled from the minimum to the maximum value of the corresponding dimension. Each data item is presented as a chain of connected line segments, intersecting each of the axes at a location corresponding to the value of the considered dimensions (see Figure 6.2) (Keim & Ward, 2002).

6.1.3.2 Iconic Displays

Another class of visual data exploration techniques is the iconic display method. The idea is to map the attribute values of a multidimensional data item to the features of an icon. Icons can be arbitrarily defined; they may be little faces (Chernoff, 1973), needle icons as used in Massive Graph Visualizer (MGV), star icons (Ward, 1994), stick figure icons (Pickett & Grinstein, 1988), color icons (Keim & Kriegel, 1994; Levkowitz, 1991), or TileBars (Hearst, 1995), for example. The visualization is generated by mapping the attribute values of each data record to the features of the icons. In case of the stick figure technique, for example, two dimensions are mapped to the display dimensions, and the remaining dimensions are mapped to the angles and/or limb length of the stick figure icon. If the data items are relatively dense with respect to the two display dimensions, the resulting visualization presents texture patterns that vary according to the characteristics of the data and are, therefore, detectable by pre-attentive perception.

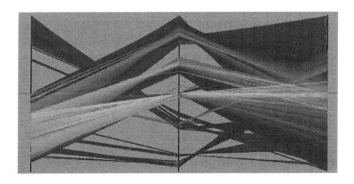

FIGURE 6.2 Parallel coordinates visualization (Keim & Ward, 2002).

Three-Dimensional Visualization

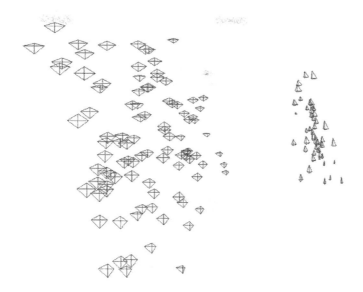

FIGURE 6.3 Iris dataset, displayed using star glyphs positioned based on the first two principal components (from XmdvTool; see Ward, 1994).

Figure 6.3 shows an example of this class of techniques. Each data point is represented by a star icon/glyph, where each data dimension controls the length of a ray emanating from the center of the icon. In this example, the positions of the icons are determined using principal component analysis (PCA) to convey more information about data relations. Other data attributes could also be mapped to an icon position (Keim & Ward, 2002).

6.1.3.3 Dense Pixel Displays

The basic idea of dense pixel techniques is to map each dimension value to a colored pixel and group the pixels belonging to each dimension into adjacent areas (Keim, 2000a). Since dense pixel displays generally use one pixel per data value, the techniques allow the visualization of the largest amount of data possible on current displays (up to about 1,000,000 data values). If each data value is represented by one pixel, the main question is how to arrange the pixels on the screen. Dense pixel techniques use different arrangements for different purposes. When the pixels are arranged in an appropriate way, the resulting visualization provides detailed information on local correlations, dependencies, and hot spots (Keim & Ward, 2002).

Well known examples are the recursive pattern technique (Keim, Kriegel, & Ankerst, 1995) and the circle segments technique (Ankerst, Keim, & Kriegel, 1996).

The recursive pattern technique is based on a generic recursive back-and-forth arrangement of the pixels and is particularly aimed at representing datasets with a natural order according to one attribute (e.g., time series data). The user may specify parameters for each recursion level and thereby control the arrangement of the pixels to form semantically meaningful substructures. The base element on each recursion level is a pattern of height h_i and width w_i as specified by the user. First, the elements correspond to single pixels arranged within a rectangle of height h_1 and width w_1 from left to right, then backward from right to left, then again forward from left to right, and so on. The same basic procedure is applied to all recursion levels; the only difference is that the basic elements on level i are in the pattern resulting from the level $(i-1)$ arrangements. Figure 6.4 shows a sample recursive pattern visualization of financial data (Keim & Ward, 2002). The visualization shows 20 years (January 1974–April 1995) of daily prices of the 100 stocks contained in the Frankfurt Stock Index (FAZ).

The idea of the circle segments technique (Ankerst, Keim, & Kriegel, 1996) is to represent the data in a circle divided into segments, one for each attribute. Within the segments, each attribute value is again

FIGURE 6.4 Dense pixel displays: recursive pattern technique (Keim & Ward, 2002).

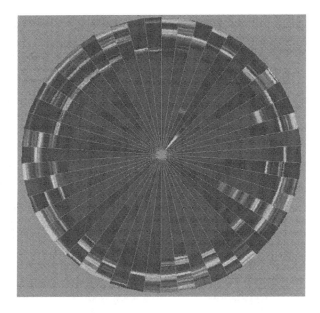

FIGURE 6.5 Dense pixel displays: circle segments technique (Keim & Ward, 2002).

visualized by a single colored pixel. The arrangement of the pixels starts at the center of the circle and continues to the outside on a line orthogonal to the segment-halving line in a back-and-forth manner. The rationale of this approach is that close to the center, all attributes are close to each other, thus enhancing the visual comparison of their values. Figure 6.5 shows an example of circle segment visualization using the same data (this time 50 stocks) as Figure 6.4 (Keim & Ward, 2002).

6.1.3.4 Stacked Displays

Stacked display techniques are tailored to present data partitioned in a hierarchical fashion. In the case of multidimensional data, the data dimensions to be used for partitioning the data and building the hierarchy have to be selected appropriately. An example of a stacked display technique is dimensional stacking (LeBlanc, Ward, & Wittels, 1990). The basic idea is to embed one coordinate system inside another coordinate system; i.e., two attributes form the outer coordinate system, two other attributes are

Three-Dimensional Visualization

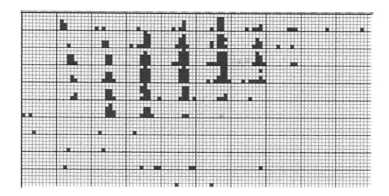

FIGURE 6.6 Dimensional stacking visualization of drill hole mining data (Keim & Ward, 2002).

embedded into the outer coordinate system, and so on. The display is generated by dividing the outermost level coordinate system into rectangular cells, and within the cells, the next two attributes are used to span the second-level coordinate system. This process may be repeated multiple times. The usefulness of the resulting visualization largely depends on the data distribution of the outer coordinates; therefore, the dimensions used to define the outer coordinate system have to be selected carefully. A rule of thumb is to choose the most important dimensions first.

A dimensional stacking visualization of mining data with longitude and latitude mapped to the outer x and y axes, as well as ore grade and depth mapped to the inner x and y axes, is shown in Figure 6.6 (Keim & Ward, 2002).

6.1.4 Goals of Visualization Techniques

6.1.4.1 Explorative Analysis

- Starting point: Data without hypotheses about the data.
- Process: Interactive, usually undirected search for structures, trends, etc.
- Result: Visualization of the data, providing hypotheses about the data (Keim, 2000).

6.1.4.2 Confirmative Analysis

- Starting point: Hypotheses about the data.
- Process: Goal-oriented examination of the hypotheses.
- Result: Visualization of the data, allowing the confirmation or rejection of the hypotheses (Keim, 2000).

6.1.4.3 Presentation

- Starting point: Facts to be presented are fixed *a priori*.
- Process: Choice of an appropriate presentation technique.
- Result: High-quality visualization of the data presenting the facts (Keim, 2000).

6.1.5 Data Type to Be Visualized

In information visualization, the data usually consist of a large number of records, each consisting of a number of variables or dimensions. Each record corresponds to an observation, measurement, transaction, etc. Examples are customer properties, e-commerce transactions, and physical experiments. The number of attributes can differ from dataset to dataset. One particular physical experiment, for example, can be described by five variables, while another may need hundreds of variables. We call the number of variables the dimensionality of the dataset. Datasets may be one-dimensional, two-dimensional, or multidimensional, or they may have more complex data types such as text/hypertext or hierarchies/graphs. A distinction is sometimes made between dense (or grid) dimensions and the dimensions which may have arbitrary values. Depending on the number of dimensions with arbitrary values, the data are sometimes also called univariate, bivariate, etc. (Keim, 2002).

6.1.5.1 One-Dimensional Data

One-dimensional data usually have one dense dimension. A typical example of one-dimensional data is temporal data. Note that with each point of time, one or multiple data values may be associated. Examples are time series of stock prices (see Figures 6.3 and 6.4) and the time series of news data used in the ThemeRiver examples (see Figures 6.1, 6.3–6.5) (Keim, 2002).

6.1.5.2 Two-Dimensional Data

Two-dimensional data have two distinct dimensions. A typical example is geographical data, where the two distinct dimensions are longitude and latitude; x-y plots are a typical method to show two-dimensional data, and maps are a special type of x-y plots to show two-dimensional geographical data. Examples are the geographical maps used in Polaris and in MGV. Although it seems easy to deal with temporal or geographic data, caution is advised. If the number of records to be visualized is large, temporal axes and maps get quickly glutted and may not help to understand the data (Keim, 2002).

6.1.5.3 Multidimensional Data

Many datasets consist of more than three attributes and do not allow a simple visualization as 2D or 3D plots. Examples of multidimensional (or multivariate) data are tables from relational databases, which often have tens to hundreds of columns (or attributes). Since there is no simple mapping of the attributes to the two dimensions of the screen, more sophisticated visualization techniques are needed. An example of a technique which allows the visualization of multidimensional data is the parallel coordinate technique (Inselberg & Dimsdale, 1990) (see Figure 6.1), which is also used in the scalable framework (see Figure 6.7). Parallel coordinates display each multidimensional data item as a polygonal line which intersects the horizontal dimension axes at the position corresponding to the data value for the corresponding dimension (Keim, 2002).

6.1.5.4 Text and Hypertext

Not all data types can be described in terms of dimensionality. In the age of the World Wide Web, one important data type is text and hypertext and another is multimedia Web page contents. These data types differ from others in that they cannot be easily described by numbers and, therefore, most of the standard visualization techniques cannot be applied. In most cases, the data must be transformed into description vectors before visualization techniques can be used. An example of a simple transformation is word counting (Nowell, Havre, Hetzler, & Whitney, 2001), often combined with a PCA or multidimensional scaling (Keim, 2002).

6.1.5.5 Hierarchies and Graphs

Data records often have some relationship to other pieces of information. Graphs are widely used to represent such interdependencies. A graph consists of set of objects, called nodes, and connections between these objects, called edges. Examples are the email interrelationships among people, their shopping behavior, the file structure of the hard disk, or the hyperlinks in the World Wide Web. A number of specific visualization techniques deal with hierarchical and graphical data (Keim, 2002).

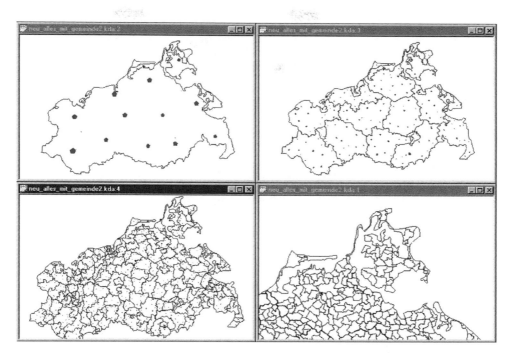

FIGURE 6.7   Refinement of geographical granularity (Kreuseler, Lopez, & Schumann, 2001).

6.1.5.6 Algorithms and Software

Another class of data is algorithms and software. Coping with large software projects is a challenge. The goal of visualization is to support software development by helping to understand algorithms, e.g., by showing the flow of information in a program, to enhance the understanding of written code, e.g., by representing the structure of thousands of source code lines as graphs, and to support the programmer in debugging the code, e.g., by visualizing errors. A large number of tools and systems support these tasks (Keim, 2002).

6.1.6 Visualization Operations

Visualization operations create renditions of given scenes or object systems. Their purpose is to facilitate the visual perception of object information. They can be scene-based or object-based (Udupa, 1999).

6.1.6.1 Scene-Based Visualization

In scene-based visualization, renditions are created directly from given scenes. Within this approach, two further subclasses may be identified: section mode and volume mode (Udupa, 1999).

6.1.6.1.1 Section Mode

Opinions differ on what constitutes a "section" and how this information is displayed. Natural sections may be axial, coronal, or sagittal; oblique or curved sections are also possible. Information is displayed as a montage with the use of roam through (fly through) and gray scale and pseudo-color.

Figure 6.8 shows a montage display of the natural sections of a computed tomography (CT) scan (Udupa, 1999).

Figure 6.9a demonstrates a 3D display-guided extraction of an oblique section from a CT scan of a pediatric patient's head. This re-sectioning operation illustrates how visualization is needed

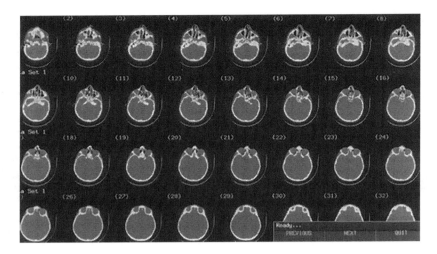

FIGURE 6.8 Montage display of a 3D CT scan of the head (Udupa, 1999).

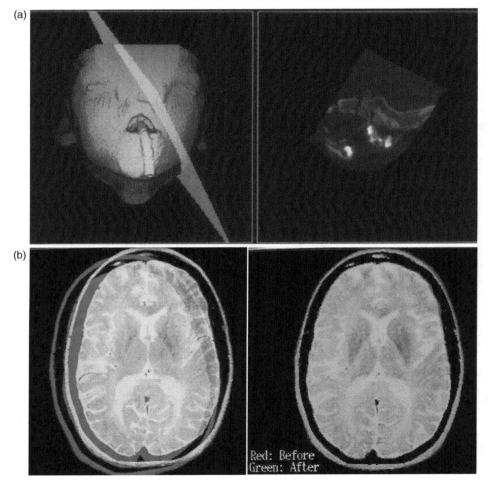

FIGURE 6.9 (a) Three-dimensional display-guided extraction of an oblique section from CT data obtained in a patient with a craniofacial disorder; (b) Pseudo-color display (Udupa, 1999).

Three-Dimensional Visualization

to perform visualization itself. Figure 6.9b illustrates pseudo-color display with two sections from a brain magnetic resonance (MR) imaging study in a patient with multiple sclerosis. The two sections, representing approximately the same location in the patient's head, were taken from 3D scenes obtained at different times and subsequently registered. The sections are assigned red and green hues. The display shows yellow (produced by a combination of red and green hues) where the sections match perfectly or where there has been no change (e.g., in the lesions). At other places, either red or green appears (Udupa, 1999).

6.1.6.1.2 Volume Mode

In volume mode visualization, information may be displayed as surfaces, interfaces, or intensity distributions with the use of surface rendering, volume rendering, or maximum intensity projection (MIP). A projection technique is always needed to move from the higher-dimensional scene to the 2D screen of the monitor. For scenes of four or more dimensions, 3D "cross sections" must first be determined, after which a projection technique can be applied to move from 3D to 2D. Two approaches may be used: first, ray casting (Levoy, 1988) consists of tracing a line perpendicular to the viewing plane from every pixel in the viewing plane into the scene domain; second, voxel projection (Frieder, Gordon, & Reynolds, 1985) consists of directly projecting voxels encountered along the projection line from the scene onto the viewing plane (see Figure 6.10). Voxel projection is generally considerably faster than ray casting; however, either of these projection methods may be used with any of the three rendering techniques (MIP, surface rendering, volume rendering) (Udupa, 1999).

In MIP, the intensity assigned to a pixel in the rendition is simply the maximum scene intensity encountered along the projection line (see Figure 6.11a) (Brown & Riederer, 1992; Schreiner, Paschal, & Galloway, 1996). MIP is the simplest of all 3D rendering techniques. It is most effective when the objects of interest are the brightest in the scene and have a simple 3D morphology and a minimal gradation of intensity values. Contrast material-enhanced CT angiography and MR angiography are ideal applications for this method, and MIP is commonly used (Napel, Marks, & Rubin, 1992; Hertz et al., 1993). Its main advantage is that it requires no segmentation. However, the ideal conditions mentioned earlier frequently go unfulfilled, due (for example) to the presence of other bright objects, such as clutter from surface coils in MR angiography, bone in CT angiography, or other obscuring vessels that may not be of interest. Consequently, some segmentation eventually becomes necessary (Udupa, 1999).

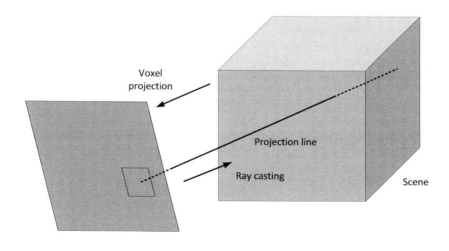

FIGURE 6.10 Schematic of projection techniques for volume mode visualization (Udupa, 1999).

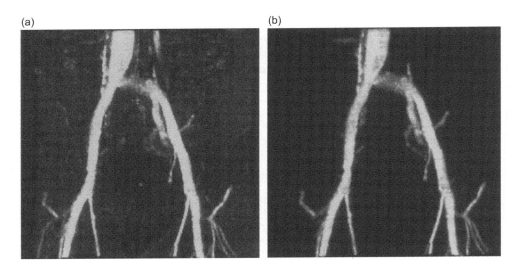

FIGURE 6.11 Fuzzy connected segmentation: (a) Three-dimensional maximum intensity projection (MIP) rendition of an MR angiography scene; (b) MIP rendition of 3D fuzzy connected vessels detected in the scene in (a). Fuzzy connectedness has been used to remove the clutter that obscures the vasculature (Udupa, 1999).

In surface rendering (Goldwasser & Reynolds, 1987), object surfaces are portrayed in the rendition. A threshold interval must be specified to indicate the object of interest in the given scene. Clearly, speed is of the utmost importance in surface rendering because the idea is that object renditions are created interactively directly from the scene as the threshold is changed. Instead of thresholding, any automatic, hard, boundary- or region-based method can be used. In such cases, however, the parameters of the method will have to be specified interactively, and the speed of segmentation and rendition must be sufficient to make this mode of visualization useful. Although rendering based on thresholding can presently be accomplished in about 0.03–0.25 s on a Pentium 300 with the use of appropriate algorithms in software (Udupa, Odhner, & Samarasekera, 1994), more sophisticated segmentation methods (e.g., kNN) may not offer interactive speed.

The actual rendering process consists of three basic steps: projection, hidden part removal, and shading. These steps are needed to impart a sense of three dimensionality to the rendered image. Additional cues for three dimensionality may be provided by techniques such as stereoscopic display, motion parallax by rotation of the objects, shadowing, and texture mapping (Udupa, 1999).

If ray casting is used as the method of projection, hidden part removal is performed by stopping at the first voxel encountered along each ray that satisfies the threshold criterion (Höhne & Bernstein, 1986). The value (shading) assigned to the pixel in the viewing plane that corresponds to the ray is determined, as described later. If voxel projection is used, hidden parts can be removed by projecting voxels from the farthest to the closest (with respect to the viewing plane) and always overwriting the shading value in one of a number of computationally efficient ways (Frieder, Gordon, & Reynolds, 1985; Reynolds, Gordon, & Chen, 1987; Herman & Liu, 1977; Udupa & Odhner, 1991).

The shading value assigned to a pixel p in the viewing plane depends on the voxel v that is eventually projected onto p. The faithfulness with which this value reflects the shape of the surface around v largely depends on the surface normal vector estimated at v. Two classes of methods are available for this purpose: object-based and scene-based methods.

In object-based methods (Chen, Herman, & Reynolds, 1985; Gordon & Reynolds, 1985), the vector is determined purely from the geometry of the shape of the surface in the vicinity of v.

In scene-based methods (Höhne & Bernstein, 1986), the vector is considered to be the gradient of the given scene at v; i.e., the direction of the vector is the same as the direction in which scene intensity changes most rapidly at v. Given the normal vector N at v, the shading assigned to p is usually determined as $[f_d(v, N, L) + f_s(v, N, L, V)] f_D(v)$, where f_d is the diffuse component of reflection, f_s is the specular component, f_D is a component that depends on the distance of v from the viewing plane, and L and V

are unit vectors indicating the direction of the incident light and the viewing rays. The diffuse component is independent of the viewing direction but depends solely on L (as a cosine of the angle between L and N). It captures the scattering property of the surface, whereas the specular component captures surface shininess. The specular component is highest in the direction of ideal reflection R whose angle with N is equal to the angle between L and N. This reflection decreases as a cosine function on either side of R. By weighting the three components in different ways, different shading effects can be achieved (Udupa, 1999).

In scene-based surface rendering, a hard object is implicitly created and rendered "on the fly" from the given scene. In scene-based volume rendering, a fuzzy object is implicitly created and rendered on the fly from the given scene. Clearly, surface rendering becomes a special case of volume rendering. Furthermore, volume rendering in this mode is generally much slower than surface rendering, typically requiring 3–20 s, even on specialized hardware rendering engines.

The basic idea in volume rendering is to assign an opacity from 0% to 100% to every voxel in the scene. The opacity value is determined on the basis of the objectness value at the voxel and how prominently we want to portray this particular grade of objectness in the rendition. This opacity assignment is specified interactively by way of an opacity function (see Figure 6.12), wherein the vertical axis indicates percentage of opacity. Every voxel is now considered to transmit, emit, and reflect light. The goal is to determine the amount of light reaching every pixel in the viewing plane. The amount of light transmitted depends on the opacity of the voxel. Light emission depends on objectness and, hence, on opacity: the greater the objectness, the greater the emission. Similarly, reflection depends on the strength of the surface; the greater the strength: the greater the reflection (Udupa, 1999).

Like surface rendering, volume rendering consists of three basic steps: projection, hidden part removal, and shading or compositing. The principles underlying projection are identical to those described for surface rendering (Udupa, 1999).

Hidden part removal is much more complicated for volume rendering than for surface rendering. In ray casting, a common method is to discard all voxels along the ray from the viewing plane beyond a point at which the "cumulative opacity" is above a high threshold (e.g., 90%) (Levoy, 1990). In voxel projection, a voxel can also be discarded if the voxels surrounding it in the direction of the viewing ray have "high" opacity (Udupa & Odhner, 1993).

The shading operation, which is more appropriately termed compositing, is also more complicated for volume rendering than for surface rendering. Compositing must take into account all three components: transmission, reflection, and emission. We can start from the voxel farthest from the viewing plane along each ray and work towards the front, calculating the output light for each voxel. The net light output by the voxel closest to the viewing plane is assigned to the pixel associated with the ray. Instead of using this back-to-front strategy, we could also make calculations from front to back, and this has actually been shown to be faster (Udupa & Odhner, 1993).

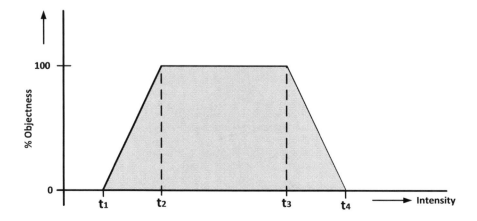

FIGURE 6.12 Diagram of fuzzy thresholding (Udupa, 1999).

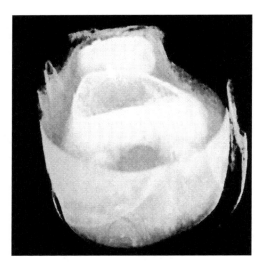

FIGURE 6.13 Scene-based volume rendering with voxel projection. Rendition of knee CT data from Figure 6.14 shows bone, fat, and soft tissue (Udupa, 1999).

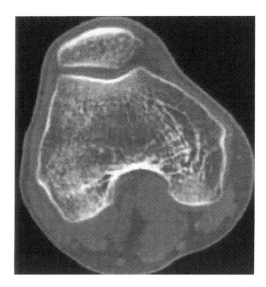

FIGURE 6.14 Graded composition and hanging togetherness. CT scan of the knee illustrates graded composition of intensities and hanging togetherness. Voxels within the same object (e.g., the femur) are assigned considerably different values. Despite this gradation of values, it is not difficult to identify the voxels as belonging to the same object (hanging togetherness) (Udupa, 1999).

In volume rendering (as in surface rendering), voxel projection is substantially faster than ray casting. Figure 6.13 shows the CT knee dataset illustrated in Figure 6.14 rendered with this method. Three types of tissue—bone, fat, and soft tissue—are identified (Udupa, 1999).

6.1.6.2 Object-Based Visualization

In object-based visualization, objects are first explicitly defined and then rendered. In difficult segmentation situations, or when segmentation is time consuming or involves too many parameters, it is impractical to perform direct scene-based rendering. The intermediate step of completing object definition then becomes necessary (Udupa, 1999).

6.1.6.2.1 Surface Rendering

Surface rendering methods take hard object descriptions as input and create renditions. The methods of projection, hidden part removal, and shading are similar to those described for scene-based surface rendering, except that a variety of surface description methods have been investigated using voxels, points, voxel faces, triangles, and other surface patches (Frieder, Gordon, & Reynolds, 1985; Reynolds, Gordon, & Chen, 1987; Udupa & Odhner, 1991). Therefore, projection methods that are appropriate for specific surface elements have been developed. Figure 6.15a shows a rendition, created with the use of voxel faces on the basis of CT data, of the craniofacial skeleton in a patient with agnathia (Udupa, 1999).

Figure 6.16 shows renditions of the bones of the foot created using the same method on the basis of MR imaging data (Udupa, 1999).

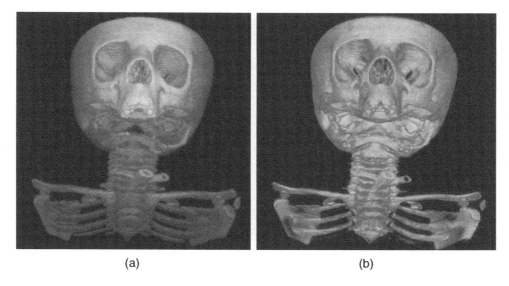

FIGURE 6.15 Object-based visualization of the skull in a child with agnathia: (a) Surface-rendered image; (b) Subsequent volume-rendered image preceded by the acquisition of a fuzzy object representation with use of fuzzy thresholding (see Figure 6.12) (Udupa, 1999).

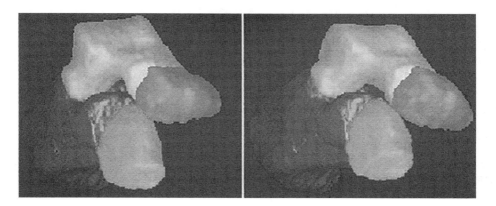

FIGURE 6.16 Rigid object-based registration. Sequence of 3D MR imaging scenes of the foot allows kinematic analysis of the midtarsal joints. The motion (i.e., translation and rotation) of the talus, calcaneus, and navicular and cuboid bones from one position to the other is determined by registering the bone surfaces in the two different positions (Udupa, 1999).

6.1.6.2.2 Volume Rendering

Volume rendering methods take as input fuzzy object descriptions in the form of a set of voxels, wherein values for objectness and a number of other parameters (e.g., gradient magnitude) are associated with each voxel (Udupa & Odhner, 1993). Because the object description is more compact than the original scene, and additional information for increasing computation speed can be stored as part of the object description, volume rendering based on fuzzy object description can be performed at interactive speeds, even on personal computers, such as the Pentium 300, entirely in software. In fact, the rendering speed (2–15 s) is now comparable to that of scene-based volume rendering with specialized hardware engines.

Figure 6.15b shows a fuzzy object rendition of the dataset in Figure 6.15a. Figure 6.17a shows a rendition of craniofacial bone and soft tissue, both of which were defined separately using the fuzzy connected methods described earlier. Note that if we use a direct scene-based volume rendering method with the opacity function illustrated in Figure 6.12, the skin becomes indistinguishable from other soft tissues and always obscures the rendition of muscles (see Figure 6.17b) (Udupa, 1999).

6.1.6.3 Misconceptions in Visualization

Several inaccurate statements concerning visualization frequently appear in the literature. The following statements are seen most often (Udupa, 1999).

6.1.6.3.1 Surface Rendering Is the Same as Thresholding

Clearly, thresholding is only one—albeit the simplest—of the many available hard region- and boundary-based segmentation methods, the output of any of which can be surface rendered (Udupa, 1999).

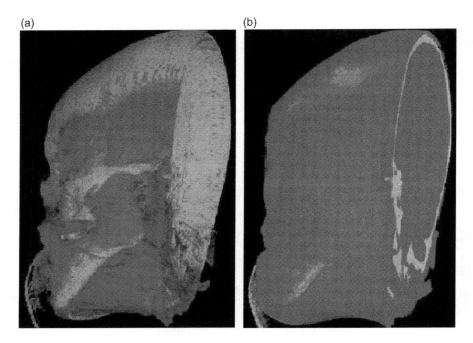

FIGURE 6.17 Visualization with volume rendering: (a) Object-based volume-rendered image demonstrates bone and soft-tissue structures (muscles) detected earlier as separate fuzzy connected objects in a 3D craniofacial CT scene. The skin is essentially "peeled away" because of its weak connectedness to muscles; (b) Scene-based volume-rendered version of the scene in (a) was acquired with use of the opacity function (see Figure 6.5) separately for bone and soft tissue. The skin has become indistinguishable from muscles because they have similar CT numbers, thus obscuring the rendition of the muscles (Udupa, 1999).

Three-Dimensional Visualization

6.1.6.3.2 Volume Rendering Does Not Require Segmentation

Although volume rendering is a general term and is used in different ways, the statement is false. The only useful volume rendering or visualization technique that requires no segmentation is MIP. The opacity assignment schemes illustrated in Figure 6.5 and described in the section entitled "Scene-based visualization" (Section 6.1.6.2) are clearly fuzzy segmentation strategies and involve the same problems encountered by any segmentation method. It is untenable to hold that opacity functions such as the one shown in Figure 6.5 do not represent segmentation while maintaining the manifestation that results when $t_1 = t_2$ and $t_3 = t_4$ (corresponding to thresholding) does represent segmentation (Udupa, 1999).

6.1.6.3.3 The Term Volume Rendering May Be Used to Refer to Any Scene-Based Rendering Technique, as Well as Object-Based Rendering Techniques

The use of the term "volume rendering" varies widely. In one sense, it can also apply to the section mode of visualization. It is better to use volume rendering to refer to fuzzy object rendering, whether performed with scene-based or object-based methods but not to refer to hard object rendering methods.

There are many challenges associated with visualization. First, preprocessing operations (and, therefore, visualization operations) can be applied in many different sequences to achieve the desired result. For example, the filtering-interpolation-segmentation-rendering sequence may produce renditions that are significantly different from those produced by interpolation-segmentation-filtering-rendering. With the large number of different methods possible for each operation and the various parameters associated with each operation, there are myriad ways of achieving the desired results. Figure 6.18 shows five images derived from CT data that were created by performing different operations. Systematic study is needed to determine which combination of operations is optimal for a given application. Normally, the fixed combination provided by the 3D imaging system is assumed to be the best for that application.

Second, the objective comparison of visualization methods becomes an enormous task in view of the vast number of ways we can reach the desired goal. Third, it can be challenging to achieve a realistic tissue display that includes color, texture, and surface properties (Udupa, 1999).

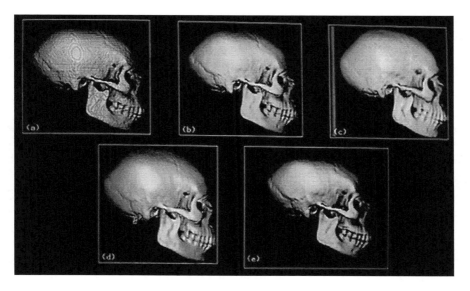

FIGURE 6.18 Preprocessing and visualization operations. Renditions from CT data created using five preprocessing and visualization operations (Udupa, 1999).

6.2 Line Security Diagnosis for Multisource Data Fusion

6.2.1 Multisource Data Fusion

6.2.1.1 Introduction

Data fusion, as a general and popular multidiscipline approach, combines data from multiple sources to improve the potential values and interpretation performances of the source data and to produce a high-quality visible representation of the data. Fusion techniques are useful for a variety of applications, ranging from object detection, recognition, identification, and classification, to object tracking, change detection, decision-making, etc. It has been successfully applied in the space and earth observation domains, computer vision, medical image analysis, and defense security.

Research on data fusion has a long history in the remote sensing community because fusion products are the basis for many applications. Researchers and practitioners have made great efforts to develop advanced fusion approaches and techniques to improve performance and accuracy. Fusing remotely sensed data, especially multisource data, however, remains challenging for many reasons, such as the various requirements, the complexity of the landscape, the temporal and spectral variations within the input dataset, and accurate data co-registration (Zhang, 2010).

In general, sensing fusion techniques can be classified at three different levels: the pixel/data level, the feature level, and the decision level (Pohl & van Genderen, 1998):

- Pixel-level fusion is the combination of raw data from multiple sources into single resolution data, which are expected to be more informative and synthetic than the input data and/or to reveal the changes between datasets acquired at different times.
- Feature-level fusion extracts various features, for example, edges, corners, lines, texture parameters, from different data sources and combines them into one or more feature maps that may be used instead of the original data for further processing. This is particularly important when the number of available spectral bands becomes so large that it is impossible to analyze each band separately. Methods used to extract features usually depend on the characteristics of the individual source data and, therefore, may be different if the datasets used are heterogeneous. In image processing, such fusion typically requires a precise (pixel-level) registration of the available images. Feature maps, thus, obtained are then used as input to preprocessing for image segmentation or change detection.
- Decision-level fusion combines the results from multiple algorithms to yield a final fused decision. When the results from different algorithms are expressed as confidences (or scores) rather than decisions, this is called soft fusion; otherwise, it is hard fusion. Methods of decision fusion include voting methods, statistical methods, and fuzzy logic-based methods.

6.2.1.2 Current Status

In this section, we address the current status of fusion methods at pixel level and at high level. Since the pixel-level fusion methods are mainly applied to optical images, we will concentrate on the methods of fusing optical panchromatic (PAN) and multispectral (MS) images. High-level fusion includes feature-level and decision-level fusion of multisource data, such as Synthetic Aperture Radar (SAR), optical images, light detection and ranging (LiDAR), Geographic Information System (GIS) data, and ground data (Zhang, 2010).

6.2.1.2.1 Pixel-Level Fusion

The main purpose of pixel-level fusion of optical images is to simultaneously improve spatial resolution, enhance structural and textural details, and retain the spectral fidelity of the original MS data. For this reason, it is also called pan sharpening. For multi-temporal data, the purpose of pixel-level fusion is to highlight the informative changes between different times, using either the same or different sensors. Although there is a trend to apply fusion techniques to change detection and change analysis, it is beyond the scope of this chapter.

Three-Dimensional Visualization

In many pixel-level fusion methods for remote sensing images, the structural and textural details of the MS image are enhanced by adopting higher resolution. The algorithms for pixel-level fusion of remote sensing images can be divided into three categories: component substitution (CS) fusion techniques, modulation-based fusion techniques, and multi-resolution analysis (MRA)-based fusion techniques (see Figure 6.19) (Zhang, 2010).

6.2.1.2.1.1 Component Substitution (CS) Fusion Techniques. The CS fusion techniques consist of three steps. First, forward transform is applied to the MS bands after they have been registered to the PAN band. Second, one component of the new data space similar to the PAN band is replaced with the higher resolution band. Third, the fused results are constructed by means of inverse transform to the original space (Zhang, 2010).

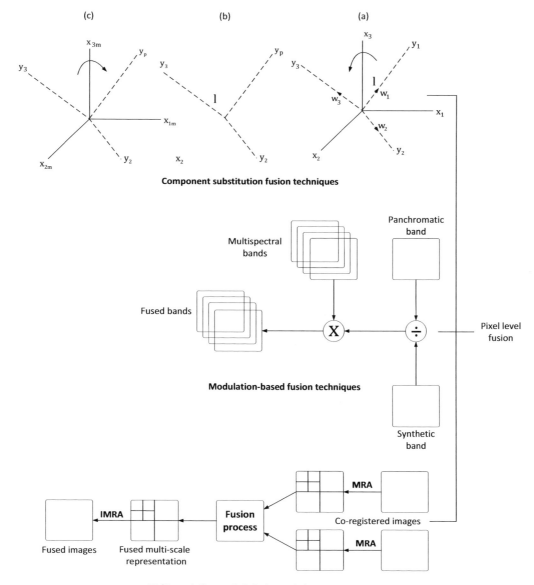

FIGURE 6.19 Three categories of pixel-level image fusion (Zhang, 2010).

6.2.1.2.1.2 Modulation-Based Fusion Techniques. The modulation-based fusion techniques use the idea that the spatial details are modulated into the MS images by multiplying the MS images by the ratio of the PAN image to the synthetic image, generally a lower resolution version of the PAN image (Zhang, 2010).

6.2.1.2.1.3 Multi-resolution Analysis (MRA) Fusion Techniques. The MRA-based fusion techniques adopt multi-scale decomposition methods, such as multi-scale wavelets, Laplacian pyramids, or bidimensional empirical mode decomposition, to decompose MS and PAN images at different levels. They derive spatial details and import them into finer scales of the MS images, highlighting relationships between PAN and MS images in coarser scales, and enhancing spatial details.

MRA-based fusion techniques consist of three main steps (Zhang, 2010):

1. MRA: Wavelet multi-resolution decomposition
2. Fusion: Replacement of approximation coefficients of PAN by those of the MS band
3. Inverse multi-resolution analysis (IMRA) transform.

6.2.1.2.2 High-Level Fusion

6.2.1.2.2.1 Fusion of Optical Images and Synthetic Aperture Radar (SAR). An undesirable property when applying the available pixel-level fusion techniques to the fusion of SAR data and optical images is that either spectral features of the optical imagery or the microwave backscattering information is destroyed or both are destroyed simultaneously. Therefore, it is necessary to develop a specifically tailored SAR-optical image fusion technique, which can fully utilize those two types of image sources. Up to now, the classifier combination has been an effective measure; it not only chooses the basic classifier corresponding to the SAR and optical imagery, respectively, but it also integrates fusion results from different basic classifiers (Zhang, 2010).

6.2.1.2.2.2 Fusion of LiDAR Data and Images. A LiDAR sensor delivers 3D point clouds with the intensities of the returned signals. In some cases, multiple pulses or full waveform signals can be provided by certain hardware systems (Wagner, 2006). As in aerial or satellite optical imagery, extensive post-processing is required to extract accurate terrain or semantic information from the LiDAR point cloud. Although the shape of topographic objects can be easily recognized by humans in 3D point clouds or optical imagery, this is not a straightforward task for computer algorithms.

The fusion of LiDAR data and imagery has been widely explored for a variety of applications, ranging from digital surface model (DSM)/digital elevation model (DEM) generation, to 3D object extraction, modeling, and land cover mapping (Zhang, 2010).

6.2.1.2.2.3 Fusion of Optical Images and GIS Data. GIS data, such as topography, land use, road, and census data, may be combined with remotely sensed data to improve the accuracy of image classification, object recognition, change detection, and 3D reconstructions. Previous research has shown that the fusion of remote sensing and GIS plays a critical role in map updating. The integration of remote sensing and GIS is emerging as a new research field (Zhang, 2010).

6.2.1.2.2.4 Data Fusion of Satellite, Aerial, and Close-Range Images. With the development of multi-view and multi-resolution earth observation systems, data fusion of satellite, aerial, and close-range images is necessary for some specific applications, such as environmental monitoring, road mapping, archaeology, building detection, and reconstruction. The cross-sensor platforms could be satellites, aircraft, unmanned aerial vehicles (UAVs), and vehicles; the surveillance range is hundreds of kilometers or only a few kilometers for site monitoring (Zhang, 2010).

As the resolution of images from space-borne, airborne, or ground-based platforms ranges from coarse to fine, data fusion using those images reflects the specific properties of the individual sensors at that resolution. For satellite images, automatic high-resolution satellite image georeferencing can be implemented by fusing existing digital orthophotos derived from aerial images. Studying and monitoring environments in urban areas requires highly accurate satellite images, but this kind of application involves an accurate georeferencing processing of the images to a given geodetic reference system. Taking advantage of the existing digital orthophoto maps and photo planes derived from aerial photogrammetry, Gianinetto and Scaioni (2003) have proposed a methodology for automatic high-resolution satellite image georeferencing (Zhang, 2010).

Fusion of aerial images and vehicle-borne sensor data can be applied for improved semantic mapping. Ground-based data are obtained by a mobile vehicle equipped with a calibrated or omnidirectional camera, differential Global Positioning System (GPS), terrestrial laser scanning, etc. This semantic information is used for local and global segmentation of an aerial image. After being matched with features detected in an aerial image, the data are used as input to a region- and boundary-based segmentation algorithm for building detection and 3D reconstruction in the aerial image. Persson, Duckett, and Lilienthal (2008) investigated the use of semantic information to link ground-level occupancy maps and aerial images. The result is a map where the semantic information has been extended beyond the range of the sensors to predict where the mobile device can find buildings (Zhang, 2010).

6.2.1.3 Trends and Challenges

Multisource data fusion (MSDF) is an evolving technology, which fuses data from multiple heterogeneous sensors to acquire enhanced information for decision-making. Applications of data fusion cross many disciplines, including environmental monitoring, ecological modeling, automatic target detection and classification, battlefield surveillance, global awareness, etc. For specific purposes, ancillary and terrestrial data, such as laser scanner data, GIS data, distributed Web-sensor data, field surveys, meteorological data, and economic census data, may be combined with remote sensing data to improve the performance of data fusion. The emerging trends are broadening the applications of MSDF (Zhang, 2010).

6.2.2 Kinds and Uses of Multisource Data

6.2.2.1 Published Paper Map

The published paper map is a basic source of multisource data. Examples include: series scale maps, geographical maps, administrative maps, circulation maps, thematic maps, and atlases. Databases are built by vectorizing these paper maps. For a region which has no vector data, the raster to vector method using the paper map is still feasible (Chen, Sun, Xu, & Xiong, 2013).

6.2.2.2 Built Spatial Database

Basic geospatial databases are a main source of data fusion. Various kinds of thematic data are also important; these include traffic data, administrative data, vegetation data, and so on. Because these databases are built at a particular point in time, the accuracy and currency of the data will decrease over time. To guarantee the quality of the built databases, a real-time updating mechanism is required (Chen, Sun, Xu, & Xiong, 2013).

6.2.2.3 Remote Sensing Images

Modern detecting technology offers abundant data which can be used to improve the currency and reliability of spatial data. Remote sensing images are widely used in spatial data updating. Russia uses RS images of Peace to update topographic maps of 1:50,000–1:100,000. Norway uses SPOT images to revise 1:50,000 traffic maps. France uses SPOT data to map 1:100,000 topographic maps. The United States uses Landsat data to revise medium- and small-scale maps. China uses TM images to update the data of 1:250,000 maps and SPOT images to update the data of 1:50,000 and 1:100,000 maps. With the startup of high-resolution detecting projects, the use of remote sensing images is the main way to update geospatial information and produce large-scale spatial data (Chen, Sun, Xu, & Xiong, 2013).

6.2.2.4 GPS Data

A GPS has the advantage of high precision, low cost, and high agility. GPS data can be used to set up geodetic control networks, level networks, and other spatial data and can assist in positioning spatial data. For example, vehicle-mounted GPS data can be used to update traffic data (Chen, Sun, Xu, & Xiong, 2013).

6.2.2.5 Field Data

Collecting field data is a traditional method of gathering geospatial information. Total station theodolites, distance meters, and other instruments are set on the known points in field work; the three-dimensional coordinates of the target points are then determined by surveying the direction, distance, and altitude differences between target points and known points (Chen, Sun, Xu, & Xiong, 2013).

6.2.2.6 Relevant Statistics and Literature

Statistics and literature are used to get attribute information. For example, national administrative codes and gazetteers can be used to update the attribute data of administrative and settlement features (Chen, Sun, Xu, & Xiong, 2013).

6.2.3 Multisource Data Integration

Multisource data differ in source, acquisition approach, data model, data format, coordinates and projection, semantic expression, and other aspects. So the multisource data can't be used to directly fuse data. The task in multisource data integration is to eliminate the differences among multisource spatial data. According to certain standards, those multisource data are integrated into one uniform system, through which the direct management and structure of the data can be realized (Chen, Sun, Xu, & Xiong, 2013).

6.2.3.1 Content and Techniques of Multisource Data Integration

6.2.3.1.1 Unified Coordinates

Different data use different coordinates. In China, for example, some current data coordinates include: old BJ54, new BJ54, CD80, WGS84, and the more recent national GPS control network, CGCS2000. The use of CGCS2000 will influence the geometrical precision of the previous coordinates, so unified coordinates are necessary, with the old coordinates transferred into the newer system. The coordinates can be transferred through three-parameter or seven-parameter models (Chen, Sun, Xu, & Xiong, 2013).

6.2.3.1.2 Projection Transformation

Depending on the mapping purposes and locations, existing multisource data use different projections. For projection transformation, it is necessary to set up a corresponding relationship between the points on different planes. For vector data, the projection can be transferred through certain equations. But for raster data, the new image can't be built through point-by-point conversion, because this may lead to slot and overlap in the new image. So the projection transformation of image uses reversed calculation (Shaomei, 2004).

6.2.3.1.3 Data Format Exchange

Different software and systems use different data models to describe the world, leading to different data formats. Commonly used data models are: Arc/Info, MapInfo, MapGIS, AutoCAD, SuperMap, and so on. Each data model has its own data format (Chen, Sun, Xu, & Xiong, 2013).

6.2.3.1.4 Data Compression

The sampling interval of GPS data is short, so there are large numbers of redundant points in the data. This means the lines are not smooth after symbolization and are hard to edit. The redundant points need to be compressed, simplified, and generalized according to the standards of a certain scale. Common compression algorithms include Ramer-Douglas-Peucker and vertical distance (Chen, Sun, Xu, & Xiong, 2013).

6.2.3.1.5 Image Enhancement

Image enhancement is used to reduce or give prominence to the features of certain elements of remotely sensed images. Image enhancement includes gray-level transformation, histogram transformation, PCA, spatial filtering, texture analysis, image correlation, and so on. The selected method should reflect the practical situation (Chen, Sun, Xu, & Xiong, 2013).

6.2.3.2 Technical Workflow of Multisource Data Integration

In the processing of multisource data integration, the geometry data and attribute data are organized separately. The former are stored in the design document, and the latter are managed by the database. The geometry data and attribute data of the same feature have a one-to-one relationship. The attribute data are divided into two parts, namely, new database and additional database, and are connected through an index table. The technical workflow of multisource data integration is shown in Figure 6.20 (Chen, Sun, Xu, & Xiong, 2013).

For the vector data, data of the same or larger scale are processed through data format transfer, coordinate unification, projection transformation, and semantic transformation. The processed vector data are integrated into a uniform digital mapping system and symbolized. The geometry data are written into the design document, and the corresponding relationship is built into the record in the database. The attribute data of the same scale are stored in the new database directly. The attribute data of the larger scale are stored in the additional database as an alternative; these data can be extracted into the new database depending on the data fusion scheme.

For the raster data, after coordinate unification, projection transformation, and image processing, the processed raster data are integrated into a uniform digital mapping system and overlaid with the symbolized geometry data in the form of a reference document (Chen, Sun, Xu, & Xiong, 2013).

6.2.4 Multisource Data Fusion (MSDF) in Scientometrics

6.2.4.1 Basic Types of Relations

There are two kinds of MSDF in scientometrics: fusion of data types and fusion of data relations.

The fusion of data types merges different data types into a single object of analysis. Data types include journal articles, conference information, dissertations, patent information, project information, book information, and so on. Hua (2013) divided multisource data into homogeneous information with a heterologous source, heterogeneous information, and multilingual information. He argued the fusion of data types was basic work, involving field mapping, field splitting, filtering repeated data, and weighting heterogeneous data (Xu, Wang, Pang, Ru, & Fang, 2016).

The fusion of data relations merges different data relations into a new relationship to characterize the relations among entities. Shibata, Kajikawa, Takeda, and Matsushima (2009) compared three networks: co-citation networks, bibliographic coupling networks, and citation networks. They found citation networks could find new topics earlier and was the most effective way to identify research fronts. Co-citation networks were the worst. They also found content clustering based on the citation network had the least risk of omitting a new research field. Klavans and Boyack (2006) examined the clustering network built by direct citations; they found its content was similar to that of a co-citation network. Couto (2006) found the bibliographic coupling approach was more effective than a textual approach in the empirical analysis of computer-related literature. However, Ahlgren and Colliander (2009) considered the textual approach was better in information retrieval (Xu, Wang, Pang, Ru, & Fang, 2016).

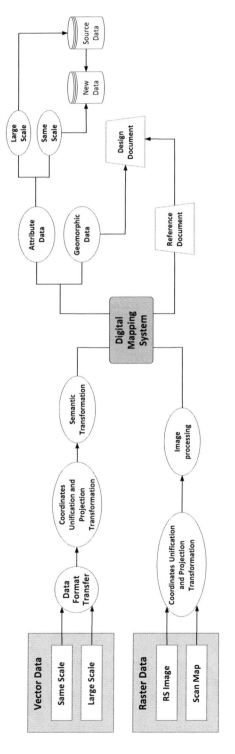

FIGURE 6.20 Technical workflow chart of multisource data integration (Chen, Sun, Xu, & Xiong, 2013).

Each of the above analyses has its own field of application, making it hard to say which method is the best. Each has advantages and disadvantages. As Morris and Van der Veer Martens (2008) point out, the scientometrics methods based on a certain relationship can reflect only a limited understanding of science from one perspective. In other words, every relationship can help researchers analyze only the partial characteristics of a research field from a certain perspective. It is only when we observe a research field from different perspectives that we can fully understand it. The integration of multiple relationships contributes to a more comprehensive understanding of a problem (Xu, Wang, Pang, Ru, & Fang, 2016).

6.2.4.2 Fusion of Multisource Data

The fusion of multisource data can be divided into the cross-integration of multimode data and the matrix fusion of multi-relational data.

The cross-integration of multimode data shows the associations among different types of entities, taking advantage of cross co-occurrence techniques but ignoring the same types of entities.

The matrix fusion of multi-relational data merges similar matrixes or distance matrixes of multisource data into a new matrix to present all relations; it then performs multivariate statistical analysis, such as cluster analysis and factor analysis.

Both the cross-integration of multimode data and the matrix fusion of multi-relational data eventually form a comprehensive matrix. For the former, the dimension of the newly formed matrix is the sum of dimensions of all the original matrixes, but for the latter, the dimension of the new matrix does not change (Xu, Wang, Pang, Ru, & Fang, 2016).

6.2.4.2.1 Cross-Integration of Multimode Data

In social network analysis, mode refers to a collection of actors, specifically, the number of the types of these collections. The network of relationships between two collections is called a two-mode crossing network (two-mode network). Similarly, the network of relationships among three collections is called a three-mode crossing network (three-mode network), and a multimode network corresponds to more than three collections.

The cross-integration of multimode data can be defined as the process of combining multisource data to form a multimode matrix, where connections only exist among different types of nodes and vice versa. Taking a two-mode network as an example, two-mode network data can be used to generate a bipartite graph whose vertices can be divided into two disjoint sets. In mathematical language, this can be expressed as the following: let $G = <V, E>$ be a graph, and V be the set of vertexes, such that $V_1 \cup V_2 = V$, and $V_1 \cap V_2 = \Phi$. G is called a bipartite graph, if for two end points of any edge in G, $(x, y) \in E$, giving us $x \in V_1$ and $y \in V_2$ or $x \in V_2$ and $y \in V_1$ (Xu, Wang, Pang, Ru, & Fang, 2016).

Some researchers combine bibliographies and words to discover research topics. One way to combine them is to use the bibliography as a qualification of the relationship between the words; another is to use citations to build the relationship between the bibliography and the words. When using indexing terms in the study of domain description, we may encounter some inherent problems because of linguistic phenomena, such as polysemy. However, a bibliography provides a specific context for indexing terms, and this, to a certain extent, can avoid these problems. Based on this consideration, Van Den Besselaar and Heimeriks (2006) used the method of word-reference co-occurrence to describe research objects and cluster literature collections in a method they call a "research front" (Xu, Wang, Pang, Ru, & Fang, 2016).

Leydesdorff (2010) applied the theory of a heterogeneous network to a three-mode network. He linked feature items of authors, journals, and keywords and showed how different types of nodes in the same network could analyze relationships between nodes to more realistically reflect research networks. Morris, DeYong, Wu, Salman, and Yemenu (2002) and Morris and Yen (2004) used the association of two co-occurrence matrixes with same feature items to conduct an application study of the cross-chart and timeline method; they solved the problem of visualization by revealing an association between two feature items. Pang (2012) improved Morris' cross-chart. The improved cross-chart not only reveals the association between two feature items but also presents the co-occurrence among three-feature items. During a patent-based tech mining analysis, Xu and Fang (2014) used association rules for the technological and functional dimensions in a patent technology-effect matrix to acquire the correlative degree of the technology subject and the effect subject in a certain field. They identified core patents or patent clusters of the same technology effect, same technology, or same effect in a two-mode network consisting of technology-effect subject terms. During this analysis, they merged technology subjects and effect subjects; they used a kind of three-mode network analysis to process the data. Using knowledge management as an example, Wei, Li, and Liu (2014) and Wei and Li (2014) built a three-mode network including authors, keywords, and periodicals and revealed the inherent law and development trend of knowledge management through empirical analysis and visualization. To address the problem of excessive outliers in authors' co-occurrence analysis, Teng (2015) analyzed the co-occurrence of authors, institutions, and countries and built a mixed author-institution-country co-occurrence network based on the theory of a super network. His empirical research showed the method could eliminate isolated nodes in the authors' co-occurrence network and enrich the amount of information in the network by adding the coupling relations between authors and institutions or countries. Finally, Pang (2012) indicated the analysis of cross co-occurrence in a three-mode network could reveal not only the knowledge of three-mode but also the knowledge of one-mode or two-mode, yielding deeper and broader knowledge than co-occurrence (Xu, Wang, Pang, Ru, & Fang, 2016).

6.2.4.2.2 Matrix Fusion of Multi-relational Data

Braam, Moed, and Van Raan (1991) combined co-citation cluster analysis with content word analysis to identify research topics. Based on co-citation clusters, they counted the frequency of words contained in references to test whether the cluster could gather literature possessing similar terms into a class. Calero-Medina and Noyons (2008) analyzed the knowledge creation and flow process between scientific publications by combining word co-occurrence and citation network analysis. They used word co-occurrence to find related terms and theories and citation network analysis to discover the key literature in this field. Zhang, Wang, and Cui (2007) found co-word clustering and strategic coordinates analysis were powerful tools to research discipline hotspots, and the combination of co-words and cited frequency could lead to better results (Xu, Wang, Pang, Ru, & Fang, 2016).

A. *Fusion of two types of relations*

In the field of information retrieval, Weiss, Vélez, and Sheldon (1996) developed a prototype system of hierarchical network search engine, HyPursuit system. It is used for retrieval and browsing by detecting the content-link clustering of hypertext documents. The clustering algorithm of content link is based on a literature similarity function which considers the similarity of the terms and the hyperlinks' similarity factor. The result is the maximum value of similarities. Using an approach based on reference links and co-words, Small (1998) identified the direct and indirect relationship between literatures. Janssens (2007) and Janssens, Glänzel, and De Moor (2008) examined the method of combining Web content and hyperlinks and merged the relationship based on words, with the relationship based on bibliographic coupling. They used Fisher's inverse chi-square method to construct new relational datasets and conducted an empirical study, the result of which showed the method could be used to find the structure of research on bioinformatics and information science. Finally, Janssens, Zhang, De Moor, and Glänzel (2009) integrated cross-citations of journals with text mining, then validated and improved the existing classification scheme of topics (Xu, Wang, Pang, Ru, & Fang, 2016).

B. *Fusion of triple relations*

Wang and Kitsuregawa (2002) proposed a clustering algorithm based on content-link coupling to retrieve Web pages. They integrated outbound links, inbound links, and terms to improve retrieval performance. He, Ding, Zha, and Simon (2001) and He, Zha, Ding, and Simon (2002) proposed a Web text clustering method merging the structure of text-based hyperlinks, co-citations, and text content. They used the structure of text-based hyperlinks to calculate similarities whose intensity was moderated by the text similarities and then integrated the hyperlink structure similarities and text similarities with co-citations using linear weighting to build a weighted adjacency matrix (Xu, Wang, Pang, Ru, & Fang, 2016).

C. *Evaluation of relational fusion results*

The findings for single relationship clustering differ. An experiment conducted by Calado (2006) found Web page retrieval based on link relations was superior to text classification, but other experiments showed similarity clustering based on words was superior to similarity clustering based on citations. However, all experiments indicated the clustering results after fusion were better than the results based on single-type relationships.

In the field of scientometrics, linear fusion is the most commonly used algorithm in relational fusion research. But MSDF is complex, and the three main types of data relationships are often not independent but correlated. Hence, a simple linear operation is not enough to solve the problem of data fusion. Still, we can learn from other research fields, such as sensors, automation, and so on, to improve and enrich the MSDF methods in scientometrics analysis (Xu, Wang, Pang, Ru, & Fang, 2016).

6.2.5 MSDF and Decision Support in Military Applications

MSDF is essential for decision-making support in the military. A sample decision-making model is depicted in Figure 6.21 (Waltz & Buede, 1986a). The model comprises four dynamically interacting elements: stimulus, hypotheses, options, and response (SHOR) (Tagarev and Ivanova, 1999).

- Stimulus: Initiation of the decision-making process to provide information on the current situation and the associated uncertainties.
- Hypotheses: A set of perception alternatives explaining the real-world situation.
- Options: Response alternatives available to the decision maker.
- Response: Selected action.

Waltz and Buede divide the "information" part of the model into two distinct subsystems: one for data fusion and the other for decision support (Waltz & Buede, 1986b). Data fusion collects information from various sensors and sources to develop the best possible perception of the situation. The situation is described by friendly and enemy orders of battle, locations, and movements of weapons and equipment, events, and intelligence as it relates to past, present, and predicted behavior of the enemy. In the fusion process, the authors include the collection, association, aggregation, and merging of data to create and display current and past situations. The decision support function creates and evaluates alternative estimates of the real situation and the responses available to the commander. Both functions are performed interactively, and the results of the military response are included in the model through a feedback loop (Tagarev & Ivanova, 1999).

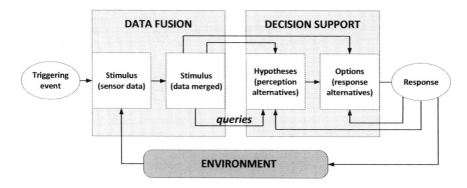

FIGURE 6.21 Data fusion and decision support in the SHOR model (Tagarev & Ivanova, 1999).

6.3 Three-Dimensional Visualization Applications

6.3.1 Application of 3D Visualization for Geographic Information Systems (GIS) Data

An online application of 3D visualization for GIS data is of interest not only to professionals, such as cartographers, geographers, geologists, and psychologists, but also to ordinary people.

The GIS is a conventional server/client-based architecture. This architecture is the main platform for online systems using a tiered distributing system. The tiers are normally set by developers to separate the various tasks. Each tier has its own responsibility, and any changes in any tier will not affect any other tier.

The three-tier framework is the best-known architecture for GIS applications, but the framework has drawbacks. More processing power is needed in the middle tier to process requests from multiple users to visualize the system. In addition, the three-tier framework is used for 2D visualization and is not appropriate for online application of 3D visualization. Applications such as GIS systems generate a massive amount of data. Consequently, the use of a three-tier framework for online applications of 3D visualization is not sufficient to process requests from users, and the system performs poorly (Che Mat, Mohamed Shariff, Mahmud, Pradhan, & Mohd Rahim, 2011).

6.3.1.1 A new Four-Tier Framework

A four-tier framework may overcome the problems of the three-tier framework and guide developers to design better online applications of 3D visualization. The system generated by this framework can help potential users visualize online applications of 3D visualization according to their needs. An online 3D visualization design using this framework will have the capability of fly through, walk through, pan, and slide and will also have a gravity setting. The data of 3D objects could be edited and controlled by users. When the data are edited, the system will automatically visualize the 3D applications with new updated data in 3D environments. Anyone who needs to monitor the updated data can access the system online. Individuals can access the system anywhere and make decisions on the spot, if required.

The four tiers are client, logic, visualization process, and database. Figure 6.22 shows the architecture. The functions of each tier are described below (Che Mat, Mohamed Shariff, Mahmud, Pradhan, & Mohd Rahim, 2011).

1. Database tier: This tier stores the data in the database of the system or database server. Only the data related to the application are stored. In this case, data on 3D objects are stored in text files or image files. When the user requests the 3D scene, the system will retrieve the related data from the database. In a three-tier framework, the middle tier will directly retrieve the data from the database tier. However, in the four-tier framework, the visualization process tier will retrieve the data from the database and perform the visualization process. The data are then sent back to the logic tier for interpretation. Thus, this system is more effective, as the visualization process tier processes the data separately without disturbing the logic tier. The data inside the database can be edited and keyed in by the users. The data can be edited based on the requirements of the application.

2. Logic tier: This is the most important tier. It functions as the interpreter of the process in the system. The entire system depends on this tier to translate the process into a form which can be understood by each tier. It will make translations between different protocols, optimize the load balance, and manage the connection between each tier. Most importantly, this tier will manage the activities or tasks which need to be performed by other tiers, such as the visualization process tier and the database tier. The entire request will be queued in this tier. After the first request is finished, the next waiting request will be processed.

3. Visualization process tier: This is the new tier added to the three-tier architecture. Its main task is to perform the visualization process on the data requested by the client tier. The process

Three-Dimensional Visualization

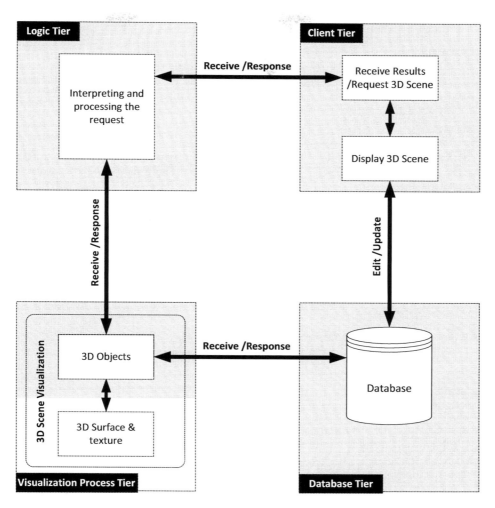

FIGURE 6.22 Architecture of four-tier framework (Che Mat, Mohamed Shariff, Mahmud, Pradhan, & Mohd Rahim, 2011).

starts when the tier receives a command from the logic tier. The command requests the visualization tier to retrieve the data from the database tier. The data on the characteristics of 3D objects are stored in the database tier. These data are submitted by database tier to the visualization process tier for processing. When it receives the data, it begins the process of visualizing the 3D objects. When it is done, the 3D objects will be placed automatically on top of a 3D surface because one of the data items stored in the database is the location of 3D objects. At the same time, the command requests the tier to process the 3D surface. The visualization process tier will process the 3D surface by overlaying it with the textured image. The processes run synchronously until the process is completed. The 3D scene is then generated and the results submitted back to the logic tier.

4. Client tier: The function of this tier is to request the 3D scene from the system. It acts as the user interface. In normal architecture, this tier is generally represented by the Web browser or plug in. In this framework, the plug in of a VRML viewer is needed to render a 3D scene. The entire process starts at this tier and ends when the generation of the 3D scene is completed. When this tier receives the response from the logic tier, it will render the 3D scene requested by the users. The users can interact with the 3D scene by using the functions of fly through, walk through, pan, slide, zoom in, and zoom out.

6.3.1.2 Interactions between Tiers

As illustrated in Figure 6.23, the tiers are represented by t_i which contains $t_1 - t_4$. Each t_i has its own functions: t_1 is the *client tier*; t_2 is the *logic tier*; t_3 is the *visualization process tier*; and t_4 is the *database tier*. Each t_i needs to interact with the others to process the request from t_1. The relationship between the tiers is very important because each t_i is dependent on the others (Che Mat, Mohamed Shariff, Mahmud, Pradhan, & Mohd Rahim, 2011).

The process starts at t_1 when the user requests the 3D data from t_4 through request a_i and the response through b_j. The logic tier t_2 receives the request a_1 from t_1 and processes the request by interpreting and making decisions on the request. When the process is completed, logic tier t_2 will make a request a_2. The request a_2 will ask visualization tier t_3 to retrieve the data from t_4. Then, visualization tier t_3 will make a request a_3. The request asks database tier t_4 to send the necessary data to visualization tier t_3. Database tier t_4 will carry out the corresponding operation and give the result back to visualization tier t_3. The result of the request will be response b_1 which will be sent from database tier t_4 to visualization process tier t_3 for the visualization process. Visualization tier t_3 will visualize the data, perform the necessary actions, and send response b_2 to logic tier t_2. At logic t_2, the results will be processed and interpreted in a form that can be understood by client tier t_1. Finally, response b_3 to the request for information will be sent from logic tier t_2 to client tier t_1. The user will get the output in the form of 3D scenes, and the scene outputs can interact.

With the four-tier framework, the processing power is reduced during the process of visualizing the 3D scene, and the performance of the system in an online environment is improved. The uniqueness of this proposed framework is that it can be applied to any type of online 3D visualization applications without changing the function of any tier (Che Mat, Mohamed Shariff, Mahmud, Pradhan, & Mohd Rahim, 2011).

6.3.2 Applications of 3D Landscape Visualization

Three-dimensional visualization tools are increasingly important in many fields of study. In geography, 3D visualization is considered an essential informational approach for such purposes as communicating existing conditions and creating alternative landscape scenarios for research, education, and consultation (Priestnall & Hampson, 2008). In landscape planning, 3D visualization can help build consensus on public issues by transforming large amounts of data into understandable images (Bishop & Lange, 2005). It can also serve as an engagement medium in public participatory decision-making processes (Stock et al., 2009; Wu, He, & Gong, 2010). Despite its utility, 3D landscape visualization has a relatively short history in landscape architecture (Ervin & Hasbrouck, 1999). Historically, landscape professions have not been thought to benefit from the use of 3D software (Hehl-Lange, 2001). The earliest effort to place 3D symbols in a landscape image was in 1969 in work by Harvard Spatial Analysis Laboratory. By 1985, 3D computer tools had been adopted more widely (Ervin & Hasbrouck, 1999).

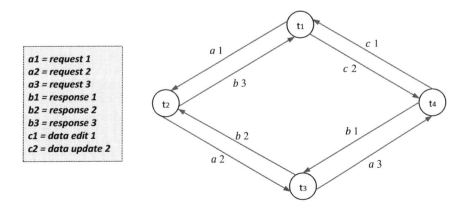

FIGURE 6.23 Interactions between tiers (Che Mat, Mohamed Shariff, Mahmud, Pradhan, & Mohd Rahim, 2011).

Three-Dimensional Visualization 219

The development of 3D software was accelerated by the booming growth of virtual reality (VR) technology in the digital gaming industry (Herwig & Paar, 2002). VR began to be used in spatial modeling, and the technique of merging image processing with geometric modeling opened the door for real-time rendering of virtual models (Danahy, 2001). However, in the early years of 3D use, many forms of information, such as leaf texture in 3D models, were not able to be synthesized by computers and had to be sampled from the real world (Ervin & Hasbrouck, 1999).

Over the past 20 years, significant research has explored the applications of 3D technology in visualization. A status report on computer use in landscape architecture in 1993 found few professionals used GIS or other 3D software in the design process (Palmer & Buhmann, 1994). A survey of projects using GIS and VR technology from 1993 to 1998 showed a rapid increase in 1994 and steady increases until 1998, after which there was a sharp decrease until 2001 (Haklay, 2002). The study concluded that the decline was due to the integration of VR technology into standard software, which reduced the justification for specialized research projects. At that time, using VR and GIS as tools for research projects in the field of urban and regional planning made up 29% of the total use of 3D software (Haklay, 2002). Before 1993, a limited number of government agencies required 3D landscape simulations as a standard procedure for landscape planning and management (Lange, 1994; Sheppard, 1989). The similarity between the actual landscape and the landscape representation was questioned (Daniel, 1992). Today, with significant improvements in image quality and related computer technologies, there is an increased level of detail in 3D visualizations. He, Yang, Shifley, and Thompson (2011), found the improved detail has helped eliminate ambiguity and increased the validity of visualization results (Yan, 2014).

A limited body of work deals with user evaluations of landscape visualization. In 1997, a nationwide survey on the computer skills and training of landscape architecture professionals found two-thirds of the respondents' computer skills were self-taught, and a lack of standard training in landscape architecture was one of the biggest obstacles to computer applications in landscape architecture (Palmer, 1997). Paar (2006) surveyed environmental planners, landscape practitioners, and other related professionals in Germany in 2006. He concluded that 3D software had a positive future in landscape visualization. In the United States, little research has been conducted since 1997 to evaluate the conditions of 3D landscape visualization. This gap is also apparent within the field of landscape architecture education where few researchers have explored the role of 3D computer technology on teaching and learning within the field of landscape architecture (Sheppard, 2001; Nielsen, Fleming, Kumarasuriyar, & Gard, 2010). This could change, however, as a criterion for evaluating the overall landscape visualization quality was recently designed by Sheppard and Cizek (2009). The criterion established six categories: accuracy, representativeness, visual clarity, interest, legitimacy, and access.

6.3.3 Three-Dimensional Visualization Software Applications

In the development process, 3D visualization software can be used to (Lifecycle Insights):

- Enable non-engineering roles to participate in design: 3D visualization software would allow those in other roles to view and interrogate 3D models, including those in manufacturing, service, procurement, and other areas.
- Allow collaboration across the supply chain: Other organizations could view and interrogate a 3D model even if they did not have the MCAD software used to design the model.

6.3.4 Future of Three-Dimensional Visualization for Industrial Applications

VR/AR (augmented reality) plus mixed reality is a hot topic. The use cases for industrial applications include: sales tools for potential customers to get familiar with the product offerings, field evaluation, design review, collaboration with customers, service calls, and training.

In a typical industrial application, engineers could use CAD systems (such as AutoCAD, Solidworks, ProE, UG-NX, Sketchup, and OnShape) to draft the base geometry in 2D, then extrude or rotate along a path into 3D objects. These 3D objects could be visualized outside the CAD system inside a Web browser without requiring the downloading of viewers onto a computer or a mobile device and without the need for immersive 3D devices, such as AR/VR or mixed reality.

The challenges of 3D visualization technologies for industrial applications include the creation of models and the fast response required to update the display content when design changes occur (Wu, 2017).

6.3.4.1 Cost and Cycle Time Challenges

Each device requires a different process and different model format. For example, individual models have to be created separately for phone, Web, VR, and AR displays. It is very expensive and time consuming to make the 3D visualization available on all devices.

Users want to see fast responses to their interactions in virtual 3D models. Dynamic contents can include the turning of blades on an engine or the animation of 3D objects along a timeline. Setting up dynamic contents currently requires intense manual efforts (Wu, 2017).

6.3.4.2 Some Short-Term and Long-Term Trends in 3D Visualization Technology

- More and more CAD applications are going online and appearing on smartphones under a business model called SaaS. The same CAD model is available on multiple devices, and data can be synchronized across devices. CAD visualization is starting to display not only 3D geometry information but also the related technical or service documents alongside the 3D geometry.
- CAD companies have started to enable the conversion from a CAD model to a 3D AR/VR model inside their CAD software. There are other software players in the market whose software can work with different CAD formats to convert them into VR/AR models and provide basic model navigation and collaboration functions. The first approach is more user friendly, as a CAD model and a VR/AR model can be designed inside CAD software without other software involved.
- Many VR software interfaces currently allow viewers' collaboration from multiple locations but are constrained by the need to use the same software and same type of device for all collaborators. In the future, multi-site collaboration will not require use of the same type of device but will be possible across multiple types of devices. A customer in Paris wearing her Oculus VR headset will view the same 3D model as a colleague in New York City looking at the model through an MS HoloLens or as the design team standing in front of a VR CAVE. They will all see each other inside the virtual 3D world, knowing where each person is inside the same 3D model.
- The bottleneck for 3D visualization technology is the hardware at this point. In the long term, all devices will eventually be consolidated, once a powerful computer can be built in a significantly smaller size to the point where it can be worn or embedded everywhere, conveniently projecting images in front of our eyes wherever we go. We could see anything as 3D or 4D (3D+time) and discuss things in the 3D virtual world with anyone anywhere in the world (Wu, 2017).

REFERENCES

Ahlgren P., Colliander C., 2009. Document–document similarity approaches and science mapping: Experimental comparison of five approaches. *Journal of Informetrics*, 3(1), 49–63.

Andrews D. F., 1972. Plots of high-dimensional data. *Biometrics*, 29:125–136.

Ankerst M., Keim D., Kriegel H. P., 1996. Circle segments: A technique for visually exploring large multidimensional data sets. In *Proceedings of Visualization 96, Hot Topic Session*, San Francisco, CA, USA, 1996.

Berthold M., Hand D. J., 2002. Chapter 11: Visualization. In *Intelligent Data Analysis: An Introduction* (pp. 403–427). Berlin: Springer.

Bishop I. D., Lange E., 2005. *Visualization in Landscape and Environmental Planning, Technology and Applications*. London: Taylor and Francis.

Braam R. R., Moed H. F., Van Raan A. F., 1991. Mapping of science by combined cocitation and word analysis, I. Structural aspects. *JASIS*, 42(4), 233–251.

Brath R., Peters M., Senior R., 2005. Visualization for communication: The importance of aesthetic sizzle. In *Proceedings of the Ninth International Conference on Information Visualisation (IV'05)*, Washington, DC, USA: IEE Computer Society, 2005, pp. 724–729.

Bowman D. A., Hodges L. F., 1995. User interface constraints for immersive virtual environment applications. Technical Report 95-26, Graphics, Visualisation and Usability Center, Georgia Institute of Technology, Atlanta, GA, USA.

Brown D. G., Riederer S. J., 1992. Contrast-to-noise ratios in maximum intensity projection images. *Magnetic Resonance in Medicine*, 23, 130–137

Calado P., 2006. Link-based similarity measures for the classification of web documents. *Journal of the American Society for Information Science and Technology*, 57(2), 208–222.

Calero-Medina C., Noyons E. C., 2008. Combining mapping and citation network analysis for a better understanding of the scientific development: The case of the absorptive capacity field. *Journal of Informetrics*, 2(4), 272–279.

Carson E., Parberry I., Jensen B., 2007. Algorithm explorer: Visualizing algorithms in a 3D multimedia environment. In *Proceedings of the 38th SIGCSE technical Symposium on Computer science education (SIGCSE '07)*, New York, USA: ACM Press, 2007, pp. 155–159.

Che Mat R., Mohamed Shariff A. R., Mahmud A. R., Pradhan B., Mohd Rahim M. S., 2011. A new four tier framework for online application of 3D visualization. In *2011 International Conference on Future Information Technology*, IPCSIT Vol. 13 (2011) © (2011) IACSIT Press, Singapore, November 2011.

Chen L. S., Herman G. T., Reynolds R. A., 1985. Surface rendering in the cuberille environment. *IEEE Computer Graphics and Applications*, 5, 33–43.

Chen H., Sun Q., Xu L., Xiong Z., 2013. Application of data fusion in the production and updating of spatial data. In *International Archives of the Photogrammetry, Remote Sensing and Spatial Information Sciences*, Volume XL-7/W1, 3rd ISPRS IWIDF 2013, Antu, Jilin Province, PR China, 20–22 August 2013.

Chernoff H., 1973. The use of faces to represent points in k-dimensional space graphically. *Journal of the American Statistical Association*, 68, 361–368.

Chuah M. C., Roth S. F., Mattis J., Kolojejchick J., 1995. SDM: Selective dynamic manipulation of visualizations. In *Proceedings of the ACM Symposium on User Interface Software and Technology, ser. 3D User Interfaces*, Pittsburgh, PA, USA, 1995, pp. 61–70.

Cleveland W. S., 1993. *Visualizing Data*. Murray Hill, NJ: AT&T Bell Laboratories; Summit, NJ: Hobart Press.

Cockburn A., McKenzie B., 2002. Evaluating the effectiveness of spatial memory in 2D and 3D physical and virtual environments. In *Proceedings of the SIGCHI Conference on Human factors in Computing Systems (CHI '02)*, New York, USA: ACM, 2002, pp. 203–210.

Couto T., 2006. A comparative study of citations and links in document classification. In *6th ACM/IEEE-CS Joint Conference on Digital Libraries*, Chapel Hill, NC, USA. pp. 75–84.

Danahy J. W., 2001. Technology for dynamic viewing and peripheral vision in landscape visualization. *Landscape and Urban Planning*, 54(14), 125–137.

Daniel T. C., 1992. Data visualization for decision support in environmental management. *Landscape and Urban Planning*, 21(4), 261–263.

Diehl S., 2007. *Software Visualization: Visualizing the Structure, Behaviour, and Evolution of Software*. New York: Springer-Verlag.

Ervin S. M., Hasbrouck H., 1999. 30 years of computing in landscape architecture. *Landscape Architecture*, 89(11), 54–56.

Frieder G., Gordon D., Reynolds R., 1985. Back-to-front display of voxel-based objects. *IEEE Computer Graphics and Applications*, 5, 52–60.

Furnas G. W., Buja A., 1994. Prosections views: Dimensional inference through sections and projections. *Journal of Computational and Graphical Statistics*, 3(4):323–353, 1994.

Gall H., Jazayeri M., Riva C., 1999. Visualizing software release histories: The use of color and third dimension. In *Proceedings of the IEEE International Conference on Software Maintenance (ICSM '99)*, Washington, DC, USA: IEEE Computer Society, 1999, p. 99.

Gianinetto M., Scaioni M., 2003. Fusion of aerial and satellite imagery over the city of Venezia. In *Proceedings of 2nd GRSS/ISPRS joint workshop on remote sensing and data fusion over urban areas*, Berlin: GRSS/ISPRS, pp. 216–219.

Goldwasser S., Reynolds R., 1987. Real-time display and manipulation of 3-D medical objects: The voxel machine architecture. *Computer Vision, Graphics, and Image Processing*, 39, 1–27.

Gordon D., Reynolds R. A., 1985. Image-space shading of three-dimensional objects. *Computer Vision, Graphics, and Image Processing*, 29, 361–376.

Haklay M. E., 2002. Virtual reality and geographical information systems: Analysis and trends. In P. Fisher & D. Unwin (Eds.), *Virtual Reality and Geography* (pp. 47–57). London: Taylor and Francis.

He X., Ding C. H., Zha H., Simon H. D., 2001. Automatic topic identification using webpage clustering. Data mining, 2001. ICDM 2001. In *IEEE International Conference*, San Jose, CA, USA. pp. 195–202.

He X., Zha H., Ding C. H., Simon H. D., 2002. Web document clustering using hyperlink structures. *Computational Statistics & Data Analysis*, 41(1), 19–45.

He H. S., Yang J., Shifley S. R., Thompson F. R., 2011. Challenges of forest landscape modeling-simulating large landscapes and validating results. *Landscape and Urban Planning*, 100(4), 400–402.

Hearst M., 1995. Tilebars: Visualization of term distribution information in full text information access. In *Proceedings of ACM Human Factors in Computing Systems Conference (CHI '95)*, Denver, CO, USA, 1995, pp. 59–66.

Hehl-Lange S., 2001. Structural elements of the visual landscape and their ecological functions. *Landscape and Urban Planning*, 54(1–4), 105–113.

Herman G., Liu L., 1977. Display of three-dimensional information in computed tomography. *Journal of Computer Assisted Tomography*, 1, 155–160.

Hertz S. M., Baum R. A., Owen R. S., Holland G. A., Logan D. R., Carpenter J. P., 1993. Comparison of magnetic resonance angiography and contrast arteriography in peripheral arterial stenosis. *American Journal of Surgery*, 166, 112–116.

Herwig A., Paar P., 2002. Game engines: Tools for landscape visualization and planning? In E. Buhmann, U. Nothelfer, & M. Pietsch (Eds.), *Trends in GIS and Virtualization in Environmental Planning and Design* (pp. 162–171). Heidelberg, Germany: Anhalt University of Applied Sciences.

Höhne K. H., Bernstein R., 1986. Shading 3D images from CT using gray-level gradients. *IEEE Transactions on Medical Imaging*, 5, 45–47.

Hua B. L., 2013. Research on the methods of multi-source fusion. *Information Studies: Theory & Application*, 36(11), 16–19.

Huber P. J., 1985. The annals of statistics. *Projection Pursuit*, 13(2), 435–474.

Inselberg A., Dimsdale B., 1990. Parallel coordinates: A tool for visualizing multidimensional geometry. In *Proceedings of Visualization 90*, San Francisco, CA, USA, pp. 361–370.

Irani P., Ware C., 2003. Diagramming information structures using 3D perceptual primitives. *ACM Transactions on Computer-Human Interaction*, 10(1), 1–19.

Janssens F., 2007. Clustering of scientific fields by integrating text mining and bibliometric.

Janssens F., Glänzel W., De Moor B., 2008. A hybrid mapping of information science. *Scientometrics*, 75(3), 607–631.

Janssens F., Zhang L., De Moor B., Glänzel W., 2009. Hybrid clustering for validation and improvement of subject-classification schemes. *Information Processing & Management*, 45(6), 683–702.

Keim D., Kriegel H. P., 1994. Visdb: Database exploration using multidimensional visualization. *Computer Graphics & Applications*, 6, 40–49.

Keim D., Kriegel H. P., Ankerst M., 1995. Recursive pattern: A technique for visualizing very large amounts of data. In *Proceedings of Visualization 95*, Atlanta, GA, USA, pp. 279–286.

Keim D., 2000. *An Introduction to Information Visualization Techniques for Exploring Large Databases*. Halle: Institute for Computer Science University of Halle.

Keim D., 2000a. Designing pixel-oriented visualization techniques: Theory and applications. *Transactions on Visualization and Computer Graphics*, 6(1), 59–78.

Keim D., 2002. Information visualization and visual data mining. *IEEE Transactions on Visualization and Computer Graphics*, 8(1), 1–8.

Keim D., Ward M., 2002. Visual data mining techniques. In: M. Berthold & D. J. Hand (Eds.), *Intelligent Data Analysis: An Introduction* (pp. 2–27). Berlin: Springer.

Klavans R., Boyack K. W., 2006. Quantitative evaluation of large maps of science. *Scientometrics*, 68(3), 475–499.

Koschke R., 2002. Software visualization for reverse engineering. In S. Diehl (Ed.), *Software Visualization*. Berlin, Heidelberg: Springer Verlag, Vol. 2269 of LNCS State-of-the-Art Survey.

Kreuseler M., Lopez N., Schumann H., 2000. A scalable framework for information visualization IEEE Symposium on Information Visualization 2000. INFOVIS 2000, Salt Lake City, UT, USA. 11 pages.

Lange E., 1994. Integration of computerized visual simulation and visual assessment in environmental planning. *Landscape and Urban Planning*, 30(1–2), 99–112.

LeBlanc J., Ward M. O., Wittels N., 1990. Exploring n-dimensional databases. In *Proceedings of the Conference on Visualization '90*, San Francisco, CA, USA, 1990, pp. 230–239.

Levkowitz H., 1991. Color icons: Merging color and texture perception for integrated visualization of multiple parameters. In *Proceedings of the Conference on Visualization '91*, San Diego, CA, USA, 1991, pp. 22–25.

Levoy M., 1988. Display of surfaces from volume data. *IEEE Computer Graphics and Applications*, 8, 29–37.

Levoy M., 1990. Display of surfaces from volume data. *ACM Transactions on Graphics*, 9, 245–271.

Leydesdorff L., 2010. What can heterogeneity add to the scientometric map? Steps towards algorithmic historiography. arXiv preprint arXiv. 1002.0532.

Lifecycle Insights. What is 3D visualization? Lifecycle Insights (LC-Insights LLC). www.lifecycleinsights.com/tech-guide/3d-visualization/. Viewed: June 03, 2019.

Mili R., Steiner R., 2002. Software engineering - Introduction. In *Revised Lectures on Software Visualization, International Seminar*, London, UK: Springer-Verlag, 2002, pp. 129–137.

Morris S., DeYong C., Wu Z., Salman S., Yemenu D., 2002. DIVA: A visualization system for exploring document databases for technology forecasting. *Computers & Industrial Engineering*, 43(4), 841–862.

Morris S. A., Yen, G. G., 2004. Crossmaps: Visualization of overlapping relationships in collections of journal papers. *Proceedings of the National Academy of Sciences*, 101(suppl 1), 5291–5296.

Morris S. A., Van der Veer Martens B., 2008. Mapping research specialties. *Annual Review of Information Science and Technology*, 42(1), 213–295.

Mullet K., Schiano D. L., Robertson G., Tesler J., Tversky B., Schiano D. J., 1995. 3D or not 3D: More is better or less is more? In *Proceedings of ACM CHI'95 Conference on Human Factors in Computing Systems, ser. Panels*, Denver, CO, USA. 1995; Vol. 2, pp. 174–175.

Napel S., Marks M. P., Rubin G. D., 1992. CT angiography with spiral CT and maximum intensity projection. *Radiology*, 185, 607–610.

Nielsen D., Fleming M. J., Kumarasuriyar A. C., Gard S., 2010. Digital design communication: Measuring learner technological prowess and self-efficacy in problem resolution. In *Paper presented in INTED 2010 International Technology, Education and Development Conference*, Valencia, Spain.

Nowell L., Havre S., Hetzler B., Whitney P., 2001. Themeriver: Visualizing thematic changes in large document collections. *IEEE Transactions on Visualization and Computer Graphics*, 8(1), 9–20.

Paar P., 2006. Landscape visualizations: Applications and requirements of 3D visualization software for environmental planning. *Computer, Environment and Urban Systems*, 30(6), 815–839.

Palmer J. F., Buhmann E., 1994. A status report on computers. *Landscape Architecture*, 84(7), 54–55.

Palmer J. F., 1997. The 1996 status report on computing skills and training in landscape architecture. *Newsletter of the ASLA Open Committee on Computing*, 9(2), 2–5.

Pang H. S., 2012. *Research on Knowledge Discovery Method Based on Multiple Co-Occurrence*. Beijing: University of Chinese academy of science.

Persson M., Duckett T., Lilienthal A., 2008. Fusion of aerial images and sensor data from a ground vehicle for improved semantic mapping. *Robotics and Autonomous Systems*, 56(6), 483–492.

Pickett R. M., Grinstein G. G., 1988. Iconographic displays for visualizing multidimensional data. In *Proceedings of IEEE Conference on Systems, Man and Cybernetics*, Piscataway, NJ, USA: IEEE Press, 1988, pp. 514–519.

Plaisant C., Grosjean J., Bederson B. B., 2002. Spacetree: Supporting exploration in large node link tree, design evolution and empirical evaluation. *Information Visualization*, 00, 57.

Pohl C., van Genderen J. L., 1998. Multisensor image fusion in remote sensing: Concepts, methods and applications. *International Journal of Remote Sensing*, 19(5), 823–854.

Priestnall G., Hampson D., 2008. Landscape visualizations: Science and art. In M. Dodge, M. McDerby, & M. Turner (Eds.), *Geographic Visualization: Concepts, Tools and Applications* (pp. 241–258). Chichester: Wiley.

Reynolds R., Gordon D., Chen L., 1987. A dynamic screen technique for shaded graphics display of slice-represented objects. *Computer Vision, Graphics, and Image Processing*, 38, 275–298.

Rilling J., Mudur S. P., 2002. On the use of metaballs to visually map source code structures and analysis results onto 3d space. In *WCRE'02*, Washington, DC, USA: IEEE Computer Society, 2002, p. 299.

Robertson G., Card S. K., Mackinlay J. D., 1993. Information visualization using 3D interactive animation. *Communications of the ACM*, 36(4), 57–71.

Schreiner S., Paschal C. B., Galloway R. L., 1996. Comparison of projection algorithms used for the construction of maximum intensity projection images. *Journal of Computer Assisted Tomography*, 20, 56–67.

Shaomei L., 2004. *Study on Digital Relief Shading Theories and Technologies*. Zhengzhou: Information Engineering University.

Sheppard S. R. J., 1989. *Visual Simulation: A User's Guide for Architects, Engineers, and Planner*. New York: Van Nostrand Reinhold.

Sheppard S. R. J., 2001. Guidance for crystal ball gazers: Developing a code of ethics for landscape visualization. *Landscape and Urban Planning*, 54(1–4), 183–199.

Sheppard S. R. J., Cizek P., 2009. The ethics of Google Earth: Crossing thresholds from spatial data to landscape visualization. *Journal of Environmental Management*, 90(6), 2102–2127.

Shibata N., Kajikawa Y., Takeda Y., Matsushima K., 2009. Comparative study on methods of detecting research fronts using different types of citation. *Journal of the American Society for Information Science and Technology*, 60(3), 571–580.

Shneiderman B., 1996. The eye has it: A task by data type taxonomy for information visualizations. In *IEEE symposium on Visual Languages*, Boulder, CO, USA. 1996.

Shneiderman B., 2003. Why not make interfaces better than 3D reality? *IEEE Computer Graphics and Applications*, 23(6), 12–15.

Small H., 1998. A general framework for creating large-scale maps of science in two or three dimensions: The SciViz system. *Scientometrics*, 41(1–2), 125–133.

Spence R., Tweedie L., Dawkes H., Su H., 1995. Visualization for functional design. In *Proceedings of International Symposium on Information Visualization (InfoVis '95)*, Atlanta, Georgia, USA, 1995. pp. 4–10.

Stasko J. T., Wehrli J. F., 1993. Three-dimensional computation visualization. In *Proceedings of IEEE Symposium on Visual Languages*, IEEE Press, 24–27 August 1993, Bergen, Norway. pp. 100–107.

Stock C., Bishop I. D., O'Connor A. N., Chen T., Pettit C. J., Aurambout J. P., 2009. SIEVE: Collaborative decision-making in an immersive online environment. *Cartography and Geographic Information Science*, 35(2), 133–144.

Tagarev T., Ivanova P., 1999. Computational intelligence in multi-source data and information fusion. *Information & Security. An International Journal*, 2, 33–49.

Teng L., 2015. Research on hybrid co-occurrence network of Author-Institution-country based on super network. *Journal of the China Society for Scientific and Technical Information*, 34(1), 28–36.

Teyseyre A., Campo M., 2008. An overview of 3D software visualization. *IEEE Transactions on Visualization and Computer Graphics*, 15(1), 87–105.

Udupa J., Odhner D., 1991. Fast visualization, manipulation, and analysis of binary volumetric objects. *IEEE Computer Graphics and Applications*, 11, 53–62.

Udupa J., Odhner D., 1993. Shell rendering. *IEEE Computer Graphics and Applications*, 13, 58–67.

Udupa J. K., Odhner D., Samarasekera S., 1994. 3DVIENIX: An open transportable, multidimensional, multimodality, multiparametric imaging software system. *SPIE Proceedings*, 2164, 58–73.

Udupa J. K., 1999. Three-dimensional visualization and analysis methodologies: A current perspective. *Imaging & Therapeutic Technology*, 19(3): 783–806. https://pubs.rsna.org/doi/full/10.1148/radiographics.19.3.g99ma13783#F13. Viewed: May 23, 2019.

Van Den Besselaar P., Heimeriks G., 2006. Mapping research topics using word-reference co-occurrences: A method and an exploratory case study. *Scientometrics*, 68(3), 377–393.

Van Wijk J. J., Van Liere R. D., 1993. Hyperslice. In *Proceedings of the Conference on Visualization '93*, San Jose, CA, USA, 1993, pp. 119–125.

Wagner W., 2006. Gaussian decomposition and calibration of a novel small-footprint full-waveform digitizing airborne laser scanner. *ISPRS Journal of Photogrammetry and Remote Sensing*, 60(2), 100–112.

Waltz E. L., Buede D. M., 1986a. Data fusion and decision support for command and control. *IEEE Transactions on Systems, Man, and Cybernetics* SMC-16, 6 (November-December 1986), 865–879.

Waltz E. L., Buede D. M., 1986b. Data fusion and decision support.

Wang Y., Kitsuregawa M., 2002. Evaluating contents-link coupled web page clustering for web search results. In *11th International Conference on Information and Knowledge Management*, pp. 499–506.

Ward M. O., 1994. Xmdvtool: Integrating multiple methods for visualizing multivariate data. In *Proceedings of the Conference on Visualization '94*, Washington, DC, USA, 1994, pp. 326–336.

Ware C., Franck G., 1996. Evaluating stereo and motion cues for visualizing information nets in three dimensions. *ACM Transactions on Graphics*, 15(2), 121–140.

Weiss R., Vélez B., Sheldon M. A., 1996. HyPursuit: A hierarchical network search engine that exploits content-link hypertext clustering. In *7th ACM conference on Hypertext*, Cambridge, MA, USA. pp. 180–193.

Wei X. Q., Li C. L., 2014. The research of science collaboration behavior based on author year keyword network—The case of the library and information science. *Journal of Intelligence*, 11, 117–123.

Wei X. Q., Li C. L., Liu F. F., 2014. Construction and visualization of 3-mode network: A literature study of knowledge management of library and information science. *Information Studies: Theory & Application*, 08, 74–78+89.

Wu H., He W., Gong J., 2010. A virtual globe-based 3D visualization and interactive framework for public participation in urban planning processes. *Computer, Environment and Urban Systems*, 34(4), 291–298.

Wu Y., 2017. Future of 3D visualization for industrial applications. Technology Leader for Data Analytics | 3D&4D Data Visualization.

Xu H. Y., Fang S., 2014. Core patents mining based on cross co-occurrence analysis to patent technology-effect subject terms and citations. *Library and Information Service*, 33(2), 158–166.

Xu H. Y., Wang C., Pang H., Ru L., Fang S., 2016. Multi-source data fusion study in scientometrics. *Qualitative and Quantitative Methods in Libraries (QQML)*, 5, 611–626. ISSN 2241-1925.

Yan J., 2014. An evaluation of current applications of 3D visualization software in landscape architecture. Master of Landscape Architecture. UTAH State University, Logan, UT, USA.

Zhang J., 2010. Multi-source remote sensing data fusion: Status and trends. *International Journal of Image and Data Fusion*, 1(1), 5–24. doi:10.1080/19479830903561035.

Zhang H., Wang X. Y., Cui L., 2007. Trend: Co- word analysis method combined with literature CI research thematic areas. *Information Studies: Theory & Application*, 30(3), 378–380.

7
Communications

7.1 Communication Methods

7.1.1 Introduction

We all know the importance of communication in our daily lives. Nothing can take place without some method of communication being used to express ourselves for whatever purpose. Communication is even more valuable in a business environment as there are several parties involved. Various stakeholders, whether customers, employees, or the media, are always sending important information to each other at all times. Without the various methods of communication available today, it would be very difficult to carry out business efficiently and quickly (Tutorials Point, 2019).

7.1.2 Communication Types and How to Improve Them

1. Verbal (in-person) communication. Face-to-face communication in the workplace can eliminate many misunderstandings. Speaking directly with someone allows both of you to see and hear the bigger picture. You can see the other person's facial expressions and body language. The emphasis on focusing on one another and the conversation promotes a feeling of value and credibility. Having one quick conversation rather than a multitude of emails can be more efficient in the end.
2. Body language and facial expressions. As mentioned above, body language and facial expressions play a vital role in how effective or ineffective a person's communication will be. Eye contact makes people feel acknowledged as they talk and listen. A relaxed stance with your arms by your sides encourages others to feel comfortable in a meeting. If you rehearse what you're going to say before a meeting, do so in front of a full-length mirror so you can see if what your body language says matches up with your words. People will receive your words more positively if your facial expressions and body language don't put them on edge.
3. Phone conversations. The phone is a common part of most business days and, as such, is a vital aspect of communication. Some people enjoy talking by phone. Others dislike it so much that they put it off until the last minute. You can positively impact the effectiveness of your phone communication by doing three simple things. Smile while you talk, as it gives your voice a friendlier tone. Take notes before the conversation and use them to ensure you cover all points. Finally, speak clearly and slowly so your words are understood.
4. Written communication. Emails, memos, and notes are common forms of written communication. Of all four modes of communication, this one can lead to the most misunderstandings. People often read between the lines or feel there are implied messages or emotions in written forms of communication. Before sending an important email, have an objective pair of eyes read it first and provide constructive criticism (Miller, 1996).

7.1.3 Why Do We Need Different Communication Methods?

Communication plays a pivotal role in the effective functioning of a business, and the various modes are suitable for different purposes. For example, email allows a customer to send an important proposal quickly and directly to the right person. Similarly, if a person is not in the country but must give a presentation, he or she can turn to video conferencing. The most effective businesses will tailor their methods of communication to the context (Tutorials Point, 2019).

7.1.3.1 Choosing the Right Method

It is important that the most cost-effective methods of communication are chosen for any organization. Simply choosing a method of communication because it is a well-known type is not going to help.

To select a method, ask the following questions:

- What is our target audience?
- How much are we willing to spend?
- Will it increase productivity in the long run?
- What kind of information do we send out most often?

There may be more questions based on the type of work and the message sent. There is no "right" method of communication. Different methods suit different purposes and tasks (Tutorials Point, 2019).

7.1.4 Effective Methods of Communication

Most of the time, when the word communication comes to the mind, people think about exchanging ideas and information by means of words, but this is a small component of communication. In the 21st century, businesses have access to a number of methods of communication with internal and external audiences. These include traditional business meetings, print, videoconferencing, and social media. While a lot of attention is paid to newer methods, traditional methods have value and relevance. The determination of the method that is right for a certain occasion is a vital decision (Rohn, 2017).

7.1.4.1 Ways to Communicate Effectively in the Workplace

This section looks at types of communication commonly used in the workplace (Rohn, 2017).

7.1.4.1.1 Oral Communication

Oral communication involves the use of words to deliver a message. It can be one-on-one, over the phone, in group settings, etc. The oral form is personal and effective. Seeing the person with whom face-to-face communication is taking place helps to gauge the response of that person by understanding his or her body language and assists in his/her active participation in the dialogue (Rohn, 2017).

Oral communication can be informal, such as the rumor mill or the grapevine, or formal such as conferences and lectures. How effective it is will depend on the speed, volume, pitch, voice modulation, and clarity of speech, as well as the various nonverbal visual cues and body language. While using verbal communication, the person needs to be aware of his or her tone of voice and inflection. Use of sarcasm and angry tones should be avoided.

Oral communication makes the conveying of thoughts faster and easier and is the most successful method of communication. However, it makes up just 7% of all human communication. Some important oral methods of communication are the following (Rohn, 2017):

1. Face-to-face communication

    This is a preferred method of communication although it is not realistic all the time, especially in organizations with several locations around the globe. However, tools like videoconferencing make face-to-face communication possible and help large organizations create personal connections between management and staff.

2. Meetings

Meetings are common in most business settings; it is now possible to augment them using technology. People in remote locations can participate in discussions, although they are not physically present.

3. Storytelling

Storytelling is an effective form of oral communication that helps in the construction of common meanings. Stories assist in clarifying key values and demonstrating the way things are done. The tone, strength, and frequency of the story have a relationship with higher organizational commitment. The quality of stories told is often related to the capability of securing capital. Stories can also reinforce and perpetuate the culture of an organization.

4. Crucial conversations

Even though the process is similar, more skill, reflection, and planning are needed for high-stakes communications than for daily interactions at work. Examples of communication involving high stakes include presenting a venture capitalist with a business plan or asking for a raise. Apart from these, at times in our professional lives, we have conversations that are crucial—discussions in which stakes are high, emotions are strong, and opinions may differ.

7.1.4.1.2 Written Communication

Written communication cannot be avoided in the workplace. It comes in the form of traditional paper and pen documents and letters, text chats, emails, typed electronic documents, reports, SMS, and anything else that might be conveyed by the use of written symbols like language. Written communication includes formal business proposals, press releases, memos, contracts, brochures, handbooks, and the like. Such methods of communication are indispensable for any organization.

The effectiveness of the written communication depends on the style of writing, vocabulary, grammar, and clarity. Written communication is most suitable in cases requiring detailed instructions, when something has to be documented, or when someone is too far away to speak in person or over the phone (Rohn, 2017). Two increasingly common forms of written communication are social media and email.

1. Social media

Social media tools such as Facebook can be used by organizations to communicate between locations, divisions, departments, and employees. Certain tools such as Yammer are specifically designed to create social media networks in the confines of a particular domain of business.

2. Email

Email communication in organizations is very common and has become an essential tool for sharing information with one or thousands of persons. It is usually a significant component of an organization's communication plan, and even though it may not be a replacement for face-to-face or other methods of written communication, it is easily accessible and inexpensive.

7.1.4.1.3 Nonverbal Methods of Communication

The most important thing in communication is hearing what isn't said.

Communication by sending and receiving wordless messages, for example, through facial expressions, is a part of nonverbal communication. Such messages are a vital part of any communication and generally reinforce oral communication (Rohn, 2017). Types of nonverbal communication include the following:

1. Physical nonverbal communication

Physical nonverbal communication or body language includes body posture, facial expressions, eye contact, gestures like a pointed finger, a wave and the like, touch, tone of voice, overall movements of the body, and others.

Movements, eye contact, gestures, and the way you sit and stand all convey a message to the person with whom you are communicating. Gestures should be used appropriately or the hands should be left at your sides. Avoid fidgeting, as it is distracting, and don't cross your arms, as it sends a signal of being closed off or angry. Always look in the eyes of the person to whom you are speaking and never stare or roll your eyes when he or she is talking.

The most common physical nonverbal communication method is facial expression. For example, a frown or a smile conveys emotions that are distinct but difficult to express otherwise. Our emotions are conveyed by our facial muscles. A silent message can be sent without uttering a word. A change in our emotional state is visible in the change of our facial expression. For example, if we focus on being confident prior to an interview, the confidence will be conveyed to the interviewer by our face. Smiling even when feeling stressed can reduce the stress levels of the body (Rohn, 2017).

2. Paralanguage

A significant component of nonverbal communication is how something is said, not what is said. This consists of style of speaking, tone, emotion, stress, pitch, intonation, and voice quality. It helps in the communication of interest, approval, or the absence of either. The following nonverbal communication forms generally communicate the personality of a person:

- Aesthetic communication or creative expressions such as painting and dancing.
- Personal appearance or style of grooming and dressing.
- Space language (e.g., landscapes or paintings).
- Ego-building, status, and religious symbols (Rohn, 2017).

7.1.4.1.4 Visual Communication

Visual communication takes place with the help of visual aids such as color, illustration, graphic design, drawing, typography, signs, and other electronic resources. Visual communication comprising charts and graphs generally reinforces written methods of communication and sometimes replaces written communication completely. As the adage says, "A picture is worth a thousand words," and developments in technology have made visual communication easier today than ever before (Rohn, 2017).

7.2 Radio Communication

7.2.1 Introduction

Radio communication requires the use of both transmitting and receiving equipment. The transmitting equipment, which includes a radio transmitter and a transmitting antenna, is installed at the point from which messages are transmitted. The receiving equipment, which includes a radio receiver and a receiving antenna, is installed at the point at which messages are received.

In the transmitter, sinusoidal oscillations are generated at a carrier frequency and modulated in accordance with the information being transmitted. The modulated radio frequency (RF) oscillations constitute the radio signal. The signal passes from the transmitter to the transmitting antenna, which excites correspondingly modulated electromagnetic waves in the surrounding space.

The radio waves travel to the receiving antenna, in which they excite electrical oscillations. These oscillations are passed on to the receiver. The signal received is very weak, since only a small fraction of the radiated energy reaches the receiving antenna. For this reason, in the receiver, the signal is fed into an amplifier and then demodulated, or detected. As a result, a signal analogous to the signal used to modulate the carrier frequency oscillations in the transmitter is obtained. Usually after an additional amplification, the signal is converted by an appropriate reproduction device into a message equivalent to the original message.

At the reception point, electromagnetic oscillations from extraneous sources of radiation may be superposed on the signal. These superposed oscillations can interfere with the correct reproduction of a message and are, therefore, called radio interference. There are two other types of radio interference. First, the quality of radio communication can be adversely affected by variations over time in the attenuation of radio waves on the path from the transmitting antenna to the receiving antenna. Second, distortion can result when radio waves propagate simultaneously along two or more paths of different lengths.

In this case, the electromagnetic field at the reception point is the sum of radio waves that are displaced in time. The interference causes distortion of the radio signal; the influence on the reception of radio signals is particularly great in long-distance communications. The growth of radio communications and the use of radio waves in, for example, radar and radio navigation are based on the principle of electromagnetic compatibility—i.e., the requirement that different systems and equipment using radio waves function simultaneously without undesired mutual interference.

The transmission of radio signals through open space in the form of radio waves means that, theoretically, the signals can be received by persons other than those for whom they are intended. In other words, radio signals are subject to interception or monitoring. This disadvantage of radio communication is not shared by telecommunication through closed lines, such as cables or radio waveguides. The privacy of telephone conversations and telegraph messages is provided by the regulations of individual countries and international agreements. When necessary, secrecy is ensured by automatic equipment for disguising radio signals. Coding is one method of providing secrecy.

Radio services are the organizational-technical measures and means for establishing and maintaining the systematic functioning of radio communications. They can be classified according to, for example, their purpose, their range, and their structure. Examples of services are Earth Space Service and Space Service, which encompass all types of radio communications involving the use of one or more satellites or other space vehicles. Other examples are the following: fixed service, where communication is between permanently located stations; mobile service, where communication is between a mobile station and a land station or between mobile stations; radio broadcasting; and television. To meet industrial and official needs, special services exist in some ministries and organizations. Thus, there are services for civil aviation, the railroads, river and marine transport, and such agencies as the fire department and hospitals. Another example is the use of radio for internal communication in industrial and agricultural enterprises and in some institutions. Radio communications are of great importance in the armed forces (FARLEX).

7.2.2 What Is Radio Communication?

Radio communication is the transmission of signals by the modulation of electromagnetic waves with frequencies below those of visible light. Electromagnetic radiation travels by means of oscillating electromagnetic fields that pass through the air and the vacuum of space. Information is carried by a systematically changing (modulating) property of the radiated waves, such as amplitude, frequency, or phase (Vázquez, Herrero, Herrero, & Gómez, 2010).

Radio communication was one of the first wireless technology developed, and it is still in use. Portable multichannel radios allow users to communicate over short distances, whereas citizen band and maritime radios provide communication services over long distances for truckers and sailors (see Figure 7.1) (Agarwal, 2018).

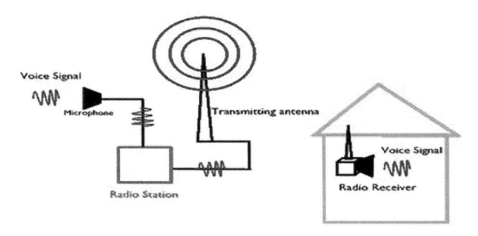

FIGURE 7.1 Radio transmission (Agarwal, 2018).

Radio broadcasts mostly sound through the air as radio waves. Radio has a transmitter which transmits the data in the form of radio signals to the receiver antenna.

To broadcast common programming, stations are associated with radio networks. The broadcast happens either in simulcast or syndication or both. Radio broadcasting may be done via cable FM and satellites over long distances at up to 2 Mbps (Agarwal, 2018).

7.2.3 Importance of Radio Communication

Radio communication has been of great importance to mankind since its inception, and its significance has continued to grow with 21st century technological advancements (Vick, 2019).

7.2.3.1 Two-Way Communication

Police, fire departments, and aircraft, among others, use two-way radio communication to ensure public safety. This medium is particularly important in emergency situations (Vick, 2019).

7.2.3.2 Broadcast Radio

Radio broadcasts on amplitude modulation (AM) and frequency modulation (FM) have been informing and entertaining the public for nearly a century (Vick, 2019).

7.2.3.3 Short-Wave Radio

Short-wave radio, while somewhat similar to broadcast media, is designed for international broadcast (Vick, 2019).

7.2.3.4 Cell Phones

Cell phones are actually sophisticated radios that receive signals from low-power stations operating on similar frequencies over a given area. Cities are usually divided into areas called cells, each of which contains a transmitting station (Vick, 2019).

7.2.3.5 New Directions

Radio technology now changes channels on TV sets, opens garage doors, provides wireless Internet service, and is part of every spacecraft's navigation system. Radio communication continues to evolve, keeping it an essential part of modern society (Vick, 2019).

7.2.4 Definition of Radio Broadcasting

Radio broadcasting refers to using radio waves to send audio signals to large groups of listeners, typically for talk or music programming.

Early use of radio focused on using the technology as an alternative to the telegraph, but by the 1920s, broadcast radio had become a major telecommunications medium. Today, in addition to traditional AM and FM radio, we have satellite radio, digital radio broadcasts, and radio-style stations available through the Internet (Melendez, 2019).

Radio stations broadcast at different frequencies, sometimes called wavelengths, as the frequency of a radio wave and its wavelength are mathematically related. To pick up a radio broadcast, you must use a radio tuned to the right frequency.

Commercial and nonprofit radio broadcasters around the world broadcast on frequencies reserved for AM and FM signals. FM is more widely used for music because it delivers better sound quality. Modern radio broadcasts include news, talk, sports, religious programming, and wide varieties of musical styles.

Communications

In the past, fictional dramas and comedies were often broadcast on the radio, but these have largely moved to television and Internet (Melendez, 2019).

7.2.5 Radio Waves

7.2.5.1 About Radio Waves

Radio signals are transmitted by an antenna with a certain output signal strength or effect. As the radio signal spreads out from the antenna, it becomes weaker. Eventually, the signal strength is so weak that the signal can no longer be understood by a receiver.

There are many reasons for the weakening of the radio signal. Here are a few examples:

- As the signal spreads out, the effect of the signal also spreads out. It is much like throwing a pebble in a pond—the height of the ripple decreases with distance from the impact.
- The effect or energy of the signal is simply absorbed by the air itself, the water in the air, particles or other things, so that the energy of the signal is lost.
- The signal is scattered, split up after colliding with particles, objects, or edges. One example of scattering is the creation of a rainbow when light is split up into different wavelengths after hitting water droplets in the air.

Radio signals are electromagnetic signals, just like light. And similar to light, radio signals can be reflected by or bounce on things. This can be either positive or negative depending on the circumstances. In a small home, it might not matter if radio signals bounce around a bit, and sometimes it can even be beneficial, but at other times, it can be a bother.

A radio wave can act much like a regular wave in the ocean. If two waves meet up and their crests and troughs line up, they will form a new much bigger wave. If two waves meet up and the troughs of one wave line up with the crests of the other wave, the two waves will cancel each other out. If there's a perfect match between the two waves for both height and wavelength, they could cancel each other out more or less completely.

Radio waves act the same way. Two radio waves that meet up will interact with each other to increase or decrease the signal strength. Figure 7.2 shows two radio waves (dashed lines) with different wavelengths and wave heights (amplitude) that meet up; together, they form a new combined radio wave (solid line) that appears very different from the two original radio waves (HOMENET HOWTO, 2016).

Even with a single radio transmitter, the radio waves spread out in all directions and bounce around the room. Basically, everything you see around you is something that radio waves can interact with by

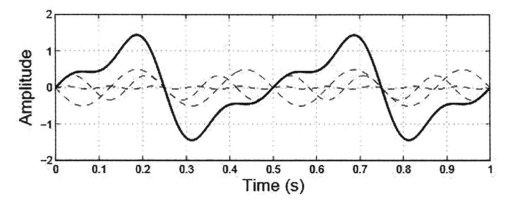

FIGURE 7.2 Two radio waves (dashed lines) with different wavelengths and wave heights (amplitude) meet up and form a new combined radio wave (solid line) (HOMENET HOWTO, 2016).

being reflected, diffracted, or scattered. The result is that the air is filled with radio signals bouncing around in all directions. So even with a single transmitter sending out a single signal, you could still get interference between the radio signals that are bouncing around.

Imagine putting a Wi-Fi router in the room. When the radio signal encounters wall, furniture, ceiling etc., it will be affected by the materials it encounters. Some of the signal effect will pass through the items, some will be absorbed, and some will reflect to bounce back into the room. This happens on every surface, so the room is filled with bouncing radio signals.

The bouncing signals are often much weaker than the original signal directly from the antenna, as not all of the signal will be reflected. If you were to put the Wi-Fi router right next to a concrete wall or steel plate, many of the signals would bounce on the hard surface right after leaving the antenna and would still be strong enough to interfere with the antenna signals.

The strongest signal strength with the best signal quality requires a completely free line of sight to the antenna of the Wi-Fi router. Anything blocking the signal will always affect it in some way. For that reason, it is always absolutely best to place a Wi-Fi router in an open space with a free line of sight to as many places as possible (HOMENET HOWTO, 2016).

7.2.6 Radio Communication Systems

A typical radio communication system used by a broadcasting station is a transmitter. The broadcasting station is allocated a unique RF carrier wave, along with a well-defined channel width. The transmitter transmits the modulated carrier wave into space through an antenna. These waves propagate through space. A receiver receives the modulated carrier through the receiving antenna with the help of a tuning circuit. It demodulates the modulated carrier and converts it into speech or intelligence. Figure 7.3 is a block diagram of a typical radio system consisting of a transmitter and receiver (Mahesh Lohith, 2014).

7.2.6.1 Transmission

The radio transmitter consists of a transducer which converts speech or intelligence into audio frequency (AF) electrical signals. These amplified AF signals modulate the RF carrier. The modulated RF carrier is then amplified and transmitted through an antenna (Mahesh Lohith, 2014).

7.2.6.2 Reception

The radio receiver consists of an antenna connected to a tuning circuit. The received-modulated RF carrier is amplified and passed through the demodulator to extract the AF signals. The AF signal is then amplified and fed to a transducer which converts it into speech or intelligence (Mahesh Lohith, 2014).

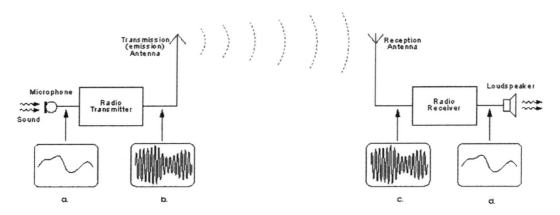

FIGURE 7.3 Radio communication system—transmission and reception (Mahesh Lohith, 2014).

7.3 Mid-Air Collision (MAC) Avoidance

7.3.1 Introduction

A mid-air collision (MAC) is an aviation accident in which two or more aircraft come into unplanned contact during flight. Because of the high velocities involved and the likelihood of subsequent impact with the ground or sea, very severe damage or the total destruction of at least one of the aircraft usually results.

The potential for an MAC is increased by miscommunication, mistrust, error in navigation, deviations from flight plans, lack of situational awareness, and the lack of collision-avoidance systems. Although a rare occurrence in general due to the vastness of open space available, collisions often happen near or at airports, where large volumes of aircraft are spaced more closely than in general flight (Wikipedia).

7.3.2 Mid-Air Collision Causes

What causes in-flight collisions? Increasing traffic and higher closing speeds represent the potential an MAC. For instance, a jet and a light twin plane have a closing speed of about 750 mph. It takes a minimum of 10 s, says the Federal Aviation Administration (FAA), for a pilot to spot traffic, identify it, realize it is a collision threat, react, and have the aircraft respond. But two planes converging at 750 mph will be less than 10 s apart when the pilots detect each other.

The main causes of MACs are (Kremer, 2012):

- Traffic congestion
- Aircraft speeds
- Failure to see other aircraft in time (Federal Aviation Administration, 2017).

The reason most often noted in the statistics is the failure of the pilot to see other aircraft, which means that the see-and-avoid system broke down. In most cases, at least one of the pilots involved could have seen the other in time to avoid contact, if he or she had just been using the visual senses properly. In sum, the human eye is the leading cause of in-flight collisions (Federal Aviation Administration, 2017).

7.3.2.1 Limitations of the Eye

The eye and consequently vision are vulnerable to just about everything: dust, fatigue, emotion, germs, fallen eyelashes, age, optical illusions, and alcohol. In flight, vision is altered by atmospheric conditions, windshield distortion, too much (or too little) oxygen, acceleration, glare, heat, lighting, aircraft design and so forth. Most of all, the eye is vulnerable to the vagaries of the mind. We can "see" and identify only what the mind lets us see. For example, a daydreaming pilot staring out into space sees no approaching traffic and is the number one candidate for an in-flight collision (Federal Aviation Administration, 2017).

7.3.2.1.1 Accommodation

One function of the eye that is a source of constant problems to the pilot (though he or she is probably never aware of it) is the time required for accommodation. Our eyes automatically accommodate for or refocus on near and far objects. But the change from something up close, like a dark panel 2 ft away, to a well-lighted landmark or aircraft target a mile or so away, takes 1–2 s or longer for eye accommodation. That can be a long time when we consider that we need 10 s to avoid in-flight collisions (Federal Aviation Administration, 2017).

7.3.2.1.2 Empty-Field Myopia

Another focusing problem usually occurs at very high altitudes, but it can happen even at lower levels on vague, colorless days above a haze or cloud layer when no distinct horizon is visible. If there is little or nothing to focus on at infinity, we do not focus at all. We experience something known as "empty-field myopia": "we stare, but we see nothing, even opposing traffic, if it should enter our visual field" (Federal Aviation Administration, 2017).

7.3.2.1.3 Binocular Vision

The effects of what is called "binocular vision" have been studied seriously by the National Transportation Safety Board (NTSB) during investigations of in-flight collisions. To actually accept what we see, we need to receive cues from both eyes. If an object is visible to one eye, but hidden from the other by a windshield post or other obstruction, the total image is blurred and not always acceptable to the mind (Federal Aviation Administration, 2017).

7.3.2.1.4 Tunnel Vision

Another inherent eye problem is a narrow field of vision. Although our eyes accept light rays from an arc of nearly 200°, they are limited to a relatively narrow area, approximately 10°–15°, in which they can actually focus and classify an object. Though we can perceive movement in the periphery, we cannot identify what is happening out there, and we tend not to believe what we see out of the corner of our eyes. This, aided by the brain, often leads to "tunnel vision" (Federal Aviation Administration, 2017).

7.3.2.1.5 Blossom Effect

At a distance, an aircraft on a collision course will appear to be motionless. It will remain in a seemingly stationary position, without appearing either to move or to grow in size for a relatively long time, and then suddenly bloom into a huge mass filling the window. This is known as the "blossom effect." Since we need motion or contrast to attract our eyes' attention, this effect becomes a frightening factor when we realize that a large bug smear or dirty spot on the windshield can hide a converging plane until it is too close to be avoided (Federal Aviation Administration, 2017).

7.3.2.1.6 Environmental Effects

In addition to the built-in problems, the eye is severely limited by environment. Optical properties of the atmosphere alter the appearance of traffic, particularly on hazy days. "Limited visibility" actually means "limited vision." The visual flight rule (VFR) of three miles may apply, but at that distance on a hazy day, opposing traffic is not easy to detect. At a range closer than three miles, opposing traffic may be detectable but no longer avoidable.

Lighting also affects our vision stimuli. Glare, usually worse on a sunny day over a cloud deck or during flight directly into the sun, makes objects hard to see and scanning uncomfortable. An object that is well lighted will have a high degree of contrast and will be easy to detect, while one with low contrast at the same distance may be impossible to see. For instance, when the sun is behind us, an opposing aircraft will stand out clearly, but when we are looking into the sun and traffic is "backlighted," it's a different story.

Another contrast problem is trying to find an airplane over a cluttered background. If it is between us and terrain that is varicolored or heavily dotted with buildings, it will blend into the background until it is quite close (Federal Aviation Administration, 2017).

7.3.2.1.7 Human Factors

The mind can distract us to the point of not seeing anything at all or lull us into cockpit myopia—staring at one instrument without even "seeing" it. It is fine to depend on instruments but not to the exclusion of the see-and-avoid system. An air traffic control (ATC) system is not infallible, even when it comes to providing radar separation between aircraft.

7.3.2.1.8 Conclusion

Visual perception is affected by many factors. It all boils down to the fact that pilots, like anyone else, tend to overestimate their visual abilities and to misunderstand the limitations of their eyes. Since the number one cause of in-flight collisions is the failure to properly adhere to the see-and-avoid concept, the best way to avoid them is to learn how to use our eyes in an efficient external scan (Federal Aviation Administration, 2017).

7.3.2.2 Situational Awareness

The prevention of MACs requires good situational awareness. Situational awareness includes the following (Kremer, 2012):

- Effective visual scanning
- Ability to gather information from radio transmissions from ground stations and other aircraft
- Ability to create a mental picture of the traffic situation
- Development of "good airmanship."

7.3.3 Mid-Air Collision Effects

An possible consequence of MACs is a temporary or permanent loss of control as a result of damage, an avoidance maneuver, or mishandling, potentially resulting in collision with terrain or an emergency landing as a result of damage to the aircraft and/or injuries to crew and passengers.

It is commonly assumed that any MAC will cause loss of both aircraft and all people on board. In fact, accident and serious incident reports show there have been a few nonfatal MAC accidents. However, in most cases, total loss is the result.

A crash following an MAC may also cause fatalities on the ground (Skybrary, 2018).

7.3.4 Mid-Air Collision Defenses

The main barriers to MACs are the following (Skybrary, 2018):

1. Strategic conflict management, including:
   - Airspace design, including the classification of airspace, the route structure, flight levels, and the airport's SIDs and STARs (procedures and checkpoints used to enter and leave the airway system).
   - Air traffic flow and capacity management (ATFCM), including capacity planning, flexible use of airspace, and flow management.
   - Traffic synchronization, including sector planning, multi-sector planning, and arrival/departure sequencing.
2. Tactical conflict management, including:
   - ATC conflict management, in which ATC officers (ATCOs) provide separation between aircraft.
   - Pilot conflict management, in which pilots are responsible for avoiding other aircraft, sometimes with the assistance of information from ATC.
   - Strategic lateral offset procedures (SLOPs), in which aircraft offset the centerline of an airway or flight route by a small amount, normally to the right, so that collision with opposite direction aircraft becomes unlikely.
3. ATC collision avoidance, including:
   - Short-term conflict alerts (STCAs).
   - Warning from ATCOs not directly responsible for separation. Although this is not a planned barrier, this type of ad hoc assistance sometimes helps avoid collisions.
4. Airborne collision avoidance, including:
   - Airborne collision avoidance systems (ACASs).
   - Visual airborne collision avoidance ("see and avoid").

Luck can also be considered a preventative barrier. A loss of separation does not necessarily lead to a collision, even if all the managed collision avoidance barriers are unsuccessful (Skybrary, 2018).

7.3.5 Typical Scenarios

Because of the multiple barriers, most collisions do not have a single cause but multiple causes, typically one for each unsuccessful barrier. "Unsuccessful" is a general term covering all types of failure causes, including technical ones, human error (e.g., lack of response or misjudgment), impracticability (e.g., not enough time), or lack of coverage (e.g., equipment not fit). Barriers may also be bypassed (Skybrary, 2018).

Common causes of unsuccessful collision avoidance are the following (Skybrary, 2018):

1. Unsuccessful STCA warning:
   - No STCA coverage of area of conflict.
   - STCA failure to give warning in time, for example, due to transponder failures, surveillance failures, STCA software failures, STCA parameters detuned to minimize false alarms.
   - ATCO failure to respond in time; for example, the ATCO is distracted and misses the warning or believes the warning is incorrect.
   - ATCO failure to recover separation in time, for example, due to inadequate communication with the pilot or inadequate response from the pilot.
2. Unsuccessful warning from other ATCOs not directly responsible for separation:
   - No independent ATCO monitoring of area of conflict.
   - Other ATCO failure to detect conflict in time, for example, for reasons as above.
   - Other ATCO failure to communicate warning to responsible ATCO in time.
   - Failure responsible ATCO to recover separation in time.
3. Unsuccessful ACAS warning:
   - ACAS not installed on the aircraft.
   - ACAS failure to detect the conflicting aircraft or issue a resolution advisory (RA) in time.
   - Pilot failure to respond with appropriate timely collision avoidance maneuver, for example, does not respond, or incorrectly prioritizes ATC instructions.
   - Avoidance action invalidated by incorrect opposing action from the other pilot.
4. Unsuccessful visual warning:
   - Other aircraft concealed, for example, by instrument meteorological conditions (IMCs), darkness, flight deck surfaces, or empty-field myopia.
   - Flight crew failure to observe the other aircraft in time to make avoidance action.
   - Pilot failure to respond with appropriate timely collision avoidance maneuver.

The causes of barriers being unsuccessful are not necessarily independent. In fact, the most important causes include ones that make several barriers unsuccessful (known as common-cause failures).

7.3.6 Reduction of Risk by Visual Scanning

- Before takeoff, check the runway visually to ensure there are no aircraft or other objects in the takeoff area.
- Check the approach and circuit to be sure of the position of other aircraft. Assess the traffic situation from radio reports.
- After takeoff, continue to scan to ensure there will be no obstacles to safe departure.
- During the climb and descent, be aware of the blind spot under the nose; maneuver the aircraft to check.
- Look behind while climbing in case the plane is in a faster aircraft's blind spot (Kremer, 2012).

7.3.7 Reduction of Risk by Flight Data Monitoring (FDM)

One way to improve operational safety is a proactive use of digital flight data from routine operations to enhance the safety and efficiency of operation, by providing greater insight into the total flight operations environment. The best potential source of operational data is the operator's own flight data monitoring (FDM), flight data analysis (FDA), or flight operations quality assurance (FOQA) programs. This information can be used by the operator to discover and identify underlying issues with the potential to negatively affect aviation safety and to allow the operator to take appropriate action.

FDM programs generally involve systems that capture flight data, transform them into an appropriate format for analysis, and generate reports and visualization to assist in assessing the data. The following capabilities are required for an effective FDM program (IATA, 2019):

- An onboard device to capture a wide range of in-flight parameters and record data on those parameters using flight data recorders (FDRs).
- A means to transfer the data recorded on board the aircraft to a ground-based computer system.
- A means for the ground-based computer system to analyze the data, identify deviations from expected performance, and generate reports to assist in interpreting the reports.
- An optional software for a flight animation capability to integrate all data, presenting them as a simulation of in-flight conditions, thereby facilitating visualization of actual events.

On some aircraft, dedicated traffic collision avoidance system (TCAS) recorders provide accurate information which can be added or matched to the FDR data should more detailed information be needed for a particular investigation.

FDM is an essential part of a well-functioning safety management system (SMS) for an aircraft operator; it is one of the main data sources for monitoring the operational safety level (IATA, 2019).

7.3.7.1 Analysis and Trends of FDM Data

FDR data must be transferred to an analysis platform. Various methods are available for downloading flight data to the analysis platform or offset for third-party analysis. The data analysis system should have the following (IATA, 2019):

- Ability to display detailed information such as where, when, and what;
- Ability to display information in a logical and user-friendly way;
- Ability to program a range of alert detection thresholds to generate events when parameters exceed present values;
- Ability to enable detailed analysis of flight data;
- Ability to provide long-term trend analysis of events.

Most FDM systems have the ability to record a TCAS's RA or warning. This means data can indicate whether a TCAS RA was issued, its duration, and its type (e.g., Climb, Descend, Level Off).

Alert detection thresholds are set to generate events when the value of the parameter exceeds a predetermined level or threshold. Exceedance detection is used to identify and assess operational risks and draw the attention of the data analyst. Alert detection thresholds are set to generate events for trending or aggregating over a period of time and to enable pilots to be alerted to their own events. They are tailored to SOPs, aircraft type, and specific operating scenarios (IATA, 2019).

7.3.8 MAC Risk Monitoring and FDM

The characteristics of the data recorded on-board aircraft and analyzed within the FDM framework strongly influence possible investigations. For an MAC, many contributing factors cannot be identified if only FDM data are available. For example, the actions of the air traffic controller, i.e., the transfer and the perception of the instructions to the flight deck crew, are essential but not traceable from FDM data. Furthermore, more than one aircraft is always involved in a loss of separation event. In general, FDM data are only available for one of the affected aircraft.

A recent problematic development is the increasing use of drones. In addition, general aviation aircraft, including gliders, often carry minimal equipment on board so that not all recovery barriers are available. Finally, because of the different nature of operation, MAC analyses of helicopters might require additional perspectives (EOFDM Working Group A, 2018).

7.3.9 MAC Risk Modeling

Figure 7.4 shows a method of modeling MAC risk—a "bow-tie model." The hazardous event to be avoided, "Aircraft in close proximity," appears in the center of the figure. Four hazards (or "triggering events") are listed on the left. To reduce the probability that these events occur and lead to the loss of separation event, certain avoidance barriers have been set up. The right side of the figure shows two potential outcomes of a loss of separation event. To prevent them from happening, certain recovery barriers are in place (EOFDM Working Group A, 2018).

A central recovery barrier of MAC events in civil aviation is the ACAS. Information about triggered alerts such as RAs and related details are often captured in the FDM data stream and can be analyzed. In addition, based on these triggers, ACAS alert hotspots can be identified by every airline if enough data are available.

Once a loss of separation event is identified, more detailed analyses outside FDM can be conducted on an individual and manual basis, potentially considering other sources of information such as the following (EOFDM Working Group A, 2018):

- Readouts of on-board ACAS memories
- Air safety reports of flight deck crew
- Feedback or reports from ATC
- Ground-based STCA data from air navigation service providers (ANSPs).

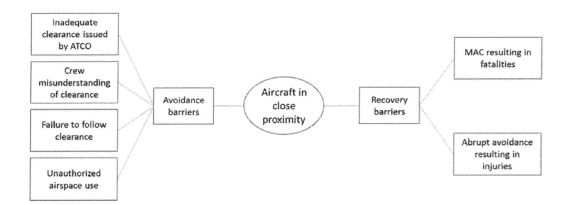

FIGURE 7.4 Simplified bow-tie model for MAC (EOFDM Working Group A, 2018).

7.4 Communications Data Rate and Bandwidth Usage

7.4.1 Data Rate

The data rate denotes the transmission speed or the number of bits per second transferred. The useful data rate for the user is usually less than the actual data rate transported on the network. Additional bits are required for the recovery of timing information at the receiver or error correction to compensate for possible transmission errors.

In telecommunications, it is common use to express the data rate or "bitrate" in bits per seconds (bit/s). In data communication, the data rate is often expressed in bytes per second (B/s) (TELECOM ABC (a)).

Prefixes are used for high bitrates. Table 7.1 shows some binary prefixes

The bitrate does not necessarily have to be constant (CBR: constant bitrate). Video, audio, or speech coding can result in a variable bitrate (VBR). A special form of VBR is average bitrate (ABR) in which the encoder tries to maintain a specific (average) bitrate. ABR can also be seen as the average bitrate of a VBR encoded bitstream (TELECOM ABC (b)).

7.4.1.1 Data Rate Definition

Data rate can be defined as (PC, 2019):

- The speed at which data are transferred within a computer or between a peripheral device and the computer, measured in bytes per second.
- The speed at which audio and video files are encoded (compressed), measured in bits per second.
- The transmission speed of a network; for example, 100Base-T Ethernet is rated at 100 Mbps (megabits per second).

7.4.1.2 Data Transfer Rate (DTR)

The data transfer rate (DTR) is the amount of digital data moved from one place to another in a given time. In general, the greater the bandwidth of a given path, the higher the DTR.

As noted above, in telecommunications, data transfer is usually measured in bits per second. For example, a typical low-speed connection to the Internet may be 33.6 kilobits per second (Kbps). On Ethernet local area networks, data transfer can be as fast as 10 Mbps. Network switches are being designed to transfer data in the terabit range. In earlier telecommunication systems, data transfer was sometimes measured in characters or blocks (of a certain size) per second. Data transfer time between the microprocessor or RAM and devices such as the hard disk and CD-ROM player is usually measured in milliseconds.

As also noted above, in computers, data transfer is often measured in bytes per second. The highest DTR to date is 14 Tbps over a single optical fiber, reported by Japan's Nippon Telegraph and Telephone (NTT DoComo) in 2006 (Rouse, 2011).

TABLE 7.1

Binary Prefixes (TELECOM ABC (b))

kilo	2^{10}	1,024
Mega	2^{20}	1,048,576
Giga	2^{30}	1,073,741,824
Tera	2^{40}	1,099,511,627,776

7.4.2 Bandwidth

7.4.2.1 Introduction

Bandwidth is the capacity of a wired or wireless network communications link to transmit the maximum amount of data from one point to another over a computer network or Internet connection in a given amount of time (usually 1 s). Synonymous with capacity, bandwidth describes the DTR. Bandwidth is not a measure of network speed (a common misconception).

While bandwidth is traditionally expressed in bits per second (bps), modern network links have greater capacity, typically measured in millions of bits per second (megabits per second, or Mbps) or billions of bits per second (gigabits per second, or Gbps).

Bandwidth connections can be symmetrical, which means the data capacity is the same in both directions to upload or download data, or asymmetrical, which means download and upload capacity are not equal. In asymmetrical connections, upload capacity is typically smaller than download capacity (Rouse, 2014).

While bandwidth is used to describe network speeds, it does not measure how fast bits of data move from one location to another. Since data packets travel over electronic or fiber-optic cables, the speed of each bit transferred is negligible. Instead, bandwidth measures how much data can flow through a specific connection at one time.

Data often flow over multiple network connections, which means the connection with the smallest bandwidth acts as a bottleneck. The Internet backbone and connections between servers generally have the most bandwidth, so they rarely become bottlenecks. Instead, the most common Internet bottleneck is the connection to an Internet Service Provider (ISP).

Note: Bandwidth also refers to a range of frequencies used to transmit a signal. This type of bandwidth is measured in hertz and is often referenced in signal processing applications (TECHTERMS, 2012).

7.4.2.2 How Bandwidth Works

The more bandwidth a data connection has, the more data it can send and receive at one time. Bandwidth can be compared to the amount of water that can flow through a water pipe. The bigger the pipe, the more water can flow through it at one time. Bandwidth works on the same principle. The higher the capacity of the communication link, or pipe, the more data can flow through it per second. End users pay for the capacity of their network connections, so the greater the capacity of the link, the more expensive it is.

The maximum capacity of a network connection is only one factor affecting network performance. Packet loss, latency, and jitter can all degrade network throughput and make a high-capacity link perform like one with less available bandwidth. An end-to-end network path usually consists of multiple network links, each with different bandwidth capacity. As a result, the link with the lowest bandwidth is often described as the bottleneck, because the lowest bandwidth connection can limit the overall data capacity of all the connections in the path (Rouse, 2014).

7.4.2.3 Considerations When Calculating Bandwidth

Technology advances have made some bandwidth calculations more complex, and they can depend on the type of network link being used. For example, optical fiber using different types of light waves and time-division multiplexing can transmit more data through a connection at one time, effectively increasing its bandwidth. In wireless networks, bandwidth is defined as the spectrum of the frequency operator's license from the Federal Communications Commission (FCC) and the National Telecommunications and Information Administration for use in mobile services in the United States.

Effective bandwidth is the highest reliable transmission rate a link can provide. It can be measured using a bandwidth test in which the link's capacity is determined by repeatedly measuring the time required for a specific file to leave its point of origin and successfully download at its destination.

In addition to testing, organizations need to calculate how much bandwidth they need to run all applications on their networks. To find out how much capacity they need, organizations need to calculate the maximum number of users who might be using the network connection at one time and multiply that number times the bandwidth capacity required by each application.

To calculate needed bandwidth for the cloud, it's important to know the capacity needed to send and receive traffic from public clouds. Capacity can be affected by any congestion on the connections used to reach public cloud providers, particularly if data are traveling over the Internet (Rouse, 2014).

7.4.2.4 Bandwidth Utilization

Bandwidth utilization is a well-known but not a well-understood network statistic. There are a couple of factors to take into account when monitoring bandwidth utilization.

The first is understanding the difference between a bandwidth utilization statistic from a single segment and that from an aggregate of segments. Administrators often want to know, or are asked the question, "What is the bandwidth utilization of our network?" The problem is that even on the most basic of networks, this is almost an impossible statistic to determine.

Even on a flat network, there are typically multiple segments, sites, and links that cannot realistically be totaled. For example, if you have ten segments for which you have gathered bandwidth utilization information for a number of weeks, averaging those data will result in a meaningless number that does not reflect any real statistics. For instance, if nine of the segments have an average utilization of 10% and one segment reaches 95% utilization, the "network utilization" will be 18.5%. This number is of no value because it hides the problem of the single site and increases the utilization of other sites. However, if each segment is monitored separately, the administrator will be able to stay on top of the network, identifying and resolving issues before they become a problem. Proactive utilization management can save any company significant time and money.

The second factor is the source of bandwidth consumption. Voice over internet protocol (VoIP), database applications, and graphic design programs are all examples of common applications known to consume fairly large amounts of network bandwidth. When considering deployment, an administrator often feels obligated to increase available bandwidth so that the application will not experience issues. However, the current budget may be tight, and purchasing additional bandwidth may require solid justification.

The first step in justifying additional bandwidth is to determine what the current utilization level is on a network segment or link; the second is to trend the actual requirements of the application in a test lab, simulating usage loads and comparing that information to actual findings to determine if an increase in available bandwidth will be required.

Those results may show that it is pertinent to increase bandwidth capacity to keep the applications and network running smoothly. Or there may only be a negligible increase in utilization, and an upgrade is unnecessary (THE EYE from Network Instrument LLC, 2005).

7.4.3 Bandwidth and Data Rate. What's the Difference?

In the wireless industry, bandwidth and data rate are used interchangeably and can sometimes be confused, as both help define the speed at which a device or network sends and receives data.

Simply stated, bandwidth is the measurement of the ability of an electronic communications device or system to send and receive information. Data rate is the speed at which data are transferred between two devices, measured in megabits per second (Mbps or mbps).

To understand bandwidth, it may be useful to think in terms of the water pipe mentioned above (Section 7.4.2.2). Bandwidth is the maximum amount of water that can travel through the pipe; i.e., the maximum about of data that can be transferred through the RF channel(s).

The bandwidth will always be greater or equal to the throughput (the amount of data that enter and go through a system). Keeping with the water pipe analogy, throughput is the actual amount of water that travels through the pipe. External things can affect the throughput, such as a physical object inside the pipe. In the same way, real-world interferences can affect the amount of data sent wirelessly. Some typical things affecting throughput are RF interference and physical obstructions.

Say we have an AW58100 Ethernet radio with an advertised 100 Mbps bandwidth. Its throughput may be 60 Mbps instead of 100 Mbps. The 40% loss is the amount of bandwidth it takes the radio to transmit the data, or the RF bandwidth (Hamby, 2016).

7.5 Antenna Types

7.5.1 Introduction

The most fundamental part of any electrical framework is an antenna. It connects the free space and transmitter or free space and the recipient. Antennas are employed in different systems in different forms. That is, the operational characteristics of some systems are designed around the directional properties of the antennas; in other systems, the antennas are used simply to radiate electromagnetic energy omni-directionally. In still other systems, they are used for point-to-point communication requiring increased gain and reduced wave interference (Dhande, 2009; Jolani, Dadgarpar, & Hassani, 2008).

Antennas convert RF signals or electrical signals into electromagnetic or wave signals. They also receive electromagnetic signals and change them into electrical signals. Antennas are useful modes of communication in different fields, as they can be used to communicate in the form of audio, video, and graphics. They are designed for different applications, including radio, television, satellite, broadcasting, and cellular system communications.

7.5.2 Antenna Definitions

Definitions of an antenna include the following:

- An antenna is a means for radiating or receiving radio waves (IEEE Std 145-1983).
- "An antenna is any device that converts electronic signals to electromagnetic waves (and vice versa) effectively with minimum loss of signals" (Dhande, 2009), as shown in Figure 7.5.
- An antenna is a transforming device that will convert impedance of transmitter output (50/75 Ω) into free space impedance (120π or 377 Ω).
- An antenna is a region of transition between guided and free space propagation.
- An antenna concentrates an incoming wave on a sensor (receiving case).
- An antenna launches waves from a guiding structure into space or air (transmitting case).
- An antenna is often part of a signal transmitting system over some distance.

The radiation pattern and radiation resistance of an antenna are the same when it transmits and when it receives, if no nonreciprocal devices are used, so the same antenna can be used for transmission and reception of electromagnetic waves. Note that an antenna is a passive device. It does not amplify signals; it only directs the signal energy in a particular direction.

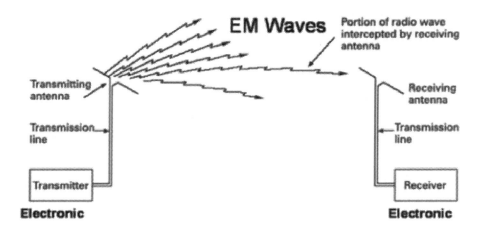

FIGURE 7.5 Wireless communication system (Dhande, 2009).

Communications

7.5.3 Properties of Antennas

7.5.3.1 Antenna Gain

The parameter that measures the degree of directivity of an antenna's radial pattern is known as gain. An antenna with a higher gain is more effective in its radiation pattern. Antennas are designed in such a way that power rises in the desired direction and decreases in the undesired direction (ElProCus, 2016).

$$G = \frac{\text{power radiated by an antenna}}{\text{power radiated by reference antenna}} \tag{7.1}$$

7.5.3.2 Aperture

The aperture actively participates in the transmission and reception of electromagnetic waves. The collective area of an antenna is known as the effective aperture (ElProCus, 2016).

$$P_r = P_d * A, \tag{7.2}$$

$$A = \frac{P_r}{P_d}, \tag{7.3}$$

7.5.3.3 Directivity and Bandwidth

The directivity of an antenna is defined as the measure of concentrated power radiation in a particular direction. It may be considered as the capability of an antenna to direct radiated power in a given direction. It can also be considered the ratio of the radiation intensity in a given direction to the average radiation intensity. Bandwidth is one of the desired parameters when choosing an antenna. It can be defined as the range of frequencies over which an antenna can properly radiate and receive energy (ElProCus, 2016).

$$D = \frac{1}{\frac{1}{4\pi} \int_0^{2\pi} \int_0^{\pi} |F(\theta,\phi)|^2 \sin\theta \, d\theta \, d\phi} \tag{7.4}$$

7.5.3.4 Polarization

An electromagnetic wave launched from an antenna may be polarized vertically or horizontally (see Figure 7.6). If the wave is polarized in the vertical direction, the E vector is vertical, and a vertical antenna is required. If the E vector is horizontal, a horizontal antenna is needed to launch the wave. Circular polarization is sometimes used; this is a combination of horizontal and vertical (ElProCus, 2016).

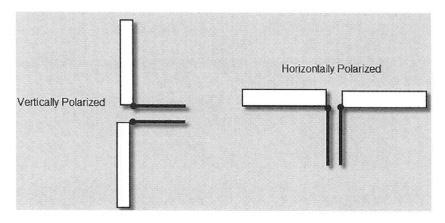

FIGURE 7.6 Antenna polarized vertically and horizontally (ElProCus, 2016).

7.5.3.5 Effective Length

The effective length is the parameter of the antenna that characterizes its efficiency in transmitting and receiving electromagnetic waves. Effective length can be defined for both transmitting and receiving antennas. The ratio of EMF at the receiver input to the intensity of the electric field on the antenna is known as the receiver's effective length. The effective length of the transmitter can be defined as the length of the free space in the conductor; current distribution across its length generates the same field intensity in any direction of radiation (ElProCus, 2016).

$$\text{Effective Length} = \frac{\text{Area under non-uniform current distribution}}{\text{Area under uniform current distribution}} \quad (7.5)$$

7.5.3.6 Polar Diagram

The most significant property of an antenna is its radiation pattern or polar diagram. In the case of a transmitting antenna, the polar diagram indicates the strength of the power field radiated by the antenna in various angular directions. A plot can be obtained for both vertical and horizontal planes; these are called the vertical and horizontal patterns, respectively (see Figure 7.7) (ElProCus, 2016).

7.5.4 How Antennas Work

Suppose you're the boss of a radio station, and you want to transmit your programs to the wider world. How do you go about it? You use microphones to capture the sounds of people's voices and turn them into electrical energy. You take that electricity and, loosely speaking, make it flow along a tall metal antenna (boosting it in power many times so it will travel just as far as you need into the world). As the electrons (tiny particles inside atoms) in the electric current wiggle back and forth along the antenna, they create invisible electromagnetic radiation in the form of radio waves. These waves, partly electric and partly magnetic, travel out at the speed of light, taking your radio program with them. What happens when I turn on my radio in my home a few miles away? The radio waves you sent flow through the metal antenna and cause electrons to wiggle back and forth (Woodford, 2019) (Figure 7.8).

Transmitter and receiver antennas are often very similar in design. For example, for a satellite phone that can send and receive a video-telephone call to any other place on Earth using space satellites, all signals transmitted and received pass through a single satellite dish—a special kind of antenna shaped like a bowl (technically known as a parabolic reflector, because the dish curves in that shape).

TV or radio broadcasting antennas are huge masts, sometimes stretching hundreds of meters/feet into the air, because they have to send powerful signals over long distances. For a TV or radio at home, however, a much smaller antenna is fine (Woodford, 2019).

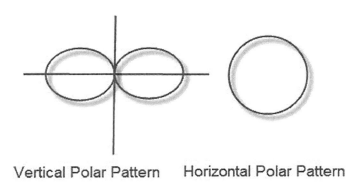

FIGURE 7.7 Antenna with radiation pattern or polar diagram (ElProCus, 2016).

Communications 247

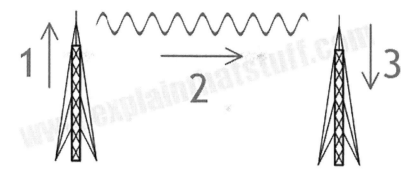

FIGURE 7.8 How a transmitter sends radio waves to a receiver: 1) Electricity flowing into the transmitter antenna makes electrons vibrate up and down it, producing radio waves. 2) Radio waves travel through the air at the speed of light. 3) Waves arrive at the receiver antenna, making electrons vibrate inside it and producing an electric current that recreates the original signal (Woodford, 2019).

Waves don't always move from transmitter to receiver. Depending on what kinds (frequencies) of waves we want to send, how far we want to send them, and when we want to do it, waves can travel in one of the following three ways (see Figure 7.9) (Woodford, 2019):

1. Via line of sight: They can shoot by a "line of sight" in a straight line—just like a beam of light. In old-fashioned long-distance telephone networks, microwaves were used to carry calls this way between very high communications towers (fiber-optic cables have largely made this obsolete).
2. Via ground waves: They can speed round the Earth's curvature in what's known as a ground wave. AM (medium-wave) radio tends to travel this way for short-to-moderate distances. This explains why we can hear radio signals beyond the horizon (when the transmitter and receiver are not within sight of each other).
3. Via the ionosphere: They can shoot up to the sky, bounce off the ionosphere (an electrically charged part of Earth's upper atmosphere), and come back down to the ground. This effect works best at night, which explains why distant (foreign) AM radio stations are much easier to pick up in the evenings. During the daytime, waves shooting to the sky are absorbed by lower layers of the ionosphere. At night, that doesn't happen. Instead, higher layers of the ionosphere catch the radio waves and fling them back to Earth—giving a very effective "sky mirror" that can help to carry radio waves over very long distances.

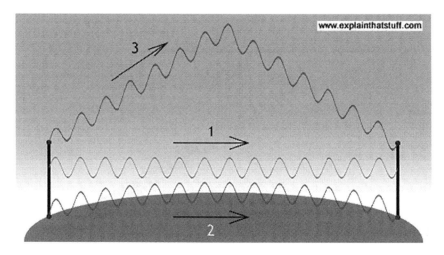

FIGURE 7.9 How a wave travels from a transmitter to a receiver: 1) Line of sight; 2) Ground wave; 3) Ionosphere (Woodford, 2019).

7.5.5 Antenna Classification

Antennas can be classified on the basis of:

1. Frequency: Very low frequency (VLF), low frequency (LF), high frequency (HF), very high frequency (VHF), ultra-high frequency (UHF), microwave, millimeter wave antenna.
2. Aperture: Wire, parabolic dish, microstrip patch antenna.
3. Polarization: Linear (vertical/horizontal), circular polarization antenna.
4. Radiation: Isotropic, omnidirectional, directional, hemispherical antenna.

7.5.5.1 Frequency Basis

Examples of antennas by frequency type include the following:

1. VLF and LF antennas: Vertical radiators, top-loaded monopoles, T and Inverted L antennas, triatic antennas, trideco antennas, valley-span antennas.
2. Medium frequency (MF) antennas: Radiators (monopoles and dipoles), directional antennas.
3. HF antennas: Log-periodic antennas, conical monopole and inverted cone antennas, vertical whip antennas, rhombic antennas, fan dipole antennas.
4. VHF) and UHF antennas: Yagi–Uda antennas, log-periodic antennas, helical antennas, panel antennas, corner reflector antennas, parabolic antennas, discone antennas.
5. Super high frequency (SHF) and extremely high frequency (EHF) antennas: Parabolic antennas, pyramidal horn antennas, discone antennas, monopole and dipole antennas, microstrip patch antennas, fractal antennas (Dhande, 2009) (Table 7.2).

7.5.5.2 Aperture Antennas

Aperture antennas transmit and receive energy from their aperture (Dhande, 2009). Types of aperture antennas are:

- Wire antennas
- Horn antennas
- Parabolic reflective antennas
- Cassegrain antennas.

TABLE 7.2

Antenna Frequency Types

Frequency Band	Designation	Typical Service
3–30 kHz	VLF	Navigation, sonar.
30–300 kHz	LF	Radio beacons, navigational aids.
300–3,000 kHz	MF	AM broadcasting maritime radio, coast guard communication, direction finding.
3–30 MHz	HF	Telephone, telegraph and facsimile, amateur radio, ship-to-coast and ship-to-aircraft communication.
30–300 MHz	VHF	Television, FM broadcast, ATC, police, navigational aids.
300–3,000 MHz	UHF	Television, satellite communication, radiosonde, surveillance radar, navigational aids.
3–30 GHz	SHF	Airborne radar, microwave links, satellite communication.
30–300 GHz	EHF	Radar, experimental.

7.5.5.2.1 Wire Antennas

A wire antenna is simply a straight wire of length ë/2 (dipole antenna) and ë/4 (monopole antenna), where ë is the transmitted signal wavelength. A wire antenna can be a loop antenna, such as circular loop, rectangular loop, etc. Basically all vertical radiators are in wire antenna categories. A whip antenna is the best example of a wire antenna (Dhande, 2009).

7.5.5.2.2 Vertical Monopole Antennas

- Length < 0.64λ (see Figure 7.10).
- Self-impedance: $ZS = Z_{ANT} + R_{GND} + R_{REF}$.
- Efficiency: $\eta = |Z_{ANT}|/|ZS|$ η ranges from <1% to >80% depending on antenna length and ground system.
- Efficiency improves as monopole gets longer and ground losses are reduced (see Figure 7.11).

FIGURE 7.10 Vertical monopole antenna (Dhande, 2009).

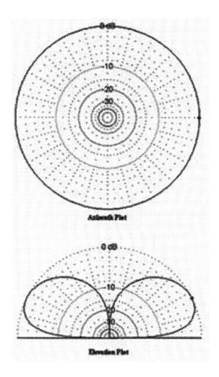

FIGURE 7.11 ë/4 vertical monopole (Dhande, 2009).

- Length ~0.25λ.
- Self-impedance: ZS ~36–70 W.
- The λ/4 vertical requires a ground system, which acts as a return for ground currents. The "image" of the monopole in the ground provides the "other half" of the antenna.
- The length of the radials depends on how many there are.
- Takeoff angle ~25°.
- Length is approximately 0.48λ.
- Self-impedance ~2,000 W.
- Antenna can be matched to 50 Ω coax with a tapped tank circuit.
- Takeoff angle ~15°.
- Ground currents at base of antenna are small; radials are less critical for λ/2 vertical (Dhande, 2009).

7.5.5.2.3 Horn Antennas

Horn radiators act as reflector antennas, illuminators, and as antennas in their own right. These kinds of antennas are not best matched to the waveguide, yet at the same time, they can accomplish standing wave proportions of 1.5:1 or less. The increase of a horn radiator corresponds to the region of the flared open rib and is conversely relative to the square of the wavelength (Kampeephat, Krachodnok, & Wongsan, 2014). Figure 7.12 demonstrates the structure of the horn antenna (Khan, Riaz, & Bilal, 2016).

A horn antenna may be regarded as a flared out or opened out waveguide. A waveguide is capable of radiating radiation into open space, provided it is excited at one end and open at the other end. If flaring is done in one direction, a sectorial horn is produced. With flaring in the direction of the electric vector and magnetic vector, the sectorial *b*-plane horn and sectorial *a*-plane horn are obtained, respectively.

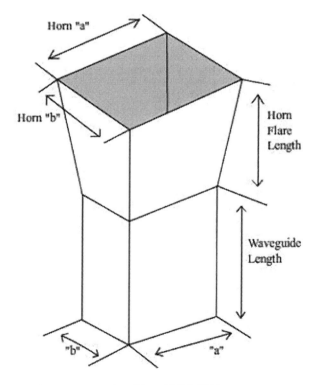

FIGURE 7.12 Representation of horn antenna (Khan, Riaz, & Bilal, 2016).

If flaring is done along both walls (*b* and *a*) of the rectangular waveguide, a pyramidal horn is obtained. By flaring the walls of a circular waveguide, a conical horn is formed (Dhande, 2009).

7.5.5.2.4 Parabolic Reflective Antennas

A parabola is a two-dimensional plane curve. A practical reflector is a three-dimensional curved surface. Therefore, a practical reflector is formed by rotating a parabola about its axis. The surface so generated is known as "paraboloid," often called a "microwave dish" or "parabolic reflector." The paraboloid reflector antenna consists of a primary antenna, such as a dipole or horn, situated at the focal point of a paraboloid reflector. The important practical implication of this property is that a reflector can focus parallel rays onto the focal point or, conversely, it can produce a parallel beam from radiations originating from the focal point (Dhande, 2009).

There are two types of parabolic reflective antennas: right cylinder and paraboloid. Linear dipole, linear array, slotted waveguide antennas, etc. are used to feed the cylinder type. Conical or pyramidal horn antennas are used to feed the paraboloid type. Parabolic reflectors gather and concentrate the parallel incoming radio wave beams and emphasize them on the actual antenna at its focal point (Ujwala, Namrata, Pooja, & Shraddha, 2014). Figure 7.13 demonstrates the structure of the parabolic reflector (Khan, Riaz, & Bilal, 2016).

7.5.5.2.5 Prime Focus Paraboloid Reflector Antennas

The prime focus paraboloid reflector antenna can be a parabolic dish or cylindrical (Dhande, 2009). The reflector acts as a large collecting area and concentrates power onto a focal region where the feed is located (see Figure 7.14).

7.5.5.2.6 Cassegrain Antennas

In a Cassegrain antenna (see Figure 7.15), the primary feed radiator is positioned around an opening near the vertex of the paraboloid instead of at the focus. A Cassegrain feed system employs a hyperboloid secondary reflector, one of whose foci coincides with the focus of the paraboloid. The feed radiator

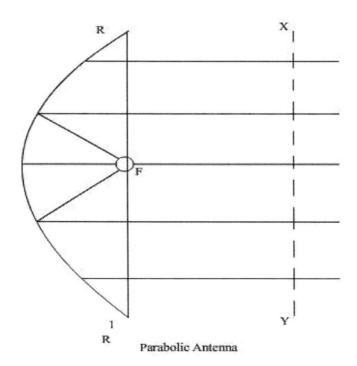

FIGURE 7.13 Parabolic reflector antenna (Khan, Riaz, & Bilal, 2016).

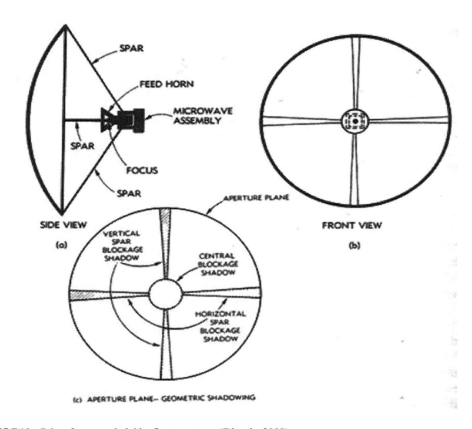

FIGURE 7.14 Prime focus paraboloid reflector antenna (Dhande, 2009).

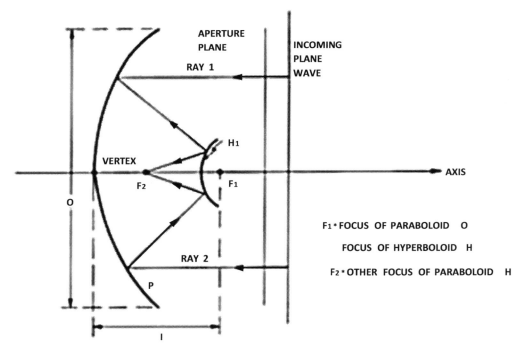

FIGURE 7.15 Cassegrain antenna (Dhande, 2009).

is aimed at the secondary hyperboloid reflector or sub-reflector. The radiations emitted from the feed radiator are reflected from the Cassegrain secondary reflector, illuminating the main paraboloid reflector as if they had originated from the focus. Then the paraboloid reflector colliminates the rays as usual (Dhande, 2009).

7.5.5.2.7 Microstrip Patch Antennas

Spacecraft or aircraft applications, where size, weight, cost, performance, ease of installation, and aerodynamic profile are constraints, require low-profile antennas. Microstrip patch antennas are used to meet these specifications (see Figure 7.16). These antennas can be flush mounted to metal or other surfaces, and they only require space for the feed line which is normally placed behind the ground plane. The major disadvantages of patch or microstrip antennas are their inefficiency and very narrow bandwidth which is typically only a fraction of a percent or, at the most, a few percent (Dhande, 2009) (Figure 7.17).

7.5.5.3 Antenna Classification Based on Polarization

Antenna polarization is governed by the polarization of electromagnetic waves. An antenna is either

1. Linearly (vertically/horizontally) polarized; or
2. Circularly polarized (Dhande, 2009).

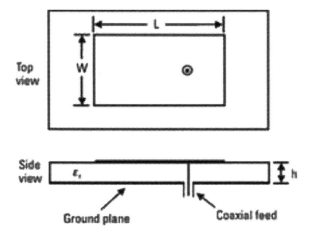

FIGURE 7.16 Microstrip patch antenna (Dhande, 2009).

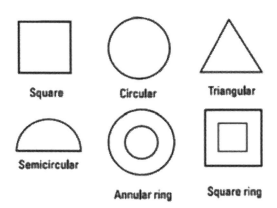

FIGURE 7.17 Various shapes of patch antenna (Dhande, 2009).

7.5.5.3.1 Linearly (Vertically/Horizontally) Polarized Antennas

If an antenna is transmitting/receiving a vertical E field vector, it is a vertically polarized antenna. If an antenna is transmitting/receiving a horizontal E field vector, it is said to be horizontally polarized (Dhande, 2009; ElProCus, 2016). See Figure 7.18 and Section 7.5.3.4.

7.5.5.3.2 Circularly Polarized Antenna

If the antenna is able to transmit or receive E field vectors of any orientation, it is said to be circularly polarized (Dhande, 2009; ElProCus, 2016).

7.5.5.4 Antenna Classification on the Basis of Radiation Pattern

On the basis of its radiation pattern, an antenna can be classified as (Dhande, 2009):

1. Isotropic,
2. Omnidirectional,
3. Directional, or
4. Hemispherical (Dhande, 2009).

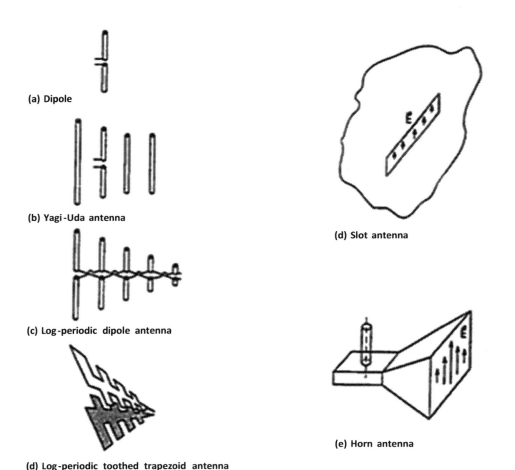

FIGURE 7.18 Linearly polarized antennas (Dhande, 2009).

7.5.5.4.1 Isotropic Antennas

An isotropic antenna is a hypothetical antenna which radiates uniformly in all directions. It is also called as isotropic source or omnidirectional antenna, or simply unipole. An isotropic antenna is a hypothetical lossless antenna, with which the practical antennas are compared. Thus, an isotropic antenna is used as reference antenna. A half-wave dipole antenna is sometimes used as a reference antenna, but an isotropic antenna is preferred (Dhande, 2009).

7.5.5.4.2 Omnidirectional Antennas

Omnidirectional antennas cover all directions equally well. Examples are whip antennas, dipole antennas, etc. The radiation patterns of omnidirectional antennas are shown in Figure 7.19 (Dhande, 2009).

7.5.5.4.3 Directional Antennas

Sending and receiving in a particular direction requires an antenna with high directivity; this is called a directional antenna. Directional antennas have very high gain and directivity to cover a large wireless distance. Examples are a paraboloid reflector antenna, Yagi–Uda antenna, and log-periodic antenna. The radiation pattern of these antennas is shown in Figure 7.20 (Dhande, 2009).

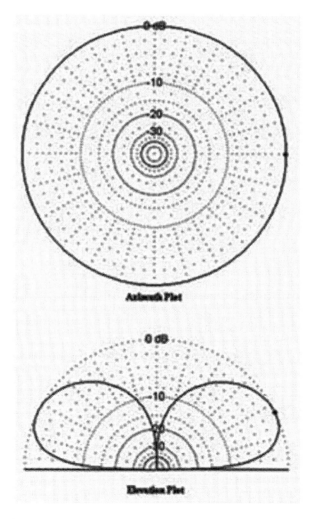

FIGURE 7.19 Omnidirectional antenna (also see Figure 7.11) (Dhande, 2009).

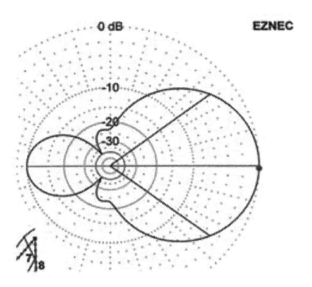

FIGURE 7.20 Directional radiation pattern (Dhande, 2009).

In directional antennas, two or more antennas are combined to make an array. This increases gain and directivity. Arrays can be parasitic or driven (RF Wireless World, 2012).

7.5.5.4.4 Hemispherical Antennas

An antenna whose radiation pattern will cover half of the hemisphere, either upper or lower, is called a hemispherical antenna. In the lower hemisphere, these antennas are implemented on an aircraft body for data link purposes. Examples are monopole antennas (Dhande, 2009).

7.5.6 Antenna Types

7.5.6.1 Log-Periodic Antennas

The most common type of antenna used in wireless communication technology is a log-periodic dipole antenna (or array). It fundamentally comprises a number of dipole elements. These dipole-array antennas reduce in size from the back end to the front end. The leading beam of this RF antenna comes from the smaller front end.

The element at the back end of the array is large, with the half wavelength operating in a low-frequency range. The spacing of the element is reduced towards the front end of the array where the smallest arrays are placed. During this operation, as the frequency varies, a smooth transition occurs along the array of the elements, forming an active region (ElProCus, 2016) (Figure 7.21).

The log-periodic array antenna operates in the VHF frequency range, from 30 to 300 MHz (Khan, Riaz, & Bilal, 2016) (Figure 7.22).

Basically, a log-periodic antenna is a broadband, multielement, tight-pillar, directional narrow beam antenna that works on a wide range of frequencies. Such antennas are used in a wide range of applications where variable bandwidth is required, along with antenna gain and directivity. They are useful for a region which requires greater frequency ranges. (ElProCus, 2016; Wong et al., 2012).

The antenna is made of a series of dipoles placed along the antenna axis at different space intervals of time followed by a logarithmic function of frequency. The antenna's radiations and impedance properties repeat regularly as a logarithmic function of the frequency excitation. A logarithmic increase in length and space of element occurs from one terminal to the other (Wong et al., 2012). The successive dipoles are alternately attached to a balanced transmission line called a feeder (Harchandra & Singh, 2014) (Figures 7.23 and 7.24).

Communications

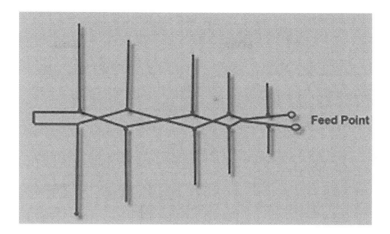

FIGURE 7.21 Log-periodic dipole antenna (ElProCus, 2016).

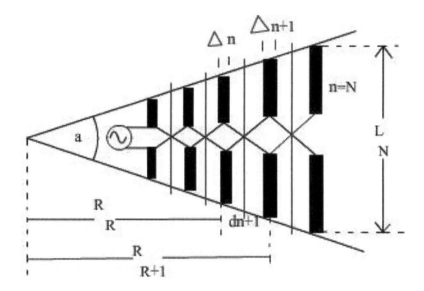

FIGURE 7.22 Log-periodic dipole array antenna (Khan, Riaz, & Bilal, 2016).

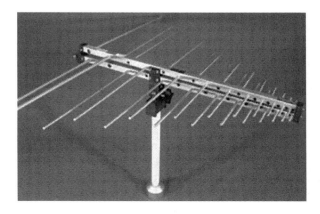

FIGURE 7.23 Log-periodic antenna (ElProCus, 2016).

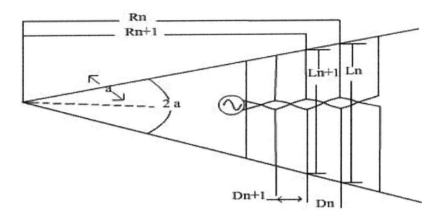

FIGURE 7.24 Representation of log-periodic dipole antenna (Kibona, 2013).

7.5.6.2 Bow-Tie Antennas

A bow-tie antenna is suited for multiband applications requiring light, low cost, dense, moveable, and easy working antennas. However, output is limited in terms of place, distance, and alignment. This type of antenna has two mirrors placed on a rectangular patch. The antenna uses a coplanar waveguide (CPW) lumped port. Applications include mobile communication networks and wireless systems (Kumar & Saini, 2013). Figure 7.25 shows a bow-tie antenna (Khan, Riaz, & Bilal, 2016).

7.5.6.3 Wire Antennas

Wire antennas are also known as linear or curved antennas. These antennas are very simple, cheap, and useful in a wide range of applications (ElProCus, 2016). There are four types of wire antennas (Figure 7.26).

7.5.6.3.1 Short Dipole Antennas

This is the simplest of all types of antennas. It is an open circuited wire in which short denotes "relative to a wavelength," so this antenna gives priority to the size of the wire relative to the wavelength of the

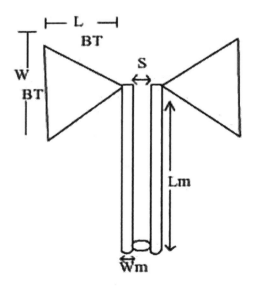

FIGURE 7.25 Representation of bow-tie antenna (Khan, Riaz, & Bilal, 2016).

Communications 259

FIGURE 7.26 Wire antenna (ElProCus, 2016).

frequency of operation. The short dipole antenna is made up of two colinear conductors placed end to end, with a small gap between conductors by a feeder. A dipole is considered short if the length of the radiating element is less than a tenth of the wavelength (ElProCus, 2016).

$$L < \lambda/10$$

The short dipole antenna can be unsatisfactory from an efficiency viewpoint because most of the power entering this antenna is dissipated as heat, and resistive losses gradually become high (ElProCus, 2016) (Figure 7.27).

7.5.6.3.2 Monopole Antennas

A monopole antenna is half of a simple dipole antenna located over a grounded plane (ElProCus, 2016). Figure 7.28 gives a representation of a monopole antenna.

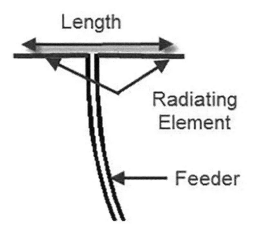

FIGURE 7.27 Representation of short dipole antenna (ElProCus, 2016).

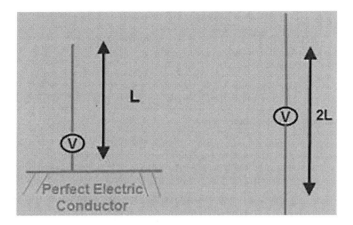

FIGURE 7.28 Representation of monopole antenna (ElProCus, 2016).

The radiation pattern above the grounded plane will be same as that for the half-wave dipole antenna, but the total power radiated will be half that of a dipole; the field is radiated only in the upper hemisphere region. The directivity of these antennas is double that of the dipole antennas.

Monopole antennas can be used as vehicle-mounted antennas, as they provide the required ground plane for antennas mounted above the Earth (ElProCus, 2016).

7.5.6.3.3 Dipole Antennas

A dipole antenna is one of the most straightforward antenna alignments. It consists of two thin metal rods with a sinusoidal voltage difference between them. The length of the rods is chosen in such a way that they a have quarter length of the wavelength at operational frequencies. These antennas are simple to construct and use.

Current and frequency flow through the metal rods. This current and voltage flow makes an electromagnetic wave, and the radio signals are radiated. A radiating element splits the rods and makes current flow through the center by means of a feeder at the transmitter that takes from the receiver (see Figure 7.29).

Types of dipole antennas include half wave, multiple, folded, and nonresonant (ElProCus, 2016).

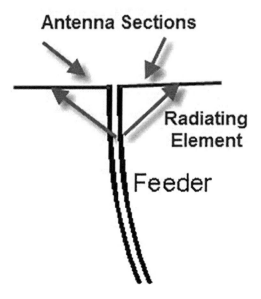

FIGURE 7.29 Representation of dipole antenna (ElProCus, 2016).

Antennas radiate effectively when the length of antenna is directly related to transmitted signal wavelength. Dipole antennas are available in half wave or quarter wavelength sizes. A half-wave dipole antenna called a doublet will have length equal to half a wavelength a operating frequency. RG-59/U is generally used for a 73 Ω coax line, and RG-11/U is used for a 75 Ω line.

$$\text{Length of this dipole, } L = 468/\text{Freq}$$

The radiation pattern of the half-wave dipole antenna is shaped like a doughnut (RF Wireless World, 2012).

7.5.6.3.4 Loop Antenna

Loop antennas are similar to dipole and monopole antennas in that they are simple and easy to construct. Loop antennas are available in different shapes, including circular, elliptical, and rectangular. The fundamental characteristics of the loop antenna are independent of its shape, however. These antennas are widely used in communication links with the frequency of around 3 GHz. They can also be used as electromagnetic field probes in the microwave bands (ElProCus, 2016) (Figure 7.30).

The circumference of the loop antenna determines the efficiency of the antenna. These antennas are further classified into two types: electrically small and electrically large, based on the circumference of the loop.

$$\text{Electrically small loop antenna} \rightarrow \text{Circumference} \leq \lambda/10$$

$$\text{Electrically large loop antenna} \rightarrow \text{Circumference} \approx \lambda$$

Electrically small loops of a single turn have small radiation resistance compared to their loss resistance. The radiation resistance of small loop antennas can be improved by adding more turns. Multi-turn loops have better radiation resistance, even if they have less efficiency (ElProCus, 2016) (Figure 7.31).

Small loop antennas are mostly used as receiving antennas where losses are not mandatory. They are not used as transmitting antennas because of their low efficiency.

Large loop antennas are directed by the operation of wavelength and are used at higher frequencies, such as VHF and UHF, where their size is convenient. They are also known as resonant antennas. Like folded-dipole antennas, they can be spherical, square, etc., and have high radiation efficiency (ElProCus, 2016).

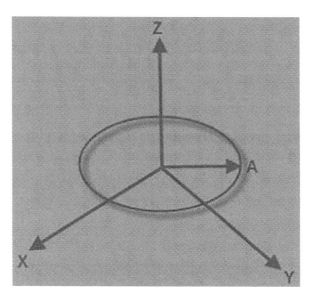

FIGURE 7.30 Loop antenna (ElProCus, 2016).

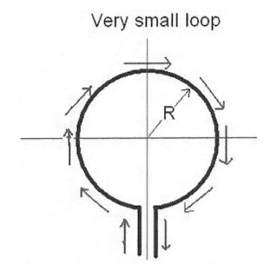

FIGURE 7.31 Small loop antenna (ElProCus, 2016).

7.5.6.4 Traveling Wave Antenna

There are several types of traveling wave antennas.

7.5.6.4.1 Helical Antennas

Helical antennas comprise a single wire or limited tape wound like a right- or left-hand screw, self-supporting, or turned on a dielectric cylinder (Kraus, 1947). Because they are practical and easy to use, helical antennas are extensively used in practice. They are widely used to get microwaves from VHF. They are also used in satellite communication where high gain is required. In a parabolic dish, higher gain is needed, so helical antennas are installed for this application as well.

Figure 7.32 shows the helical antenna. It comprises one empty dielectric chamber with relative permittivity 2.1 and cross-distance 61.33 mm. There is a generator at the base, between the antenna and the ground plane. The feed is situated at the base (Azdanboost & Kohno, 2005).

7.5.6.4.2 Yagi–Uda Antennas

The Yagi–Uda antenna is sometimes called a Yagi antenna or Yagi. These antennas contain reflectors, a dipole, and directors. They are made with aluminum tubes and an aluminum cross member.

This type of antenna is used in UHF/VHF radar, phased Doppler radar, and wind profiler systems (Kraus, 1988). Yagi antennas are widely used for TV reception, but as they are designed for one frequency only, they are not suited for a wide frequency range (RF Wireless World, 2012).

Figure 7.33 demonstrates the structure of the Yagi–Uda antenna (Khan, Riaz, & Bilal, 2016).

7.5.6.5 Microwave Antennas

Antennas operating at microwave frequencies are called microwave antennas. These antennas have a wide range of applications (ElProCus, 2016).

7.5.6.5.1 Rectangular Microstrip Antennas

For spacecraft or aircraft applications—based on specifications such as size, weight, cost, performance, and ease of installation—low-profile antennas are preferred. These antennas are known as rectangular microstrip antennas or patch antennas; they only require space for the feed line which is normally placed behind the ground plane. The major disadvantage of using these antennas is their inefficient and very narrow bandwidth, typically a fraction of a percent or, at the most, a few percent.

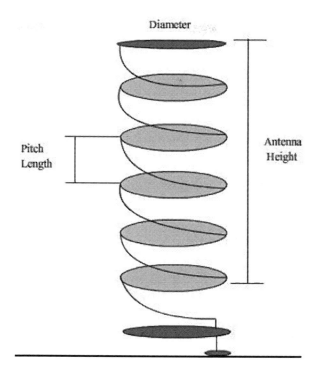

FIGURE 7.32 Representation of helical antenna (Khan, Riaz, & Bilal, 2016).

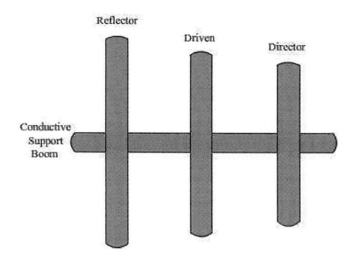

FIGURE 7.33 Representation of Yagi–Uda antenna (Khan, Riaz, & Bilal, 2016).

7.5.6.5.2 Planar Inverted-F Antennas

A planar inverted-F antenna (IFA) can be considered a type of linear IFA in which the wire radiating element is replaced by a plate to increase the bandwidth. The advantage of these antennas is that they can be hidden in the housing of the mobile, unlike whip, rod, or helical antennas, etc. The other advantage is that they can reduce the backward radiation towards the top of the antenna by absorbing power, thus enhancing the efficiency. They provide high gain in both horizontal and vertical states. This feature is most important for any kind of antenna used in wireless communications (ElProCus, 2016).

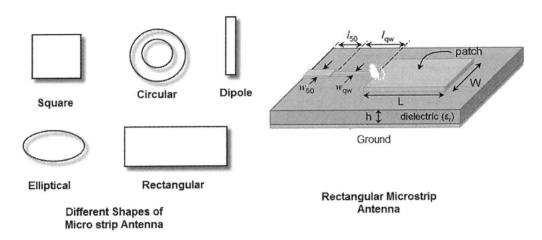

FIGURE 7.34 Rectangular microstrip antennas (ElProCus, 2016).

Small antennas are in demand in many commercial communication networks. For example, the IFA is used in Bluetooth technology. It is dense, with a simple construction, effective radiation, and an omnidirectional radiation pattern; it operates a bandwidth of 250 MHz (Vinoy, 2002). The antenna gives adaptability in impedance coordinating, and it makes an even and vertical electric field (Ramadan, Kabalan, El-Hajj Khoury, & Al-Husseini, 2009). It is useful for indoor communication (Vinoy, 2002). Figure 7.35 shows the structure of an IFA (Khan, Riaz, & Bilal, 2016).

7.5.6.6 Reflector Antennas

There are two types of reflector antenna: corner reflector and parabolic reflector.

7.5.6.6.1 Corner Reflector Antennas

An antenna that comprises one or more dipole elements placed in front of a corner reflector is known as a corner reflector antenna. These antennas are simple, effective, and highly efficient. The antenna is made up of a dipole element and two plane reflector panels. It is more directorial, as this arrangement prohibits radiation in the back and side directions. A dipole or a variety of the collinear dipole is put parallel to the vertex. The feed elements are bicolical dipoles or thick cylindrical wire rather than thin wire for greater bandwidth (Rao & Jones, 1991). The directivity of any antenna can be increased by using reflectors. In the case of a wire antenna, a conducting sheet is used behind the antenna to direct radiation in the forward direction (ElProCus, 2016).

Figure 7.36 explains the structure of the corner reflector antenna (Khan, Riaz, & Bilal, 2016).

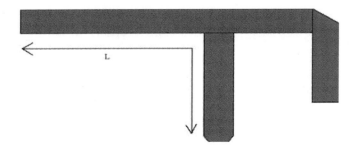

FIGURE 7.35 Representation of IFA (Khan, Riaz, & Bilal, 2016).

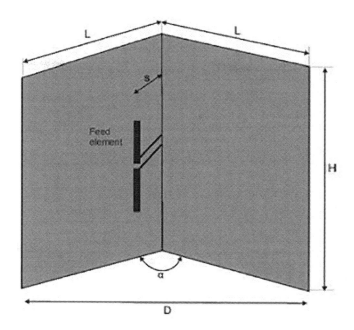

FIGURE 7.36 Corner reflector antenna (Khan, Riaz, & Bilal, 2016).

7.5.6.6.2 Parabolic Reflector Antennas

This antenna is discussed in Section 7.5.5.2.4.

7.6 Tracking with Multiple Autonomous Vehicles

Tracking with an autonomous vehicle involves accurately identifying and localizing dynamic objects in the environment surrounding the vehicle. Tracking surrounding vehicles is essential for many tasks crucial to truly autonomous driving, such as obstacle avoidance, path planning, and intent recognition. To be useful for such high-level reasoning, the generated tracks should be accurate, long, and robust to sensor noise (Rangesh & Trivedi, 2019).

7.6.1 Multi-Object Tracking (MOT) with Autonomous Vehicles

Traditional multi-object tracking (MOT) techniques for autonomous vehicles can roughly be categorized into three groups based on the sensory inputs they use (Rangesh & Trivedi, 2019):

1. Dense point clouds from range sensors
2. Vision sensors
3. A fusion of range and vision sensors.

Some studies used dense point clouds created by 3D light detection and ranging (LiDAR) systems like the Velodyne HDL-64E. Such sensors, although bulky and expensive, are capable of capturing finer details of the surroundings because of their high vertical resolution. Trackers can create suitable mid-level representations, such as 2.5D grids and voxels that retain unique statistics of the volume they enclose, and group such units together to form coherent objects that can be tracked. It must be noted, however, that these approaches rely on having dense point representations of the scene and do not scale well to LiDAR sensors with fewer scan layers.

Other studies use stereo vision to perform tracking. The pipeline usually involves estimating the disparity image and optionally creating a 3D point cloud, followed by similar mid-level representations such as stixels and voxels which are then tracked from frame to frame. These sensors are limited by the quality of disparity estimates and the field of view (FoV) of the stereo pair. Unlike 3D LiDAR-based systems, they are unable to track objects in full surround. There are other single camera approaches to surround vehicle behavior analysis, but they are limited in their FoVs and localization capabilities.

Finally, some approaches based in fusion make use of LiDAR systems, stereo pairs, monocular cameras, and radars in a variety of configurations. These techniques perform either early or late fusion based on their sensor setup and algorithmic needs. However, none seems to offer full-surround solutions for vision sensors, and they are ultimately limited to fusion only in the FoV of the vision sensors (Rangesh & Trivedi, 2019).

7.6.1.1 Two-Dimensional MOT for Autonomous Vehicles

Recent research in MOT has focused on tracking-by-detection, where the main challenge is data association to link object detections to targets. The majority of batch (offline) methods formulate MOT as a global optimization problem in a graph-based representation, while online methods solve the data association problem either probabilistically or deterministically (e.g., Hungarian algorithm or greedy association). A core component in any data association algorithm is a similarity function between objects. Both batch methods and online methods have explored the idea of learning to track, where the goal is to learn a similarity function for data association from training data (Rangesh & Trivedi, 2019).

7.6.1.2 Three-Dimensional MOT for Autonomous Vehicles

There is considerable ongoing research on the use of 3D MOT for autonomous vehicles. For example, Pfeiffer and Franke (2010) use a stereo rig to calculate the disparity using semi-global matching (SGM). This is followed by height-based segmentation and free-space calculation to create a mid-level representation using stixels that encode the height within a cell. Each stixel is then represented by a 6D state vector, which is tracked using the extended Kalman filter (EKF). Broggi, Cattani, Patander, Sabbatelli, and Zani (2013) use a voxel-based representation instead and cluster neighboring voxels based on color to create objects that are then tracked using a greedy association model. For their part, Vatavu, Danescu, and Nedevschi (2015) use a grid-based representation of the scene, where cells are grouped to create objects, each of which is represented by a set of control points on the object surface. This creates a high-dimensional state-space representation, accounted for by a Rao-Blackwellized particle filter.

Osep et al. (2016) propose semantic segmentation on the disparity image, which is then used to generate generic object proposals by creating a scale-space representation of the density, followed by multi-scale clustering. The clusters are tracked using a quadratic pseudo-Boolean optimization (QPBO) framework. Dueholm, Kristoffersen, Satzoda, Moeslund, and Trivedi (2016) use a camera setup, but the authors propose an offline framework for tracking, hence limiting their use to surveillance-related applications (Rangesh & Trivedi, 2019).

Some approaches make use of dense point clouds generated by LiDAR rather than creating point clouds from a disparity image. Choi, Ulbrich, Lichte, and Maurer (2013) carry out ground classification based on the variance in the radius of each scan layer, followed by a 2.5D occupancy grid representation of the scene. The grid is segmented, and regions of interest (RoIs) identified within, each of which is tracked by a standard Kalman filter (KF). Data association is achieved by simple global nearest neighbor. Similar to this, Asvadi, Peixoto, and Nunes (2015) use a 2.5D occupancy grid-based representation, but augment this with an occupancy grid map which accumulates past grids to create a coherent global map of occupancy by accounting for ego motion. Using these two representations, they create a 2.5D motion grid by comparing the map with the latest occupancy grid; this isolates and identifies dynamic objects in the scene. Although Asvadi, Premebida, Peixoto, and Nunes (2016) follow the same general idea, they propose a piecewise ground plane estimation scheme capable of handling nonplanar surfaces. In a departure from grid-based methods, Song, Xiang, and Liu (2015) project the 3D point cloud onto a virtual image plane, creating an object appearance model based on four image-based cues for each template of the desired target. A particle-filtering framework is implemented; the particle with least reconstruction error with respect to the stored template is chosen to update the tracker. Background filtering and occlusion detection are implemented to improve performance.

7.6.2 Tracking Multiple Autonomous Underwater Vehicles (AUVs)

7.6.2.1 Introduction

Autonomous underwater vehicles (AUVs) are becoming a reliable and cost-effective solution for performing a variety of underwater tasks in a fully automated way. Among the main tasks performed by AUVs are bathymetric surveys and environmental inspections, surveillance and patrolling, or even mine countermeasure operations. The use of such vehicles means the assigned tasks can usually be performed in a cost-effective way. It also enables operations in challenging scenarios that otherwise would not be safe or even possible for human intervention. While most of these tasks are traditionally performed using only a single vehicle, significant research efforts are focused on the development of algorithms that allow fleets of AUVs, navigating in a coordinated fashion, to achieve a common goal (Melo & Matos, 2018).

The use of multiple vehicles allows the parallelization of tasks that otherwise would not be possible, thus reducing operations time. The potential for efficiency gains is even greater if the various vehicles collaborate in the completion of a task. The use of teams of collaborating AUVs has been foreseen for different applications, for example, mine countermeasure missions (Prins & Kandemir, 2008) or archaeological missions. Even though there is an extensive and growing literature on cooperative control theory, there are only a few approaches demonstrating complete multi-AUV cooperation in field trials in water. An example is the efficient mapping of a given area using multiple vehicles by Paull, Huang, Seto, and Leonard (2015). Teams of cooperating AUVs have also been reported to perform adaptive environmental sampling tasks by having multiple vehicles performing plume tracking quickly and with high temporal and spatial resolution (Schulz et al., 2003). In a somewhat similar mission, sea trials with a fleet of ten autonomous underwater gliders deployed as an adaptive, coordinated ocean sampling network have also been reported (Fiorelli et al., 2006; Leonard et al., 2010). A possible application within the framework of environmental sampling is a distributed multi-vehicle patrolling approach (Marino, Antonelli, Aguiar, Pascoal, & Chiaverini, 2015).

With such developments, it is reasonable to expect that in the near future, new applications will appear requiring the operation of multiple AUVs concurrently, cooperating to achieve a desired goal. With the increase of multi-vehicle missions for underwater vehicles, the problem of tracking multiple AUVs in real-time becomes even more relevant. In fact, for most of the classical missions for AUVs, the ability to track the vehicles during operations is not only desirable but even critical for some more uncertain and hazardous scenarios, such as military or oil industry applications.

Tracking AUVs can be done by listening to the acoustic signals exchanged between the vehicle and a set of acoustic beacons deployed in the area of operations. The AUVs need to emit an acoustic signal that is detected by each beacon at different times, according to its distance from the vehicle. By combining the time of flight (ToF) of the signals as detected by the beacons, the position of a vehicle can be computed using multi-lateration techniques.

Out of all the literature considering the tracking of AUVs, only a few authors have addressed the problem of tracking multiple vehicles. Scenarios with multiple targets are seldom considered or experimentally validated. The main obstacle is the ability to uniquely associate acoustic signals with the source emitting them. Tracking more than one vehicle usually requires that each vehicle emits a signal that can be easily distinguishable. A natural solution is to have the vehicles use different frequency-modulated signals. Alternatively, time-division multiplexing schemes can be derived. While these approaches are proven to work, they are far from being optimal.

Neither option is scalable, particularly in situations with several vehicles. This is even more problematic if, as frequently happens, the acoustic beacons are also required to emit their own acoustic signals, in order to provide navigational aids to the vehicles. When using time multiplexing schemes, time slots are attributed to each device operating on the network, so it can emit acoustic signals. For operations with multiple vehicles, the number of time slots is increased, and this, in turn, increases the time interval between two consecutive signal emission slots for a given vehicle. As the number of vehicles increases, the performance of the trackers can significantly degrade.

However, increasing the number of distinct frequency signals is cumbersome and costly, as it requires the development of specific hardware for emission and detection of the signals. This is even more complicated if we consider that the acoustic signals are usually in a very confined band, from approximately 10 to 30 kHz, and this limits the number of available frequencies (Melo & Matos, 2018).

REFERENCES

Agarwal T., 2018. Different types of wireless communication technologies. EDGEFX.IN. www.edgefx.in/different-types-wireless-communication-technologies/. Viewed: June 17, 2019.

Asvadi A., Peixoto P., Nunes U., 2015. Detection and tracking of moving objects using 2.5D motion grids. In *2015 IEEE 18th International Conference on Intelligent Transportation Systems (ITSC)*, Las Palmas, Spain. IEEE, 2015, pp. 788–793.

Asvadi A., Premebida C., Peixoto P., Nunes U., 2016. 3D LIDAR-based static and moving obstacle detection in driving environments: An approach based on voxels and multi-region ground planes. *Robotics and Autonomous Systems*, 83, 299–311.

Azdanboost K. Y., Kohno R., 2005. Ultra wideband L-loop antenna. In *Ultra-Wideband, ICU. IEEE International Conferences*, Zurich, Switzerland, September 2005, pp. 201–205.

Broggi A., Cattani S., Patander M., Sabbatelli M., Zani P., 2013. A full 3D voxel-based dynamic obstacle detection for urban scenario using stereo vision. In *2013 16th International IEEE Conference on Intelligent Transportation Systems-(ITSC)*, The Hague, Netherlands. IEEE, 2013, pp. 71–76.

Choi J., Ulbrich S., Lichte B., Maurer M., 2013. Multi-target tracking using a 3D-LIDAR sensor for autonomous vehicles. In *2013 16th International IEEE Conference on Intelligent Transportation Systems-(ITSC)*, The Hague, Netherlands. IEEE, 2013, pp. 881–886.

Dhande P., 2009. Antennas and its applications. *Science and Spectrum*, 2, 66–78.

ElProCus, 2016. Here's a quick way to know about different types of antennas. www.elprocus.com/different-types-of-antennas-with-properties-and-their-working/. Viewed: June 25, 2019.

EOFDM Working Group A, 2018. Review of Mid Air Collision (MAC) precursors from an FDM perspective. EOFDM Working Group A, Version 1. May 18, 2018.

FARLEX. *The Free Dictionary. Radio Communication. The Great Soviet Encyclopedia*, Third Edition (1970–1979). ©2010 Farmington Hills: The Gale Group, Inc. All rights reserved. https://encyclopedia2.thefreedictionary.com/Radio+Communication. Viewed: June 17, 2019.

Federal Aviation Administration. How to avoid a mid air collision - P-8740-5. 2017. www.faasafety.gov/gslac/alc/libview_normal.aspx?id=6851. Viewed: June 16, 2019.

Fiorelli E., Leonard N., Bhatta P., Paley D., Bachmayer R., Fratantoni D., 2006. Multi-AUV control and adaptive sampling in Monterey bay. *IEEE Journal of Oceanic Engineering*, 31(4), 935–948.

Hamby C., 2016. Bandwidth. Data rate. Throughput. What's the difference? AvaLAN wireless. August 9, 2016. www.avalan.com/blog/bandwidth.-data-rate.-throughput.-whats-the-difference. Viewed: June 23, 2019.

Harchandra B., Singh R., 2014. Analysis and design of bowtie antenna with different shapes and structures. *International Journal of Engineering Trends and Technology*, 18(4), 171–175.

HOMENET HOWTO, 2016. Radio communication basics. www.homenethowto.com/advanced-topics/radio-communication-basics/. Viewed: June 18, 2019.

IATA, 2019. Assessment of Pilot Compliance to TCAS. Guidance Material. Performance assessment of pilot compliance to Traffic Alert and Collision Avoidance System (TCAS) using Flight Data Monitoring (FDM). EUROCONTROL, Edition 1, January 2019.

Jolani E., Dadgarpar A. M., Hassani H. R., 2008. Progress in electromagnetic research letter. Vol. 3, pp. 35–42.

Kampeephat S., Krachodnok P., Wongsan R., 2014. Efficiency improvement for conventional rectangular horn antenna by using EBG technique. *International Journal of Electrical, Computer, Electronics and Communication Engineering*, Vol. 8, No. 7, pp. 904–909.

Khan A. Q., Riaz M., Bilal A., 2016. Various types of antenna with respect to their applications: A review. *International Journal of Multidisciplinary Sciences and Engineering*, Vol. 7, No. 3, pp. 1–8.

Kibona L., 2013. Gain and directivity analysis of the log periodic antenna. *International Journal of Scientific Engineering and Research (IJSER)*, 1(3), 14–18.

Kraus J. D., 1947. Helical beam antennas. *Electronics*, 20, 109–111.

Kraus J. D., 1988. *Antennas, Network*. New York: McGraw-Hill.

Kremer P., 2012. Collision avoidance. UPL Safety Seminar 2012. https://upl-aopa.lu/wp-content/uploads/Collision-Avoidance.pdf. Viewed: June 16, 2019.

Kumar N., Saini G., 2013. A novel low profile Planar Inverted-F Antenna (PIFA) for mobile handsets. *International Journal of Scientific and Research Publications*, Vol. 3, No. (3), pp. 1–4.

Leonard N. E., Paley D. A., Davis R. E., Fratantoni D. M., Lekien F., Zhang F., 2010. Coordinated control of an underwater glider fleet in an adaptive ocean sampling field experiment in monterey bay. *Journal of Field Robotics*, 27(6), 718–740.

Mahesh Lohith K. S., Maratha Mandal Engineering College, 2014. Communication Systems a Lecture Notes by Mahesh Lohith K. S. Department of Applied Science. 10ELN15/25- Part B - Unit 7- Communication Systems. ML, Maratha Mandal Engineering College, Belgaum.

Marino A., Antonelli G., Aguiar A., Pascoal A., Chiaverini S., 2015. A decentralized strategy for multirobot sampling/patrolling: Theory and experiments. *IEEE Transactions on Control Systems Technology*, 23(1), 313–322.

Melendez S., 2019. Definition of radio broadcasting. TECHWALLA. www.techwalla.com/articles/importance-of-radio-communication. Viewed: June 18, 2019.

Melo J., Matos A. C., 2018. Tracking multiple Autonomous Underwater Vehicles. *Autonomous Robots*, 43(1), 1–20. ©Springer Science+Business Media, LLC, part of Springer Nature 2018. January 29, 2018.

Miller R., 1996. 4 different types of workplace communication and how to improve in each area. Sandler training. www.sandler.com/blog/4-different-types-workplace-communication-and-how-improve-each-area. Viewed: June 23, 2019.

Osep A., Hermans A., Engelmann F., Klostermann D., Mathias M., Leibe B., 2016. Multi-scale object candidates for generic object tracking in street scenes. In *2016 IEEE International Conference on Robotics and Automation (ICRA)*, Stockholm, Sweden. IEEE, 2016, pp. 3180–3187.

Paull L., Huang G., Seto M., Leonard J., 2015. Communication constrained multi-AUV cooperative slam. In *2015 IEEE international conference on robotics and automation (ICRA)*, Seattle, WA, USA. pp. 509–516.

PC, 2019. Definition of: Data rate. Encyclopedia. www.pcmag.com/encyclopedia/term/40833/data-rate. Viewed: June 23, 2019.

Pfeiffer D., Franke U., 2010. Efficient representation of traffic scenes by means of dynamic stixels. In *Intelligent Vehicles Symposium (IV)*, San Diego, CA, USA. IEEE, 2010, pp. 217–224.

Prins R., Kandemir M., 2008. Time-constrained optimization of multi-AUV cooperative mine detection. In *Proceedings of the MTS/IEEE Oceans'08 conference*, Quebec, Canada, pp. 1–13.

Ramadan A. H., Kabalan K. Y., El-Hajj Khoury A., Al-Husseini M., 2009. A reconfigurable U-Koch microstrip antenna for wireless applications. *Progress in Electromagnetics Research*, 93, 355–367.

Rangesh A., Trivedi M. M., 2019. No blind spots: Full-surround multi-object tracking for autonomous vehicles using cameras & LiDARs.

Rao B. R., Jones D. N., 1991. Characterization of a high frequency Beverage antenna using a fiber-optic measurement technique. In *Antennas and Propagation Society International Symposium*, London, Ontario, Canada. IEEE, 1991, pp. 1190–1193.

RF Wireless World, 2012. Types of Antenna | Antenna types for wireless communication. RF & Wireless Vendors and Resources. www.rfwireless-world.com/Terminology/Types-of-Antenna.html. Viewed: June 28, 2019.

Rohn J., 2017. Different effective methods of communication (useful). EDUCBA. www.educba.com/different-methods-of-communication/. Viewed: June 16, 2019.

Rouse M., 2011. Data transfer rate (DTR). SearchNetworking. TECHTARGET. May 2011. https://searchunifiedcommunications.techtarget.com/definition/data-transfer-rate. Viewed: June 23, 2019.

Rouse M., 2014. Bandwidth. SearchNetworking. TECHTARGET. https://searchnetworking.techtarget.com/definition/bandwidth. Viewed: June 20, 2019.

Schulz B., Hobson B., Kemp M., Meyer J., Moody R., Pinnix H., St Clair M., 2003. Field results of multi-UUV missions using ranger micro-UUVS. In *Proceedings of the MTS/IEEE Oceans'03 conference*, San Diego, CA, USA; Vol. 2, pp. 956–961.

Skybrary, 2018. Mid-air collision. February 18, 2018. www.skybrary.aero/index.php/Mid-Air_Collision#Definition. Viewed: June 16, 2019.

Song S., Xiang Z., Liu J., 2015. Object tracking with 3D LIDAR via multitask sparse learning. In *2015 IEEE International Conference on Mechatronics and Automation (ICMA)*, Beijing, China. IEEE, 2015, pp. 2603–2608.

TECHTERMS, 2012. Bandwidth. May 16, 2012. https://techterms.com/definition/bandwidth. June 20, 2019.

TECHTERMS, 2019. Data transfer rate. https://techterms.com/definition/datatransferrate. Viewed: June 20, 2019.

TELECOM ABC (a). www.telecomabc.com/d/data-rate.html. Viewed: June 27, 2019.

TELECOM ABC (b). www.telecomabc.com/b/bitrate.html. Viewed: June 27, 2019.

THE EYE from Network Instrument LLC, 2005. Bandwidth utilization. June 2005. https://insight.viavisolutions.com/en/2005/jun_en_web.html. Viewed: June 20, 2019.

Tutorials Point, 2019. Communication methods. Telangana, INDIA-50008. 2019. https://www.tutorialspoint.com/management_concepts/communication_methods.htm. Viewed: June 26, 2019.

Ujwala P., Namrata M., Pooja K., Shraddha M., 2014. Performance analysis of corner reflector antenna. *International Journal of Innovative Research in Computer and Communication Engineering*, 201–205.

Vatavu A., Danescu R., Nedevschi S., 2015. Stereovision-based multiple object tracking in traffic scenarios using free-form obstacle delimiters and particle filters. *IEEE Transactions on Intelligent Transportation Systems*, 16(1), pp. 498–511.

Vázquez A. R., Herrero J., Herrero D., Gómez J., 2010. Description and analysis of an indoor positioning system that uses wireless zigbee technology. doi:10.4018/978-1-61520-701-5.ch019.

Vick A., 2019. Importance of radio communication. TECHWALLA. www.techwalla.com/articles/importance-of-radio-communication. Viewed: June 18, 2019.

Vinoy K. J., 2002. Fractal shaped antenna elements for wide-and multi-band wireless applications. *Doctoral Dissertation*, The Pennsylvania State University, Pennsylvania, 2002.

Wikipedia. Mid-air collision. https://en.wikipedia.org/wiki/Mid-air_collision#cite_note-1. Viewed: June 16, 2019.

Wong H., Luk K. M., Chan C. H., Xue Q., So K. K., Lai H. W., 2012. Small antennas in wireless communications. In *2012 Proceedings of the IEEE*, Kowloon, Hong Kong. Vol. 100, No.7, pp. 2109–2121.

Woodford C., 2019. Antennas and transmitters. EXPLAINTHATSTUFF! April 21, 2019. www.explainthatstuff.com/antennas.html. Viewed: June 26, 2019.

8

Autonomous Vehicles for Infrastructure Inspection Applications

8.1 Power Line Inspection

8.1.1 Introduction

The inspection of power lines is critical for the safe operation of power transmission grids (Miralles, Pouliot, & Montambault, 2014). The monitoring of power lines incorporates two aspects: power line components and their surrounding objects, especially vegetation. The conditions of the components require regular checking to detect faults that are caused, for example, by corrosion and mechanical damage. There is also a need for the regular inspection of vegetation both inside and near the power line corridor. Trees or tree branches that are too close to power lines should be trimmed; vegetation is conductive, and this could lead to electric arcs. To ensure the safe operation of power lines, it is necessary to have a range of empty space without conductive objects around an extra high-voltage (EHV) power line. However, vegetation within the power line corridor will naturally grow after a power transmission grid becomes operational. A discharge may be generated when the distance between the vegetation and the power line is less than the safety threshold, thereby endangering the safe operation of the power transmission grid (Ahmad, Malik, Abdullah, Kamel, & Xia, 2015). Therefore, much research has been focusing on finding a highly efficient method of detecting obstacles along a power line corridor across a large area.

Seven types of data can be used for the inspection of power line corridors (Matikainen et al., 2016):

1. Synthetic aperture radar (SAR) images
2. Optical satellite images
3. Optical aerial images
4. Thermal images
5. Airborne laser scanner (ALS) data
6. Land-based mobile mapping data
7. Unmanned aerial vehicle (UAV) images.

SAR image pixels include the radar backscattering intensity, the phase of the backscattering signal, and the range from the sensor to the target. This information can be easily used to map power lines and towers and to monitor disasters that can harm power lines, such as earthquakes and typhoons.

Optical satellite images can be used to monitor vegetation in power line corridors, although the detected results are rather coarse because of the lower spatial and temporal resolution of satellite images compared to aerial visible wavelength images. The advantages of aerial images are their high resolution and availability; thus, they are regularly used for the reconstruction of power line corridors. Thermal images are sometimes used to inspect electrical faults in high-voltage electric utility transmission and distribution lines. Even at short distances, accurate temperature measurements of electrical faults are impossible to quantify. ALS is an active remote sensing technique that can be applied to power line mapping, vegetation mapping, and power line monitoring. ALS-based inspection methods use a laser scanner mounted on an aircraft to scan the power line corridor, obtaining a 3D point cloud to conduct a 3D reconstruction of the power lines and the ground in order to detect obstacles within the corridor.

At present, this method is applied only within a certain range, and for these applications, the point cloud density is the key factor in the 3D reconstruction of power lines. However, the ALS technique has not been popular because of the high costs associated with laser scanning equipment possessing large scanning ranges and high scanning frequencies. Because of its large size and heavy weight, such equipment is usually mounted on a manned helicopter; thus, the cost is high, making it more difficult to arrange inspection work. Although small laser scanner equipment can be mounted on UAVs, the small scanning range and slow scanning frequency of this equipment make it difficult to efficiently obtain sufficiently dense 3D point clouds. With the development of ALS equipment in the direction of miniaturization, a number of new types of ALS have appeared in the market.

A land-based mobile mapping technology is based on the integration of various positioning, navigation, and imaging data collection sensors that constitute a mobile mapping system (MMS), mounted on a kinematic platform, such as a car or a human. Unfortunately, many EHV transmission lines are located in areas that lack a transportation network.

A UAV equipped with a digital camera is a new and convenient method for power line inspection. UAVs have significantly lower operating costs than helicopters; thus, power line corridors across large areas can be inspected by installing a digital camera on a UAV. Different UAV platforms have been applied. Generally, fixed-wing UAVs can fly higher and faster, and they are more suitable for vegetation monitoring and the rough inspection of long power lines, while helicopter and multi-rotor UAVs can be used to acquire detailed pictures by hovering in the air close to the objects (Zhang, Yuan, Li, & Chen, 2017). The automatic measurement of power lines using stereo images is similar to automatic stereo mapping for a specific target in a particular scene. When performing an inspection, a UAV equipped with a lightweight digital camera flies along the power line while acquiring stereo images of the corridor according to a certain overlap. A 3D reconstruction of the power lines and the ground can be performed using photogrammetric methods, and the obstacles can be automatically identified and located by calculating the spatial distance between the power line and the ground. Zhang, Yuan, Fang, and Chen (2017) proposed a dense matching algorithm for the 3D reconstruction of the ground in power line corridors, known as SPMEC; the proposed method has accomplished the automatic 3D reconstruction of objects such as the crowns of trees and roofs. For a 3D reconstruction of power lines, the researchers adopted the traditional manual stereo measurement method, but it restricts the level of automation of inspection by using UAV images and needs further improvement (Zhang, Yuan, Li, & Chen, 2017).

8.1.2 Inspection with a UAV

Inspection with a UAV is an upgrade of automated helicopter inspection, so both concepts have common problems. An evaluation of using a UAV for power line inspection (Jones & Earp, 1996) found this inspection method could be faster than foot patrol and would yield the same or better accuracy than costly helicopter inspection. The system was concluded to be feasible from a technical point of view. The concept was further investigated by Jones (2005). This study proposed a small electrically driven rotorcraft which could pick up energy from power lines. This vehicle would be equipped with gyro-stabilized cameras, navigation and position regulation, a computer for image and other sensor data processing, a communication link, and a system for electric power pickup. Power would be obtained from the power line using a pantograph mechanism. Other research has considered the development of a vision system for power line tracking and image quality assurance. Good power line tracking is important for visual position control and navigation, while image quality is of utmost importance for inspection (Katrasnik, Pernus, & Likar, 2008b).

8.1.2.1 Position Control

Since power lines have to be inspected from a small distance but must under no circumstances get damaged even in strong wind, position control of the UAV is very important but difficult. Because conductors have to be in the field of view of the camera almost all the time, determining the position of the helicopter visually from the images of the conductors seems a possibility (Campoy et al., 2001;

Golightly & Jones, 2005; Jones, Golightly, Roberts, & Usher, 2006). Position control is thus closely related to the automatic tracking of power lines. The helicopter is a very complex, unstable, and nonlinear system with cross couplings. Campoy et al. (2001) chose a linear quadratic Gaussian (LQG) controller for roll and pitch control and a proportional-integral-derivative (PID) controller for yaw control. The controllers were implemented on the basis of the measured dynamic characteristics of the helicopter. Because only the position of the helicopter could be measured, all other required variables were estimated by the Kalman filter. The robustness was tested when the helicopter was in hover by pulling it with a cable. The regulation worked well in the presence of such external disturbances (Katrasnik, Pernus, & Likar, 2008b).

A rotorcraft model and a position control system for a power line inspection robot were presented by Jones, Golightly, Roberts, and Usher (2006). They derived a mathematical model of a ducted-fan rotorcraft with the center of gravity above the aircraft center and used it to develop a control system. The control is achieved by moving a mass, positioned above the center of gravity, left or right. When the mass is moved, the craft tilts in the same direction and accelerates in that direction. The control system is closely linked with the visual tracking of power lines and controls the height and lateral position of the craft to the lines. Lateral position and height are both measured with image analysis (Katrasnik, Pernus, & Likar, 2008b).

8.1.2.2 Automatic Power Line Tracking

The visual tracking of power lines with a UAV is similar to visual tracking with a helicopter. The only major difference is that the UAV can get closer to the lines. The tracking methods are therefore a little different. Jones, Golightly, Roberts, and Usher (2006) developed a simple tracking algorithm that can track a power line with three lines based on the Hough transform. The main purpose of this tracking algorithm is to provide height and lateral displacement of the vehicle to the control system. The method was tested on a scaled model and was proven to be successful even when the background was cluttered. Another study on visual power line tracking (Campoy et al., 2001) utilized a vector-gradient Hough transform for line detection. Only one line was tracked and simultaneously inspected. The position of the helicopter in relation to the line was determined with stereovision (Katrasnik, Pernus, & Likar, 2008b).

8.1.2.3 Obstacle Avoidance

A problem related to robot mobility is obstacle avoidance and path planning. The space around power lines is usually obstacle free; nevertheless, the robot must be able to avoid obstacles in its way, when it is not controlled by a human operator. A computer vision solution to this problem was proposed by Williams, Jones, and Earp (2001). The researchers determined the positions of the obstacles using optical flow. The obtained positions were used in a path planning algorithm based on the distance transform. The algorithms were tested in a laboratory environment using a test rig with a scaled version of a power line. It was established that the principles used were correct, but the method was sensitive to the variations in background, lighting, and perspective. Another important problem was the computing power because image analysis demands were high and rapid obstacle detection was required (Katrasnik, Pernus, & Likar, 2008b).

8.1.2.4 Power Supply

An important characteristic of an inspection vehicle is the duration of its power supply. The longer the craft can stay operational, the more lines can be inspected. Current battery technology does not permit long flights for small electrically driven helicopters. Power lines are an abundant source of energy, but obtaining that power is not trivial. A concept of a power line power pickup device was proposed by Jones (2005). The power would be acquired by touching two lines of different phases with a special

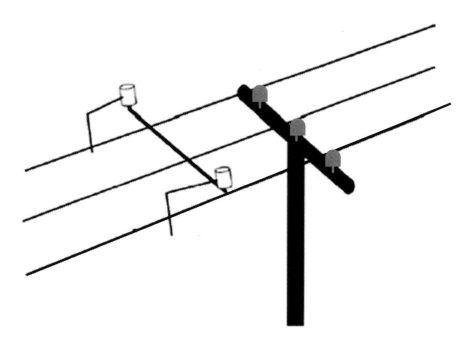

FIGURE 8.1 Proposed pantograph power pickup mechanism (Katrasnik, Pernus, & Likar, 2008b).

pantograph mechanism (see Figure 8.1). For this concept to work, however, line tracking and position control algorithms have to be highly reliable (Katrasnik, Pernus, & Likar, 2008b).

8.1.2.5 Other Problems

A problem that has not been researched thoroughly is automatic power line fault detection. It would be convenient if a robot could automatically detect faults on-site, because it could then inspect them more thoroughly. Automatic fault detection could be done in the ground station after the inspection; this would be easier to implement but would not provide detailed information about the defects. A big difficulty with fault detection is the quality of images taken from the UAV. Because of the distance from the line and constant movement of the craft, the quality of images is usually poor, making automatic fault detection especially demanding.

With UAV inspections, almost every system on the robot (position control, obstacle avoidance, fault detection, and power pickup) depends on the visual tracking of power lines, and this is not very reliable. Although visual power line tracking is successful in the laboratory, the real environment is much more demanding. Contrast between the lines and the background is usually very low. Lighting varies a great deal and depends on unpredictable weather conditions. The UAV is in constant motion and vibrates, so the images acquired are poor quality, and the faults are difficult to detect, even for a human. Unintentional detection of other straight lines on the image, such as other power lines or railroad tracks, could also pose a serious problem (Katrasnik, Pernus, & Likar, 2008a,b).

8.1.3 Climbing Robots

An alternative approach to power line inspection is to use a climbing robot. The robot must be able to climb on the conductor and overcome all the various obstacles on the power lines. The main advantage of this concept is the inspection accuracy. Namely, close proximity to the line and low vibrations increase the quality of image acquisition. Unfortunately, the development of a robot mechanism for overcoming

Autonomous Vehicles for Inspection Applications

obstacles on the line is extremely difficult. The main research problem with climbing robots is the development of a robot mechanism and a control system for obstacle crossing. The proximity of the conductor also brings problems related to electromagnetic shielding; sensitive electronics and sensors have to be protected from the electric and magnetic fields of the conductor (Katrasnik, Pernus, & Likar, 2008a).

8.1.3.1 Robot Mechanisms and Obstacle Traversing

One of the first operational robot mechanisms for power line inspection was devised by Sawada, Kusumoto, Maikawa, Munakata, and Ishikawa (1991). The robot consisted of a drive, an arc-shaped rail, a guide rail manipulator, and a balancer with a controller. It could travel on slopes of up to 30°. When the robot came across an obstacle, it would unpack its rail and mount it on the conductor on both sides of the obstacle. Then the drive mechanism would release the conductor and travel on the rail to the other side. The robot was able to negotiate towers and other equipment on overhead ground wires. As it did not have proper shielding and mechanisms for overcoming obstacles, it could not travel on phase conductors (Katrasnik, Pernus, & Likar, 2008a).

A more complex robot mechanism presented by Tang, Wang, and Fang (2004) had two arms (front arm and rear arm) and a body. Each arm had 4° of freedom and a gripper with a running wheel. The body also had a running wheel with a gripper. When overcoming obstacles, the robot would release the conductor with the front arm, elongate it over the obstacle, and grasp the conductor on the other side. Then the body would release, and the two arms would move it across the obstacle, where it would grip the conductor again. Finally, the rear arm would move across the obstacle. This robot could overcome all standard obstacles on phase conductors of overhead power lines. However, it could not travel on bundled conductors (Katrasnik, Pernus, & Likar, 2008a).

The robot configuration in Xinglong, Hongguang, Lijin, Mingyang, and Jiping (2006) had two arms and a special gripper, combined with a driving wheel. The specialty of this mechanism was that the gripper could always grasp the conductor when it was in contact with the running wheel. The gripper pressed on the conductor from the left and right side of the wheel. The main disadvantage was that the gripper could not handle large torque, which can easily occur when crossing obstacles. For that reason, a special obstacle crossing strategy that also simplified the design of the robot was designed (see Figure 8.2). When the robot detected an obstacle ahead, it would stop, grasp the conductor with the front arm, and move its body under the front arm to minimize the torque when crossing the obstacle (see Figure 8.2a). Next, the rear arm would lift the running wheel up, and the front arm would rotate the robot around its own axis. Finally, the rear arm would lower the wheel on the conductor (see Figure 8.2b). The same process would then be repeated with the arms' roles changed. Because of this obstacle traversing strategy, the robot arms needed only two degrees of freedom; in addition, the torques in the joints and on the conductor were small and, consequently, the motors did not need to be as powerful and heavy (Katrasnik, Pernus, & Likar, 2008a).

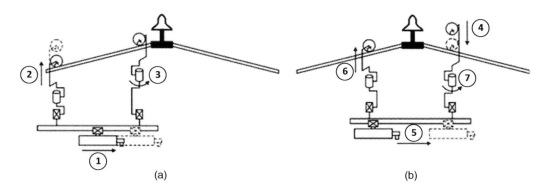

FIGURE 8.2 (a and b) Obstacle traversing strategy (Xinglong, Hongguang, Lijin, Mingyang, & Jiping, 2006).

8.1.3.2 Robot Control System

The main purpose of the robot control system is to navigate the robot over obstacles on the line. One of the first robot control algorithms for power line inspection was the one described above, created by Sawada, Kusumoto, Maikawa, Munakata, and Ishikawa (1991). A more complex control system using a distributed expert system divided between the robot and the ground station, also mentioned previously, was designed by Tang, Wang, and Fang (2004). The latter robot control system ran on an embedded PC/104 based computer, connected to the ground station with a wireless data link and a separate image transmission channel. The robot expert system consisted of an inference engine, knowledge base, static database, external information input module, and decision-making module. The inference engine would decide what commands to execute on the basis of sensor information and information in the static database. Sensors provided information about the current position of the robot and the obstacles around it, while the static database contained data on towers and other obstacles on the line. The robot expert system would plan the path of the robot arms so that the robot would overcome obstacles successfully. The ground station was used for monitoring and guiding the robot, as well as for detecting faults on the power line from the images sent by the robot. Similar distributed expert system designs were proposed by Ludan, Hongguang, Lijin, and Mingyang (2006) and are described above.

8.1.3.3 Obstacle Detection and Recognition

Obstacle detection is usually done with a proximity sensor; the method is simple yet effective, but the detection of the obstacle is usually not enough to overcome it. In most cases, the type of the obstacle has to be known before it is overcome. Zhang et al. (2006) presented a computer vision method for obstacle recognition and distance measurement. The method determines the obstacle types from the shapes on the image. An ellipse represents a suspension insulator string with two circles left and to the right of the conductor a strain insulator string. After the obstacle is recognized, its position is located with stereovision. The method was tested on a real power line, and an accuracy of 7% or better was reported.

Another important problem associated with visual obstacle detection and recognition is the elimination of motion blur from the captured images (Fu et al., 2006). Although a climbing robot is fixed on the conductor, it swings in wind and when traveling along the line (Katrasnik, Pernus, & Likar, 2008a,b).

8.1.3.4 Power Supply

Power lines could provide the inspection robot with energy for its operation. Energy could be extracted from the magnetic field of the line. This concept was proposed by Peungsungwal, Pungsiri, Chamnongthai, and Okuda (2001). A magnetic iron core was placed around the conductor. Current induced in the secondary coil around the core was measured at different numbers of secondary windings. The researchers found the current reached its maximum value at a certain number of secondary windings, and the power transferred to the secondary coil increased with the current of the power line (Katrasnik, Pernus, & Likar, 2008a).

8.1.4 Climbing-Flying Robot

The advantageous features of flying and climbing robots can be combined (see Figure 8.3). The proposed robot would combine a helicopter for flying over the obstacles and a special drive mechanism for traveling on the conductor. During inspection, the robot would travel on the conductor up to an obstacle. Then it would fly off the conductor over the obstacle, land on the other side, and continue traveling along the conductor. Traveling on the conductor would be automated, while flying over the obstacles would likely have to be done manually. Some of the problems to be solved are similar to those described in previous sections, for instance, the power pickup system, obstacle detection and recognition, and drive mechanism for traveling on the conductor (Katrasnik, Pernus, & Likar, 2008b).

Autonomous Vehicles for Inspection Applications 277

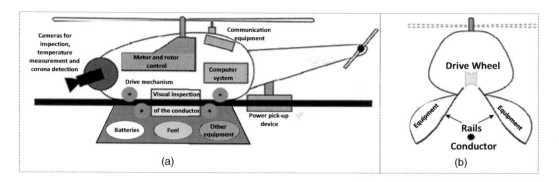

FIGURE 8.3 Proposed robot: (a) Illustration of the proposed climbing-flying robot and its components; (b) Sketch of the robot from the front, showing the rails for easier landing on conductors and equipment placement for stability (Katrasnik, Pernus, & Likar, 2008b).

8.1.4.1 Robot Design

Designing a climbing-flying robot is much more difficult than designing a flying robot, although not as difficult as a climbing robot. A design for a climbing-flying robot must consider the weight limitations of the helicopter, which are much stricter than for the flying robot. The addition of the drive mechanism and the electromagnetic shielding significantly increases the weight of the robot.

Another major problem is the weight distribution in the robot. To achieve a good degree of stability on the power line, the center of gravity of the robot must be below the conductor. This conflicts with the design of the helicopter, where the majority of the weight is placed directly below the rotor to achieve good maneuverability. The parts of the robot must therefore be carefully positioned to achieve the optimal position of the center of gravity. A rough distribution of robot parts inside the robot is proposed in Figure 8.3.

Setting the weight limitation and distribution problems aside, the most important problem of the climbing-flying robot is the design of individual systems, i.e., the helicopter, drive mechanism, visual inspection system, power pickup device, and communication system. The design of these systems is discussed in the following subsections (Katrasnik, Pernus, & Likar, 2008a).

8.1.4.2 Helicopter

When choosing the rotor configuration of the helicopter for the climbing-flying robot, we have three choices: Sikorsky configuration, tandem rotor configuration, and coaxial configuration.

The most common is the Sikorsky configuration. Ninety percent of all the helicopters in the world are made in this configuration. It is simple to produce and has good maneuverability and sufficient lift.

The tandem rotor configuration has worse maneuverability but produces more lift, as no power is needed to balance the main rotor torque. This configuration is also more difficult to make and maintain, as it has more moving parts and a more complex design.

The coaxial rotor configuration is more expensive to build and maintain, but it requires less space, while producing the same amount of lift as the other two configurations. This results in a smaller and more maneuverable helicopter for the same payload limitations. The coaxial configuration has better maneuverability than the Sikorsky configuration but is more expensive and has more frequent maintenance. For the climbing-flying robot, the coaxial configuration is, therefore, the best choice (Katrasnik, Pernus, & Likar, 2008a).

8.1.4.3 Drive Mechanism

The drive mechanism consists of the front and the rear drive mechanism. Each of the two drive mechanisms has two wheels (Figure 8.4). The upper wheel is the drive wheel, while the lower wheel provides

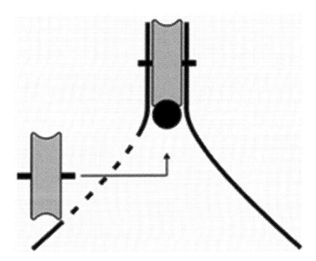

FIGURE 8.4 Part of the proposed drive mechanism. After the robot lands on the conductor, the lower wheels grasp the conductor from the sides (Katrasnik, Pernus, & Likar, 2008a).

stability for the robot on the power line. The drive wheel is connected to an electrical motor with a drive chain, whereas the lower support wheel runs freely. The wheels are made of aluminum, and the conductor contact surfaces of the wheels are covered with conductive rubber to increase traction, damp vibrations, and keep the robot on the same electric potential as the conductor. The support wheel will grasp the conductor. At landing, the robot will sit down on the drive wheels with the help of special rails (Figures 8.3b and 8.4). After the robot is positioned on the drive wheels, the support wheels are moved into position with servomotors. The contact force with the conductor is applied with springs. Before take-off, the support wheels are retracted, and the robot is free to lift off the conductor (Katrasnik, Pernus, & Likar, 2008a).

8.1.4.4 Power Pickup Device

The power pickup device consists of two parts, the toroidal core and the clasping mechanism. The toroidal core (Figure 8.5) is made from a ferromagnetic iron core and is split into two halves. On each half is a winding that transforms the energy of the magnetic field in the iron core to electrical energy, which is then further treated with a special converter circuit to obtain a useable voltage to power the systems on board the robot. The converter must be capable of handling a large range of input voltages, as the voltage in the winding changes linearly with the power line current. The clasping mechanism takes care of

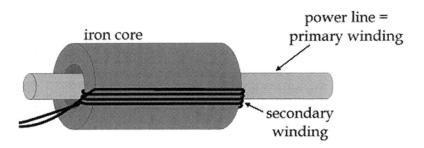

FIGURE 8.5 Power pickup device (Katrasnik, Pernus, & Likar, 2008a).

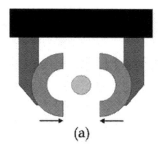 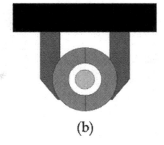

FIGURE 8.6 Power pickup device with clasping mechanism: (a) Open power pickup device; (b) Closed power pickup device (Katrasnik, Pernus, & Likar, 2008a).

the closing and opening of the toroidal core after landing and before takeoff (Figure 8.6). It is extremely important that the clasping mechanism closes the two halves of the toroidal core as closely together as possible, as even a small slit between the two halves significantly affects the efficiency of the power pickup device. Its precision is, therefore, of great importance.

A very important parameter of the power pickup device is its power-to-weight ratio. It is crucial that a power pickup device is as light as possible, as weight is limited on the robot. The power produced by the power pickup device depends on the power line current and on the geometry of the toroidal core. A preliminary analysis showed that such a power pickup device is feasible, as a power-to-weight ratio of more than 250 W/kg can be achieved for a relatively small 400 A power line current (Katrašnik, 2007; Katrasnik, Pernus, & Likar, 2008a).

8.1.4.5 Communication

To guide the robot over the obstacles effectively, the operator needs real-time visual feedback on the robot's surroundings, while the guiding data have to be sent with minimal latency. For these reasons, a reliable high-bandwidth wireless data link with very low latency is required. Another requirement, which conflicts with the high-bandwidth requirement, is long communication distance that should reach at least 5 km for an efficient operation (Katrasnik, Pernus, & Likar, 2008b).

8.1.4.6 Visual Inspection System

Visual inspection of the conductor consists of visual inspection of the power line using infrared (IR) and ultraviolet cameras at the front of the robot and visual inspection using three-line scan cameras.

Visual inspection of the power line includes looking for defects on the conductor, insulator, supporting tower, and other equipment. Visual detection of defects on all these systems is not very reliable; the conductor is more accurately inspected with the second visual inspection system, while defects on other equipment are detected with IR and ultraviolet cameras, a part of the front visual inspection system. IR cameras can easily detect overheating of any part of the power line equipment. Ultraviolet cameras detect coronas, usually a sign of a defect, fairly straightforwardly.

The second visual inspection system performs a more accurate visual inspection of the conductor. This visual inspection system consists of three-line scan cameras placed around the conductor 120° apart (Figure 8.7b). The conductor is illuminated with two LED-based lights for each camera (Figure 8.7a). The lights are placed on both sides of the cameras. This lighting configuration provides diffuse illumination of the conductor to enable efficient visual defect detection. An incremental encoder on one of the wheels of the drive mechanism is used to trigger the line scan cameras (Katrasnik, Pernus, & Likar, 2008a).

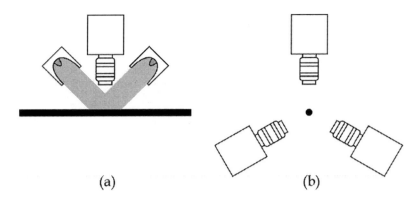

FIGURE 8.7 Conductor visual inspection system: (a) Illumination, camera, and conductor configuration for one camera; (b) Camera configuration around the conductor (Katrasnik, Pernus, & Likar, 2008a).

8.2 Building Monitoring

8.2.1 Introduction

A major concern in the inspection of building infrastructure is the aging process and life cycle management. A conventional means of monitoring the condition of buildings is a man-driven visual inspection, possibly supported by some tap testing. This way of monitoring provides essential information about cracking conditions and possibly the detachment of the covering layers of concrete or stone-based structures.

Much more effort is required when considering such structures as a dam, a cooling tower, a church, or even a simple multistoried building, since significant lifting equipment is required to perform the inspection. A way to circumvent this is to use unmanned aerial vehicles (UAVs) or micro aerial vehicles (MAVs) as airborne sensor systems to capture the required data.

The potential applications for unmanned aircraft for nondestructive testing (NDT) include state detection, damage analysis, and condition monitoring.

Unmanned aircraft are categorized according to size restrictions, weight limits, or the area of operation (operating radius, flight duration). These categories include both aircraft and helicopters, as well as any other type of aircraft. For insurance-related reasons, the use of UAVs outside registered (model) airfields is limited to a maximum takeoff weight of 5 kg. The UAVs used for building inspection cannot exceed the classification category micro UAV (μUAV, MAV) which also has a weight limit of 5 kg.

In terms of requirements and goals for selective damage detection and in conditions of high building density, especially urban areas, vertical takeoff and landings (VTOLs) seem to be capable platforms. In areas of high building density and the associated traffic, the risks for personal injury and infrastructural damage must be kept as low as possible. Consequently, the redundancy factor of the system and subsystems, thus the reliability of flight functions, is of particular significance. Another major concern is the need for stable hovering characteristics to ensure detailed damage inspection (Eschmann, Kuo, Kuo, & Boller, 2012).

8.2.2 Building Inspection

8.2.2.1 Data Acquisition

The main focus is clearly on acquiring data. A preliminary flight track plan is needed to fly around an object; this is usually done by using a common software based on Global Positioning System (GPS) waypoint navigation. However, GPS navigation is insufficient for inspecting a building because of the

Autonomous Vehicles for Inspection Applications 281

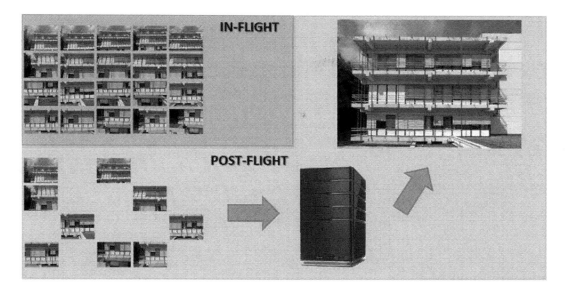

FIGURE 8.8 Two-step process for facade modeling (Eschmann, Kuo, Kuo, & Boller, 2012).

precision required for a facade to be monitored and the threat of the shadowing effects of nearby buildings. Moreover, GPS does not allow accurate flight altitude control, an essential factor in flight planning. A combination of collision and navigation sensors must be developed for an autonomous flight program (under pilot control) to become feasible in the long term. Manual flight control is still the only option when flying close to a building (see Figure 8.8) (Eschmann, Kuo, Kuo, & Boller, 2012).

Obtaining images and videos using UAV cameras is one method of data acquisition. When the building is very big, data acquisition can be divided into three phases: right side, front, and left side. At least two persons must be involved in flying the UAV. One remotely controls the UAV and the other monitors the smartphone and gives directions to the controller (Kaamin et al., 2017).

For images to be easily allocated to the real object in a structured way, there are two flight pattern options (Figure 8.9). The flight path can be allocated horizontally as a story-wise scanning of the building, or it can follow vertically aligned slices (Eschmann, Kuo, Kuo, & Boller, 2012).

The second option can be eliminated, as the main vertical movement increases the lens-induced effects negative for stitching. However, the horizontal speed has to be quite limited while recording images,

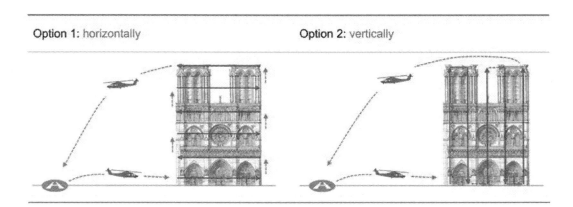

FIGURE 8.9 Options for on-site flight pattern (Eschmann, Kuo, Kuo, & Boller, 2012).

so that fast bank angle changes affecting images are reduced and not leveled by the automatic stabilization of the camera pod.

The integrated digital camera is controlled by an automatic photo-firing sequence, which can be set to a frequency of up to three pictures per second. The camera can be controlled manually to set zoom, focus, and shutter release if necessary. For optimal in-flight detection of damage, in both automatic recording and manual focusing modes, the real-time video link (low definition) can be used by the pilot or another person for camera orientation. It is not necessary for this application to transmit high-quality images in real time, as the digital building reconstruction remains time consuming. Therefore, the data stored on the camera are read out after landing. Because of the automatic triggering of the camera, each flight generates a large amount of data, for example, in a 15-min flight, normally more than 1,200 photos. This amount is far more than what is needed for inspection, but as a result of the not completely stable hover, a relatively high number of unusable images is produced, a consequence of the not fully filtered vibrations from the platform or external influences, such as wind gusts. In addition, there is often a very high overlap of the area captured on each image, which varies depending on the hover speed parallel to the building façade. Accordingly, unnecessary records are eliminated when there is too much overlapping to avoid double or multiple information within the images and to keep the image database as small as possible without loss of quality (Eschmann, Kuo, Kuo, & Boller, 2012).

8.2.2.2 Image Processing

After the aerial survey is completed, the next step is the digital post-processing of the selected images. Programmed for applications such as airborne terrestrial mapping or panoramic photography, several experimental and commercial software solutions are available to reassemble the individual images. These stitching or mosaicking methods are based on pattern recognition techniques which analyze similar image content structures, called matching points, in two or more images and link them together based on these points. The panorama creation software analyzes the input data under the assumption that images are recorded only by pivoting without changing the camera's position. Since the aerial survey with UAVs generates images from different positions, the above algorithms are not suitable. In contrast, the software for the mapping of landscape or similar mainly 2D objects can handle images from different locations. However, these algorithms are based on a precise georeferencing procedure, made possible by noticeable GPS location changes, together with inertial systems (inertial measurement units, IMUs) of high accuracy.

To this point, the stitching of up to 15 images has been successful, but generating a full facade reconstruction consisting of several hundred images is only possible by stitching the pre-stitched parts, resulting in a multilevel stitching process. Because the edge areas of the facade parts created by the software are currently not suitable for fully automated stitching, stitching has to be done mainly by hand, using programs such as Corel Photo-Paint, with which every single image has to be distorted and resized until it is adapted and integrated in the collage (Eschmann, Kuo, Kuo, & Boller, 2012).

8.2.2.3 Image-Based Digital Inspection and Monitoring

For the building shown in Figure 8.10, more than 12,000 images were taken over four days of flight but only several hundred images were used in the 2D model shown. The digital facade reconstruction has an overall resolution of about 1.27 gigapixel. However, the picture size at this resolution is very hard to handle. To make the inspection more user-friendly, the model is separated into floors which are then separated into parts of ten window frames each (Eschmann, Kuo, Kuo, & Boller, 2012).

The inspection then can be done directly on these window sections. In areas of special interest, high-resolution detailed photos can be linked to the sections for even better monitoring.

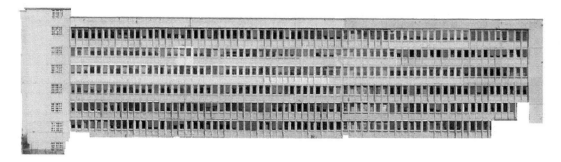

FIGURE 8.10 Digital facade reconstruction (Eschmann, Kuo, Kuo, & Boller, 2012).

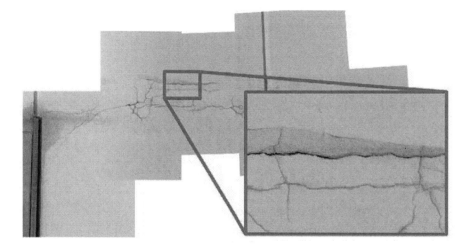

FIGURE 8.11 High-resolution crack inspection in submillimeter range (Eschmann, Kuo, Kuo, & Boller, 2012).

In the case of damage inspection, tests on high resolution areas have characterized crack sizes down to the millimeter range (Figure 8.11). The aim is to have this feature integrated in the full building reconstruction wherever significant damage appears. Therefore, some filtering algorithms have been programmed to detect cracks automatically (Eschmann, Kuo, Kuo, & Boller, 2012).

8.2.3 Infrared Building Inspection with UAVs

8.2.3.1 Thermal Imaging

There are many techniques for investigating the energy efficiency of a building which do not interfere with its structure. Some are simple, such as equipment to measure the moisture. Others are more complicated and expensive, for example, thermography. In any case, it is important to make an expert selection of appropriate methods of measurement and an expert interpretation of results.

IR is invisible radiant energy emitted by every object with the temperature above absolute zero. IR or thermal imaging can be used in various applications, including detecting and assessing heat losses in insulated systems, observing changing flows in pipe networks, or detecting overheating of electrical apparatus. Thermal imaging can be used in the construction and building sector, mostly to detect defects in insulation layers, thermal bridges, and heating networks.

Thermography, or thermal imaging, is photography using a camera that captures IR light rather than the visible light captured by a standard camera. IR rays are outside the visible spectrum and are invisible to the naked eye. All objects that are warmer than absolute zero (−273°C) emit IR light. The warmer an object is, the more IR light it emits. IR cameras record the amount of IR light and transfer it into temperature, which is indicated by the scale bar or thermogram.

As buildings are typically large constructions, it is convenient to perform thermal imaging inspections, keeping the proper distance from the object. A perfect solution is the use of remotely operated UAVs equipped with IR cameras. This also allows roofs or other tall constructions that cannot be easily accessed by a human to be inspected.

Thermal imaging cameras are able to capture even very small differences in temperature, in the order of 0.1°C. The image presented by a thermal imaging camera is multicolored, with each color representing a different temperature. Different scales may be used, depending on the presented objects. Thermal imaging studies have many uses in many fields and can be a useful diagnostic tool in the analysis of the state of a building. Noninvasive methods are very important, for example, in historic buildings.

Current solutions for IR inspections combine hardware (camera, lens, detectors, etc.) and software. Smart use of advanced software can give information on the quantity of energy loss, the risk of the dew point, and other valuable data.

Thermal imaging studies can identify problems with a building surface. Thermal imaging may be used to identify and locate areas of moisture, thinner wall elements, cracks, and cavities. Specialization is required to make decisions about how to take pictures and how to interpret them later. Unfortunately, objects that have high or low emissivity, such as metal, do not provide accurate temperature measurements. Elements such as weather conditions, orientation, and time of day also affect the results. Information collected from thermal images can be properly evaluated only in conjunction with the data collected in the framework of a comprehensive study of the state of the building (Krawczyk, Mazur, Sasin, & Stokłosa, 2015).

8.2.3.2 Classification of UAVs for Thermal Imaging

UAVs (or drones) can fly without an onboard human operator (Thompson II, 2013). An unmanned aircraft system (UAS) consists of the following main components: an unmanned aircraft (UAV), a communication data link, a ground control station (GCS), and an operator. UAVs can be ground or remotely controlled, partially or fully autonomous (Gupta, Ghonge, & Jawandhiya, 2013).

A wide range of UAV types can be classified according to their operational range, weight, missions, performance, configuration, complexity, and altitude. Unfortunately, there is no widely used, universal, and common UAV classification. The categories are classified according to the range (long, short, and close), altitude (high, medium, low, and very low), and purpose (target and decoy, reconnaissance, combat, research and development, civil and commercial) (Gupta, Ghonge, & Jawandhiya, 2013; Sawicki, 2012; Thapa & Shrestha, 2014).

According to Gupta, Ghonge, and Jawandhiya (2013), the UAV classification can also be based on the application: dull, dirty, or dangerous missions. The main civilian applications are border security, environmental monitoring and agriculture, remote sensing, aerial mapping, and meteorology.

The following classifications, which can vary with national legal restrictions, can be taken into account when selecting a UAV for thermal imaging (Krawczyk, Mazur, Sasin, & Stokłosa, 2015):

- NAVs (nano air vehicles): Usually used in swarms of UAVs for radar confusion for ultra-short range surveillance. All sensors, including cameras, propulsion system, and control subsystems, need to be small (Austin, 2010).
- Micro UAVs: Have a close range, are portable, can be hand launched, and fly at very low altitude. Payload weight is less than 2 kg. Takeoff weight is less than 5 kg. Total power is lower than 100 W. Total time less than 1 h (Dalamagkidis, Valavanis, & Piegl, 2008).
- Mini UAVs: Have a close range and fly at low altitudes. Payload weight is 1–3 kg. Takeoff weight is less than 20 kg. Total power is less than 10 kW. Total time is shorter than 2 h (Dalamagkidis, Valavanis, & Piegl, 2008).

- Regular/small UAVs: Have medium range, medium altitude. These include launch systems. Payload weight is less than 150 kg. Takeoff weight is less than 400 kg. Total power is 10–50 kW. Total time less than 6 h.
- MALE (medium-altitude long-endurance) UAVs: Perform long flights at medium altitudes (Dalamagkidis, Valavanis, & Piegl, 2008).
- HALE (high-altitude long-endurance) UAVs: Perform long flights at high altitudes. Total flight time is 24–48 h (Dalamagkidis, Valavanis, & Piegl, 2008).

A lot of research has sought to increase the flight endurance and payload of UAVs. UAVs with different aerodynamic configurations and sizes with a variety of capabilities, endurance, and flight levels can be classified as follows:

- Fixed-wing UAVs: In general, these need a runway to take off and land or a catapult launching system. Fixed-wing UAVs have long endurance and high cruising speed.
- Rotary-wing UAVs: These are also called rotorcraft or VTOL systems. Rotary-wing UAVs may be in different configurations, including helicopters, coaxial, tandem, and multi-rotors. Their main advantages are high maneuverability and hovering ability to enable operation in closed spaces (Hrishikeshavan, 2011).
- Blimps: These are usually balloons and airships. They are lighter than air and are large, have long endurance, and can fly at low speeds (Gupta, Ghonge, & Jawandhiya, 2013).
- Flapping-wing UAVs: These UAVs have flexible and/or morphing small wings inspired by birds and flying insects (Gupta, Ghonge, & Jawandhiya, 2013).
- Others: Other hybrid or convertible configurations include convertible rotor aircraft, tilt-wing-body aircraft, ducted-fan aircraft, and jet-life aircraft (Austin, 2010).

8.2.3.3 Advantages of UAV for Thermal Imaging

A number of applications are directly related to thermal imaging, including building thermal diagnosis, inspection of industrial facilities, such as chimneys, cooling towers, water towers, tanks, and pipelines, detection of water deficit in agriculture, temperature measurements for production lines and industrial facility monitoring, search and rescue missions, monitoring of areas after fires or other natural or industrial disaster, and many others.

Depending on their characteristics and performance, UAVs can be used to perform all sorts of tasks that require special sensors. For example, well-designed and equipped UAVs can be used for building thermal inspections. Aerial thermal imaging performed by UAVs can be very advantageous in the building sector because it can identify thermal losses and bridges, find construction defaults, detect defects, lack of insulation, or air leakage, control energy losses, and provide monitoring.

The use of UAVs in thermal imaging is beneficial for a number of technical and economic reasons, including the following: they can take pictures in small, difficult areas; they have relatively no time and space constraints; they can provide a view of a wide area in minimum time and at minimum cost; they can support digital processing of aerial and satellite images; they offer an environmentally friendly application; and they have relative independence from aviation weather conditions (Krawczyk, Mazur, Sasin, & Stokłosa, 2015).

For civil structures, the main advantages of using UAVs for thermal imaging are the following (Krawczyk, Mazur, Sasin, & Stokłosa, 2015):

- Increased capabilities, such as endurance, real-time deployment, and full spectrum coverage compared to satellite-based remote sensing applications and manned aircraft.
- Better investment and operation costs.
- Technology maturation due to military applications.

Even so, there are significant challenges associated with the development of UAVs, both technical and regulatory (Krawczyk, Mazur, Sasin, & Stokłosa, 2015).

UAVs with the greatest potential for thermal imaging applications have rotary-wing (multi-rotor) and fixed-wing structures which can be assimilated into mini UAVs. These structures have relatively low manufacturing, investment, and operation costs and minimal onboard equipment, including a control system, navigation, and registration data. In addition to the cost, these UAVS have advantages in various critical technical parameters, such as altitude, range, flight time, payload, takeoff weight, and external dimensions; they have automatic and autonomous control and navigation and offer different applicability (Sibilski, Żyluk, Kowalski, & Wiśniowski, 2015; Mieloszyk & Tarnowski, 2015). UAVs allow fast, relatively cheap high-resolution data and image acquisition from any orientation (Sawicki, 2012).

UAVs with vertical takeoff can carefully examine all elements, such as vertical walls, and can accurately diagnose roofs and facades of buildings without needing people to work at heights. This method provides fast and precise control of both the roof surface and the side surfaces; it is the most accurate method and also very economical. Regular monitoring can lead to significant cost savings and defect removal at an early stage (Krawczyk, Mazur, Sasin, & Stokłosa, 2015).

In some buildings, apart from an individual evaluation of their energy performance and roof inspection, the availability of thermal imaging information allows taking into account the spatial circumstances and location of the buildings and permits a general study of the surrounding area for the global performance measurement (Lagüela, Díaz–Vilariño, Roca, & Lorenzo, 2015).

Properly used thermal imaging cameras can give a fast and reliable indication of the amount of heat loss in the building as a result of leaks and defects. The use of appropriate sensors and computer systems will provide automatic and precise location of any faults, fast acquisition of thermal imaging data, high resolution of image data, highly accurate positioning, and multiple and repetitive data acquisition (Krawczyk, Mazur, Sasin, & Stokłosa, 2015).

8.2.4 Benefits of Using a Drone for Building Inspection

Proper building maintenance means doing regular, visual inspections of the exterior. The problem is that it can be difficult to see parts of the outside of a building, particularly if it is more than a few stories.

Drones offer a more thorough, safer, and faster option, especially if the building is tall. Using drones for building inspection is also a fraction of the cost of doing a survey or an inspection using a harness and line off the top of a building. Drones can also detect defects and anomalies that a naked eye cannot, which is useful in data collection. This is especially true for drones equipped with thermal imaging and IR cameras (Drone Insurance Depot, 2016).

It isn't all that easy, however. When drones are used for building inspection, transport regulations must be followed at all times, and only a veteran UAS pilot should attempt the inspection. In addition, GPS guidance systems are not reliable as the drone gets closer to a building, and manual flying will be necessary if the GPS is suddenly disabled or lost. Finally, privacy and ground security are other challenges (Drone Insurance Depot, 2016).

8.2.5 Automated Crack Detection

Since the walls of inspected buildings are mostly light in color, cracks appear as black lines in visual inspections. Two different methods can be used to highlight them (Eschmann, Kuo, Kuo, & Boller, 2012):

- Adding additional color: This method analyzes the threshold value to determine if each photo pixel needs to be given more "black" or "white," resulting in black crack areas. However, this method only works with gray or white walls.
- Edge detection: Edges are detected by applying a Gaussian blur (Equation 8.1) to the original image, then subtracting it from the image again. The edge will be displayed as almost black, while others will be almost pure white (Beard, Jones, Chacon Jr. & Ahumada, 2005).

$$G(x, y) = \frac{1}{2\pi\sigma^2} e^{\frac{x^2+y^2}{2\sigma^2}}, \qquad (8.1)$$

Autonomous Vehicles for Inspection Applications 287

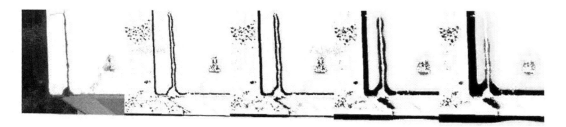

FIGURE 8.12 Influence of sigma value: Original image (left) and results with increasing sigma values (right) (Eschmann, Kuo, Kuo, & Boller, 2012).

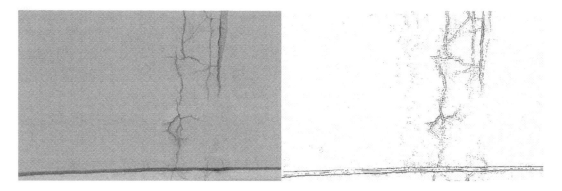

FIGURE 8.13 Original image (left) and result after crack enhancement by edge detection (right) (Eschmann, Kuo, Kuo, & Boller, 2012).

Typical damage found in buildings is shown on the left of Figure 8.12. In this study by Eschmann, Kuo, Kuo, and Boller (2012), the original photo was processed using the edge detection method to enhance the damaged areas. The four results on the right show the effect of varying the variance σ (sigma) used in the Gaussian blur (increasing sigma value to the right). Values of sigma that are too small can result in an insufficient damage enhancement, but a higher sigma does not necessarily mean better crack extraction (Eschmann, Kuo, Kuo, & Boller, 2012).

To analyze the cracking shown in Figure 8.11 in more detail, Eschmann, Kuo, Kuo, and Boller (2012) reinspected the image using edge detection software (see Figure 8.13). After the extraction, the width of the long but small cracks on the surface was filtered out for clearer visualization. However, they found the edge detection method is not perfect. It helps to extract bigger damage, but tiny surface cracks are still not very visible after image processing, whereas man-made edges resulting from the object's geometry may be mistaken as cracks (Eschmann, Kuo, Kuo, & Boller, 2012).

8.2.6 Micro Aerial Vehicles

Micro aerial vehicles (MAVs) show promise for digital building monitoring. The high resolution camera attached to the MAV has good results even under nonoptimal flight conditions, and its visual recording methods provide valuable information for infrastructural inspection purposes. However, more work needs to be done. MAVs need improved data acquisition. They require better stabilization of the flight platform, improved anti-collision and navigation systems, and better route planning algorithms to expand the automation of the process. In addition, the image post-processing has to be improved by reducing manual workflow through appropriate image stitching and mosaicking software, ideally with an integrated crack detection feature (Eschmann, Kuo, Kuo, & Boller, 2012).

8.3 Railway Infrastructure Inspection

8.3.1 Unmanned Aircraft System Technology in Railway Applications

UAS technology is having a powerful and transformative impact on the rail industry. In railroad environments, UASs are particularly suitable for the following (US Department of Transportation/Federal Railroad Administration, 2018):

- Structural monitoring, especially for critical assets like bridges and tunnels, and for fault detection (i.e., diagnostics/prognostics).
- Environmental security monitoring, such as assessments of fire, explosions, earthquakes, floods, and landslides along the track.
- Physical security monitoring, including detection of intrusions, objects stolen or moved, graffiti, etc.
- Safety monitoring, for example, to early detect failures on track elements/devices or obstacles on the track.
- Situation assessment and emergency/crisis management to monitor accident scenarios and coordinate the intervention of first responders.

The use of UAS technology offers the following direct benefits for routine inspection activities (US Department of Transportation/Federal Railroad Administration, 2018):

- Reduction of risk to staff and people and infrastructure in the project area
- Reduced planning cycle
- More efficient work processes
- More flexible, affordable verification tools
- Higher quality data available in larger quantities at lower costs.

When natural disasters strike, many railway assets can be at risk. In such situations, it is critical to determine which part of the railway needs repair prior to the movement of trains. UASs can gather information on the condition of the track or bridges, as well as the presence of debris on the right of way.

The aging of rail infrastructure poses challenges. Visual condition assessment remains the predominant input to the decision-making process. Many railways use machine-vision technology installed on rail-bound vehicles, but there are situations in which inspectors on foot or in hi-rail vehicles assess the track's surroundings. In the case of high or steep slope embankments, UASs can collect detailed information that could be missed by inspectors (US Department of Transportation/Federal Railroad Administration, 2018).

8.3.1.1 UAVs Suitable for Railway Applications

Two primary UAV types are available for railway operations: "rotary wing" aircraft, shown in the top portion of Figure 8.14, and "fixed-wing" aircraft, shown in the lower portion of Figure 8.14 (Jurić-Kaćunić, Librić, & Car, 2016).

Rotary wing UAVs share many characteristics with manned helicopters. Rather than a continuous forward movement to generate airflow, these units rely on lift from the constant rotation of the rotor blades. There is no limit on how many blades an aircraft has, but the average is between four and eight. Unlike fixed-wing units, rotary wing units have the ability to vertically take off and land, so they can be deployed virtually anywhere. This enables the aircraft to lift vertically and hover at a specific location. These UAVs can move in any direction, hovering over important areas, collecting the most intricate data. This ability makes them well suited for inspections where precision maneuvering is critical to the operation.

Autonomous Vehicles for Inspection Applications 289

FIGURE 8.14 Rotary-wing and fixed-wing UAVs (Buchanan, 2016).

Fixed-wing UAVs are designed for higher speeds and longer flight distances. This type of UAV is ideal for coverage of large areas, such as aerial mapping and surveillance applications. It can often carry heavier payloads than a rotary UAV. Fixed-wing UAVs glide efficiently, and the single fixed wing drastically reduces the risk of mechanical failure. The maintenance and repair requirements for these units are often minimal, saving time and money. However, the current beyond visual line of sight (BVLOS) regulations limit the utility of fixed-wing UAVs. Several railways are using multi-rotor or hybrid vehicles that employ multiple rotors along with fixed wings to facilitate short takeoffs. Among the various types of UAVs, the one with the highest number of units worldwide is the rotary wing followed by the fixed wing (US Department of Transportation/Federal Railroad Administration, 2018).

The nano-type UAV is becoming prevalent in the UAS market space. It is a palm-sized platform with a maximum takeoff mass of less than 30 g. It has advanced navigation systems, full-authority autopilot technology, digital data links, and multi-sensor payloads. The operational radius for this type of platform is more than 1.5 km, and it can be flown safely in strong wind. Future development is anticipated to yield even smaller and more advanced nano UAVs with high levels of autonomy (European Aviation Safety Agency, 2015).

8.3.1.2 Sensors Used in Railway Applications

Cameras are still the most common sensor used on a UAV. However, dynamic sensor technologies created for use with UAVs provide essential situational awareness and a level of detail often missed by the human eye and standard cameras. Light detection and ranging (LiDAR) sensors on UAVs, such as that shown in Figure 8.15, capture imagery which only a few years ago required an aircraft and a crew to collect (Jurić-Kaćunić, Librić, & Car, 2016). A LiDAR sensor mounted on a UAV, along with sophisticated software, can produce accurate three-dimensional images very quickly (US Department of Transportation/Federal Railroad Administration, 2018).

UAV payloads can integrate sensors of a different nature, such as temperature sensors or multispectral cameras to provide diverse functionalities, depending on energy consumption and maximum allowed weight. Self-powered chemical sensors can be mounted on the aerial platform to provide quick and safe analyses of chemical or air samples near a derailment (US Department of Transportation/Federal Railroad Administration, 2018).

FIGURE 8.15 Various UAV-mounted sensors (Jurić-Kaćunić, Librić, & Car, 2016).

Current standard UAS technology allows the registration and tracking of position with GPS or inertial navigation systems (INS) and orientation of the implemented sensors in a local or global coordinate system. UAS-based photogrammetry, or the practice of making measurements from imagery, now allows the collection of information from platforms that are remotely controlled or operated in a semiautonomous or autonomous manner, eliminating the need for a pilot sitting in the vehicle (Flammini, Pragliola, & Smarra, 2016). UAS photogrammetry can be understood as a photogrammetric measurement tool with applications in the close-range domain; it combines aerial and terrestrial photogrammetry to provide a real-time application and low-cost alternative to the classical manned aerial photogrammetry. This approach can provide both an overview of a situation and detailed area documentation (Jurić-Kaćunić, Librić, & Car, 2016).

The collection of three-dimensional data by conventional surveying methods can be time consuming, expensive, and even dangerous for the field operator, especially on steep slopes and cuts where there are potential rockfalls, landslides, or mudslides. Visual inspection of the terrain in such locations, just as geodetic data collection with classical methods, can result in incomplete and insufficiently detailed data, thus posing a risk to the railroad. The use of UAVs in such locations can complement, enhance, and even completely replace the classical methods of mapping, determining the volume, cross sections, contours, and other parameters that are necessary for remediation measures, as illustrated in Figure 8.16 (Jurić-Kaćunić, Librić, & Car, 2016).

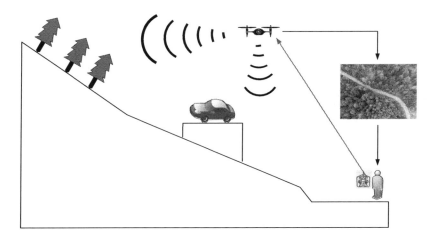

FIGURE 8.16 UAV mapping steep slopes and contours (Jurić-Kaćunić, Librić, & Car, 2016).

The present challenge is to increase the level of automation to reduce the need for human interventions with the ongoing enhancement of UAV endurance and payloads, even in critical situations. The number of scenarios in which railway UAVs would be useful will be proportional to UAS performance growth (US Department of Transportation/Federal Railroad Administration, 2018).

8.3.2 Analysis of Organizational Activities for Flight over Railway Infrastructure Using UAVs

Preparation of field tests for flight over the railway infrastructure using UAVs should start by checking the availability of the airspace above the given area and analysis of the flight route. In an uncontrolled space, it is possible to perform flights without permission from appropriate departments. In other cases, space is controlled, and permission or formal notification is required. In still others, UAV flights are banned or can be carried out only with the consent of the manager and under the conditions set out by him. In such cases, the operator is obliged to report an intention to perform a flight to the air navigation service provider and, if necessary, request consent from the facility manager.

During the preparation of the flight plan, the most sensitive elements of the infrastructure should be identified and analyzed for possible flight routes. All restrictions related to the constraints of the flight and the procedures used in the facility should be considered in the final version of the scenario. In case of flights within the railway line, special attention should be paid to the elements of the electric traction infrastructure and rolling stock. On railway sidings, the number of such threats will be even greater.

It is important to choose the right day for testing. Weather conditions have a significant impact on the success of the inspection, because they affect the safety and quality of the conducted tests. Too much sun can cause the sun's rays to reflect on the infrastructure, which can lead to a deterioration in the quality of images. Large gusts of wind and intense rainfall may prevent the UAV from staying airborne and flying in a straight line or even lead to destruction of the equipment. Even when safety and stabilizing systems are installed, a minimum, safe distance from the tested object should be maintained. The minimum distance from the elements of the power infrastructure, taking into account electromagnetic interference, control errors, or a sudden gust of wind, is 5 m (Kochan, Rutkowska, & Wójcik, 2018).

8.3.3 Conditions for Recording Video Material

8.3.3.1 Ways of Recording

Photos or videos of railway infrastructure can be recorded in (Kochan, Rutkowska, & Wójcik, 2018):

- Remotely controlled flights, in which the drone is manually controlled by operators, and the image is recorded in the form of video
- Semiautonomous flights, the purpose of which is to create an orthophotomap to map the area of the siding in the form of a cartographic map.

8.3.3.2 Remotely Controlled Flight

A remotely controlled flight is a flight mode in which the operator manually controls the drone. The method requires the creation of a precise scenario, containing all the elements of the infrastructure in detail, along with the order in which they are recorded. Performing this type of flight is complicated from the point of view of maneuvering a drone. The operator must safely avoid various types of obstacles appearing on the flight route (such as buildings, cables, high voltage lines, traction poles, trees, and rolling stock). In addition, the drone cannot move away from the operator for more than 250–300 m because it will be lost from the field of view (Kochan, Rutkowska, & Wójcik, 2018) (Figure 8.17).

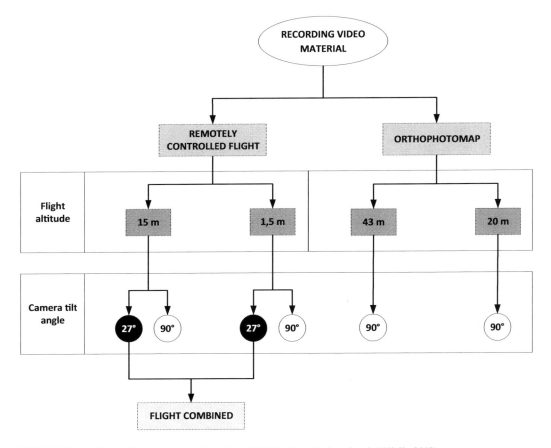

FIGURE 8.17 Methods of material recording using UAV (Kochan, Rutkowska, & Wójcik, 2018).

The parameters listed below are significant for drone positioning to record video material (Kochan, Rutkowska, & Wójcik, 2018):

- Flight altitude
- Tilt angle
- Speed.

Kochan, Rutkowska, and Wójcik (2018) studied flights over (i.e., at 15 m) and under the contact lines (i.e., at 1.5 m) to determine their relative advantages and disadvantages in video recording. Table 8.1 presents the advantages and disadvantages of flight with such parameters (Kochan, Rutkowska, & Wójcik, 2018):

8.3.3.3 Orthophotomap

Orthophotomap represents a set of processed aerial photographs, adjusted to a uniform scale and fit to geodetic control (photogrammetric) points. An orthophotomap as opposed to aerial photography is characterized by the following (Kochan, Rutkowska, & Wójcik, 2018):

- Orthogonal projection (not middle)
- Uniform scale for the entire surface area (however, the scale does not have objects protruding above the ground surface, e.g., houses, trees).

TABLE 8.1

Advantages and Disadvantages of Carrying Out Flight under and over the Overhead Contact Line (Kochan, Rutkowska, & Wójcik, 2018)

	Under Traction	**Over Traction**
Advantages	More detailed picture of the track infrastructure (track, rails, turnouts, drives).	Ability to speed up the flight (higher speed). The frame shows elements of the infrastructure that rise vertically (traction, signaling, poles).
Disadvantages	Some parts of the track may be used by rolling stock or wagons (and others) which may hinder or prevent passage over this section. It also requires operator skill. For railway traffic control devices (especially signaling devices), it is necessary to perform detailed flight routes.	The flight must be performed a few meters above the traction because of potential interference from the contact wire. The traction network and its supporting structures may obscure some elements in the track

An orthophotomap requires the following (Kochan, Rutkowska, & Wójcik, 2018):

- Internal orientation of the photos, as well as mutual and absolute orientation (aerial triangulation)
- A numerical terrain model
- Orthorectification or geometric correction of photographs (repositioning of image pixels resulting from slope and the properties of the middle projection)
- Mosaicking, i.e., combining ortho-images according to some sectional cut
- Rasterization with vector content (details, frames, and post-border descriptions).

Before starting the creation of the orthophotomap, it is necessary to determine the numerical model of coverage of the examined area and the height at which the flight can take place in a safe manner.

The flight takes place in a semiautonomous manner; the operator programs the route, entering data into the UAS onboard computer. After the start, the drone is guided by an automatic pilot and positioned using GPS, but at any moment, control of the UAS can be taken over by the operator.

Telemetry data (coordinates, altitude, etc.) are assigned to the acquired photos and videos (Kochan, Rutkowska, & Wójcik, 2018).

The benefits of using an orthophotomap are as follows (Kochan, Rutkowska, & Wójcik, 2018):

- No need to develop a scenario before the flight; it is enough to specify the area of interest
- Relatively fast flight time
- No need for direct access by the filming crew
- Unlike video material from remote flight, does not require additional processing
- Convenient document for use.

However, an orthophotomap also has disadvantages (Kochan, Rutkowska, & Wójcik, 2018):

- Limited image accuracy from safe heights (e.g., 43 m)
- Inaccuracies related to the combination of photos
- Overriding of selected elements by higher objects
- Map preparation time = a few days using expensive software and a computer with high computing power.

8.3.4 Proposed Application of UASs in Mobile Networks

In 2016, Flammini, Pragliola, and Smarra described the benefits of overall railway monitoring using UAS solutions. They noted that conventional methods for railway infrastructure surveillance employ

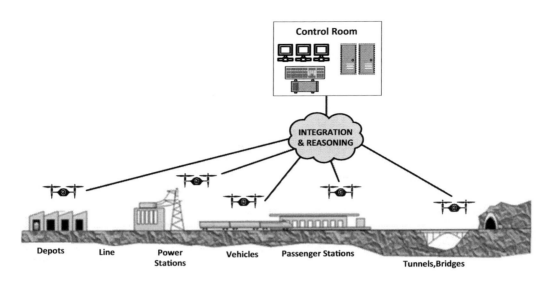

FIGURE 8.18 Communication events using UAV-based mobile terminals (Flammini, Pragliola, & Smarra, 2016).

such tools as analog or Internet Protocol (IP) cameras, wired intrusion detection, and other structural monitoring sensors, such as piezoelectric accelerometers. As each sensor/camera costs thousands of Euros, extensive surveillance of very large geographical areas, as in the case of railways, can be impractical. The researchers suggested that instead of using analog or IP cameras, interconnected UAVs and their payloads could be used as mobile terminals to communicate events to remote control centers (Flammini, Pragliola, & Smarra, 2016).

Figure 8.18 shows the concept and implementation of an overall hardware architecture of UAV-based surveillance. Distributed smart sensors are installed along the railway line, both in fixed (e.g., bridges, tunnels, stations) and mobile (e.g., passenger trains, freight cars) locations. They are integrated locally using local wireless infrastructures and data are collected by gateway nodes to be transmitted remotely by means of wide area network (WAN) infrastructures, such as 3G and 4G cellular networks, satellite links, or fiber-optic geographic networks along the line. Given the availability of large bandwidth (in the Gbps range) provided by fiber optics links, for example, it is possible to transmit high-resolution videos at very high frame rates for a superior situational awareness in the control rooms. In the case of UAVs, those links could be leveraged by fixed UAV stations to quickly download large amounts of recorded data (US Department of Transportation/Federal Railroad Administration, 2018).

Railways face challenges with the integration of UAS into their rail fleet for long endurance and forward-looking inspections (US Department of Transportation/Federal Railroad Administration, 2018).

8.4 Waterways and Other Infrastructures

8.4.1 Waterway Inspection

8.4.1.1 Aerial Inspection of Siltation in Waterways

Silt accumulation and sedimentation in canal beds lead to the deterioration of watercourses over time. Every year, a forced closure of the canals in the Indus basin is required for canal cleaning, entailing a very large-scale and costly operation. Silt removal precision is prone to inefficiencies due to subjective decision-making in the cleaning process. In this section, we lay out a theoretical framework to map the semi-structured (emptied) canal bed terrains with a UAV system for quantitative inspection of deposited silt (Anwar, Muhammad, & Berns, 2015).

Figure 8.19 shows the side view of a silted canal bed being profiled by a flying robot.

Autonomous Vehicles for Inspection Applications 295

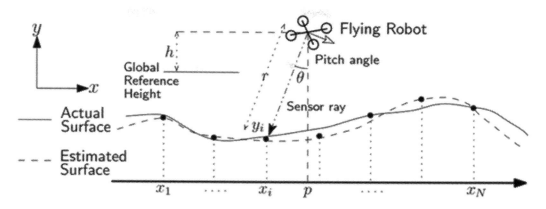

FIGURE 8.19 Side view of a silted canal bed being profiled by a flying robot (Anwar, Muhammad, & Berns, 2015).

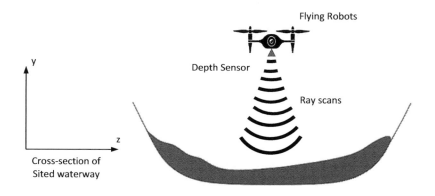

FIGURE 8.20 Cross-sectional view of a silted canal bed profiled by a flying robot (Anwar, Muhammad, & Berns, 2015).

Figure 8.20 shows a cross-sectional view of the canal channel with the flying robot scanning the surface. The robot is moving along the canal; since surface fitting is independent along orthogonal axes, the problem reduces to a single dimension (Anwar, Muhammad, & Berns, 2015).

Anwar, Muhammad, and Berns (2015) proposed a 3D perception system, to be deployed on board an aerial robot to assist the human operator in surveying and cleaning the canal effectively. It looked for at least 6-inch silt depth at specific data points. The system envisaged efficient cost-effective cleaning, reduced water discharge variability, and enhanced agricultural productivity (Anwar, Muhammad, & Berns, 2015).

8.4.2 Other Infrastructure/Construction & Infrastructure Inspection

The net market value of the deployment of UAVs in support of construction and infrastructure inspection applications is about 45% of the total UAV market. In this section, we explain the uses of UAVs for infrastructure inspection and present some ongoing challenges, research trends, and future insights (Shakhatreh et al., 2018).

8.4.2.1 Deployment of UAVs for Construction and Infrastructure Inspection Applications

Figure 8.21 illustrates the classification of the deployment of UAVs for construction and infrastructure inspection applications (Shakhatreh et al., 2018).

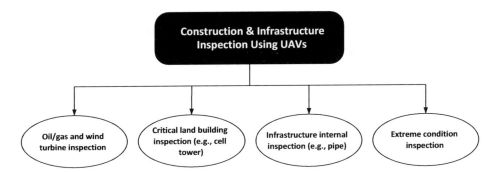

FIGURE 8.21 Deployment of UAVs for construction and infrastructure inspection (Shakhatreh et al., 2018).

8.4.2.1.1 Oil/Gas and Wind Turbine Inspection

Pacific Gas and Electric Company (PG&E) has tested the use of drones to inspect its electric and gas services (PG&E, 2018). The inspections focused on hard-to-reach areas to detect methane leaks across its 70,000-square-mile service area. In the future, PG&E plans to extend drone tests to include storm and disaster response (Shakhatreh et al., 2018).

Cyberhawk, an oil and gas company, uses UAVs for inspection (Cyberhawk, 2008). It has completed more than 5,000 structural inspections, including the following (Shakhatreh et al., 2018):

1. Oil and gas structures
2. Wind turbines
3. Live flares.

Cyberhawk's UAVs take photos and conduct close visual and thermal inspections of the assets (Shakhatreh et al., 2018).

Industrial SkyWorks (https://industrialskyworks.com/) uses drones for building inspections and oil/gas inspections in North America. A powerful machine learning algorithm, BlueVu, has been developed to handle the captured data. It provides the following solutions (Shakhatreh et al., 2018):

- Asset inspections and data acquisition
- Advanced data processing with 2D and 3D images
- Detailed reports of the inspected asset (i.e., annotations, inspector comments, and recommendations).

8.4.2.1.2 Critical Land Building Inspection (e.g., Cell Tower)

AT&T owns about 65,000 cell towers that need to be inspected, repaired, or installed. The video analytic team at AT&T Labs has collaborated with other forces (e.g., Intel, Qualcomm) to develop faster, better, more efficient, and fully automated cell tower inspection using UAVs (Gilbert, 2017). One approach is to employ deep learning algorithms on high definition (HD) videos to detect defects and anomalies in real time.

Honeywell InView inspection service has been launched to provide industrial critical structure inspections (Honeywell, 2017). The service combines the Intel Falcon 8+ UAV system with Honeywell aerospace and industrial technology solutions. Specifically, the Honeywell InView inspection service can achieve (Shakhatreh et al., 2018):

1. Worker safety
2. Improved efficiency
3. Advanced data processing.

8.4.2.1.3 Infrastructure Internal Inspection

Maverick has provided industrial UAV inspection services for equipment, piping, tanks, and stack internals in western Canada since 1994 (www.maverickinspection.com). It offers a dedicated service for the internal inspection of assets using Flyability ELIOS. Maverick also provides post-data processing that analyzes data using measurement software and CAD modeling (Shakhatreh et al., 2018).

8.4.2.1.4 Extreme Condition Inspection

Bluestream offers UAV inspection services for onshore and offshore assets (www.bluestreamoffshore.com). Its services are particularly suitable for the following applications (Shakhatreh et al., 2018):

1. Onshore and offshore live flare inspections
2. Topside, splash zone, and under deck inspections
3. Hard-to-access infrastructure inspection.

8.4.2.2 Challenges

At the moment, the use of UAVs for construction and infrastructure inspection faces several challenges. These include (Shakhatreh et al., 2018):

- Limited energy, short flight time, and limited processing capabilities (Gheisari, Irizarry, & Walker, 2014).
- Limited payload capacity. The onboard loads could include optical wavelength range cameras, thermal InfraRed (TIR) cameras, color and stereo vision cameras, and different types of sensors, such as gas detection and GPS (Bretschneider & Shetti, 2015).
- Lack of research on multi-UAV cooperation for construction and infrastructure inspection applications. Multi-UAV cooperation could provide wider inspection scope, higher error tolerance, and faster task completion time.
- Need for autonomous UAVs that can maneuver an indoor environment with no access to GPS signals (Dupont, Chua, Tashrif, & Abbott, 2017).

8.4.2.3 Research Trends and Future Insights

8.4.2.3.1 Machine Learning

Machine learning has become an increasingly important artificial intelligence (AI) approach to allow UAVs to operate autonomously. Applying advanced machine learning algorithms (e.g., deep learning algorithm) could help the UAV system draw better conclusions. For example, with improved data processing models, deep learning algorithms could help to obtain new findings from existing data, leading to more concise and reliable results. The UAV inspection program at AT&T uses deep learning algorithms on HD videos to detect defects and anomalies in real time (Gilbert, 2017). Industrial SkyWorks has introduced advanced machine learning algorithms to process 2D and 3D images (https://industrialskyworks.com/).

More specifically, deep learning is useful for feature extraction from the raw measurements provided by sensors on board a UAV. Convolutional neural networks (CNNs) are deep learning feature extractors used in the area of image recognition and classification and have been proven very effective (Carrio, Sampedro, Rodríguez-Ramos, & Campoy, 2017). Figure 8.22 illustrates one example of how CNNs works (Shakhatreh et al., 2018).

8.4.2.3.2 Image Processing

Construction and infrastructure inspection using UAVs equipped with onboard cameras and sensors can employ image processing techniques. The employment of image processing techniques allows

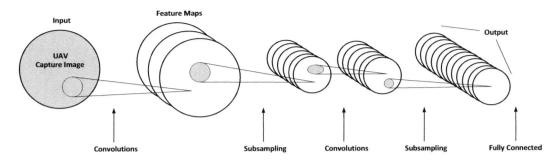

FIGURE 8.22 Illustration of how CNNs work (Carrio, Sampedro, Rodríguez-Ramos, & Campoy, 2017).

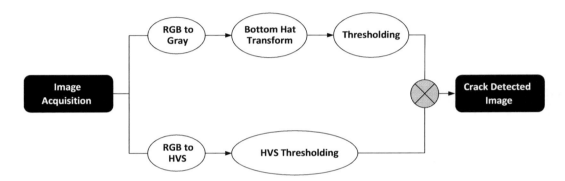

FIGURE 8.23 Proposed approach for crack detection algorithm (Sankarasrinivasan, Balasubramanian, Karthik, Chandrasekar, & Gupta, 2015).

construction projects to be monitored and assessed. In addition, infrastructures can be inspected, for example, surveying construction sites, monitoring work progress, inspecting bridges, monitoring irrigation structures, detecting construction damage and surface degradation (Sankarasrinivasan, Balasubramanian, Karthik, Chandrasekar, & Gupta, 2015; Ham, Han, Lin, & Golparvar-Fard, 2016).

Sankarasrinivasan, Balasubramanian, Karthik, Chandrasekar, and Gupta (2015) presented an integrated data acquisition and image processing platform mounted on UAV. They proposed using it for the inspection of infrastructures and real-time structural health monitoring (SHM). In the proposed framework, a real-time image and data are sent to the GCS. Then the images and data are processed using the image processing unit in the GCS to facilitate the diagnosis and inspection process. The authors proposed combining hue, saturation, value (HSV) thresholding and the hat transform for crack detection on concrete surfaces. Figure 8.23 presents a block diagram of their crack detection algorithm (Sankarasrinivasan, Balasubramanian, Karthik, Chandrasekar, & Gupta, 2015).

Luque-Vega, Castillo-Toledo, Loukianov, and González Jiménez (2014) proposed the deployment of UAV with a vision-based system that consists of a color camera, TIR camera, and a transmitter to send the captured images to the GCS. These images are processed in the GCS and used in the inspection and estimation of the real temperature of power lines joints (Shakhatreh et al., 2018).

Larrauri, Sorrosal, and González (2013) proposed a fully automatic system to determine the distance between power lines and trees, buildings, and any other obstacles. The authors designed and developed vision-based algorithms for processing the HD images obtained using an HD camera installed on a UAV. The captured video is sent to the computer in the GCS. At the GCS, the video is converted into the consecutive images and processed to calculate the distance between the power line and the obstacles (Shakhatreh et al., 2018).

8.4.2.3.3 Future Insights

Possible research directions include the following (Shakhatreh et al., 2018):

- More research is required to improve UAVs battery life time to allow longer distances and increased flight time (Gheisari, Irizarry, & Walker, 2014).
- An accurate, autonomous, and real-time power line inspection approach using UAVs, including ultrasonic sensors, TIR or color cameras, image processing, and data analysis tools, is needed. More specifically, techniques to monitor, detect, and diagnose defects automatically are called for (Pagnano, Höpf, & Teti, 2013).
- Research on advanced data collection, sharing, and processing of algorithms for multi-UAV cooperation is required to achieve faster and more efficient inspections.
- Autonomy and safety when UAVs maneuver in congested and indoor environments with no or weak GPS signals is a final research concern (Dupont, Chua, Tashrif, & Abbott, 2017).

8.4.3 Autonomous Vehicles: The Future of the Construction Industry

8.4.3.1 Connected and Autonomous Vehicles (CAVs)

Connected and autonomous vehicles (CAVs), or "driverless cars," promise to transform the way we travel. They will move people around roads, public spaces, airports, hospitals, business parks, shopping centers, and tourist areas (Gard, 2017).

8.4.3.2 Opportunities for Construction Industry

The connected road system that will be required to accommodate autonomous vehicles presents a challenge and an opportunity for the construction industry. Highways and transport infrastructure will face increasing pressure to keep up with the rapidly changing technologies. This could lead to a period of rapid infrastructure upgrades, resulting in a boon for construction and engineering firms. But more than this, autonomous vehicles and robotics have the potential to revolutionize the construction industry itself and transform building sites (Gard, 2017).

8.4.3.3 Use of Autonomous Vehicles/Equipment in Construction Industry

To date, the main focus for developers of autonomous vehicles has been personal transport and freight. However, there are many other potential applications. Similar technology has already been embraced by the mining industry, for example; it has adopted autonomous vehicles to extract vital rocks and minerals. Automation of construction vehicles (bulldozers, cranes etc.) which are currently manually operated could provide similar benefits for the construction industry (Gard, 2017).

The growth of the construction industry is frustrated by a number of factors, including (Gard, 2017):

- An aging workforce
- Labor shortages
- Concerns about efficiency, productivity, and innovation
- Narrow profit margins.

There is also increasing pressure on contractors and developers to reduce emissions and air pollution from construction sites. Construction permits are increasingly subject to environmental and social impact studies. Furthermore, use of vehicles and heavy equipment remains a major source of construction fatalities and injuries.

Embracing new technologies could alleviate a number of these issues (Gard, 2017).

8.4.3.4 Benefits

When combined with other new forms of technology, including robotics, AI, 3D printing, UAVs (drones), building information modeling (BIM), and so on, the benefits of automated vehicles in the construction industry include (Gard, 2017):

- Improved safety and working conditions
- Reduced costs and greater efficiency: combined with drones, autonomous vehicles can be guided in more efficient patterns and provide better fuel usage and shortened schedules
- Lower emissions and congestion due to more efficient vehicle use
- Greater productivity
- Reduction in project delays
- Savings of time, labor, and cost for contractors and developers
- Overcoming the growing shortage of heavy equipment operators
- Remaining competitive in an industry operating with very thin margins.

A number of contractors are already using driver-assisted or semiautonomous functionality in heavy equipment on-site (e.g., many bulldozers use machine learning for blade control), and UAVs are being used to carry out site surveys and to monitor during construction.

In Japan, a company known as Komatsu, has taken this a stage further with the introduction of Smart Construction for excavation and earthmoving operations. Smart Construction uses drones to map sites in 3D and guide robotic vehicles by providing real-time data on the amount of Earth being moved around (Gard, 2017).

8.4.4 Relation to Other Fields

Generating as-is 3D models of the built environment using visual data collected via UAVs, transforming the outcome into useful information about the scene, and guaranteeing the accuracy and completeness during the data collection process have all received significant attention from the computer vision and robotics communities. However, direct application of these methods to construction monitoring is challenging because enormous amounts of visual data need to be frequently captured and rapidly processed; for example, thousands of images are collected per flight on a construction site, resulting in 3D point clouds with more than a 100 million points (Ham, Han, Lin, & Golparvar-Fard, 2016):

In the construction domain, BIM can provide strong a priori information about geometry and the appearance of scenes. Therefore, collection of visual data and detection and analysis of individual elements are simpler problems than the generic problems encountered in the computer vision and robotics fields. This makes the problem domain of UAV-driven construction monitoring interesting, as it can lead to domain-specific findings which can be generalized to solve the generic problems. The UAV problems are closely related to several research fields, encouraging interdisciplinary and multidisciplinary efforts (Ham, Han, Lin, & Golparvar-Fard, 2016).

8.4.4.1 UAV-Driven Visual Monitoring for Construction

A UAV-driven visual performance monitoring procedure will do the following (Ham, Han, Lin, & Golparvar-Fard, 2016):

1. Collect images or videos from the most informative views on a project site
2. Analyze them with or without a priori BIM to reason about performance deviations during construction (e.g., progress and quality)
3. Monitor ongoing operations for productivity and safety
4. Quickly and frequently visualize and communicate the most updated state of work in progress with onsite and offsite project participants.

Autonomous Vehicles for Inspection Applications 301

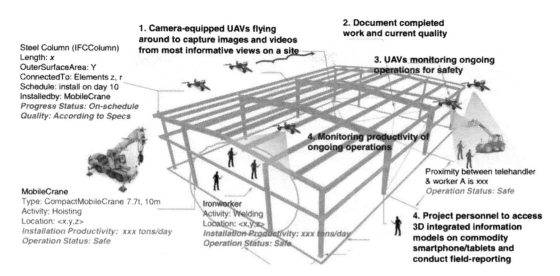

FIGURE 8.24 A vision for the next-generation construction site where camera-equipped UAVs autonomously monitor work in progress for improving safety, quality, and productivity (Ham, Han, Lin, & Golparvar-Fard, 2016).

Figure 8.24 illustrates an example of a next-generation construction site where camera-equipped UAVs autonomously monitor the construction performance. The following sections describe the most recent research in each procedure, discuss their challenges, and suggest future directions for research (Ham, Han, Lin, & Golparvar-Fard, 2016).

8.4.4.2 Collecting Informative Visual Data

Providing accurate performance information about the state of construction requires UAVs to collect visual data in the form of images and videos (e.g., digital: RGB; thermal: T; depth: D; digital+depth: RGB+D) from the most relevant locations and views. To streamline this process, research in UAV-driven visual data collection needs to address the following challenges (Ham, Han, Lin, & Golparvar-Fard, 2016):

1. Autonomous or semiautonomous path planning, navigation, and takeoff and landing procedures
2. Characterization of the criteria necessary for data collection, including the configurations among the images to guarantee complete information
3. Identification of the most informative views for data collection (e.g., canonical view, top-down view) for appearance-based recognition of work in progress (Han, Lin, & Golparvar-Fard, 2015).

In practice, UAV-driven data collection still relies on experienced pilots navigating the UAVs on and around project sites, although the research community has done some recent work on simultaneous localization and mapping (SLAM) techniques. For example, researchers have proposed autonomous navigation and data collection capabilities using GPS waypoints and predetermined flight paths. These methods have benefits, but a GPS-driven flight planning method that builds on existing maps has the following limitations (Ham, Han, Lin, & Golparvar-Fard, 2016):

1. Does not account for dynamics on a construction site and their impact on safety (e.g., the location and orientation of temporary resources such as cranes)
2. Can be negatively affected by loss/interference in GPS signal at interiors or caused by the shadowing effects of nearby buildings or other structures in densely populated metropolitan areas and high-rise buildings
3. Can be subject to navigational hazards due to the loss of calibration on magnetometer sensors in proximity to structural and nonstructural steel components.

Research on SLAM techniques for UAVs and ground robots has become mainstream in the robotics community. Using high-resolution laser scanners, monocular cameras, and RGB-D cameras, these new techniques generate 3D maps of unknown scenes and localize the robot in that environment. The latest efforts (e.g., Michael et al., 2014) have focused on experimenting with a variant of SLAM using RGD-B to scan post-disaster buildings, yet there is little to no work reported on actively testing these algorithms to produce 3D maps for construction monitoring. Methods that account for evolving structures and dynamic objects (i.e., equipment and human) are not reported in the literature either.

Identifying the most informative views for observing different tasks with or without BIM and collecting visual data to monitor locations, activities of equipment, and workers on a jobsite are not well studied. There is an opportunity to leverage a priori information about geometry and the appearance of a site via BIM. When integrated with schedules, BIM can also report on the most likely locations for expected changes on the site to steer data collection. Together with 4D (3D+time) point clouds, BIM definitely has the potential to minimize many technical challenges facing fully autonomous navigation and data collection for the built environment (Ham, Han, Lin, & Golparvar-Fard, 2016).

8.4.4.3 Information Visualization

Achieving an effective flow of information to and from project sites and conducting actionable analytics for construction monitoring require intuitive visualization of the information produced throughout the

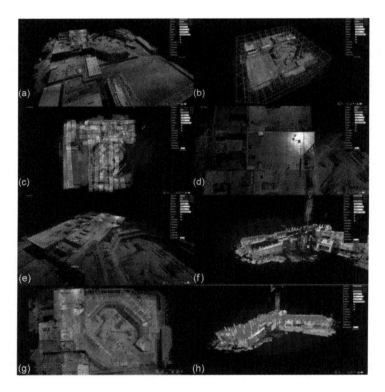

FIGURE 8.25 Web-based visualization and navigation of work in progress on construction sites using UAV-collected images, 4D point clouds, and 4D BIM: (a) image-based 3D point cloud models; (b) nested octree structure for mapping and interactions with point cloud data; (c) location and orientation of images taken via the UAV mapped in 3D using pyramids as camera frusta; (d) texture mapping onto the frontal part of the camera frusta, allowing users to interact with both images and point cloud data; (e) conducting linear, area-based, volumetric, and angular measurements; (f) color-coding changes on the construction site via point cloud data; (g) 4D BIM superimposed on point cloud data wherein a location at risk is highlighted in yellow; (h) 4D BIM superimposed on the point cloud model and performance deviations color-coded. All images are from the ongoing Flying Superintendents project at University of Illinois at Urbana-Champaign (Ham, Han, Lin, & Golparvar-Fard, 2016).

process, in addition to the UAV visual data images and point clouds. While attention to visual sensing and analytics has been the mainstream of the literature, less work has been conducted on interactive visualization. Recent work by Zollmann et al. (2014) introduced an interactive multilayer 4D visualization of information captured and analyzed through a UAV-driven procedure in the form of mobile augmented reality. Their system, similar to that proposed by Karsch, Golparvar-Fard, and Forsyth (2014) overlays color-coded 3D construction progress information on the physical world and adopts filtering methods to avoid information clutter and issues associated with displaying detailed augmented information.

Other recent examples by Han, Lin, and Golparvar-Fard (2015) and Lin, Han, Fukuchi, Eda, and Golparvar-Fard (2015) introduce Web-based tools with scalable system architectures for visualizing and manipulating large-scale 4D point cloud data, large collections of images, 4D BIM, and other project information. These methods account for the level of detail in data representation and dynamically consider the limited computational power and connection bandwidth when visualizing data on commodity smartphones. Since BIM is hosted on the server side of their system architecture, these tools can support the pull and push of the geometry and other semantic information from BIM. This provides access to the most updated information and does not require storing BIM locally on the client's device (Figure 8.25).

More research is needed to map, visualize, and explore the modalities of user interaction with operation-level construction performance data (e.g., locations and activities of workers and equipment) both on- and off-site (Ham, Han, Lin, & Golparvar-Fard, 2016).

REFERENCES

Ahmad J., Malik A. S., Abdullah M. F., Kamel N., Xia L., 2015. A novel method for vegetation encroachment monitoring of transmission lines using a single 2D camera. *Pattern Analysis and Applications*, 18, 419–440.

Anwar H., Muhammad A., Berns K., 2015. A theoretical framework for aerial inspection of siltation in waterways. LUMS Faculty Initiative Fund (FIF) and German Academic Exchange Service (DAAD) for the project titled RoPWat (Robotic Profiling for Clearing Watercourses).

Austin R., 2010. *Unmanned Aircraft Systems: UAVS Design, Development and Deployment*. Hoboken, NJ: John Wiley & Sons Ltd.

Beard B. L., Jones K. M., Chacon C., Jr., Ahumada A. J., 2005. Detection of blurred cracks: A step towards an empirical vision standard. Final Report for FAA Agreement DTFA-2045.

Bluestream. Alongside UAV inspection services. www.bluestreamoffshore.com/site/services/uav-inspection.html#. Viewed: July 14, 2019.

Bretschneider T. R., Shetti K., 2015. UAV-based gas pipeline leak detection. In *Proceedings of ARCS*, Singapore, 2015.

Buchanan K., 2016. Regulation of drones: New Zealand. April 2016.

Campoy P., Garcia P. J., Barrientos A., Cerro J. D, Aguirre I., Roa A., García R., Muñoz J. M., 2001. A stereoscopic vision system guiding an autonomous helicopter for overhead power cable inspection. In *Proceedings, Robot Vision: International Workshop RobVis 2001*, Auckland, New Zealand, 2001, pp. 115.

Carrio A., Sampedro C., Rodríguez-Ramos A., Campoy P., 2017. A review of deep learning methods and applications for unmanned aerial vehicles. *Journal of Sensors*, 2017, 2017.

Cyberhawk, 2008. Aerial inspection and survey using UAVs.

Dalamagkidis K., Valavanis K. P., Piegl L. A., 2008. Current status and future perspectives for unmanned aircraft system operations in the US, Springer Science and Business Media B. V.

Drone Insurance Depot, 2016. The benefits of using a drone for building inspection. August 9, 2016. http://blog.droneinsurancedepot.com/the-benefits-of-using-a-drone-for-building-inspection. Viewed: July 13, 2019.

Dupont Q. F., Chua D. K., Tashrif A., Abbott E. L., 2017. Potential applications of UAV along the construction's value chain. *Procedia Engineering*, 182, 165–173.

Eschmann C., Kuo C. M., Kuo C. H., Boller C., 2012. Unmanned aircraft systems for remote building inspection and monitoring. In *6th European Workshop on Structural Health Monitoring, Dresden, Germany*, January 2012.

European Aviation Safety Agency, 2015. Notice of Proposed Amendment 2017-05 (B).

Flammini F., Pragliola C., Smarra G., 2016. Railway infrastructure monitoring by drones. In *International Conference on Electrical Systems for Aircraft, Railway, Ship Propulsion and Road Vehicles & International Transportation Electrification Conference (ESARS-ITEC)*, Toulouse, France.

Fu S., Zhang Y., Zhao X., Liang Z., Hou Z., Zou A., Tan M., Ye W., Bo L., 2006. *Motion Deblurring for A Power Transmission Line Inspection Robot. Computational Intelligence.* Berlin, Heidelberg: Springer.

Gard W., 2017. Autonomous vehicles: The future of the construction industry. Burges Salmon. July 28, 2017. www.burges-salmon.com/news-and-insight/legal-updates/autonomous-vehicles-the-future-of-the-construction-industry/. Viewed: July 14, 2019.

Gheisari M., Irizarry J., Walker B. N., 2014. UAS4SAFETY: The potential of unmanned aerial systems for construction safety applications. In *Construction Research Congress 2014: Construction in a Global Network*, Atlanta, Georgia, USA, 2014, pp. 1801–1810.

Gilbert M., 2017. Drones & AI: The next phase of automation. Technology Blog. August 15, 2017. http://about.att.com/innovationblog/drones_automation. Viewed: July 13, 2019.

Golightly I., Jones D., 2005. Visual control of an unmanned aerial vehicle for power line inspection. In *Presented at Proceedings 12th International Conference on Advanced Robotics (ICAR '05)*, Seattle, WA, USA, 2005.

Gupta S. G., Ghonge M. M., Jawandhiya P. M., 2013. Review of unmanned aircraft system (UAS). *International Journal of Advanced Research in Computer Engineering & Technology (IJARCET)*, 2(4), 1646–1658.

Ham Y., Han K. K., Lin J. J., Golparvar-Fard M., 2016. Visual monitoring of civil infrastructure systems via camera-equipped unmanned aerial vehicles (UAVs): A review of related works. *Visualization in Engineering*, 4(1), 1.

Han K., Lin J., Golparvar-Fard M., 2015. A formalism for utilization of autonomous vision-based systems and integrated project models for construction progress monitoring. *Proceedings of 2015 Conference on Autonomous and Robotic Construction of Infrastructure*, Ames, Iowa, USA.

Honeywell, 2017. Honeywell launches UAV industrial inspection service, teams with Intel on innovative offering. https://aerospace.honeywell.com/en/press-release-listing/2017/october/honeywell-launches-uav-industrial-inspection-service-teams-with-intel-on-innovative-offering.

Hrishikeshavan V., 2011. Experimental investigation of a shrouded rotor micro air vehicle in hover and in edgewise gusts. *Doctoral dissertation*, University of Maryland. https://drum.lib.umd.edu/handle/1903/12351.

Industrial Skyworks. Drone inspections services. https://industrialskyworks.com/.

Jones D. I., Earp G. K., 1996. Requirements for aerial inspection of overhead electrical power lines. In *12th International Conference on Remotely Piloted Vehicles*, Bristol, England. Bristol, 1996.

Jones D., 2005. Power line inspection - A UAV concept. In *Presented at The IEE Forum on Autonomous Systems 2005* (Ref. No. 2005/11271), London, UK, 2005.

Jones D., Golightly I., Roberts J., Usher K., 2006. Modelling and control of a robotic power line inspection vehicle. In *Presented at IEEE International Conference on Control Applications (CCA '06)*, Munich, Germany, 2006.

Jurić-Kaćunić D., Librić L., Car M., 2016. Application of unmanned aerial vehicles on transport infrastructure network. *Građevinar*, 68(4), 287–300.

Kaamin M., Azyyati Idris N., Mohd Bukari S., Zaurin Ali Z., Samion N., Anjang Ahmad M., 2017. Visual inspection of historical buildings using micro UAV. In *MATEC Web of Conferences*; Vol. 103, p. 07003. doi:10.1051/matecconf/20171030.

Karsch K., Golparvar-Fard M., Forsyth D., 2014. Construct aide: Analyzing and visualizing construction sites through photographs and building models. In *Proceedings of ACM Transactions on Graphics*, Illinois, USA.

Katrašnik J., 2007. *Robotic Power Line Inspection.* Ljubljana, Slovenia: Faculty of Electrical Engineering, University of Ljubljana.

Katrasnik J., Pernus F., Likar B., 2008a. A climbing-flying robot for power line inspection. In *2008 IEEE Conference on Robotics, Automation and Mechatronics*, Ljubljana, Slovenia. ©2008 IEEE.

Katrasnik J., Pernus F., Likar B., 2008b. *New Robot for Power Line Inspection.* Ljubljana, Slovenia: Laboratory of Imaging Technologies. University of Ljubljana, Faculty of Electrical Engineering. 978-1-4244-1676-9/08 /$25.00 ©02008 IEEE.

Kochan A., Rutkowska P., Wójcik M., 2018. *Inspection of the Railway Infrastructure with the Use of Unmanned Aerial Vehicles.* Warszawa: Warsaw University of Technology, Faculty of Transport, Ośrodek Certyfikacji Transportu.

Krawczyk J. M., Mazur A. M., Sasin T., Stokłosa A. W., 2015. Infrared building inspection with unmanned Aerial Vehicles. *Transactions of the Institute of Aviation*, 3(240), 32–48. doi 10.5604/05096669.1194965.

Lagüela S., Díaz–Vilariño L., Roca D., Lorenzo H., 2015. Aerial thermography from low cost UAV for the generation of thermographic digital terrain models, applied Geotechnologies Research Group, University of Vigo, Lab 22, ETSE Minas, Campus Universitario Lagoas Marcosende, Pontevedra, Spain.

Larrauri J. I., Sorrosal G., González M., 2013. Automatic system for overhead power line inspection using an unmanned aerial vehicle. Trelifo project. In *International Conference on Unmanned Aircraft Systems (ICUAS)*, Atlanta, GA, USA. IEEE, 2013, pp. 244–252.

Lin J., Han K., Fukuchi Y., Eda M., Golparvar-Fard, M., 2015. Model based monitoring of work in progress via images taken by camera equipped UAV and BIM. In *Proceedings of 2015 International Conference on Civil and Building Engineering Informatics*, Tokyo, Japan.

Ludan W., Hongguang W., Lijin F., Mingyang Z., 2006. Research on obstacle-navigation control of a mobile Robot for inspection of the power transmission lines based on expert system. In *Proceedings of the 8th International Conference on Climbing and Walking Robots and the Support Technologies for Mobile Machines (CLAWAR 2005)*, Berlin, Heidelberg: Springer.

Luque-Vega L. F., Castillo-Toledo B., Loukianov A., González Jiménez L. E., 2014. Power line inspection via an unmanned aerial system based on the quadrotor helicopter. In *17th Mediterranean Electrotechnical Conference (MELECON)*, Beirut, Lebanon. IEEE, 2014, pp. 393–397.

Matikainen L., Lehtomäki M., Ahokas E., Hyyppä J., Karjalainen M., Jaakkola A., Kukko A., Heinonen T., 2016. Remote sensing methods for power line corridor surveys. *Journal of Photogrammetry and Remote Sensing*, 119, 10–31.

Maverick Industrial UAV Inspection. www.maverickinspection.com/services/industrial-drone-inspection/. Viewed: July 14, 2019.

Michael N., Shen, S, Mohta K., Kumar V., Nagatani K., Okada Y., Kiribayashi S., Otake K., Yoshida K., Ohno K., Takeuchi E., Tadokoro S., 2014. Collaborative mapping of an earthquake damaged building via ground and aerial robots. In K. Yoshida & S. Tadokoro (Eds.), *Field and Service Robotics* (pp. 33–47). Berlin, Heidelberg: Springer.

Mieloszyk J., Tarnowski A., 2015. Mass, time and cost reduction in MAV manufacturing. *Transactions of the Institute of Aviation*, 1(238), pp. 22–34.

Miralles F., Pouliot N., Montambault S., 2014. State-of-the-art review of computer vision for the management of power transmission lines. In *Proceedings of the IEEE International Conference on Applied Robotics for the Power Industry (CARPI)*, Foz do Iguassu, Brazil, 14–16 October 2014.

Pagnano A., Höpf M., Teti R., 2013. A roadmap for automated power line inspection. Maintenance and repair. *Procedia CIRP*, 12, 234–239.

Peungsungwal S., Pungsiri B., Chamnongthai K., Okuda M., 2001. Autonomous robot for a power transmission line inspection. In *Proceedings of the 2001 IEEE International Symposium on Circuits and Systems (ISCAS 2001)*, Sydney, NSW, Australia.

PG&E, 2018. Testing safety drones to inspect electric and gas infrastructure. PG&E External Communications (415) 973-5930. May 18, 2016. www.pge.com/en/about/newsroom/newsdetails/index.page?title=20160518_pge_testing_safety_drones_to_inspect_electric_and_gas_infrastructure. Viewed: July 13, 2019.

Sankarasrinivasan S., Balasubramanian E., Karthik K., Chandrasekar U., Gupta R., 2015. Health monitoring of civil structures with integrated UAV and image processing system. *Procedia Computer Science*, 54, 508–515.

Sawada J., Kusumoto K., Maikawa Y., Munakata T., Ishikawa Y., 1991. A mobile robot for inspection of power transmission lines. *IEEE Transactions on Power Delivery*, 6(1), 309–315.

Sawicki P., 2012. Unmanned aerial vehicles in photogrammetry and remote sensing – state of the art and trends, Katedra Fotogrametrii i Teledetekcji, uniwersytet Warmińsko Mazurski w olsztynie, archiwum Fotogrametrii, Kartografii i Teledetekcji, 23, 365–376.

Shakhatreh H., Ahmad Sawalmeh A., Al-Fuqaha A., Dou Z., Almaita E., Issa Khalil I., Shamsiah Othman N., Khreishah A., Mohsen Guizani M., 2018. Unmanned aerial vehicles (UAVs): A survey on civil applications and key research challenges. *IEEE Access*, 7, 48572–48634. doi:10.1109/ACCESS.2017.

Sibilski K., Żyluk A., Kowalski W., Wiśniowski W., 2015. Simulation studies of micro air vehicle. *Journal of KONES Powertrain and Transport*, 22(4), 243–252.

Tang L., Wang H., Fang L., 2004. Development of an inspection robot control system for 500KV extra-high voltage power transmission lines. In *Proceedings of SICE 2004 Annual Conference*, Sapporo, Japan.

Thapa M., Shrestha R., 2014. Autonomous UAV: General and case study of reaper MQ-9, Tribhuvan University Institute of Engineering Central Campus Pulchowk Department of Mechanical Engineering.

Thompson R. M., II, 2013. Drones in domestic surveillance operations: Fourth amendment Implications and legislative responses, legislative attorney.

U.S. Department of Transportation/Federal Railroad Administration, 2018. Unmanned aircraft system applications in international railroads. Final Report, Office of Research, Development and Technology Washington, DC, USA, February 2018.

Williams M., Jones D. I., Earp G. K., 2001. Obstacle avoidance during aerial inspection of power lines. *Aircraft Engineering and Aerospace Technology*, 73, 472–479.

Xinglong Z., Hongguang W., Lijin F., Mingyang Z., Jiping Z., 2006. A novel running and gripping mechanism design based on centroid adjustment. In *Proceedings of the 2006 IEEE International Conference on Mechatronics and Automation*, Luoyang, Henan, China.

Zhang Y., Fu S., Zhao X., Liang Z., Tan M., Zhang, Y., 2006. *Visual Navigation for a Power Transmission Line Inspection Robot. Computational Intelligence*. Berlin, Heidelberg: Springer.

Zhang Y., Yuan X., Fang Y., Chen S., 2017. UAV low altitude photogrammetry for power line inspection. *International Journal of Geo-Information*, 6, 14.

Zhang Y., Yuan X., Li W., Chen S., 2017. Automatic power line inspection using UAV images. *Remote Sensing*, 9, 824. doi:10.3390/rs9080824.

Zollmann S., Hoppe C., Kluckner S., Poglitsch C., Bischof H., Reitmayr G., 2014. Augmented reality for construction site monitoring and documentation. *Proceedings of the IEEE*, 102(2), 137–154.

9

Failure Detection Application in Autonomous Vehicles

9.1 Repeated Inspections and Failure Identification

9.1.1 Definition of Faults and Failures

9.1.1.1 Fault

A good definition of fault is the following: "A fault is an unpermitted deviation of at least one characteristic property (feature) of the system from the acceptable, usual, standard condition" (Ducard, 2009).

Based on this definition, a fault corresponds to an abnormal behavior of the system, which may not affect the overall functioning but may eventually lead to a failure (defined below). A fault may be smaller than the detectable value or can be hidden. Detecting or estimating such a fault is difficult. Actuators are used on an aircraft's control surfaces, such as ailerons, elevators, and rudders; they are also used to actuate the engine throttle or the landing gear mechanism. An actuator is considered faulty if it behaves abnormally; this can be caused by a loss of effectiveness or a bias of the actuator (Sadeghzadeh & Zhang, 2011).

9.1.1.2 Failure

A useful definition of failure is the following: "A failure is a permanent interruption of a system's ability to perform a required function under specified operating conditions" (Ducard, 2009).

One or more faults can cause a failure which leads to the end of the system functioning. To continue the previous example of an actuator is declared failed when it can no longer act in a controlled manner. The actuator may become ineffective or float at the zero-moment position. The control surface can be locked in any position or reach and stay at the saturation position. Mechanical failure can also occur during a flight, such as a mechanical break between the control surface and its actuator. The engine and system structure may fail as well (Sadeghzadeh & Zhang, 2011).

Many sources of possible irreversible damage to the aircraft may be classified as structural failures. They correspond to scenarios where a piece of the aircraft is missing, such as an aileron, a tail rudder, an elevator, or a part of a wing (Chowdhary, Johnson, & Kimbrell, 2013; DeBusk, Chowdhary, & Johnson, 2009). An active (reconfigurable) flight control system is capable of detecting faults in the system (which are more difficult to detect than failures) and is able to adequately compensate for failures (Sadeghzadeh & Zhang, 2011).

9.1.2 Fault Detection, Identification, and Accommodation Techniques for Unmanned Airborne Vehicle (UAV)

Unmanned airborne vehicles (UAVs) are assuming prominent roles in both the commercial and military aerospace industries. The promise of reduced costs and reduced risk to human operators is one of their major attractions, but these low-cost systems have yet to gain acceptance as a safe alternative to manned solutions. The absence of a thinking, observing, reacting, and decision-making pilot reduces the capability of UAVs to manage adverse situations, such as faults and failures.

This section reports on the fault detection and accommodation techniques for low-cost UAV systems. It begins by highlighting current fault detection and accommodation approaches for UAVs, with a focus on sensor failures. Sensor failures are critically important as they can lead to unrecoverable flight. As reduced complexity, lower costs, and weight optimization are major design specifications, traditional approaches, such as built-in tests and multiple redundancies, are no longer appropriate. One method employed to combat sensor failures is the use of model-based techniques to produce parameter estimates that can be used for both fault detection and fault accommodation (Cork, Walker, & Dunn, 2005).

The approach presented in this section uses neural networks to provide analytical redundancy from sensors already existing onboard a UAV. It investigates the neural network-based sensor fault detection, isolation, and accommodation (SFDIA) process proposed by Napolitano, An, and Seanor (2000), with a focus on UAV-specific sensor models and the closed-loop aircraft performance (Cork, Walker, & Dunn, 2005).

9.1.2.1 Fault Detection and Accommodation Techniques in UAVs

The aerospace industry has used fault detection for a variety of applications, such as detecting structural failures, engine failures, and avionics/power system failures, but few of these applications have specifically been for UAVs (Cork, Walker, & Dunn, 2005).

9.1.2.1.1 Types of Faults and Failures

Faults in closed-loop systems are commonly represented as actuator, plant, or sensor failures (see Figure 9.1). The fault detection and accommodation techniques required will depend on the type of failure experienced. Plant and actuator failures in UAVs generally result from mechanical or structural failures and often require adaptation of the control system. Providing redundancy or other types of fault accommodation is normally not feasible. Sensor failures can be a major source of error in UAVs, particularly when a UAV uses less reliable components because of design restrictions. To detect and manage these three types of failures, a combination of techniques is required, but in this section, we focus on sensor failures and sensor fault detection and accommodation (SFDA) (Cork, Walker, & Dunn, 2005).

9.1.2.1.2 Fault Detection Methods

Fault detection procedures can be divided into the following (Cork, Walker, & Dunn, 2005):

1. Knowledge-based techniques
2. Signal processing approaches
3. Model-based approaches.

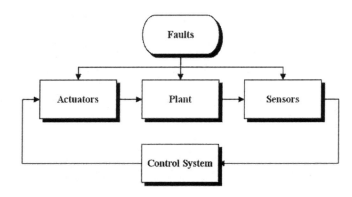

FIGURE 9.1 Representation of fault locations in a closed-loop system (Cork, Walker, & Dunn, 2005).

Knowledge-based techniques can be used for fault detection, identification, and accommodation (FDIA). They use artificial intelligence approaches, such as neural networks or fuzzy decision logic, to detect and classify faults.

Signal processing techniques use signal features (spectrum information, statistical information etc.) to generate signals that indicate the existence of a failure.

Model-based techniques are similar to signal processing techniques, except that a model is used to estimate the values, from which error signals (residuals) can be used to give an indication of the existence of a failure. Neural networks, model-based techniques form a considerable portion of FDIA research (Cork, Walker, & Dunn, 2005).

Neural Network Approaches to Fault Detection The neural network approach is an extremely powerful method for fault detection. Neural network FDIA solutions can be knowledge-based, where neural networks are trained to recognize faults, based on certain criteria/features, or model-based, where neural networks are used to provide analytical redundancy for fault detection purposes (Cork, Walker, & Dunn, 2005). An example of work on neural network fault detection for aircraft systems is by Napolitano, An, and Seanor (2000).

9.1.2.1.3 Fault Detection Performance

The performance of a fault detection procedure is measured by its percentages of successful detections, as well as its percentage of false alarms. There are four possible outcomes (two successful, two unsuccessful) for a fault detection system. A successful outcome is one which determines the correct health status. A false alarm is where a fault is declared when no fault exists, whereas a missed alarm is where a fault is not detected when a failure occurs. Other important terms in fault detection performance are detectability, isolability, robustness, and detection delay (Cork, Walker, & Dunn, 2005).

9.1.2.1.4 Fault Accommodation Procedures

Accommodation of a sensor failure is generally achieved using one of two methods. The first method is system reconfiguration, where the system is altered to minimize the effect of a fault. A reconfigurable controller fits into this type of solution. The other method is modification or replacement of the faulty signal. This requires a form of redundancy (either hardware or analytical) that can be used to accurately estimate the faulty sensor measurement (Cork, Walker, & Dunn, 2005).

9.1.2.1.5 System Performance

The ultimate goal of a fault detection system for a UAV is to allow the aircraft to continue flying with an acceptable level of performance, for an adequate time span to either complete its mission or recover from the failure. Providing UAVs with FDIA capabilities will improve their reliability and safety, but there will always be a limit to the level of faults that can be detected and accommodated. As more components fail, the system becomes less capable and reliable. For this reason, any fault detection scheme and accommodation process must be backed up by suitable maintenance procedures and realistic expectations (Cork, Walker, & Dunn, 2005).

9.1.3 Failure Prediction for Autonomous Vehicles

Autonomous vehicles (AVs) will soon have a substantial impact on people's daily lives, both personally and professionally. For instance, automated vehicles can increase human productivity by turning driving time into working time, provide personalized mobility to nondrivers, reduce traffic accidents, or free up parking space (Sadeghzadeh & Zhang, 2011). Thus, developing automated vehicles is the core interest of many diverse industrial players. We have seen great progress in autonomous driving, with AVs being driven many thousands of miles and companies aspiring to sell such vehicles in a few years.

Yet significant technical obstacles must be overcome before assisted driving can be turned into fully fledged automated driving. To make matters worse, an automated car that from time to time will call on the driver to take over, will, by many drivers, be considered worse than having no automated driving at all. Indeed, in such a transition situation, the driver will be required to permanently pay attention to the road, so as to not be out of context when s/he suddenly needs to act. And then there is the boredom coming with not having to intervene for a long time. The more successful the automation, the worse the issue. Add legal responsibilities to the picture, and the possibility that the human driver is called upon to take decisions, however rarely, may be with us for a while.

With so much effort currently going into improving autonomous driving, such systems will certainly improve quickly. Yet during the coming years, performance will probably not be strong enough to avoid all mistakes. Driving models may still fail due to congested traffic, bad weather, frontal illumination, road construction, etc., or simply fail unexpectedly due to the idiosyncrasies of the underlying algorithms. Failures of a vehicle can be catastrophic, so it is crucial to obtain an early warning of impending trouble. Despite this importance, the community has so far paid limited attention to the automated predictions of potential failures.

One exception is work on "scene drivability": i.e., how easy a scene is for an automated car to navigate. A low drivability score means the automated vehicle is likely to fail for the particular scene. Obviously, scene drivability is dependent on the autonomous driving system at hand. Therefore, to quantify and learn this property, we first need to pick a particular autonomous driving model and train it using videos from car-mounted cameras. Via machine learning, it will automatically learn things like "when the vehicle is in the left-most lane, the only safe maneuvers are a right-lane change or keeping straight, unless the vehicle is approaching an intersection." Such learning requires the system to be exposed to a representative sample of scenarios; therefore, the model should be trained on a large, real driving dataset, containing video sequences and other time-stamped sensor measurements, such as steering angles, speeds, and Global Positioning System (GPS) coordinates. Discrepancies between the predictions by the trained driving model and the ground-truth maneuvers by human drivers can then be used to assess the likelihood of failure, i.e., the scene drivability score.

Given the success of deep neural networks in supervised learning, and especially in autonomous driving, a recurrent convolutional network (RCNet) has been tested with four convolutional neural networks (CNNs) as visual encoders and three long short-term memory (LSTMs) to integrate the visual contents, temporal relationships, and the previous driving states (steering angle and speed) into a single prediction model. The model can be trained very efficiently in an end-to-end manner, and its architecture is shown in Figure 9.2. This architecture can be used for both car driving and failure prediction.

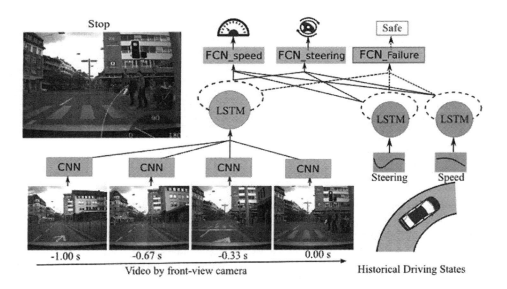

FIGURE 9.2 Architecture of driving model which provides future maneuvers (i.e., speed and steering angle) and the drivability score of the scene. The drivability scores are quantized into two levels: safe and hazardous. In this case, the coming scene is safe for the driving model, so the system does not need to alert the human driver. FCN stands for fully connected network (Sadeghzadeh & Zhang, 2011).

All layers, except the task-specific fully connected layers, are shared for computational efficiency (Sadeghzadeh & Zhang, 2011).

The emphasis of the following section is not on achieving state-of-the-art driving performance. Rather, it is to provide a sensible driving model and infer failure prediction for it, so autonomous driving can survive the risky market situation ahead (Sadeghzadeh & Zhang, 2011).

9.1.3.1 Failure Prediction

Performance-blind algorithms can be disastrous. As automated vision increasingly penetrates industrial applications, this issue is gaining attention. Notable examples in computer vision for learning model uncertainty or failure include: semantic image segmentation, optical flow, image completion, stereo, and image creation. In addition to creating warnings, performance-aware algorithms bring other benefits. For instance, they can speed up algorithms downstream by adaptively allocating computing resources based on scene difficulty. For autonomous driving, this can also mean using sensors adaptively or selectively (Sadeghzadeh & Zhang, 2011).

9.1.4 The UAV as a Complex Maintenance System

A UAV is a complex aerial system composed of, say, m subsystems (or subparts) defined as $J = \{1, 2, ..., m\}$ and consisting of l_j components. At the component level, we can continuously control and check the degradation of a defined collection of physical parameters. The physical conditions degrade monotonically during use and are restored by maintenance actions. For each component or subpart $i \in I$, $X_i(t)$ indicates the degradation trajectory in a fixed time interval $t \in [0, \infty)$. Soft failure can be defined as the ability of a component, part, subsystem, or system to continue its work even with degraded performance, i.e., up to the point when its reduced performance exceeds a specific fixed threshold H_i (with $X_i(t) > H_i$). Typically, components subjected to thermal stress or mechanical degradation are hit by soft failures.

When $X_i(t)$ exceeds H_i, a soft failure happens between two maintenance points $(n-1)\tau$ and $n\tau$. This implies an action of corrective maintenance (CM), which has a specific cost $\left(c_i^{CM}\right)$ on the critical component. This action is executed in a fixed time called maintenance point $n\tau$, as shown in Figure 9.3a (Petritoli, Leccese, & Ciani, 2018a,b).

The period between the occurrence of the soft failure point and the maintenance point $n\tau$ is defined as the "soft failure period" (see Figure 9.3). This period defines loss of quality in production or poorer performance with a cost rate indicated by c_i^P (Poole, 2011).

9.1.5 Benefits of Drone Inspection

An increasing number of companies are using drone technology for visual inspection, as it is a cost-conscious and effective way to inspect at heights and in inaccessible areas (Simonsen, 2014a). Drone inspection makes it possible to quickly and safely access areas that may pose health, safety, and environmental risks. Using drone inspection for visual inspection provides benefits such as the following (Simonsen, 2014b):

- Inspection of areas difficult to access.
- Preventive maintenance planning and optimized production.
- Access to areas that pose health, safety, and environmental risk to humans.
- Reduced downtime.
- Quick overview and evaluation of the condition.
- Sharp and detailed photographs of defects.
- High level of safety.

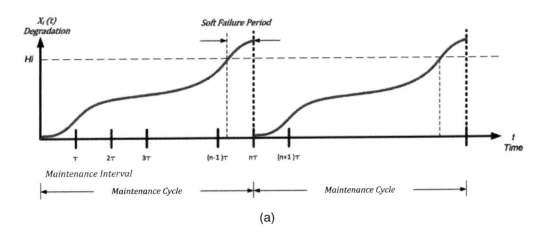

(a)

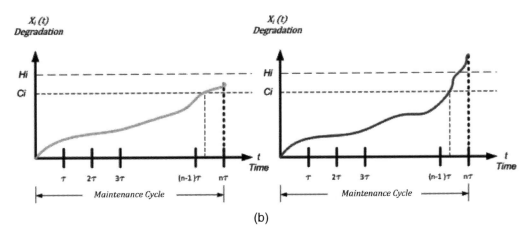

(b)

FIGURE 9.3 Degradation threshold of a system with a cycle of CM only (Petritoli, Leccese, & Ciani, 2018a). Condition-based maintenance of single components CM only (a), preventive ((b) left figure), and CM ((b) right figure) (Petritoli, Leccese, & Ciani, 2018b).

Typically, drone inspection is recommended for the following three categories (Simonsen, 2014a):

1. Offshore constructions
   The offshore environment is tough and demanding, but it is vital to keep faults and loss of production to a minimum.
   One example of a drone inspection is a flare inspection on an oil-gas platform, where the flare is still active. Here the drones collect pictures or thermographic data on the flare's condition, while production continues unaffected.
   The data collection takes place in a turbulent and flame-filled environment, while the drone pilot, inspector, and employees are at a safe distance from the inspection area and, therefore, out of the risk zone.
2. Onshore constructions
   Drones are normally restricted to flying at an altitude of 100 m, but inspections higher than 100 m may be required.
   An example is the inspection of a 225 m chimney and masts at a power station, refinery, or hydro power plant. Another area clearly suitable for drone inspection is bridges, whether on land, connecting islands, or spanning divisions in a landscape.

Drones are also delivering good results with the increasingly taller wind turbines. In particular, leading edges, lightning receptors, and nacelles are subjects for inspection. Here the drone collects high resolution pictures of possible defects, making it possible to plan repair work in detail; thus the data collected will help avoid unpleasant surprises during the maintenance period.

3. Confined spaces

Confined spaces, such as boilers at power plants and storage tanks for fuel, represent a relatively new addition to the drone inspection area but are gradually becoming more commonplace. Such inspections are advantageous if it is too expensive or time consuming to set up a scaffold or too hazardous for physical inspection.

There are considerable challenges when using drones in confined spaces; these include lack of proper light condition, color nuances, reflection of radio signals, and propeller turbulence.

9.1.5.1 Technical Possibilities

Today, there are both simple drones which can record video or take still images and advanced drones to which all types of data collecting equipment can be attached. It is especially relevant to choose the advanced drones when measuring flue gas emissions or inspecting wind turbines, as a thermographic camera can detect delaminating, that is to say, the internal composite layers separating beneath the surface of the wing. In recent years, thermographic cameras have been used to inspect house roofs and electrical and production installations, facilitating the detection of heat-related faults or defects (Simonsen, 2014a).

9.1.5.2 Data Analysis

The drone inspection itself is normally quickly finished. Most of the work lies in the analysis of the data collected and the pictures from the drones. It is essential to enlist experts in analysis to obtain full value from a drone inspection. Someone lacking the necessary qualified experience may overlook a critical defect or fault, possibly with fatal consequences. For instance, an inspection of a steel construction requires an expert metallurgist to analyze and interpret the collected data. An inspection report is produced, based on the analysis of these data, and is usually divided into categories of red, yellow, and green priority according to the seriousness of the damage. This report can include a service plan, with recommendations for cost-effective improvements (Simonsen, 2014a).

9.2 Autonomous Vehicle Emergency Inspection Applications

The automotive industry is changing with the aim of developing intelligent vehicles. The most significant changes in the future will include the following (MORSU, 2017):

- Increased vehicle safety: Systems such as obstacle detection, lane departure warning, lane keep assist, adaptive cruise control, and so on will help avoid accidents and, thus, increase vehicle safety.
- Steady traffic flow: Traffic congestion caused by driver error will be avoided by various autonomous driving applications.
- Reduced fuel consumption: Automatically adjusting speeds on highways by using an adaptive cruise control feature will have a direct effect on increased fuel efficiency.

Perception plays a crucial role in the development of any AV applications. The key challenges are to manage and combine the significant amount of data coming from the different sensors and to create a consistent model from these data that can make decisions. In developing self-driving vehicle technology, SAE International, initially established as the Society of Automotive Engineers, has classified standards of driving automation levels based on the amount of driver intervention and attentiveness required rather than the vehicle capabilities. The definitions include the following (MORSU, 2017):

- Partially automated: The driver must continuously monitor the automatic functions and cannot perform any nondriving task.
- Highly automated: The automatic system recognizes its own limitations and calls for the driver to take control when needed. The driver can perform a certain number of nondriving tasks.
- Fully automated: The system can handle all the situations autonomously, and there is no need for human intervention. Driverless driving is possible at this level.

As the technology is approaching fully autonomous driving, the number of sensors in vehicles will increase drastically, but which sensors will provide the most value is an important question for automakers. It is in their best interest to push hard for cost optimization without sacrificing safety. Some of the main sensors include the following (MORSU, 2017):

- LiDAR: LiDAR stands for light detection and ranging. The technology uses laser light to measure distances up to 100 m in all directions and generates a precise 3D map of the vehicle's surroundings. The problem with LiDAR is that it generates huge amounts of data and is expensive to implement. At the same time, fully autonomous driving requires LiDAR.
- Radar: Radar stands for radio detection and ranging. This sensor system uses radio waves to determine the velocity, range, and angle of objects. Radar is computationally lighter and uses far less data than LiDAR. However, its output is less angularly accurate.
- Camera: Cameras are the masters of classification and texture implementation. A camera sees in two dimensions with high spatial resolution. It is also possible to infer depth information from camera images. Cameras are the cheapest and most available sensors by far. Self-driving functions can be developed by using cameras with smart algorithms.

9.2.1 Emergency Control for a Fail-Safe System in Automated Vehicles

Research into fail-safe and fall back systems for AVs is important. AVs are composed of various sensors, computers, actuators, and other types of equipment, and the equipment is configured to communicate. For fault diagnosis, each piece of equipment needs real-time monitoring and maneuvers to be configured in the event of a failure. The system makes an emergency stop when a fault is determined by a fault diagnosis system, and there is no driver intervention. Even if an error occurs and the vehicle does not receive normal data from the upper controller designed to recognize and judge components of the autonomous driving system, the proposed algorithm uses the vehicle's chassis information to respond safely. The vehicle hardware configuration is divided into upper controllers responsible for recognition and judgment and lower controllers responsible for vehicle control. The lower controller has robust hardware that allows safe longitudinal and lateral control in the event of errors in the upper controller.

In autonomous driving, SAE International Level 4 suggests a specific response by an automated driving system, even if a human driver does not respond appropriately to a request to intervene (see Section 9.3.3.1) (Lee, Oh, & Yi, 2016).

Although there are not many studies on the fault diagnosis system of autonomous driving systems, Jeong and Lee (2015) developed a vehicle sensor fault diagnosis and acceptance algorithm and conducted residual and adaptive threshold fault diagnosis without additional hardware. Another study was conducted on the predictive fault diagnosis algorithm using sliding mode observers (OH, 2017).

Google Waymo self-driving vehicles have applied fallback systems. Figure 9.4 is Waymo's safety-critical system description. Waymo's redundant system is composed of backup computing, backup braking, backup steering, and a backup power system (Waymo Safety Report, 2017). Similar to Google Waymo, CRUISE has a backup computer, backup actuator, a signal communication system, and a data accumulation system (General Motors, 2018).

Failure Detection Application

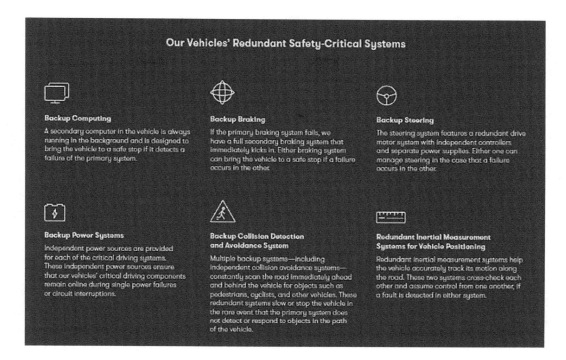

FIGURE 9.4 Waymo vehicles' redundant safety-critical systems (Waymo Safety Report, 2017).

9.2.1.1 Hardware-Based Fail Detection Classification

This section introduces the module-based classification of an AV's fail detection and maneuvering system. Hardware can be divided into actuators, sensors, upper controllers, controller area networks (CANs), and lower controllers. An actuator acts as steering or as a throttle/brake. Sensors include LiDAR, radar, and mono-vision (camera). The upper controller contains logic for perception and judgment of the autonomous driving algorithms; it calculates these at regular intervals and transmits the calculated values to the lower controller over real-time communication. CAN communication refers to the overall communication of the vehicle, including many sensors, actuators, and vehicle intercommunication systems; it conducts real-time monitoring of their values. The lower controller consists of algorithms that calculate the relative sub-controller of the overall configuration, the algorithm for path tracking, and a control input that enters longitudinal control in the event of failure (Lee, Oh, & Yi, 2016).

9.2.1.1.1 Sensor Fail Detection

The hardware fault detection method of sensors is shown in Figure 9.5a. The manufacturer sends a corresponding fault signal from the sensor itself in the event of a failure. Delphi's radar, shown in the figure, has signals that can find many faults, such as sensor communication error, sensor status failed, status blocked, and temperature status. Communication settings allow users to read the appropriate information.

Figure 9.5b shows Ibeo LUX Ridar error and warning messages. Error contents include internal error, motor error, temperature rise, data loss, internal communication error, incorrect scan data, etc. The warning signal sent to the user can include error messages, such as internal communication error and temperature increase (Lee, Oh, & Yi, 2016).

9.2.1.1.2 Communication and Controller Fail Detection

Inside the vehicle, communication is via the CAN bus. The communication protocol is a standard communication specification designed to enable multiple devices to communicate with each other without a host computer.

4E0	CAN_TX_COMM_ERROR	Indication that the sensor has detected a communication error	False/True	bool	0: False 1: True	False	50ms	MMR
4E1	CAN_TX_XCVR_OPERATIONAL	Sensor Status Radiating 0 = not radiating 1 = radiating	0 to 1	n/a	1	0	50ms	MMR
4E1	CAN_TX_INTERNAL_ERROR	Sensor Status Failed 0 = not failed 1 = failed	0 to 1	n/a	1	0	50ms	MMR
4E1	CAN_TX_RANGE_PERF_ERROR	Sensor Status Blocked 0 = not blocked 1 = blocked	0 to 1	n/a	1	0	50ms	MMR

(a)

6.1 Error register 1

Bytes	LUX error	Description	Comment
Bit 0	E-SP	internal error	contact support
Bit 1	E-Motor_1	motor fault	contact support
Bit 2	E-Buffer_1	scan buffer transmitted incompletely	decrease scan resolution/frequency/range, contact support
Bit 3	E-Buffer_2	Scan buffer overflow	decrease scan resolution/frequency/range, contact support
Bit 4	E-Meas_1	APD voltage failed	contact support
Bit 5		reserved	
Bit 6		reserved	
Bit 7		reserved	
Bit8...9	E-Temp	Bit 9: APD Over Temperature Bit 8: APD Under Temperature Bit 8 and 9: APD Temperature Sensor defect	provide cooling provide heating contact support
Bit 10	E-Motor_2	motor fault	contact support
Bit 11	E-Motor_3	motor fault	contact support
Bit 12	E-Motor_4	motor fault	contact support
Bit 13	E-Motor_5	motor fault	contact support
Bit14...15		reserved	

6.2 Error register 2

Bytes	LUX error	Description	Comment
Bit 0	E-IF_internal_1	no scan data received.	contact support
Bit 1	E-IF_internal_2	internal communication error	contact support
Bit 2	E-IF_internal_3	incorrect scan data	contact support
Bit 3	E-Configuration_1	FPGA not configurable	contact support
Bit 4	E-Configuartion_2	incorrect configuration data	load correct configuration values
Bit 5	E-Configuration_3	configuration contains incorrect parameters	load correct configuration values
Bit 6	E-Timeout_1	data processing timeout	decrease scan resolution or scan frequency
Bit 7	E-Timeout_2	reset the computation of the environmental model	contact support
Bit8...15	reserved		

6.3 Warning register 1

Bytes	LUX warning	Description	Comment
Bit0	W-CMD	internal communication error	
Bit1	W-Range_1	internal warning	
Bit2	W-Range_2	internal warning	
Bit3	W-low_temperature	temperature too low	warning of insufficient temperature
Bit4	W-high_temperature	temperature too high	warning of exceeding temperature
Bit5	W-Motor_1	internal warning	
Bit6	W-Motor_2	internal warning	
Bit 7	W-Sync	synchronization error	check synchronization- and scan frequency
Bit7...15	RES 7...15	reserved	

6.4 Warning register 2

Bytes	LUX warning	Description	Comment
Bit0	W-IF_CAN	CAN interface blocked	check CAN bus and CAN connection
Bit1	E-IF_ETH	Ethernet interface blocked	check Ethernet connection
Bit2	W-CANdata	incorrect CAN message received	check CAN data
Bit3	W-IF_internal_1	incorrect scan data	contact support
Bit4	W-ETHdata	unknown or incomplete data	check Ethernet data
Bit5	W-Command	incorrect or forbidden command received	check command
Bit6	W-Flash	memory access failure	restart Ibeo LUX, contact support
Bit7	W-Overflow_1	internal overflow	contact support
Bit8	W-EgoMotion	vehicle data update missing	check CAN vehicle data
Bit9	W-Mounting Position	incorrect mounting parameters	correct mounting position according to OM
Bit10	W-CalcFrequency	no object computation due to scan frequency.	set the scan frequency to 12.5 Hz to receive objects

(b)

FIGURE 9.5 (a) Delphi radar error message; (b) LiDAR IBEO LUX error message (Ibeo LUX, 2014).

To detect errors in CAN signals, it is important to identify the characteristics of CAN signals. The PC LabVIEW signal processing program, a higher control of an AV, can recognize an error in a CAN state. The LabVIEW program can also detect errors in the CAN signal using a virtual instrument (VI).

The upper controller is linked to a PC and to the lower controller (micro-autobox) responsible for control. The PC-to-autobox system sends and receives data over CAN communication in real time, thus allowing the fault-finding system to recognize when a system fails (Lee, Oh, & Yi, 2016).

9.2.2 Sensor Fault Detection and Diagnosis for Autonomous Vehicles

9.2.2.1 Introduction

Many AVs are being tested on public roads to see if they can be safely and reliably operated in real-world situations. Fault-tolerant architectures have been designed for steering, braking, control, and some specific sensor functions that integrate AVs. However, the long-term behavior of many sensors has not been tested, and fault-tolerant perception architectures have not yet been developed (Realpe, Vintimilla, & Vlaci, 2015b).

Fault-tolerant systems should be able to compensate for faults in order to avoid unplanned behaviors (Isermann, 2011). Thus, a fault-tolerant system should have the capability to detect and isolate the presence and location of faults and then reconfigure the system architecture to compensate for those faults (fault recovery).

Several sensor validation methods have been proposed in diverse applications. Some methods produce their own health information using a sensor alone. The sensor readings are usually compared to a pre-established nominal value, and a faulty sensor is declared whenever a threshold value is exceeded. A more common sensor validation method for complex systems is analytical validation, based on information from multiple sensors. An analytical validation requires a model of the system or of the relations between the sensors, executed in parallel to the process. This provides a group of features which are then compared with the system, forming residual values. The residuals that differ from the nominal values are called symptoms and can be subject to a symptom-fault classification to detect a fault and its location (fault diagnosis) (Isermann, 2011).

Model-based methods are categorized as parity equations (Muenchhof, 2005; Chan, 2006), parameter estimation methods (Escobet & Trave-Massuyes, 2001), and observer-based methods with Luenberger observers (Hilbert, 2013) or Kalman filters (Heredia & Ollero, 2009). These methods are very popular for fault-tolerant control systems. Nevertheless, soft computing techniques, such as neural networks, fuzzy logic, evolutionary algorithms, and support vector machines (SVMs), have been developed for fault detection and fault isolation, because it is not always possible to obtain a good model of the system (Meskin & Khorasani, 2011).

Fault diagnosis is based on observed symptoms and experience-based knowledge of the system (Isermann, 2011). One approach is to use classification methods, where the relations between symptoms and faults are determined experimentally in a previous phase of the system. Another approach is to use inference methods, where causal relations are created in the form of rules based on partially known relations between faults and symptoms (Realpe, Vintimilla, & Vlaci, 2015b).

After identifying faults, the system architecture must be reconfigured. Fault recovery can be achieved using direct redundancy or analytical redundancy (Aldridge, 1996). With direct redundancy, a spare module is employed to replace the faulty one. Although direct redundancy is effective and easy to configure, it can be very expensive and unfeasible. Analytical redundancy requires utilizing the working modules to complete the tasks which failed. For instance, if there is a fault in a laser scanner of an AV, the information from two cameras can be used to create range data and compensate for the laser scanner functions (Realpe, Vintimilla, & Vlaci, 2015b).

Only a few specific fault-tolerant perception systems have been developed for AVs. However, many researchers have implemented fault-tolerant modules in AVs in areas such as vehicle navigation sensors. Furthermore, the multi-sensor architecture of navigation systems can be compared with perception systems. In general, two different architectures are applied in navigation systems; centralized architecture, a one-level fusion process with little fault tolerance against soft sensor faults (Lawrence & Berarducci, 1994), and federated architecture, a two-level fusion method with good fault tolerance potential (Realpe, Vintimilla, & Vlaci, 2015b).

Federated architecture was proposed by Carlson (1988) to fuse decentralized navigation systems, with the objective of isolating faulty sensors before their data become integrated into the entire system. This architecture is composed of a group of local filters that operate in parallel and a master filter (Figure 9.6). A fundamental component of the federated filter is the reference sensor. Its data are frequently used to initialize sensors and set preprocessing information in local filters. Consequently, the most reliable and accurate sensor should be chosen as the reference for the local filters (Edelmayer & Miranda, 2007). Xu, Sutton, and Sharma (2007) and Xu and Weigong (2010) implemented a federated Kalman filter in a multi-sensor navigation system and applied a fuzzy logic adaptive technique to adjust the feedback signals on the local filters and their participation in the master filter. Similarly, an expert system was implemented by Fengyang (2007) to adjust the information-sharing coefficients for local filters (for a summary, see Realpe, Vintimilla, & Vlaci, 2015b).

9.2.2.2 Model Description

The Joint Directors of Laboratories (JDL) model is the most widely used model in the data fusion community. The JDL model is integrated by a common bus that interconnects five levels of data processing. A revision of the JDL model, the ProFusion2 (PF2) functional model, applies sensor fusion in

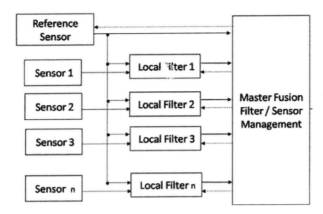

FIGURE 9.6 Federated sensor data fusion architecture (Realpe, Vintimilla, & Vlaci, 2015a).

multi-sensor automotive safety systems. It groups the original levels into three layers to add hierarchical structure. It also establishes inter-level and within-layer interactions, excluding the process refinement (level 4) from the original JDL model; it monitors the overall data fusion process and provides a feedback mechanism to each of the other layers (Realpe, Vintimilla, & Vlaci, 2015b).

There is a third possibility: the reintegration of the process refinement level into the sensor fusion to communicate with all levels, while maintaining the hierarchical structure of the PF2 functional model, as shown in Figure 9.7. In this model, the perception layer provides state estimations of the objects; the decision application layer predicts future situations and deduces the output of potential maneuvers; the action/HMI layer collects and provides information to the user. Meanwhile, the process refinement layer analyzes residuals from all the layers and provides information about faulty states to the decision application layer and feedback to each layer to minimize the effects of faults (Realpe, Vintimilla, & Vlaci, 2015b).

The perception system has been implemented by Realpe (2015a) based on the perception sensors available in the KITTI dataset (a Velodyne sensor and two pairs of stereovision cameras). The federated perception architecture suggested to fuse sensor data from the KITTI dataset is shown in Figure 9.8. The system is divided into different modules: one object detection (OD) for each sensor type, one local fusion (LF) for each support sensor, one master fusion (MF), a tracking module, and a fault detection and diagnosis (FDD) module (Realpe, Vintimilla, & Vlaci, 2015b).

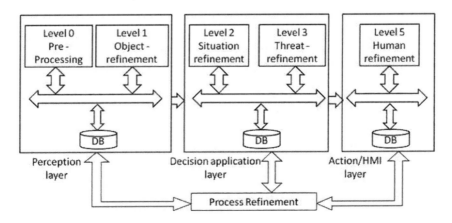

FIGURE 9.7 Data fusion model for fault-tolerant implementation (Realpe, 2015a).

Failure Detection Application

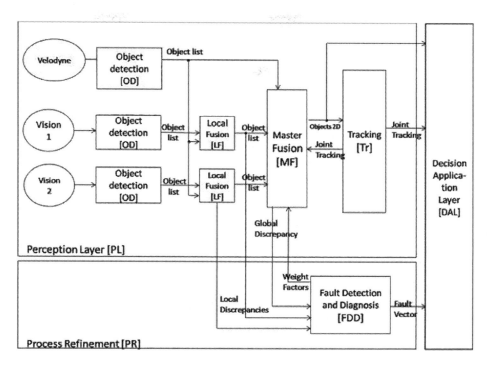

FIGURE 9.8 Fault-tolerant perception system for KITTI dataset (Realpe, 2015a).

9.2.2.2.1 Object Detection (OD) and Local Fusion (LF)

OD and LF are implemented and described by Realpe (2015a). Vision OD processes information from cameras, combining motion detection, the histogram of oriented gradients (HOG) detector, and disparity maps to detect obstacles in the frontal area of the vehicle. Velodyne OD groups the scanned points into objects, according to their distances using the nearest neighbor algorithm based on the Euclidean distance metric. The LF module creates an object list using data from a specific sensor and the reference sensor; it also creates the discrepancy values between those sensors. These values represent the residuals used by the FDD module to determine the presence of a sensor fault (Realpe, Vintimilla, & Vlaci, 2015b).

9.2.2.2.2 Master Fusion (MF)

MF combines data from the reference sensor, LF modules, and the tracking module. First, the lists of objects from all the inputs are fused based on their overlapping areas, creating candidate objects. Then, patterns in the objects' pixels and the weight from each sensor are used to validate pixels in the candidate objects. The MF discrepancy values are estimated by obtaining the difference between the numbers of pixels from the candidate objects' list, the reference sensor, and the fused objects list.

The pixel relationship of the objects is represented by a vector composed of six features. The first three features are Boolean values that represent the origin of the object (reference, LF1, LF2), and the next three features are the distance field values that show the distance of the corresponding pixel to the closest object. Three extra features representing the weight of each sensor are added to create a training vector (Table 9.1). The MF feature vector is trained offline with an SVM algorithm using positive vectors from a group of pixels that have been manually marked as detected objects and using negative vectors selected randomly from the other pixels (no objects) (Realpe, Vintimilla, & Vlaci, 2015b).

9.2.2.2.3 Fault Detection and Diagnosis (FDD)

The FDD module uses a SVM to recognize changes in the discrepancy values for MF and LF modules. The LF discrepancy values are integer numbers representing the percentage of pixels from a fusion module that are present in its associated sensor and the reference sensor. For example, Figure 9.9 shows

TABLE 9.1

MF Feature Vector (Realpe, Vintimilla, & Vlaci, 2015b)

Feature	Value
Reference sensor	True, false
Local fusion 1	True, false
Local fusion 2	True, false
Reference distance field	0–255
Local distance field 1	0–255
Local distance field 2	0–255
Weight reference	High, low, off
Weight vision 1	High, low, off
Weight vision 2	High, low, off

FIGURE 9.9 Discrepancy map from LF (Realpe, 2015a).

the discrepancies in an LF module coded by color: green represents pixels from the Velodyne OD, red represents pixels from the vision OD, and yellow represents pixels present in both. The MF discrepancy is given by the difference between the resulting fused objects and the objects detected by the reference sensor (Realpe, Vintimilla, & Vlaci, 2015b).

An SVM model is created for every sensor. Each model is trained using a vector of nine features, as shown in Table 9.2. The negative vectors are created by introducing a displacement in the calibration matrix of the associated sensor, while the positive vectors are obtained from the unaltered data (Realpe, Vintimilla, & Vlaci, 2015b).

The FDD module is trained to detect faults in a specific sensor. Thus, the system has three different models, one from each sensor, with the faulty sensor obtained directly from a specific model.

TABLE 9.2

FDD Feature Vector (Realpe, Vintimilla, & Vlaci, 2015b)

LF1	Reference
	Vision 1
	Both
LF2	Reference
	Vision 1
	Both
LF3	Reference
	Not reference
	Fused

To avoid false positives, the output from the SVM is considered only if a faulty response is given after N consecutive outputs. Then, the respective sensor is reconfigured to a lower priority (high->low->off) (Realpe, Vintimilla, & Vlaci, 2015b).

9.2.3 Landmark Assistance in Local Positioning Systems

A key tool for automated vehicle localization is the use of landmarks. Landmarks can be natural or artificially added to the vehicle's environment. When natural, the vehicle identifies shapes that are supposed to be there, such as trees, barriers, and posts. Natural landmarks are often complicated to identify, decreasing system reliability. For example, a tree can change shape, or barriers may appear different from different angles. In comparison, artificial landmarks provide an easy, consistent way for the automated vehicle to identify its position. Examples of suitable artificial landmarks include reflectors, radio frequency identification (RF IDS), bar codes, lines on the ground, panels with geometric shapes and colors, lights, and Wi-Fi or infrared (IR) beacons. When the vehicle has identified positioning via landmarks, it can update and track its position over time. In this way, reliable correction of positioning on its map becomes feasible when the vehicle is either in motion or moved in a known environment (Freelance Robotics).

9.2.4 Control of Automated Vehicles

Once the automated vehicle has acquired knowledge of its position and orientation, it can use that information to drive. The vehicle will have a command in speed and bearing to move around. The initial objective is to make the automated vehicle control its motion along a specific AB line. To do so, a proportional integral derivative (PID) is often used to get the distance and orientation difference. Essentially, the PID compares where the automated vehicle is against where it is supposed to be. The aim is to reduce this difference to the smallest amount possible. To this end, the automated vehicle applies a theoretical model to output a command. For example, to change vehicle curvature (angle of trajectory), a command may be given to the actuators driving the vehicle. The PID modifies this command by boosting or damping the model reaction; it considers previous, present, and predicted measurement differences from the desired measurement. This process allows the automated vehicle to drive towards the AB line with minimum overshoot and the quickest line acquisition the vehicle can manage. To optimize this line acquisition, the PID must be tuned (Freelance Robotics).

9.2.5 Automated Vehicle Path Planning

Once the automated vehicle controls motion along a line, the same principle can be easily extrapolated to control from one line to another. The vehicle can thereby create a path with any desired shape to which the vehicle's model can conform. This new line could be too small to be considered a curve; alternatively, Bezier curves may be used to create a path where curvature itself can be used to conform to a curved path. Thus, the vehicle can either follow a preregistered path entered by the user or determine its own path using decision rules (Freelance Robotics).

9.2.6 Obstacle Avoidance

Automated vehicles are required to navigate a number of path modifications for successful independent driving. One of the path modifications an automated vehicle often has to determine by itself occurs when an obstacle gets in the way of the original path. An obstacle could be stationary or in a collision path if it is in motion. To avoid the obstacle, the automated vehicle needs to take into consideration the vehicle model and the environment map; it has to determine the shortest way to recover the initial path considering a minimum clearance of surrounding obstacles. To do so, decision algorithms are used to split the map into a grid. Decision algorithms operate with more or less efficiency; running tree exploration solutions, for example, may help evaluate which is the more or less optimal path around the obstacle (Freelance Robotics).

9.2.7 Conclusion

Putting these together leads to the construction of an automated vehicle. It is the outcome of a series of processes, which loop from the acquisition of environmental data, to data processing and the usage of processed data to control the vehicle. Motion control can occur according to a user predetermined or vehicle-generated path by sending a command to the vehicle actuator. This loop ensures the vehicle moves in the desired direction. Motion changes the environmental data acquired by the sensors, and the series of processes is restarted. Adaptation to the environment following set objectives is fundamental to robotics, as it is to life. This behavior brings machines closer to biological systems, as the vehicle learns by reacting to its environment over time. Vehicles will become smarter, following a more complex decisional pattern and taking into account more information. Automated vehicles will be able to perform tasks of increasing complexity. Looking into the future of human–machine interaction, our vision of the machine can only evolve in tandem with machine complexity. Vehicle automation, from this viewpoint, is a window into the evolution of self and technology (Freelance Robotics).

9.3 Autonomous Vehicle Navigation Security

9.3.1 Autonomous Vehicle Security

9.3.1.1 Introduction

A typical prediction of the future of AVs includes people being relieved of the stress of daily commute driving, perhaps even taking a nap on the way to work. This is expected to be accompanied by a dramatic reduction in driving fatalities because imperfect human drivers will be replaced by (presumably) better computerized autopilots. But how to get such fully automated vehicles to actually be safe is not a simple matter. A number of areas present significant challenges to creating acceptably safe AVs.

The question is not whether AVs will be perfect (they won't). The question is when we will be able to deploy a fleet of fully autonomous driving systems that are actually safe enough to leave humans completely out of the driving loop. The challenges are significant and span a range of technical and social issues for both acceptance and deployment. A holistic solution will be needed and must include a broad appreciation of the range of challenges (and potential solutions) by all relevant stakeholders and disciplines involved.

Even understanding what "safe" really means for AVs is not simple. "Secure" means at least correctly implementing vehicle-level behaviors, such as obeying traffic laws (which can vary depending upon location) and dealing with nonroutine road hazards, such as downed power lines and flooding. But it also means things such as performing fail-over mission planning, finding a way to validate inductive-based learning strategies, providing resilience in the face of likely gaps in early-deployed system requirements, and having an appropriate safety certification strategy to demonstrate that a sufficient level of safety has actually been achieved.

Thus, achieving a safe AV is not something that can be solved with a single technological silver bullet. Rather, it is a coupled set of problems that must be solved in a coordinated, cross-domain manner. As everyone gains more experience with the technology, a few more high-level problems and many more detailed issues will emerge, but this is a starting point for understanding the bigger picture (see Figure 9.10) (Koopman & Wagner, 2017).

9.3.1.2 Security Engineering

Once we have small-scale deployment of fully autonomous Level 4 (NHTSA, 2013; see also Section 9.3.3.1) vehicles, we expect that most vehicles will work well most of the time in everyday on-road environmental conditions. Now we will want to deploy at scale. The challenge becomes managing failures that are very infrequent for any single vehicle but will nonetheless happen too often to be acceptable as exposure increases to millions of vehicles in a fleet.

Failure Detection Application

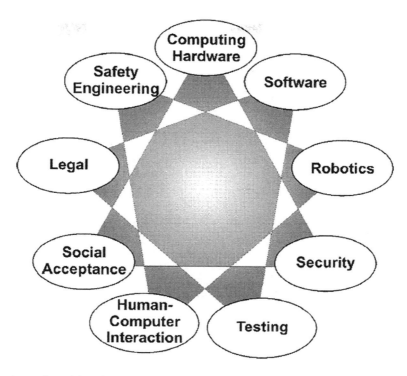

FIGURE 9.10 A coordinated, interdisciplinary approach to security (Koopman & Wagner, 2017).

There is a well-developed body of knowledge on how to make computer-based automotive systems safe and an even longer history of creating safety-critical computer-based systems for trains, chemical process components, aircraft, and so on (Koopman & Wagner, 2017).

Current accepted practice for vehicle computer-based system safety is typically based on a functional security approach, such as the automotive-specific ISO 26262 safety standard. A key issue is that ISO 26262 generally gives a system credit for a human driver ultimately being responsible for safety. But, with full automation, the human won't be responsible for driving at all. Relying on autonomy to be completely responsible for vehicle safety without driver oversight is a huge change compared to current driver-assistance systems that largely rely upon the driver to be responsible for vehicle safety. One approach to handling a lack of human driver oversight is to set ISO 26262's "controllability" aspect to zero for an autonomous system; this could dramatically increase the safety requirements of a variety of automotive functions compared to today's vehicles. Whether this will work or if ISO 26262 can even be used as is to validate AVs are interesting but open questions (Koopman & Wagner, 2016).

A significant security certification concern is validating any use of self-adaptive system behavior by AVs (de Lemos, 2013). Unconstrained adaptation, such as real-time learning of new behaviors (Rupp & King, 2010), means a vehicle can be expected to have a different behavior during operation than during testing and certification. Current certification approaches are unable to handle that situation, because they require considering all possible system behaviors up front in the design and validation process. Unless limits are somehow put on and fully explored during system design, it may be impossible to ensure the safety of such a system because the system being tested won't have the same behavior as a system that is deployed. Formal approaches may be able to prove properties about adaptive systems, but such proofs come with assumptions which are not necessarily provable or testable, and such approaches don't scale well to full-size software systems, at least not yet. Note that by "adaptive," we mean the system changes its behaviors depending upon its operational history by, for example, using online machine learning. This is a much more dynamic range of behaviors than is seen in traditional controls-based systems, such as adaptive cruise control, which can be validated using more traditional methods (Kianfar, Falcone, & Fredriksson, 2013).

A commonly mentioned approach to hedge system-level security for highly autonomous systems is to reengage the driver when there is an equipment failure, providing a human safety net for automation. For example, a human might be taking a nap and will need time to gain enough situational awareness to take over responsibility for driving. To bridge the human inattention gap, the vehicle will need have to have some sort of fail-operational autonomy capability to carry through until a human can regain control.

Fortunately, cars can typically achieve a safe state in seconds (pull to the side of the road), compared to hours for airplanes (fly to a diversion airport). Thus, an effective safety strategy might be for vehicles to change operational modes to a short duration "safe mode" or "safing mission" when a critical primary component fails. The thinking is that an autonomy system that just has to be smart enough to pull a vehicle over to the side of the road over the course of a few seconds might be designed in a less complex way than a full-driving autonomy system (Koopman & Wagner, 2017).

For example, a safing subsystem might stay in the current lane while coming to a stop, and thereby dispense with sensors and control systems needed for lane change maneuvers. Moreover, a short mission time (seconds, not hours) would likely ease reliability and redundancy requirements on the safing subsystem itself. As an added benefit, designing a safe shutdown mission capability may relax the safety requirements on primary vehicle autonomy. If a safing mission is always available, primary autonomy need not be fail operational. Instead, it might be sufficient to ensure that a safing mission is invoked whenever there is a failure of the primary autonomy system, permitting a less-than-perfect primary autonomy system so long as failures are detected quickly enough to invoke a safing mission. Relaxing the safety requirements on primary autonomy (while keeping the vehicle as a whole safe) could potentially offer a dramatic reduction in overall system cost and complexity (Koopman & Wagner, 2016).

9.3.2 Autonomous Vehicle Navigation

9.3.2.1 Introduction

An AV, also known as a driverless vehicle or a self-driving vehicle, is a vehicle capable of fulfilling the human transportation capabilities of a traditional vehicle. As an AV, it can sense its environment and navigate without human input.

AVs sense their surroundings with such techniques as radar, LiDAR, GPS, and computer vision. Advanced control systems interpret sensory information to identify appropriate navigation paths, as well as obstacles and relevant signage. Some AVs update their maps based on sensory input, allowing the vehicles to keep track of their position even when conditions change or when they enter uncharted environments.

For any mobile robot, the ability to navigate in its environment is one of its most important capabilities. In general, the navigation task can be defined as the combination of three basic competences: localization, path planning, and vehicle control. Localization denotes the robot's ability to determine its own position and orientation (pose) within a global reference frame. Path planning defines the computation of an adequate sequence of motion commands to reach the desired destination from the current robot position. Because of its planning component, path planning is typically done before motion. The planned path is followed by the robot using feedback control. This controller includes reactive obstacle avoidance and global path preplanning.

The potential application areas of the autonomous navigation of mobile robots include automatic driving, guidance for the blind and disabled, exploration of dangerous regions, transporting objects in factory or office environments, collecting geographical information in unknown terrains like unmanned exploration of a new planetary surface, and many more (Dhanasingaraja, Kalaimagal, & Muralidharan, 2014).

9.3.2.2 Methodology

The location of the vehicle is uploaded to the server through general packet radio service (GPRS). The system takes the current position as the source and gets the destination point. At the server, the latitude and longitude coordinates are obtained and displayed in the Google map for monitoring purposes; this

Failure Detection Application

means the vehicle can be monitored from anywhere in the world. The system finds the shortest path to the destination and sends the information to the vehicle. The vehicle follows the coordinates using GPS and compass. If a GPS signal is not received, an inertial navigation system (INS) is used to obtain the coordinates. Obstacles around the vehicle are sensed by a laser range finder.

A block diagram is shown in Figure 9.11 (Dhanasingaraja, Kalaimagal, & Muralidharan, 2014).

9.3.2.2.1 Waypoint Extraction

Standardized geodata from the OpenStreetMap (OSM) project can be used for autonomous robot navigation. Founded in July 2004, OSM is a collaborative project which aims to create a free-to-use and editable map of the world. Unlike commercial map distributors like Google, Navteq, and Teleatlas, the OSM map is a public domain and created by volunteers performing systematic ground surveys with a handheld GPS receiver.

The OSM map allows the user to select the vehicle's navigation path. Waypoints along the path are extracted and send to the vehicle through Internet. A basic visual application is designed to perform this task (Dhanasingaraja, Kalaimagal, & Muralidharan, 2014).

9.3.2.2.2 Perception

The perception system is responsible for providing a model of the world to the behavioral and motion planning subsystems. The model includes the moving vehicles and static obstacles; it localizes the vehicle relative to and estimates the shape of the roads it is driving on. A rotatable laser range finder is used to sense the obstacles around the vehicle (Dhanasingaraja, Kalaimagal, & Muralidharan, 2014).

9.3.2.2.3 Localization

Localization is the estimation of a robot's position relative to its environment using sensor observations. Localization is a necessity for successful mobile robot systems; it has been called "the most fundamental problem to providing a mobile robot with autonomous capabilities." To achieve autonomous navigation, the robot must maintain accurate knowledge of its position and orientation. Successful achievement of all other navigation tasks depends on the robot's ability to know its position and orientation accurately. GPS and inertial sensors are commonly used for localization.

A GPS receiver calculates its position by precisely timing the signals sent by GPS satellites high above the Earth. Each satellite continually transmits messages that include:

1. Time the message was transmitted
2. Satellite position at time of message transmission.

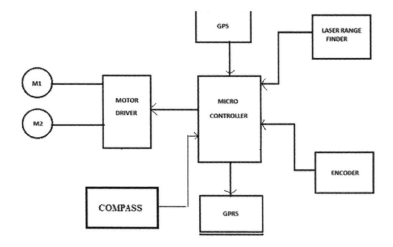

FIGURE 9.11 Block diagram of autonomous navigation (Dhanasingaraja, Kalaimagal, & Muralidharan, 2014).

The receiver uses the messages it receives to determine the transit time of each message and computes the distance to each satellite using the speed of light. Each of these distances and satellite locations defines a sphere. The receiver is on the surface of each of these spheres when the distances and the satellite locations are correct. These distances and satellite locations are used to compute the location of the receiver using the navigation equations. This location is then displayed, perhaps with a moving map display or latitude and longitude; elevation information may be included. Many GPS units show derived information, such as direction and speed, calculated from position changes.

GPS receivers also work on NMEA Standards. Every sentence begins with a "$" sign, has about 80 characters, and ends up with a carriage return/line feed. Sentences are mostly framed in single lines (occasionally multiple lines), and the data items in each sentence are separated by commas. The data received are in ASCII text, and they vary in precision. A sentence ends with checksum which consists of a "*" and two hexadecimal digits. The checksum digits represent either an eight-bit exclusive or all characters between, but not including, the $ and *.

Latitude, longitude, and number of satellites in view are extracted from the GPS data. Accuracy of the information depends on the number of satellites in view. As noted above, if the data are not accurate, localization is obtained from inertial sensors. Inertial localization works by keeping track of the positions a vehicle has been in. It uses the wheel encoder and compass to sense the robot's movements (Dhanasingaraja, Kalaimagal, & Muralidharan, 2014).

9.3.2.2.4 Navigation

The algorithm for the navigation of an autonomous robot is written below and depicted in Figure 9.12. The system gets the current position from GPS and destination coordinates from the user. It finds the shortest path between the present position and the destination. It selects and travels on the shortest path. If it senses any obstacles in the path, it selects a new path based on the next shortest path. This process is continued until the destination is reached. Once the destination is reached, the vehicle stops. At the same time, the system sends its current position coordinates to the part of the system doing the monitoring (Dhanasingaraja, Kalaimagal, & Muralidharan, 2014).

$$X = \text{Long}_{goal} - \text{Long}_{vehicle}$$

$$Y = \text{Lat}_{goal} - \text{Lat}_{vehicle}$$

$$\theta_{goal} = \tan^{-1}(y/x)$$

$$\text{Angle}(\theta) = \text{Goal}_{angel} - \text{Vehicle}_{angle}$$

where
   X: Distance to be traveled in latitude
   Y: Distance to be traveled in longitude

If the path between the goal and the vehicle is not a straight line, small goal points are considered at the various turns along the path (Figure 9.13) (Dhanasingaraja, Kalaimagal, & Muralidharan, 2014).

9.3.2.2.5 Mapping

The system allows a user to view the present and past positions of a target object on Google Map through the Internet. The system reads the current position of the object using GPS, and the data are sent via GPRS service from the GSM network to a Web server using the POST method of the HTTP protocol (see Figure 9.14) (Dhanasingaraja, Kalaimagal, & Muralidharan, 2014).

The object's position data are then stored in the database for live and past tracking. A Web application has been developed using PHP, JavaScript, Ajax, and MySQL with the Google Map embedded. Third-party server Cosm Pachube (now known as Xively) was used for the system shown in the figure (Dhanasingaraja, Kalaimagal, & Muralidharan, 2014).

Failure Detection Application

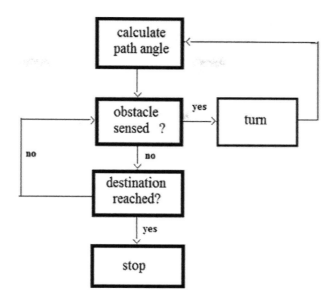

FIGURE 9.12 Depiction of navigation algorithm (Dhanasingaraja, Kalaimagal, & Muralidharan, 2014).

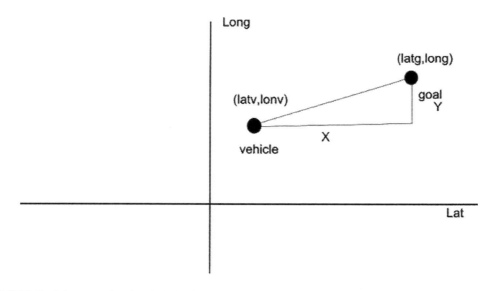

FIGURE 9.13 Distance and angle calculation for AV navigation (Dhanasingaraja, Kalaimagal, & Muralidharan, 2014).

9.3.3 Autonomous Vehicle: Security by Design

9.3.3.1 Introduction

Modern cars are fitted with a range of autonomy features. In order to distinguish cars with varying degrees of autonomy in a consistent manner, SAE International (Society of Automotive Engineers) proposed six levels of autonomy in the standard J3016 (SAE, 2016a), with Level 0 meaning no automation and Level 5 full automation. From Levels 0 to 2, a human driver monitors the driving environment, whereas from Levels 3 to 5, the driving system monitors the driving environment. A tabular description of the capabilities associated with the different levels of autonomy is provided in Table 9.3 (Chattopadhyay & Lam, 2018).

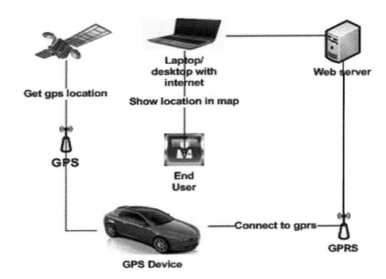

FIGURE 9.14 POST method of HTTP protocol (Dhanasingaraja, Kalaimagal, & Muralidharan, 2014).

TABLE 9.3

AVs: Levels of Autonomy (S, System; H, Human) (Chattopadhyay & Lam, 2018)

Level	Automation	Steering Cruise	Environment Monitoring	Fallback Control	Driving Modes
0	None	H	H	H	N/A
1	Supportive	H,S	H	H	Some
2	Partial	S	H	H	Some
3	Conditional	S	S	H	Some
4	High	S	S	H	Some
5	Full	S	S	S	All

Understandably, with a higher degree of autonomy, the security risks escalate. From Level 3 onwards, the car must be fit with an increased number of sensing and communicating devices to be "self-aware." AVs could be considered vehicles with autonomy features at Level 3 and above (Chattopadhyay & Lam, 2018).

A connected AV is subject to cyberattacks through its various network interfaces to the public network infrastructure, as well as its direct exposure to the open physical environment. An attack surface of a system is the sum of the different attack vectors, that is, the different points where attackers can make attempts to inject data to or extract data from the system to compromise the security control of the AV. Figure 9.15 depicts the typical attack surfaces of an AV (Intel IoT, 2016) and potential attack sources. As the figure shows, the attack sources are typically external agents/events but can also be internal components with malicious intent that attempt to compromise the expected autonomy functionality of the AV. For example, the Bluetooth interface of the AV shown in Figure 9.15 can be considered a potential attack surface that can be compromised by plugging malicious devices (attack source) into this communication channel (Chattopadhyay & Lam, 2018).

9.3.3.2 Security by Design for Autonomous Vehicles

This section describes the security assumptions, requirements, threats, and control measures of AVs and shows how these are handled by the so-called security-by-design approach advocated by the Open Web Application Security Project (OWASP). Specifically, we describe the key steps of security-by-design in

Failure Detection Application

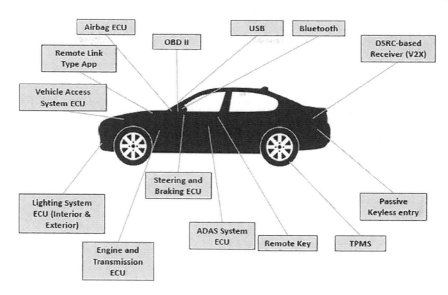

FIGURE 9.15 Potential attack sources and surfaces in cars (Intel IoT, 2016).

the context of AV by viewing it as a socio-technical system. Designing security for complex systems such as AVs typically involves the following steps (Chattopadhyay & Lam, 2018):

1. Identify the security objectives and requirements of the system.
2. Assess the value and sensitivity of the system to be protected.
3. Define security policy in accordance with the security requirements.
4. Estimate the capabilities of the adversaries.
5. Design control features commensurate with the sensitivity of the system and the risks it is exposed to.

To set the security-by-design process within a sound framework, it is important to establish the operational model of the AV so that security objectives and requirements may be analyzed in a holistic and systematic manner (Chattopadhyay & Lam, 2018).

9.3.3.2.1 AV Operation Model

For simplicity, the following basic operational model of an AV is assumed (Chattopadhyay & Lam, 2018):

- Communication: AV periodically sends operation logs to the manufacturer to allow life cycle management and maintenance.
- Communication: AV has wireless or wired interfaces to support firmware/software update/upgrade at maintenance and service workshop.
- Communication: AV supports communication with some traffic management system infrastructure for traffic flow control, as well as remote control in case of emergency.
- Sensing: AV is equipped with a variety of sensors to sense the physical environment and detect collisions.
- Decision: AV has a number of navigation-related control functions that allow Level 5 autonomy; i.e., it requires real-time updating of travel routes, enables intelligent route planning, and provides automatic steering in accordance with road conditions.
- Decision: AV has a number of safety-related control functions that allow Level 5 autonomy; i.e., it enables automatic steering, speed regulation, and braking in accordance with road conditions.

9.3.3.2.2 Security Objectives and Requirements of AV

Based on this model, the basic security objectives of an AV can be listed as follows (Chattopadhyay & Lam, 2018):

- Integrity of the remote control functions of the AV (possibly as an emergency operation from the traffic management system) so that no attacker is able to take control of the AV by tampering with the remote control system.
- Integrity of the sensor systems so that navigation and safety-related control features will not be interfered with by attackers tampering with the sensor data.
- Integrity of the safety-related control operations, such as braking and speed control, in accordance with the sensed road conditions or based on remote control instructions.
- Integrity of the navigation-related control operations, such as steering, braking, and speed control, in accordance with the sensed road conditions or based on a preprogrammed route path.
- Confidentiality of communications between AV and traffic management system so that safety-related control parameters will not be disclosed to unauthorized parties who may attempt to exploit the AV.
- Confidentiality and integrity of communications between AV and its manufacturer so that the robustness of the life cycle management and maintenance of the AV can be assured.
- Integrity and authenticity of communications between AV and maintenance workshop so that software patches and updates to the AV can be performed with high assurance.
- Confidentiality of cryptographic keying materials stored inside the AV so that attackers cannot bypass higher-level security mechanisms by siphoning the cryptographic keys.

9.3.3.2.3 Safety Standards for AV

An AV should be modeled as a socio-technical system where safety is of utmost importance because human lives are at stake. For AVs, safety-criticality directly leads to the criticality of cybersecurity, not the other way around. Hence, the AV security objectives are based on, and derived from, the relevant AV safety standards. The present safety standards for AVs are presented in the following (Chattopadhyay & Lam, 2018):

- ISO 26262 (ISO 26262 – 1: 2011): This standard is derived from IEC 61508 (International Electrotechnical Commission (IEC), 2010), which was developed for all electrical/electronic safety-related systems. ISO 26262 is specifically targeted to automotive safety. The standard provides a safety life cycle throughout all phases, including management, development, production, operation, service, and decommissioning. ISO 26262 also defines the Automotive Safety Integrity Level (ASIL). ASIL includes severity classifications (S0 – S3), exposure classifications (E0 – E4), and controllability classifications (C0 – C3) to quantify the severity of an injury, the probability of occurrence, and the controllability of a situation, respectively. ASIL is expressed as follows:

$$ASIL = Severity \times Exposure \times Controllability$$

  where the higher level of ASIL indicates a more grievous situation.
- In the context of AV, it can be noted that the controllability level is extremely high for Levels 3–5. Hazard analysis and risk assessment (HARA), fault tree analysis (FTA), and failure mode and effects analysis (FMEA) can be used to assess the ASIL.
- SAE J3016 (SAE International, 2016a): Recognizing the specific need for a standard in the wake of cybersecurity incidents, the SAE decided to move together with ISO to define the standard J3016. While it is based on ISO 26262, it identifies the growing threat landscape and tries to establish awareness and a common terminology across the AV supply chain. It establishes the terminology of threat (malicious attacker), vulnerability (unguarded gateway), and

Failure Detection Application

risk (likelihood of attack). Most importantly, J3016 delineates the scope of cybersecurity by stating that cybersecurity-critical systems are not necessarily safety-critical, but the reverse is true. It also emphasizes the distinction between system safety (fault/accident) and system cybersecurity (purposeful attack).

Figure 9.16 shows the AV security objectives and requirements, together with the relevant AV safety and security policies and standards.

9.3.3.2.4 Adversarial Models for AV Security

The following adversarial capabilities were defined for AVs by Chattopadhyay and Lam (2018).

- Property 1: An adversary is capable of intercepting and tampering with all inter-vehicle and intra-vehicle communication.
- Property 2: An adversary is capable of introducing malicious nodes into the inter-vehicle and intra-vehicle communication network.

These adversarial models have been documented in several attacks (Chattopadhyay & Lam, 2018).

9.3.3.2.5 System Security Model for AVs

The aforementioned standards can act as the original guidelines, but there is a lack of consistency in the approaches used to identify attack surfaces, identify threats, and assess risk. There is no single standard applicable to AV security because of its complex overlap across multiple technology domains, including wireless communication, electronics (ISO 26262 – 1: 2011), mechanical systems, and software development (ISO/IEC 27034, 2011).

Figure 9.17 presents a high-level model for security by design of AV. The model divides the attack surfaces of the AV into three layers (Chattopadhyay & Lam, 2018):

1. Core layer, defined by the physical enclosure of the AV.
2. Interface layer, or AV gateway layer, characterized by the collection of connectivity interfaces between the AV and the external world.
3. Infrastructure layer, composed of all the infrastructure and backend modules trusted by and connected to the AV.

The security issues are clearly distributed over these three layers: first, the AV system; second, the AV gateway; and third, the vehicle-to-vehicle (V2V)/vehicle-to-infrastructure (V2I) communications.

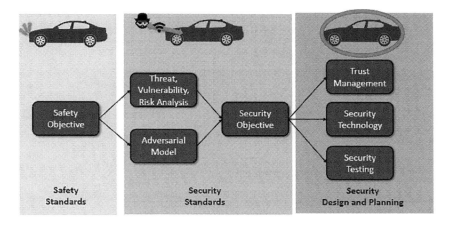

FIGURE 9.16 AV security design flow (Chattopadhyay & Lam, 2018).

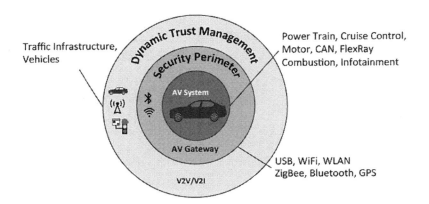

FIGURE 9.17 AV security by design (Chattopadhyay & Lam, 2018).

As it is defined by the physical enclosure of the AV, the core layer enforces its perimeter through physical security.

The AV gateway layer consists of diverse forms of communication links, including Bluetooth, Wi-Fi, and ZigBee; these links help establish and maintain communication with external vehicles and infrastructures. The gateway layer defines a network security perimeter, which is enforced by network access control mechanisms to guard against unauthorized access to the internal functionalities of the AV. The layer relies on the assumed physical security of the core layer so that any damage done to this vehicle within this layer through physical means is not considered an AV security hazard because of cyberattacks.

The outermost layer contains all external systems interacting with the AV, including vehicles, traffic infrastructures, and cloud-based navigation service providers, for example. Note that not all of these external systems can be categorized as a trusted party. Therefore, a dynamic trust management approach has to be undertaken at this layer. A threat arises when a trusted component communicates malicious packets (e.g., malware during software update) or when an untrusted component is able to bypass the secure gateway (e.g., by compromising the vehicle's sim card). Note that the AV system does not necessarily fall prey to an attacker if the security perimeter is breached. For example, Audi A8 maintains a network layout (Peters, 2017), where the wireless infrastructure is kept away from the internal AV network via a secure gateway. Thus, a cybersecurity incident may occur, but safety will not be compromised if the safety-related features are not accessible from the wireless gateway (Chattopadhyay & Lam, 2018).

9.3.4 Accurate and Secure Navigation for Autonomous Vehicles

The navigational systems of AVs integrate positioning signals from a variety of sources, each of which may have certain limitations. Internal navigation systems (INSs) are prone to accumulated uncertainty errors when operating alone but are often paired with other systems. Added sensors, such as cameras, lasers, and sonar, may be too large or expensive in some cases. GPS signals can be lost in dense urban environments or altered by a malicious attack, so other sensor-less and GPS-independent signals are needed to ensure and improve autonomous navigation (UC Riverside, 2017).

9.3.4.1 Use of Cell Phones as Navigation Systems: A Brief Description

While cellular phone networks are not designed for navigation, they are abundant in urban environments which are known to challenge GPS signals. At the University of California, Riverside, researchers integrated signals of opportunity from mobile phone networks to provide AVs with precise navigational information (UC Riverside, 2017).

9.3.4.2 Advantages

- Low cost and lightweight: Use of cellular phone signals instead of bulky sensors or cameras.
- Accurate: Real-time vehicle demonstrations achieved positioning estimation with reliable submeter-level accuracy (UC Riverside, 2017).

9.3.4.3 Applications

- Autonomous/manned vehicles: Aerial, terrestrial, naval.
- Navigational and positional systems: For example, telematics/remote, smartphones, wearables, location-based services, surveillance, mapping, precision localization (UC Riverside, 2017).

9.3.5 Towards Navigation Security for Autonomous Vehicles

Promising new technology has recently emerged to increase the level of safety and autonomy in driving, including lane and distance keeping assist systems, automatic braking systems, and highway auto-drive systems. Each of these technologies brings cars closer to the ultimate goal of fully autonomous operation. While it is still unclear if and when safe driverless cars will be on the mass market, a comparison with the development of aircraft autopilot systems can provide valuable insight. This section investigates a path towards ensuring safety for self-driving or autonomous cars by leveraging prior work in aviation. It focuses on navigation, or localization, a key aspect of automated operation.

Highly automated vehicles (HAVs) can make driving more fuel and time-efficient. They can significantly reduce traffic congestion and emissions by driving at a precise speed, minimizing lane changes, and maintaining an exact distance to neighboring cars. They can also increase accessibility and mobility for disabled and elderly persons.

Sharing an HAV instead of owning one is projected to dramatically reduce a household's yearly transportation budget, which currently ranges between approximately $8,000 and $11,000 per car in the United States. HAVs promise improved road mobility and accessibility and may generate architectural and societal changes that will make mass parking spaces and personal car ownership obsolete in urban areas. Above all, HAVs can improve road safety by preventing car accidents that cause more than 30,000 deaths/year in the United States alone, cost approximately $230 billion/year in medical and work loss costs, and are caused by humans 90% of the time.

In the 1950s and 1960s, many predicted that autonomous cars and "electronic highways" would become widely available by 1975. While this obviously didn't happen, major milestones in the use of new sensors, computation, and communication technologies have recently reenergized hope. One such milestone was the 2005 "DARPA Grand Challenge," where four different HAVs designed by teams of engineers from industry and academia completed a 132-mile trip across the Mohave Desert in less than 7.5 h with no human intervention. The 2007 DARPA "Urban Challenge" saw six teams autonomously complete a 60-mile course in an urban environment, while following traffic laws (see Chapter 1). Most teams used a combination of LiDAR, cameras, differential GPS, and computation power that was multiple orders of magnitude higher than is typically needed for a commercial passenger vehicle. In 2009, Google began designing and testing "self-driving" cars, which have since accumulated more than 3 million miles in autonomous mode (see Chapter 1).

Today, most car manufacturers have HAV prototype systems and HAV pilot testing programs, including fully autonomous systems for public transportation, which, for now, are confined to segregated lanes and geofenced areas. Multiple Tier-2 supplier companies specializing in autonomous car technology have also emerged. In early 2017, 36 companies were registered to test prototype HAV systems on public roads in the state of California.

In Figure 9.18, Gartner's "2016 hype cycle for emerging technologies" shows that HAV technology might be at the "peak of inflated expectations," approaching the "trough of disillusionment." Hype cycle curves are nonscientific tools that have been empirically verified for multiple example technologies over many years. Two emerging technologies, commercial unmanned aircraft systems (UASs) and virtual reality, are included in Figure 9.18 for illustrative purposes. The curve's time scale may differ for each technology.

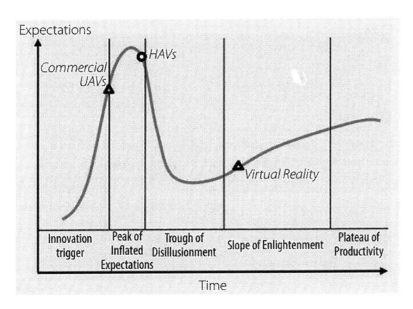

FIGURE 9.18 Gartner's "hype cycle for emerging technologies" (Gartner, 2016).

One of many indicators of decreasing expectations of HAVs is a reduction in press coverage and the emergence of the first negative news stories, especially following the May 2016 crash of a Tesla Model S whose autopilot failed to distinguish a white trailer truck from the bright Florida sky. The Model S ran under the trailer, tearing its roof off and killing the operator. The car kept going full speed through two fences until it hit a pole and came to a stop (Joerger & Spenko, 2017).

In parallel, until the end of 2016, Google was providing detailed reports of its self-driving car performance, which were designed to operate in real-world urban environments. These reports contain records of millions of miles driven autonomously, but they also acknowledge "disengagements", i.e., where the operator needed to take control to avoid collisions. The data show HAVs are much more likely to be involved in collisions, even though these collisions are often less severe than collisions caused by conventional human driving (HAVs typically get rear-ended because of their unusual road behavior). Uber's autonomous taxis in Pittsburgh have a reported rate of one disengagement per mile autonomously driven.

Another troubling sign is that the first fielded autonomous systems have revealed new safety threats. In particular, the technology's functionality, as perceived by the human operator, does not always match the intended operational domain: for example, there have been cases of highway autopilots passing red lights without slowing down.

Human–machine interaction is at the heart of role confusion (is the operator or the HAV in charge?), of mode confusion (is the HAV in autonomous or manual mode?), and of the operator's trust in this multimodal system. Misinterpretation may grow because a given functionality will not achieve the same level of performance across models and manufacturers, and operators may not be aware of the systems' independently verified safety ratings. And within the next few years, operators will be expected to anticipate hazardous situations and take over the control. This means they may need more education and different training (Joerger & Spenko, 2017).

9.3.5.1 HAV Taxonomy

Creating a path to successful automated navigation requires an overall methodology to prioritize the imminently achievable objectives and then expand to more challenging missions. A classification using the six SAE autonomy levels mentioned previously appears in Table 9.4. This classification is refined by segmenting a car's trip into basic driving competencies and by specifying the conditions under which a given HAV should achieve these competencies. A similar classification was made in the early days of

TABLE 9.4
SAE International Levels of Driving Automation (SAE International, 2016b)

SAE Level	Name	Description
		Human Driver Monitors the Driving Environment
0	No automation	The human driver performs all driving tasks at all times.
1	Driver assistance	Either steering or acceleration/deceleration task by the system; driver expected to perform all other aspects of driving.
2	Partial automation	Both steering and acceleration/deceleration tasks by the system; driver expected to perform all other aspects of driving.
		HAV Monitors the Driving Environment
3	Conditional automation	The HAV performs all driving tasks under limited, predefined conditions, and can request the human driver to intervene and take over control.
4	High automation	The HAV performs all driving tasks under limited, predefined conditions, without the expectation of any human intervention.
5	Full automation	The HAV performs all driving tasks under all conditions.

GPS-based commercial aircraft navigation safety analysis, with distinctions made between different phases of flight, weather conditions, vehicle equipment, and airport infrastructure capabilities (Joerger & Spenko, 2017).

For example, in the early 1990s, 40% of aircraft accidents occurred during final approach and landing and 26% during takeoff and initial climb, representing an average of 4% and 2% of flight time, respectively. The Federal Aviation Administration (FAA), therefore, concentrated its efforts on improving safety during these phases of flight. GPS augmentation systems were designed, with varying capabilities, depending on airborne equipment and airport infrastructure, to guide the aircraft under the cloud ceiling or to bring it all the way to touchdown. Similarly, the "first and last mile" are identified as the most challenging parts of HAV operations. In its 2016 Federal Automated Vehicles Policy, NHTSA identifies 28 HAV behavioral competencies which are particularly challenging to meet in the first and last miles of a typical trip. These competencies are basic abilities that an HAV must have to complete nominal driving tasks; they include, for example, lane keeping, obeying traffic laws, and responding to other road users.

To better describe an HAV's ability, the Federal Automated Vehicles Policy specifies that basic driving competencies should be available under an HAV's predefined operational design domain (ODD), described by its geographical location, road type and condition, weather and lighting conditions, vehicle speed, etc. The ODD captures the circumstances under which an HAV is supposed to operate safely.

Such classification is key to safety analysis. The classification can also help identify geographical areas where improved road infrastructure is needed for automated operation, similar to airports requiring equipment for instrument navigation to deal with higher traffic density.

Standards for electronic equipment, measured by ASILs, have been issued and compared with aviation's Design Assurance Levels (DAL). Overall system safety levels have been codified, which in aviation account for both the severity and probability of occurrence of an incident, and in automotive applications account, in addition to these, for "controllability," a measure of how likely an average driver is to maneuver out of a given imminent danger (Joerger & Spenko, 2017).

The following elements must be specified to carry out a formal HAV safety analysis (Joerger & Spenko, 2017):

a. HAV autonomy level
b. Basic driving competency
c. Operation design domain
d. Vehicle electronic equipment
e. Overall safety risk requirement.

Still missing are clear guidelines, or sample methods, on how to implement these safety requirements (Joerger & Spenko, 2017).

9.3.5.2 Path Towards HAV Navigation Safety

When quantifying the safety of HAV navigation systems, such as in the example displayed in Figure 9.19, every component of the system, including raw sensors, estimator and integrity monitor, and safety predictor, can potentially introduce risk. Unlike aircraft, HAVs require multiple and varied sensors to compensate for GPS signal blockages caused by buildings and trees. These sensor types must be integrated, and new methods to evaluate the integrity of multi-sensor systems must be developed. Furthermore, HAVs must have the ability to continuously predict integrity in a dynamic HAV environment (Joerger & Spenko, 2017).

In general, research on the analytical evaluation of HAV navigation safety is sparse. For example, Lee et al. (2015) use the concept of a "safe driving envelope," but they focus mostly on collision avoidance. A paper by Le Marchand, Bonnifait, Guzman, and Btaille (2009) evaluates ground vehicle navigation but shows an "approximate radial-error" of tens of meters, far exceeding the necessary submeter alert limit. A multi-sensor augmented GPS/inertial measurement unit (IMU) system is used by Toledo-Moreo, Zamora-Izquierdo, Beda Miarro, and Gómez-Skarmeta (2007) with "horizontal trust levels" of 7–10 m, still an order of magnitude higher than the required HAV alert limit.

Multi-sensor integrity is addressed by Brenner (1996) but for a sensor combination specific to aviation and insufficient for terrestrial mobile robots. Other approaches to multi-sensor integration show promise but do not provide rigorous proof of integrity. In fact, most publications use estimation error covariance as a measure of performance, which is understood as not being sufficient but is the only metric currently available. Most critically, the metric does not account for fault modes introduced by feature extraction and data association, two algorithms commonly used in mobile robot localization.

Unlike GPS, which gives absolute position fixes, IMUs, LiDAR, radar, and cameras provide relative displacements with respect to a previous time step or with respect to a map. Thus, measurement time filtering is required; this makes integrity risk evaluation more challenging, as past-time sensor errors and undetected faults can now impact current-time safety (Joerger & Spenko, 2017).

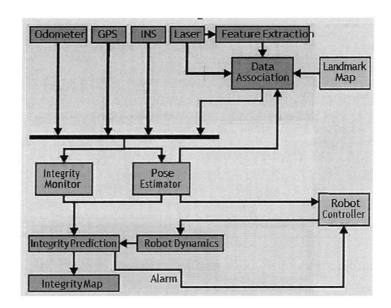

FIGURE 9.19 HAV navigation system. Key sensors are represented on top. Their measurements are processed to estimate HAV position, velocity, and orientation and then to predict safety risk and send alerts if needed (Joerger & Spenko, 2017).

REFERENCES

Aldridge H. A., 1996. Robot position sensor fault tolerance. *Ph.D.* 9713717, Carnegie Mellon University, Pittsburgh, PA, USA, 1996.

Brenner M., 1996. Integrated GPS/inertial fault detection availability. *Navigation*, 43(2), 111–130.

Carlson N. A., 1988. Federated filter for fault-tolerant integrated navigation systems. In *Position Location and Navigation Symposium, 1988. Record. Navigation into the 21st Century. IEEE PLANS '88*, Orlando, FL, USA. IEEE, 1988, pp. 110–119.

Chan C. W., 2006. Application of fully decoupled parity equation in fault detection and identification of DC motors. *IEEE Transactions on Industrial Electronics*, 53, 1277–1284.

Chattopadhyay A., Lam K.-Y., 2018. Autonomous vehicle: Security by design. October 01, 2018.

Chowdhary G., Johnson E. N., Kimbrell M. S., 2013. Guidance and control of an airplane under severe structural damage. In *AIAA Infotech@Aerospace 2010*, Atlanta, GA, USA: American Institute of Aeronautics and Astronautics, Georgia Institute of Technology, July, 2013.

Cork L. R., Walker R., Dunn S., 2005. Fault detection, identification and accommodation techniques for unmanned airborne vehicle. In *Australian International Aerospace Congress (AIAC)*, Melbourne, Australia. January 2005.

DeBusk W. M., Chowdhary G., Johnson E. N., 2009. Real-time system identification of a small multi-engine aircraft with structural damage. In *AIAA Infotech@Aerospace 2010*, Atlanta, GA, USA: Georgia Institute of Technology, 10–13 August, 2009.

de Lemos R. 2013. Software engineering for self-adaptive systems: A second research roadmap. In R. de Lemos, H. Giese, H. A. Müller, & M. Shaw (Eds.), *Software Engineering for Self-Adaptive Systems II* (pp. 1–32). Lecture Notes in Computer Science, Vol. 7475. Berlin, Heidelberg: Springer.

Dhanasingaraja R., Kalaimagal S., Muralidharan G., 2014. Autonomous vehicle navigation and mapping system. *International Journal of Innovative Research in Science, Engineering and Technology*, Vol. 3, Special Issue (3), pp. 1347–1350. 2014 International Conference on Innovations in Engineering and Technology (ICIET'14). On 21st & 22nd March Organized by K.L.N. College of Engineering and Technology, Madurai, Tamil Nadu, India.

Ducard G. J. J., 2009. *Fault-Tolerant Flight Control and Guidance Systems: Practical Methods for Unmanned Aerial Vehicles*. Switzerland: Springer.

Edelmayer A. and Miranda M., 2007. Federated filtering for fault tolerant estimation and sensor redundancy management in coupled dynamics distributed systems. In *Mediterranean Conference on Control & Automation, 2007 (MED '07)*, Athens, Greece, 2007, pp. 1–6.

Escobet T., Trave-Massuyes L., 2001. Parameter estimation methods for fault detection and isolation. Bridge Workshop Notes, 2001.

Fengyang D., 2007. Study on fault-tolerant filter algorithm for integrated navigation system. In *International Conference on Mechatronics and Automation (ICMA 2007)*, Harbin, China, 2007, pp. 2419–2423.

Freelance Robotics. Overview of techniques and applications for autonomous vehicles. Embracing the 4th Industrial Revolution. www.freelancerobotics.com.au/technological-articles/overview-techniques-applications-autonomous-vehicles/. Viewed: July 24, 2019.

Gartner, 2016. Hype cycle for emerging technologie. www.gartner.com/newsroom/id/3412017.

General Motors (GM), 2018. Self-driving safety report, 2018.

Heredia G., Ollero A., 2009. Sensor fault detection in small autonomous helicopters using observer/Kalman filter identification. In *IEEE International Conference on Mechatronics (ICM 2009)*, Malaga, Spain, 2009, pp. 1–6.

Hilbert M., 2013. Observer based condition monitoring of the generator temperature integrated in the wind turbine controller. In *EWEA 2013 Scientific Proceedings*, Vienna, Austria, 4–7 February 2013, pp. 189–193.

Ibeo LUX, 2014. Ibeo LUX and ibeo LUX systems CAN protocol description, ibeo automotive, 2014.

Intel IoT, 2016. Intel automotive security research workshops, 2016. www.intel.com/content/dam/www/public/us/en/documents/product-briefs/automotive-security-research-workshops-summary.pdf. Viewed: July 24, 2019.

International Electrotechnical Commission (IEC), 2010. Functional safety and IEC 61508, 2010. www.iec.ch/functionalsafety/. Viewed: July 24, 2019.

Isermann R., 2011. *Fault-Diagnosis Applications: Model-Based Condition Monitoring: Actuators, Drives, Machinery, Plants, Sensors, and Fault-tolerant Systems*: Berlin, Heidelberg: Springer.

ISO 26262 – 1: 2011. Road vehicles, 2011. *Functional Safety, International Standard*, First Edition. Switzerland: ISO.

ISO/IEC 27034:2011+ – Information technology Security Techniques Application security, 2011. www.iso27001security.com/html/27034.html.

Jeong Y. H., Lee K. J., 2015. Virtual sensor based vehicle sensor fault tolerant algorithm. KSME, pp. 1761–1766.

Joerger M., Spenko M., 2017. Towards navigation safety for autonomous cars. HAV Safety. Inside GNSS, November/December, 2017.

Kianfar R., Falcone P., Fredriksson J., 2013. Safety verification of automated driving systems. *IEEE Intelligent Transportation Systems Magazine*, Winter 2013, 5(3), 73–86.

Koopman P., Wagner M., 2016. Challenges in autonomous vehicle testing and validation. *SAE International Journal of Transportation Safety*, 4(1), 15–24.

Koopman P., Wagner M., 2017. Autonomous vehicle safety: An interdisciplinary challenge. *IEEE Intelligent Transportation Systems Magazine*, Spring 2017, 9(1), 90–96. Accepted Manuscript version. Official published version doi:10.1109/MITS.2016.2583491.

Oh K. S., 2017. Kinematic model-based predictive fault diagnosis algorithm of autonomous vehicles using sliding mode observer. *KSME*, 41(10), 931–940.

Lawrence P. J., Jr., Berarducci M.P., 1994. Comparison of federated and centralized Kalman filters with fault detection considerations. In *Position Location and Navigation Symposium*, Las Vegas, NV, USA. IEEE, 1994, pp. 703–710.

Lee J., Kim B., Seo J., Yi K., Yoon J., Ko B., 2015. Automated driving control in safe driving envelope based on probabilistic prediction of surrounding vehicle behaviors. *Society of Automotive Engineers International Journal of Passenger Cars - Electronic and Electrical Systems*, 8(1), 207–218.

Lee J., Oh K., Yi K., 2016. Development of an emergency control algorithm for a fail-safe system in automated driving vehicles. School of Mechanical and Aerospace Engineering, Seoul National University Korea. Paper Number 19-0101

Le Marchand O., Bonnifait P., Guzman J. I., Btaille D., 2009. Vehicle localization integrity based on trajectory monitoring. In *IEEE/RSJ International Conference on Intelligent Robots and Systems*, St. Louis, MO, USA, 2009, pp. 3453–3458.

Meskin N., Khorasani K., 2011. *Fault Detection and Isolation: Multi-Vehicle Unmanned Systems*. New York: Springer.

MORSU, 2017. Camera based moving object detection for autonomous driving. Technischen Universität Chemnitz. www.tu-chemnitz.de/etit/dst/lehre/docs/results/morsu.pdf. Viewed: July 24, 2019.

Muenchhof M., 2005. Comparison of change detection methods for a residual of a hydraulic servo-axis. *IFAC Proceedings Volumes*, 38(1), 1854–1854.

Napolitano M. R., An Y., Seanor B. A., 2000. A fault tolerant flight control system for sensor and actuator failures using neural networks. *Aircraft Design*, 3, 103–128.

NHTSA, 2013. Preliminary statement of policy concerning automated vehicles. www.nhtsa.gov/staticfiles/rulemaking/pdf/Automated_Vehicles_Policy.pdf.

Peters M.-B., 2017. Have a safe trip. https://audi-encounter.com/en/car-security. Viewed: July 24, 2019.

Petritoli E., Leccese F., Ciani L., 2018a. Reliability and maintenance analysis of unmanned aerial vehicles. *Sensors*, 18, 3171; doi:10.3390/s18093171.

Petritoli E., Leccese F., Ciani L., 2018b. Reliability degradation, preventive and corrective maintenance of UAV systems. In *Proceedings of the 5th IEEE International Workshop on Metrology for AeroSpace (MetroAeroSpace)*, Rome, Italy, 20–22 June 2018.

Poole J. A., 2011. Fast reliability analysis for an unmanned aerial vehicle performing a phased mission. *Ph.D. Thesis*, Loughborough University, Loughborough, UK, 2011.

Realpe M., 2015a. Towards fault tolerant perception for autonomous vehicles: Local fusion. In *Presented at the 7th IEEE International Conference on Robotics, Automation and Mechatronics (RAM)*, Angkor Wat, Cambodia, 2015.

Realpe M., Vintimilla B., Vlaci L., 2015b. Sensor Fault Detection and Diagnosis for autonomous vehicles. *MATEC Web of Conferences*, Singapore. ©Owned by the authors, published by EDP Sciences, 2015; Vol. 30, p. 04003.

Rupp D., King A., 2010. Autonomous driving – A practical roadmap. SAE 2010-01-2335.

SAE International, 2016a. Automated driving: Levels of driving automation as per SAE international standard J3016. www.sae.org/news/2019/01/sae-updates-j3016-automated-driving-graphic. Viewed: July 24, 2019.

SAE International, 2016b. Surface vehicle recommended practice: Taxonomy and definitions for terms related to driving automation systems for on-road motor vehicles. SAE Standard J3016, 2016.

Sadeghzadeh I., Zhang Y., 2011. A review on fault-tolerant control for Unmanned Aerial Vehicles (UAVs). In *Infotech@ Aerospace 2011*, Montreal, Canada: Concordia University, 29–31 March 2011.

Simonsen R. B., 2014a. Inspection using drone technology. Force technology. Brondby, Denmark. https://forcetechnology.com/en/articles/inspection-using-drone-technology. Viewed: July 19, 2019.

Simonsen R. B., 2014b. Drone inspection services. Force technology. Brondby, Denmark. https://forcetechnology.com/en/services/drone-inspection-services. Viewed: July 19, 2019.

Toledo-Moreo R., Zamora-Izquierdo M. A., Beda Miarro B., Gómez-Skarmeta A. F., 2007. High-integrity IMM-EKF-based road vehicle navigation with low-cost GPS/SBAS/INS. *IEEE Transactions on Aerospace and Electronic Systems*, 8(3), 491–511, 2007.

UC Riverside, 2017. Accurate and secure navigation for autonomous vehicles. Tech ID: 30212/UC Case 2017-484-0. University of California, Riverside. Office of Technology Commercialization. https://techtransfer.universityofcalifornia.edu/NCD/30212.html. Viewed: July 24, 2019.

Waymo Safety Report 2017. On the road to fully self-driving. October, 2017.

Xu L., Weigong Z., 2010. An adaptive fault-tolerant multisensor navigation strategy for automated vehicles. *IEEE Transactions on Vehicular Technology*, 59, 2815–2829.

Xu T., Sutton R., Sharma S., 2007. A multi-sensor data fusion navigation system for an unmanned surface vehicle. In *Proceedings of the Institution of Mechanical Engineers*, Plymouth, England, 2007; Vol. 221, pp. 167–175, 177–186.

10

Autonomous Inspection and Maintenance with Artificial Intelligence Infiltration

10.1 Artificial Intelligence Techniques Used in AVs

10.1.1 Artificial Intelligence and Autonomous Vehicles

Artificial intelligence (AI) has taken the automotive industry by storm to drive the development of Level-4 and Level-5 autonomous vehicles (AVs). Why has AI become so popular now, even though it has been around since the 1950s? Simply put, the reason for this explosion of AI is the enormous amount of data available today. With the help of connected devices and services, we are able to collect data in every industry, thus fueling the AI revolution. Nvidia unveiled its first AI computer in October 2017 to enable deep learning, computer vision, and parallel computing algorithms. AI has become an essential component of automated drive technology, and it is important to know how it works in autonomous and connected vehicles (Gadam, 2018).

10.1.1.1 What Is Artificial Intelligence?

John McCarthy, a computer scientist, coined the term "Artificial Intelligence" in 1955. AI is defined as the ability of a computer program or machine to think, learn, and make decisions. In general use, the term means a machine which mimics human cognition. With AI, we are getting computer programs and machines to do what humans do. We are feeding these programs and machines with a massive amount of data that are analyzed and processed to ultimately think logically and perform human actions. The process of automating repetitive human tasks is just the tip of the AI iceberg; medical diagnostics equipment and autonomous cars have implemented AI with the objective of saving human lives (Gadam, 2018).

The roots of AI go back to classic philosophers, attempts to map and describe how humans process information and think about manipulating symbols in their environment. As technology continued to advance, engineers and philosophers were able to apply programmable digital computers to address complex mathematical reasoning. The ultimate goal of AI is to get machines to do what humans do. While computer programs can develop patterns of thought similar to humans, they require massive amounts of data that are analyzed and processed through sophisticated algorithms. The process of analyzing data helps AI mimic a logical thought process and perform human actions. AI engineers look to automate repetitive human tasks, and this ambitious goal can be applied to many industries across the world (Giarratana, 2018).

10.1.1.2 How Does AI Work in Autonomous Vehicles?

AI is popular buzz word today, but how does it actually work in AVs?

Let us first look at the human perspective of driving a car with the use of sensory functions, such as vision and sound, to watch the road and the other cars on the road. When we stop at a red light or wait for a pedestrian to cross the road, we are using our memory to make quick decisions. The years of driving experience habituate us to look for the little things that we encounter often on the roads—it could be a better route to the office or just a big bump in the road.

We are building AVs that drive themselves, but we want them to drive like human drivers do. That means we need to provide these vehicles with the sensory functions, cognitive functions (memory, logical thinking, decision-making, and learning) and executive capabilities that humans use to drive vehicles. The automotive industry is continuously evolving to achieve exactly this.

According to Gartner, 250 million cars will soon be connected with each other and the infrastructure around them through various V2X (vehicle-to-everything communication) systems. As the amount of information being fed into IVI (in-vehicle infotainment) units or telematics systems grows, vehicles will be able to capture and share not only internal system status and location data but also the changes in their surroundings, all in real time. AVs are being fitted with cameras, sensors, and communication systems to enable them to generate massive amounts of data which, when applied with AI, enable them to see, hear, think, and make decisions just like human drivers do (Gadam, 2018).

10.1.2 Autonomous Vehicles and Embedded Artificial Intelligence

10.1.2.1 Artificial Driving Intelligence: Context of Autonomous Vehicle Decisions

AVs offer the opportunity to link the benefits of the latest sensory technologies with those of AI to make driving decisions which mitigate many risks associated with human driving decisions. Indeed, the focus on the AI driving of AVs gives rise to two contrasting formulations of machine decisionality and its impacts in terms of benefits and risks to society as a whole–risk and risk mitigation. Some argue that AVs eliminate decisionality problems and hence mitigate risk associated with the human frailties of fatigue, misperception, and intoxication, along with the problematic decisions humans often make in the context of driving. This safety argument identifies the welfare benefits of machine decisions and endorses claims that AVs should be supported by policy. Conversely, others highlight the risks of new errors. Overall, there is a need to define and disseminate the benefits of AV decisional intelligence to avoid underutilization of the technology due misplaced risk perceptions (Floridi et al., 2018).

10.1.2.2 Autonomous Vehicle Literature Space

AVs offer many significant societal benefits: enhancing the mobility of those who lack it, transforming urban spaces and supporting the environment, and radically improving safety and saving lives. However, since the opportunities of any substantive technology also carry both embedded and new forms of risk, any actualization of potential AV benefits also necessitates mitigation of the risks involved. Moreover, AV risk mitigation cannot be undertaken by governance regimes alone but must be a multi-stakeholder phenomenon. In this instance, traditional government and new governance models are simply outpaced, as is evident throughout the current era of digital innovation (Marchant, 2011) and highlighted by a US National Highway Traffic Safety Agency's (NHTSA) comment (2016) that "the speed with which HAVs are advancing, combined with the complexity and novelty of these innovations, threatens to outpace the Agency's conventional regulatory processes and capabilities." For these reasons, intelligence technologies can only be responded to by a shared risk mitigation process wherein numerous actors cooperate. As such, the conceptualization and framing of technology in terms of meaning, benefits, and risks will ultimately determine how stakeholders engage with the technology (Cunneen, Mullins, & Murphy, 2019).

The key consideration of AV risk mitigation discussed across the literature concerns assessment of an AV's capacity to make driving decisions. Any research which illuminates the decisionality phenomenon of AVs contributes to the multi-stakeholder risk mitigation process and promotes access to AV societal benefits. Moreover, analysis of the scope of AV decisions in terms of both benefits (risk mitigation) and potential limitations (new forms of risk) supports the dynamics of new governance relations which are both top-down and bottom-up.

Furthermore while AVs arguably afford opportunities to minimize and potentially eliminate the many risks associated with human driving, future benefits cannot be realized unless accurate and effective anticipatory risk governance research is undertaken today. A broad and immensely complex decisional context is inherent to AV technologies, such as how different governance actors and policy writers understand the decisional capacity and societal impact of AV decisions. Research should also consider the diverse ethical interpretations of AV decisions, including the need to control ethical decisions as a predetermined configuration of action or the calculation of metrics such as values and risk. Some research repeats the many abstract questions surrounding machine ethics, while other work considers meaning, conceptual confusions, and limited decisional capacity. Issues of the technical scalability of AV decisional capacity are of significance, as well, as are issues of legality and governance whose interpretation relies on an understanding and anticipation of the impacts of AV, particularly in terms of societal risk (Cunneen, Mullins, & Murphy, 2019).

Trappl (2016) underscores the need to consider the important conceptual differences between human and machine contexts of moral decisionality in the context of AVs, while Bringsjord and Sen (2016) highlight the potential confusion surrounding the differing contexts AVs' of intelligence and ethical capacities. They also point out the need to support actors in reaching more accurate and informed choices in terms of AV policy and regulation. Millar (2016) proposes the need to investigate ethical AV decision-making while Goodall (2016) shifts the emphasis from ethics to risk. Others, such as Coeklebergh (2016), elucidate the importance of changes in relations which socially embedded technologies bring about between agents and actors (Coeklebergh, 2016). This is most evident in the consideration of key legal and ethical concepts by way of the changing human phenomenological relations with AVs (Coeklebergh, 2016). However distinct these approaches may be, they are united in their attempts to fathom the decisional relations of AI and applied applications, such as AV.

Understanding the role of AV decisionality is a complex challenge which requires careful elucidation, as the basic function of AV requires the driving intelligence to make decisions affecting human welfare and life (Lin, 2016). In fact, an AV will typically make thousands of such decisions on every trip, and global deployment will translate into millions of such decisions per day. Accordingly, it is imperative to explore the many facets of the AV decisional spectrum, not merely in terms of awareness of the limitations of AV decisionality but also in terms of the key contexts wherein different actors confuse or misunderstand the meaning of AV decisions (Cunneen, Mullins, & Murphy, 2019).

10.1.2.3 Framing Artificial Intelligence and Autonomous Decisions

The technological paradigm of AVs has generated some technological disorientation, especially with respect to the decisional capacity of embodied AI products. A progression of conceptual meaning and conceptual framing begins with the research phase and culminates with how the media and society engage with the concepts relating to the technology. However, as development of innovation depends on the key metrics of governance, the media, and public perception, there is a need for closer scrutiny of how initial framing plays out in the public arena. The literature on risk amplification speaks to this issue and points to the need for debates which set a positive and inclusive tone (Pidgeon, Kasperson, & Slovic, 2003). This is true of the more general phenomenon of risk amplification and of, more discrete phenomena, such as risk dread.

Risk amplification and the fear of new and emerging technology are well documented in the literature and suggest the care required in the initial conceptual framing (Frewer, Miles, & Marsh, 2002). This aspect is taken up by Johnson and Verdicchio (2017) who maintain the need to "argue for a reframing of AI discourse that avoids the pitfalls of confusion about autonomy and instead frames AI research as what it is: the design of computational artifacts that are able to achieve a goal without having their course of action fully specified by a human programmer" (Johnson & Verdicchio, 2017). While their critical approach points to the challenges of framing embodied AI products such as AVs, they represent a minority who addresses the question.

In addition to autonomy, there are further related complex challenges specific to the framing of AI and AI decisionality. Effective ontological domains are required for individual concepts (Franklin & Ferkin, 2006), so there is a need to anticipate conceptual challenges in the initial framing and ontologies of AI products (Cunneen, Mullins, Murphy, & Gaines, 2018). This is essentially a call for temporal considerations to be captured in the concepts employed, as this field is highly dynamic, and the configuration of actors and their anticipated roles are liable to change over time (Cunneen, Mullins, & Murphy, 2019).

In short, there is a need to anticipate the societal, ethical, and legal (SEL) impacts of AVs' decisional capacity. A critical analysis suggests there has been a failure to engage at the necessary meta-level and construct informed accurate conceptual frameworks of AV decisional capacity, as well as a failure to consider the important differences between how society and users understand human and machine decision-making in more detail. In fact, the core question of the SEL impact of AVs is yoked to the meaning framework of machine driving decisions and human driving decisions. This underlines the necessity to interrogate the conceptual framing of AV driving decisions. Without accurate SEL impact analysis, the challenges of uncertainty and risk will hinder informed research, development, and societal perception (Renn, 2008: xv). And without accurate metrics of the SEL impact, systems of governance cannot provide the mechanisms which balance the need to support innovation with the duty to assess potential risks and protect society from harms. All innovation warrants a process of analysis by which to accurately frame the legal and general principles of associated societal rights to safety, freedom, equality, privacy, and welfare.

Broadly stated, some researchers focus on AV safety ; others tackle the ethical challenges inherent in using AI. Both types of analysis frame AVs by centering the decisional capacity of vehicular driving intelligence, but they offer very different matrices of the range of decisions AI must carry out to safely traverse the human road network. One claims it is a superior driving decision capacity that will save lives; the other insists it presents a risk of limited decisional capacity which could inadvertently pose significant ethical problems (Lin, 2016). Each interpretation begins with the focus on decisions but frames the decision capacity differently, and each anticipates very different accounts of the potential SEL impacts of AV decisions and governance.

Of course. diverse perspectives and interpretations are an integral aspect of developing research and knowledge contexts, but as multiple agents and actors engage with the different frameworks around AV, the potential for inaccurate framing feeding into systems of governance is a significant concern. We have two very different accounts of decisional capacity possible SEL impacts, and proposed governance of AVs. They frame the decisional capacity in dramatically opposing ways: Proper analysis clarifies the AV decision domain, and if we are to judge by the two principal framing values of the safety argument and ethical challenge, the AV decisional framework presents a technological medium that remains conceptually obscure (Cunneen, Mullins, & Murphy, 2019).

10.1.3 Drones and Robots Are Taking over Maintenance Inspections ... and That's Not a Bad Thing

Smart devices are taking over. From drones to robot vacuums to driverless cars, we're not only seeing them in household settings but also in industrial settings. The recent surge of interest and use is largely due to new advances in AI, which increases their autonomy, capabilities, and usefulness.

As smart devices get better and better, they are increasingly used to automate dangerous and laborious jobs, often completing the work more efficiently and with much higher precision than humans can.

Just one example of this is the use of service robots carrying out asset inspections. Let's take a closer look at the benefits, what we can expect to see in the near future, and what it means for the technicians who used to do these jobs (Edge 4 Industry, 2018).

10.1.3.1 Robots and Drones for Asset Inspections: The Benefits

For many organizations, human-mediated asset inspections can be high cost, risky, and time consuming. Though we're still in the early days of deploying drones and robots to alleviate the challenges of traditional asset inspections, many are looking forward to the day when deploying drones and robots is the norm. Using smart devices to conduct asset inspections has numerous benefits over traditional methods (Edge 4 Industry, 2018):

1. Increase safety and decrease risks to maintenance technicians.

    Drones reach places that are dangerous for human workers, such as tall structures and hazardous areas (e.g., areas with radiation or high-voltage lines). As such, they are much safer when it comes to inspecting refineries, mine areas, and pipelines. Likewise, drones can operate under adverse weather and physical conditions, such as wind, waves, and radiation, which are among the most common sources of safety risks for human workers in field service and enterprise maintenance. Of course, there are limitations to what they can withstand, but in many cases, they can surmount challenges that could be perilous or, at the very least a significant nuisance to human workers. They can capture images and video of assets that are in difficult or dangerous areas that are hardly accessible by human workers. For example, engineers have successfully guided robots into the destroyed Fukushima nuclear power plant, where radiation could be lethal to human workers. Overall, one of the leading benefits of using devices powered by AI to conduct asset inspection is that by putting the drones and robots at risk, human workers are safer and healthier.

2. Provide unprecedented richness and accuracy in data collection of an asset's condition.

Modern drones are able to capture high resolution images and video from the assets they inspect. The rich visualizations they provide can give context and clarity regarding the assets' conditions. For instance, they can capture images of damage and defects from multiple angles. These images can help maintenance engineers plan and execute optimal service strategies.

3. Versatile and flexible enough for a wide range of maintenance inspection tasks.

Many different types of UAVs are commercially available, including drones of many sizes that can fly in different ranges and can operate autonomously for varying amounts of time. Now more than ever before, plant owners and field service engineers have options to choose from so that they can select and deploy the UAV that is most suitable for the inspection and service tasks at their organization.

Before choosing a drone (and any applicable attachments, add-ons, and tools), enterprise maintenance professionals will need to consider their requirements, including:

- Flying altitude
- Quality of images
- Longest flight time
- Data transmission rates.

This flexibility of the available drones on the market means that there are now machines capable of performing inspections where only a few short years ago no such technology existed.

4. Make data structuring and information sharing easier.

Simply put, drones with their related attachments, software, etc. simplify the collection, organization, accessibility, sorting, sharing, processing, and interpretation of data. UAVs collect and process data in digital formats, making it much easier to store and organize the data and to produce usable reports about the inspections. Furthermore, once those data have been collected, using the right tools to make organizing, manipulating reviewing, and sharing the data easy can have benefits across maintenance stakeholders, including plant owners, maintenance and automation solution providers, maintenance engineers/technicians, and field workers.

5. Decrease maintenance equipment downtime.

Drone-based inspections can often be performed without a need to shut down systems or equipment. In some cases, machines, tools, buildings, and other systems can continue to operate while being inspected by a drone. This is not always the case with manual inspections, where some systems must be shut down in order to avoid equipment damage and injuries to human workers. Thus, UAVs lead to better OEE (Overall Equipment Efficiency) and do not disrupt services or production tasks.

6. Decrease enterprise maintenance labor and insurance costs.

There are many different ways in which UAVs contribute to cost reductions in enterprise maintenance environments. First, they reduce insurance costs for inspectors and field service personnel, who no longer engage in many dangerous tasks. Second, the actual reduction of injuries reduces costs associated with absences and healthcare. Third, enterprises save on costs for renting the equipment that supports manual inspections, such as ladders and aerial lifts.

Finally, significant cost savings stem from the fact that inspections are performed faster, i.e., more inspections can be concluded in a given timeframe.

7. Repurpose maintenance inspection UAVs for other uses.

Not only do drones make it easier and more effective to collect and analyze asset data and conduct maintenance inspections, in some circumstances, organizations may be able to leverage them for supplemental purposes. For instance, a drone might be used to capture breathtaking aerial views of facilities, buildings, plants, and their surrounding property, which can be used in marketing from social media posts to blog articles and website updates. Finding additional ways to use an inspection drone can increase the return on investment (ROI).

10.1.3.2 Examples of Robots and Drones in Enterprise Maintenance

The vision of using drones and robots in inspection tasks is already materializing. During recent years, drones have been deployed in many different industries not only for asset inspections but also for security and surveillance. Most of the deployments can be found in industries, such as utilities and power generation, insurance, oil and gas, and building construction and facility management.

Take, for example, the Deloitte Maximo Center of Excellence. It has integrated an autonomous drone as part of its Enterprise Asset Management solutions. The drone solution leverages other IBM products as well, such as Watson and Bluemix, which manage UAV data integration as part of the asset management application.

As another example, aviation companies like EasyJet and Thomas Cook Airlines are planning to deploy UAVs to inspect their aircrafts and other assets. Their strategy includes the possibility of launching a UAV every time an aircraft approaches a gate, as a means of monitoring potential damage. A more advanced deployment example is Eelume's swimming robots, deployed for subsea inspection and light intervention.

We're just now seeing the beginning of what's to come with drones, robots, and UAVs. Their use will rapidly become more popular and commonplace because of their ability to decrease costs and keep human workers safe. Moreover, drones, robots, and other smart objects are becoming more versatile and useful as their functionalities and intelligence continue to be improved.

For instance, UAV vendors are working towards releasing cognitive drones which will be able to intelligently tune the rate of their data collection depending on the context of the inspection. In particular, cognitive drones will be able to collect more images about damaged parts, by adapting their operation whenever they identify a damaged part.

In the future, it's likely that we will see inspections and maintenance tasks carried out by voice-guided robots. We'll also see a greater number of actuator robots that will complete routine field inspections, while human workers lead the way in safer, supervisory roles (Edge 4 Industry, 2018).

10.2 Artificial Intelligence Approaches for Inspection and Maintenance

10.2.1 Introduction

Over the past two decades, there has been substantial research and development in the area of operations management including maintenance. Kobbacy, Vadera, and Rasmi (2007) argue that the continuous research in these areas implies that solutions have not been found for many problems. Many proposed solutions are for well-defined problems; moreover the solutions assume accurate data are available; finally the solutions are too computationally expensive to be practical. AI is recognized by many researchers as a potentially powerful tool especially when combined with operational research (OR) techniques. This section reviews the application of AI in maintenance and inspections and introduces the concept of developing an intelligent maintenance optimization system (Kobbacy & Prabhakar Murthy, 2008).

10.2.2 AI Applications in Maintenance

10.2.2.1 Knowledge-Based Systems (KBSs)

MYCEN, the diagnostic medical expert system, was designed in the early 1970s at Stanford University and generated lots of interest even though it was never used. KBS began to be researched in AI in the 1980s and initial interest in applying AI techniques in maintenance started with KBS. An early paper by Dhaliwal (1986) argued for the use of AI techniques in the maintenance management of large-scale systems. Kobbacy (1992) proposed the use of KBS in this area, while Kobbacy, Proudlove, and Harper (1995) detailed "IMOS," an Intelligent Maintenance Optimization System using KBS (Kobbacy, 2012).

Over the years, the interest in KBSs in maintenance has varied but, in general, moderate interest has always been maintained. Applications have covered a wide range of areas. Batanov, Nagarur, and Nitikhunkasem (1993) presented a prototype KBS system for maintenance management (EXPERT-MM). The system suggests maintenance policy, provides machine diagnosis, and offers maintenance scheduling. Su, Hwang, and Liu (2000) proposed a KBS system for analyzing cognitive type recovery in preventive maintenance. Gabbar, Yamashita, Suzuki, and Shimada (2003) suggested a computer-aided Reliability-Centered Maintenance (RCM)-based plant maintenance management system. The adopted approach utilizes a commercial computerized maintenance management system (CMMS) (Kobbacy, 2012).

10.2.2.2 Fuzzy Logic (FL)

Fuzzy logic (FL) is a popular AI technique that has many applications and is frequently employed with other AI techniques in hybrid systems. Zadeh (1965) introduced the concept of fuzzy sets. Zadeh (1999) later defined narrow FL which aims at formalizing of approximate reasoning and wide FL which is coextensive with fuzzy set theory and is far broader than narrow FL. Today, the term FL is used predominantly in its wide sense (Kobbacy, 2012).

There is steady interest in applying FL to model maintenance problems because of its ability to deal with uncertainty. Mozami, Behbahani, and Muniandy (2011) used FL to prioritize maintenance activities based on pavement condition index, traffic volume, road width, and rehabilitation and maintenance cost. Al-Najjar and Alsyouf (2003) assessed the most popular maintenance approaches using a fuzzy multiple criteria decision-making (MCDM) evaluation methodology. Derigent, Thomas, Levrat, and Iung (2009) presented a fuzzy methodology to assess component proximity based on which opportunistic maintenance strategy can be implemented. They used component proximity fuzzy modeling targeting the system components "close to" a given reference component, on which a maintenance action is planned. Sasmal and Ramanjaneyulu (2008) developed a systematic procedure and formulation for condition evaluation of bridges using the analytic hierarchy process (AHP). Malik, Singh, Kr, and Jarial (2012) used UV/VIS spectroscopy-based FL to assess transformer oil health (Kobbacy, 2012).

Risk-based maintenance/inspection is becoming a popular approach to ensure safe and economically viable operations. Khan, Sadiq, and Haddara (2004) presented a structured risk-based inspection and maintenance methodology using FL to estimate risk by combining (fuzzy) likelihood of occurrence and its (fuzzy) consequence with case studies from oil and gas operations. Singh and Markeset (2009) presented a methodology for risk-based inspection program for pipes using an FL framework (Kobbacy, 2012).

10.2.2.3 Case-Based Reasoning (CBR)

Case-based reasoning (CBR) uses index schemes, similarity functions and adaptation to utilize past experiences in solving new problems. When combined with other AI techniques, it can provide powerful hybrid systems that are able to solve complex problems through the use of machine learning capability (Kobbacy, 2012).

Yu, Iung, and Panetto (2003) studied remote maintenance decision-making and identified the need for suppliers and customers to negotiated and cooperate based on knowledge. Chou (2009) outlined the development of a Web-based CBR prototype system to assist decision makers in project screening and budget allocation. The system determines preliminary project cost with information based on previous experience of pavement maintenance-related construction. Cai, Jia, Gu, and Wu (2011) proposed a framework for a command forces system, based on CBR, of equipment maintenance support. The paper analyzes the CBR-based method of representation and equipment maintenance (Kobbacy, 2012).

Finally, Cheng, Jia, Gao, and Wu (2008) presented an Intelligent Reliability-Centered Maintenance Analysis (IRCMA) in which they used CBR on the historical records of similar items to analyze a new item (Kobbacy, 2012).

10.2.2.4 Genetic Algorithms (GAs)

The genetic algorithm (GA) is an optimization technique developed in the 1970s. It is based on the principles of genetics and natural selection; i.e., solutions can evolved through mutation, and weaker solutions become extinct. GAs have superior and more versatile features than the classic optimization and search techniques. For example, a GA can work with large number of continuous or discrete variables and optimize extremely complex cost functions. However, it has the drawback of requiring high computing power and this, in turn, may require the use of parallel processing (Haupt and Haupt, 2004). Because of its advantages, GAs are used to solve many complex maintenance management problems (Kobbacy, 2012).

GA is the most popular AI technique applied in maintenance. There are many publications on applying GAs in a wide range of maintenance problems and applications, albeit mostly of an operational rather than a strategic nature. There are numerous recent developments and applications of GAs in maintenance. In what follows, we highlight the most popular maintenance problem situations and industrial applications (Kobbacy, 2012).

Maintenance scheduling problems represent large-scale and complex applications because the large number of variables involved and their interdependence. Planning preventive maintenance is a complex because of the nature of the problem and the typical lack of data. Lapa, Claudio, Pereira, and Paes de Barros (2006) presented a model for PM planning using GAs. They aimed to optimize PM schedules automatically, considering factors such as the probability of requiring repairs and their costs, the typical downtime, PM costs, maintenance impact on the systems reliability, probability of imperfect maintenance, etc. They used a case study of high-pressure injection system (HPIS) of a typical four-loop pressurized water reactor (PWR) to evaluate the methodology. Volkanovski, Mavko, Boševski, Čauševski, and Čepin (2008) proposed a GA approach to optimize the maintenance schedule of a generating unit of a power system. Tsai, Wang, and Teng (2001) optimized PM for mechanical components. They considered a choice of either a PM intervention to partially renew the component or a replacement with a new component. The optimal activity combination at each PM stage is decided by using GA to minimize the system unit-cost life. The optimal minimum cost life is reached at the point in time when the system's unit-cost life is less than its discarded life at PM intervention. For instance, Yang and Yang (2012) used GA to optimize aircraft maintenance planning (Kobbacy, 2012).

There have been a few attempts to use GAs to optimize the maintenance of parallel/series systems. For example, Levitin and Lisnianski (1999) addressed the problems of redundancy and reliability for a multi-level system with different levels of output at a point in time. The optimal system structure and replacement policy provide the desired level of system reliability with minimal cost of maintenance and failures measured by unsupplied demand. Monga and Zuo (2001) studied the optimal design of series-parallel systems considering maintenance and salvage values. They developed a reliability-based design model with deteriorating components and considering the effects of PM and asset depreciation. They used GA to perform constrained optimization of the system cost function subject to both active and non-active constraints. Nourelfath, Châtelet, and Nahas (2012) considered joint redundancy and imperfect PM optimization for a series-parallel multistate degraded system. They used a heuristic based on combined GAs and Tabu Search to solve the formulated problem (Kobbacy, 2012).

There are also a few publications on the combined use of GAs and Monte Carlo simulation. Marseguerra, Zio, and Podofillini (2002) used GA to determine the optimal degradation level beyond which preventive maintenance has to be performed for a continuously monitored multicomponent system. Marseguerra and Zio (2000) presented an interesting study on using Monte Carlo simulation in terms of economic

Artificial Intelligent Infiltration

costs and revenues of the operation of a plant. The used GA to optimize the components, maintenance periods and the number of repair teams. Chootinan, Chen, Horrocks, and Bolling (2006) introduced a pavement maintenance programming methodology that can account for uncertainty in pavement deterioration using a simulation-based GA approach. This GA approach is capable of planning the maintenance activities over a multi-year planning period. The uncertainty of future pavement conditions is simulated based on the calibrated deterioration model, and GA is used to solve the combinatorial optimization problem (Kobbacy, 2012).

Levitin and Lisnianski (2000) addressed the problem of optimizing imperfect preventive maintenance (PM) for multi-state systems, i.e., systems with different levels of output at a point of time. They used GA to find an optimal sequence of minimum cost PM actions, while providing the required level of system reliability. Nahas, Khatab, Ait-Kadi, and Nourelfath (2008) proposed an improved optimization approach to find the optimal PM actions based on the extended great deluge algorithm which is simpler and produces the best solution (Kobbacy, 2012).

Examples of applying GAs in maintenance optimization in nuclear power systems include work by Lapa, Claudio, Pereira, and Mol (2000). These authors used GA to maximize the availability of nuclear power systems through maintenance scheduling. Jiejuan, Dingyuan, and Dazhi (2004) utilized probabilistic safety assessment (PSA) technology to study the effect of testing and PM on component unavailability and built a maintenance risk-cost model for global maintenance optimization in a nuclear power plant. The used GA to obtain the optimal maintenance strategy (Kobbacy, 2012).

Tan and Kramer (1997) presented a general framework for PM optimization in chemical processing operations using GAs and Monte Carlo simulation. They argued that the proposed framework can easily be integrated with general process planning and scheduling and can be applied for opportunistic maintenance problems. Nguyen and Bagajewicz (2008) also used GA to optimize PM scheduling in a processing plant. The used Monte Carlo simulation to evaluate the maintenance cost, economic loss, and maintenance performance (Kobbacy, 2012).

Another important application area for GA in maintenance is in manufacturing. System reliability is important in manufacturing; this is even more critical for multifactory production, as a failure may cause a chain reaction. Moradi, Fatemi Ghomi, and Zandieh (2011) proposed a bi-objective model for joint production and maintenance scheduling in a flexible job-shop problem (FJSP). The model aims at minimizing unavailability using GA (Kobbacy, 2012).

10.2.2.5 Neural Networks (NNs)

Neural networks (NNs) are based on the idea of emulating the human brain. They are often used in modeling and statistical analysis and in classification and optimization (Kobbacy, 2012).

Siener and Aurich (2011) presented a concept for quality-oriented maintenance scheduling. The used artificial NN to predict the influence of individual machines on final product quality based on expected conditions of quality-relevant process objects. Mazhar, Kara, and Kaebernick (2007) estimated the remaining life of used components using a two-step approach based on integrating reliability assessment model using Weibull analysis and an artificial neural network (ANN) model using degradation and condition monitoring data. The authors showed the advantages of using this approach in the sustainable management of supply chains. Lee, Sanmugarasa, Blumenstein, and Loo (2008) proposed an approach to improve bridge management systems (BMSs) by utilizing an ANN-based prediction model to generate historical bridge condition ratings using limited bridge inspection records. They used the model to predict the missing historical condition ratings of individual bridge elements leading to improved BMS reliability in predicting future bridge conditions (Kobbacy, 2012).

10.2.2.6 Data Mining (DM)

Data mining (DM) uses statistical and machine learning techniques to automate the detection of data patterns in a database and to develop predictive models to support decision-making (Kobbacy, 2012).

10.2.2.7 Hybrid Systems

Hybrid systems are those which employ two or more AI techniques. In maintenance applications, the most frequently used AI technique in a hybrid system is FL when used with NN and GAs. Hybrid systems using other combinations in maintenance include CBR with AHP and GAs with artificial immune systems and simulated annealing (Kobbacy, 2012).

Christodoulou, Deligianni, Aslani, and Agathokleous (2009) presented a neuro-fuzzy decision support system (DSS) for urban water distribution networks. The system undertakes multi-factored risk-of-failure analysis and asset management. Huang (1998) proposed a genetic-evolved fuzzy system for the maintenance scheduling of generating units. Kim, Nara, and Gen (1994) introduced a method for maintenance scheduling using GA and simulated annealing (Kobbacy, 2012).

Kobbacy, Proudlove, and Harper (1995) presented the next generation of Integrated Modeling of Optical Systems (IMOS) discussed above under KBS. The hybrid IMOS, HIMOS, uses a hybrid approach of KBS and CBR to provide the system with greater learning capability (Kobbacy, 2012).

10.2.3 Artificial Intelligence for Industrial Inspection

10.2.3.1 Introduction

Visual inspection is routinely carried out across industries to determine whether a structure, product, component, or process meets the specified requirements. Typical examples include the detection of product defects in service or during maintenance. It is also done as part of manufacturing in-process monitoring. Such inspections are usually carried out by a trained individual who has sufficient knowledge and experience to visually identify faults and non-conforming quality and performance.

AI provides an opportunity to introduce innovation and new technology to the visual inspection process.

For example, the Centre for Modelling & Simulation (CFMS) has produced a demonstrator that uses a combination of computer vision and AI technologies to automate the manual inspection process.

Automated inspection technology is applicable across sectors and will significantly benefit high value manufacturing, maintenance, repair and operations in aerospace, automotive, construction, and energy sectors, where visual inspection is highly utilized (CFMS, 2018a).

10.2.3.2 Applicability of Artificial Intelligence for Industrial Inspection

A large number of visual inspection tasks are routinely carried out in industry; these activities are largely manual, involve recurring costs, and can have health and safety implications.

AI has the potential to transform things. Within typical inspection workflows, assessment can be carried out to verify whether a product, component, structure, or process meets specified requirements. This decision can be made by a skilled engineer, but this type of inspection is often associated with high recurring costs, increased susceptibility human error due to fatigue, and can carry high operational risk. For example, the industrial asset could be difficult to access due to confined space or a hazardous environment.

The CFMS company recently produced three demonstrators to disseminate emerging AI technologies across the aerospace industry to automate industrial inspection tasks. It hoped this kind of automation would promote a culture of data curation within the industry, leading to the identification of further opportunities for the adoption of AI technologies to automate business processes more generally.

The underlying AI model used in all the demonstrators was deep learning, consisting of NNs vaguely inspired by the biological NNs in that the connectivity pattern between neurons resembles the organization of an animal visual cortex. NNs consist of a number of connected layers wherein each layer has a number of neurons. The connections pass signals from one neuron to another. The crux of training a NN is finding out (i.e., self-learning) the strength of the various signals (also known as "weights") passed between the neurons from training examples. This process is analogous to finding out coefficients of an equation that maps input variables to corresponding outcomes in a typical empirical dataset.

Since all the demonstrators were based on image processing, a specific type of NN called convolutional neural network (CNN) was used to train the algorithm to recognize features and objects in the images. One of the key advantages of using CNN is that it uses less parameters for training which means it is computationally efficient. The video footage was split into individual frames and labeled with areas of interest for the inspection task. The labeled frames were then used to train the CNNs, i.e., train them to "self-learn" the rules for feature identification and to identify objects and/or defects irrespective of their position or orientation in a real-world inspection scenario.

The first demonstrator related to a typical manual inspection task of an aircraft wing as part of a preflight inspection at the airport or a maintenance check at an aircraft hangar. The purpose of inspection in these cases is to detect imperfections such as foreign object damage, corrosion, cracks, dents, and scratches, and the typical challenge associated with these types of inspections is accessibility due to aircraft size. The demonstrator was developed in collaboration with the National Composites Centre (NCC) using its drone to fly over a full-scale Airbus A400M Aircraft Wing at Airbus UK. The camera mounted on the drone captured the condition of the aircraft wing in a video format. This footage was then used to train the CNN to learn to detect defects (e.g., corrosion, dents) and features (bolts, brackets etc.).

In the second demonstrator, a smart phone was used to capture video footage of the inside of the A400M wing box. This area needs to be routinely inspected during aircraft maintenance cycles, but it is quite a confined space, which makes inspection physically challenging. The footage from the smartphone camera was again used to train the CNN to learn to detect the defects previously outlined. In the future, organizations could deploy a crawling robot, equipped with a camera and AI module to automate inspection in these cases.

Finally, the third demonstrator related to the inspection aspect of the composite manufacturing process involving an Auto Fiber Placement (AFP) machine. This manufacturing process needs to be monitored to ensure there are no excessive gaps between composite tapes (tows) or no undesired overlaps or defects, like wrinkles. These monitoring and documenting steps are currently manual, which not only means they involve high recurring costs but that they are also prone to variation in documentation—usually when different wording or language is used among inspectors—which, in turn, can lead to information being interpreted differently. The good news is that most of these machines are equipped with a video camera that is capable of recording video footage of the ongoing manufacturing process. However, cameras are used infrequently or switched off altogether, as the benefits of automating the inspection process using AI is not widely known. Hence, in this demonstrator, video footage from a fixed camcorder was used to train the CNN to detect defects like tow gaps or tape cutting failure and features of composite manufacturing (untrimmed) component like "bat ears."

All demonstrators aimed to raise awareness and stimulate the identification of further opportunities for AI technologies, which could revolutionize inspection activities across the aerospace industry, with possible applications in other industries. As an independent, not-for-profit specialist center, CFMS will continue to assess the latest AI developments and their ability to help organizations streamline processes, boost efficiencies, and save time and cost (CFMS, 2018b).

10.2.4 Artificial Intelligence for Predictive Maintenance

Leveraging AI models to identify anomalous behavior turns equipment sensor data into meaningful, actionable insights for proactive asset maintenance—preventing downtime or accidents. Commonly known as predictive maintenance, this intelligence forecasts when or if functional equipment will fail so that its maintenance and repair can be scheduled before the failure occurs.

Identifying causes of potential faults allows companies to deploy maintenance services more effectively, improving equipment uptime.

Critical features that help predict faults or failures are often buried in structured data, such as year of production, make, model, and warranty details, as well as unstructured data such as maintenance history and repair logs. However, emerging technologies such as the Internet of Things (IoT), Big Data analytics, and cloud data storage are enabling more equipment to share condition-based data with a centralized server, making fault detection easier, more practical, and more direct (Otto, 2019).

10.2.4.1 Predictive Maintenance Model

The underlying architecture of a preventive maintenance model is fairly uniform, irrespective of applications. Analytics usually reside on various IT platforms, with layers systematically described as (Otto, 2019):

- Data acquisition, storage: Cloud or edge systems.
- Data transformation: Conversion of raw data for machine learning models.
- Condition monitoring: Alerts based on asset operating limits.
- Asset health evaluation: Diagnostic records based on trend analysis if asset health declines.
- Prognostics: Failure predictions through machine learning models, estimate remaining life.
- Decision support system: Best action recommendations.
- Human interface layer: Information accessible in easy-to-understand format.

Failure prediction, fault diagnosis, failure-type classification, and recommendation of relevant maintenance actions are all a part of predictive maintenance methodology.

Manufacturing, energy, and utilities verticals are among the biggest demand drivers for predictive maintenance, and the technology is growing elsewhere, as manufacturers look to control maintenance and downtime costs. It is critical for equipment manufacturers and owners/operators to adopt a predictive maintenance solution to maintain a competitive advantage.

The bigger players have been using this methodology for more than a decade. Small- and medium-sized companies in the manufacturing sector also can reap its advantages by keeping repair costs low and meeting initial operational costs for new operations.

Offering more business benefits than corrective and preventative maintenance programs, predictive maintenance is a step ahead of preventive maintenance. As maintenance work is scheduled at preset intervals, maintenance technicians are informed of the likelihood of parts and components failing during the next work cycle, and can act to minimize downtime:

"It's critical for equipment manufacturers and owners/operators to adopt a predictive maintenance solution to maintain a competitive advantage" (Otto, 2019).

10.2.4.2 Performance Benefits

Predictive maintenance employs nonintrusive testing techniques to evaluate and compute asset performance trends. Additional methods used can include thermodynamics, acoustics, vibration analysis, and infrared (IR) analysis.

The continuous developments in Big Data, machine-to-machine communication, and cloud technology have created new possibilities for investigating information derived from industrial assets. Condition monitoring in real time is viable from sensors, actuators, and other control parameters. What stakeholders need is a bankable analytics and engineering service partner who can help them leverage data science to predict embryonic asset failures, eliminate them, and act in a timely manner (Otto, 2019).

10.2.5 Predictive Maintenance and Predicting Industrial Revolutions

Predictive maintenance can be tricky, but it is certainly simpler than predicting the next industrial revolution.

How transformational is AI? That seemingly simple question can be tricky to answer.

First of all, the term often has a nebulous meaning. The term is often shorthand for "artificial general intelligence." In this theoretical construct, nonorganic intelligence has the ability to reason and carry out an array of tasks. It is also this kind of AI Hollywood has referenced in films like "The Terminator," as well as the variety that elicits dystopian fears in the likes of Elon Musk and Bill Gates. But the question of when such a form of AI will exist is impossible to decidedly answer. It will probably be decades before researchers create machines capable of approximating human reasoning, even though machines trained at narrow tasks can outshine humans in intellectual games such as chess and, more recently, Go. Martin Ford, author of *Architects of Intelligence: The truth about AI from the people building it*, says most experts believe AI is a unpredictable.

So are, it seems, industrial revolutions. While concepts such as Industry 4.0 broadly hint that the convergence of AI, Industrial Internet of Things (IIoT), and other technologies could foment the next industrial revolution, productivity in Western nations has been sluggish for decades.

But while the industrial macrocosm, measured by various productivity indices, putters along, there are a growing number of success stories emerging from industrial companies embracing IIoT technologies in tandem with machine learning. The startup FogHorn, for instance, helped the Japanese industrial electronics company Daihen eliminate 1,800 h worth of manual data entry in a single factory. And a top beverage company saved the equivalent of 1 million cans of beer through predictive maintenance in one fell swoop. The firm installed machine monitoring technology from the firm Augury, which marries wireless vibration, ultrasonic, temperature, and magnetic sensors with machine learning to detect machine problems for a range of industrial machines, including those used by breweries.

Air Liquide provides another example of the power of Industry 4.0 technologies. Incorporating technology from OSIsoft, Air Liquide created a plant optimization platform known as SIO at Alizent, the company's digital subsidiary. OSIsoft's PI software served as the embedded data engine for the platform, which enabled data collection and refining for analytics. It achieved payback in three months.

Another example of a company with a rapid digital transformation ROI is White House Utility District. One of the top water and sewer utilities in Tennessee, the organization cut water leaks from approximately 32% to 15%. The savings in water resulted in millions of dollars in savings.

But the process of applying machine learning—or the application of it in predictive maintenance—in the industrial realm is rarely simple. Data can be inconsistent or missing, and sensors can give false readings. Data must be cleaned before they can be analyzed.

We are still at the infancy of applying machine learning to industrial applications.

A study from Bain & Co reached similar conclusions after surveying 600 high-tech executives. IoT in general and predictive maintenance in particular are often more challenging to deploy than expected—as is the prospect of extracting valuable insights from the data from IoT projects. The report concluded, however, that industrial IoT represents a promising opportunity."

One central challenge is that making sense of such data requires a rare blend of domain expertise and data science savvy.

It can be difficult to find this type of talent. While much has been made about the difficulty of finding an expert in, say, industrial cybersecurity or industrial data science, the problem is actually much larger.

Much of the legacy manufacturing experience in the United States will disappear as older industrial employees retire in the next 5–10 years. Meanwhile, industrial jobs are low on the list for younger employees.

It may be too early to tell if broad technologies like IoT and AI will create an era of productivity, rivaling the first industrial revolutions or whether Industry 4.0 will be something more like a software revision than a cyber-physical system-driven revolution. For now, it is a more practical question to ask how such technologies can meet their most pressing needs—getting the right people (technicians) to the right place (a potentially failing machine) at the right time (before that machine breaks) (Buntz, 2019).

10.3 Current Developments of AVs with AI

10.3.1 Introduction

Autonomous driving and AVs are currently among the most intensively researched and publicly followed technologies in the transportation domain. Before realizing the vision of fully autonomous vehicles, many technical and legal challenges remain to be solved. In addition to these technical and legal challenges, questions about user and societal acceptance come into play.

AVs involve the application of intelligent automation. In general, automation is defined as technology that actively selects data, transforms information, and makes decisions or controls processes. The decision-making process employed in the technology is based on inherent AI, hence the term "intelligent automation." The transportation industry is only one among many industries that are increasingly influenced by automation involving AI. Intelligent, personal robots have begun to noticeably appear in diverse application fields ranging from home automation to medical assistance devices.

In order to approach the topic of AI, the distinction between strong and weak AI is essential. Strong AI implies a system with human or superhuman intelligence in all facets and is pure fiction today. Currently, only weak AI is of interest for commercial applications in terms of specific tasks that require single human capabilities, for example, visual perception, understanding context, probabilistic reasoning, and dealing with complexity. In these domains, machines exceed human capabilities by far. However, intelligent technologies are not able to execute intelligent tasks, such as ethical judgments, symbolic reasoning, and managing social situations or ideation (Hengstler, Enkel, & Duelli, 2016).

10.3.2 AI Technology in the Automotive Industry

While AI has been beating humans in computing and board games for some time, AI has taken the automotive industry by storm in recent years. Since the 1950s, researchers and engineers have been working towards Level 4 and Level 5 AV development. From Tesla to Google, most industry leaders feel that fully autonomous cars are closer than ever because of the advancements in AI technology, especially deep learning. Anyone who has taken a drive on city streets can testify that there are many variables to consider during their journey. Everything from volatile weather conditions, obstructions in the road like potholes, and other drivers can make a pure drive to the store a stressful event. Humans have a complex visual system to absorb information about their surroundings and then process that information to react accordingly. Autonomous cars will need to be developed in a similar way because this technology will be interacting with their surroundings and human drivers. As more information is captured by internal systems like cameras, sensors, and communication systems, onboard AI computing programs will need to process massive amounts of data in real time (Giarratana, 2018).

10.3.3 Will Artificial Intelligence Make Autonomous Vehicles a Reality?

There are countless variables and other forces that could destroy the dream of autonomous cars. The field of self-driving vehicles is like a scientific experiment that we don't know the answer to. We have been able to apply deep learning to deliver a certain level of automated driving, but current deep learning models could impose a hard limit on how far that progress can go.

The future of AVs depends on how well onboard sensors can gather information and how accurately AI systems can process that information in real time. Some companies are testing different techniques to solve the problem posed by deep learning when applied to autonomous cars.

We will to see which technique will prove more effective (Giarratana, 2018).

10.3.4 AI Perception Action Cycle in Autonomous Vehicles

A repetitive loop, called a perception action cycle, is created when the AV generates data from its surrounding environment and feeds them into the intelligent agent, who, in turn, makes decisions and enables the AV to perform specific actions in that same environment. Figure 10.1 illustrates the data flow in AVs (Gadam, 2018).

In what follows, we break this process down into three main components (Gadam, 2018).

10.3.4.1 Component 1: In-Vehicle Data Collection & Communication Systems

AVs are fitted with numerous sensors, radar, and cameras to generate massive amounts of environmental data. All of these form a digital sensorium, through which the AV can see, hear, and feel the road, road infrastructure, other vehicles, and every other object on/near the road, just like a human driver would pay attention to the road while driving. These data are processed with supercomputers,

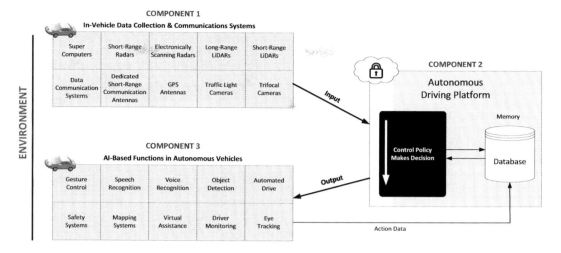

FIGURE 10.1 AI perception action cycle in autonomous cars (Gadam, 2018).

and data communication systems are used to securely communicate valuable information (input) to the autonomous driving cloud platform. The AV first communicates the driving environment and/or the particular driving situation to the autonomous driving platform (Gadam, 2018).

10.3.4.2 Component 2: Autonomous Driving Platform (Cloud)

The autonomous driving platform in the cloud contains an intelligent agent which makes use of AI algorithms to make meaningful decisions. It acts as the control policy or the brain of the AV. This intelligent agent is also connected to a database which acts as a memory where past driving experiences are stored. These data, along with the real-time input coming in through the AV and the immediate environment around it, help the intelligent agent make accurate driving decisions. The AV now knows exactly what to do in this driving environment and/or particular driving situation (Gadam, 2018).

10.3.4.3 Component 3: AI-Based Functions in Autonomous Vehicles

Based on the decisions made by the intelligent agent, the AV is able to detect objects on the road, maneuver through the traffic without human intervention, and get to the destination safely. AVs are also being equipped with AI-based functional systems, such as voice and speech recognition, gesture controls, eye tracking and other driving monitoring systems, virtual assistance, mapping and safety systems to name a few. These functions are also carried out based on the decisions made by the intelligent agent in the autonomous driving platform. These systems have been created to give customers a great user experience and keep them safe on the roads. The driving experiences generated by every ride are recorded and stored in the database to help the intelligent agent make much more accurate decisions in the future.

This data loop, called the perception action cycle, takes place repetitively. The more the number of perception action cycles take place, that much more intelligent the intelligent agent becomes, resulting in a higher accuracy of decisions, especially in complex driving situations. With connected vehicles, more driving experiences are recorded, enabling the intelligent agent to make decisions based on data generated by multiple AVs. This means that not every AV has to go through a complex driving situation before it can actually understand it.

AI, especially NNs and deep learning, has become an absolute necessity to make AVs function properly and safely. AI is leading the way for the launch of Level 5 AVs, where there will be no need for a steering wheel, accelerator, or brakes (Gadam, 2018).

10.3.5 AI in Action: Autonomous Vehicles

AVs will transform our daily lives and our communities. What seemed like science fiction a decade ago is now visible, as test vehicles gather data, tune sensors, and develop the AI to make cars self-driving and safer. Every major auto company, their suppliers, and startups across the globe are using the latest technology in a race to the future where cars drive themselves.

It isn't enough for the vehicle to navigate itself. It must also be prepared for the unexpected. There can't be instructions for every pedestrian, careless driver, obstruction, or fault. Instead, the industry is using AI frameworks to build vehicle controls that recognize and react. Taking data from dozens of sensors, the controls recognize the difference between a stationary rock and an animal crossing the street and quickly steer, brake, or accelerate to avoid it. To do this faster and more safely than a human driver, they use machine learning and deep learning to "teach" systems to classify, measure, and react to unanticipated scenarios based upon previous data. Every run of each vehicle adds more information, and the feedback improves the system.

As more vehicles travel, more data are captured. With light detection and ranging (LiDAR), radar, cameras, Global Positioning Systems (GPS), internal systems, and mechanical sensors, industry experts estimate an AV can generate close to 10 TB per hour. These data come in many forms, but what is critical is that they are all properly identified as being from that one car at that one time, and they are in proper order so that the inputs, system recognition, and reaction can be reconstructed, so humans can evaluate whether the vehicle performed optimally.

Evaluating the effectiveness of the AI programs that drive the car requires a continuous process of testing. By going through many series of scenarios, the current program is tested against the latest version. This is a virtuous cycle of champion and challenger repeated through a growing volume of data and unique test cases to improve the AI and the systems behind it. To support this testing, high-speed scalable storage drives clusters of servers, each with multiple graphics processing units (GPUs) programmed to quickly crunch through the data. Scalability in a balanced system with sufficient data bandwidth, fast storage, networks, and servers is critical (O'Flaherty, 2018).

10.3.5.1 Managing the Data Flow

Properly managing these data is not a trivial task. Initially, data were stored within the test vehicle. Cables were attached after testing, and high-speed dedicated networks moved the terabytes (TBs) of data to high-speed storage for computational analysis and human review. As development matured, and there were more cars on the road, the transfer complexity increased, and there was a shift to wireless networks. Data value will soon be strongly based on metadata–the data about the data–which will identify the vehicle, sensor, time, and location where those data were gathered. With proper metadata management, developers will be able to quickly augment specialized training and testing sets to help develop features or test a particular situation.

Long-term data retention and rapid represent another challenge for AVs. A large amount of data is needed for development, but there is a potential for massive data growth with each new car on the road. Each vehicle, each near-miss, and every accident provide raw data to improve the AV's AI systems. As systems are improved and software is developed, automakers will need to demonstrate that previous incidents will be avoided in the future. Similarly, should human error or some other unique circumstance result in an incident, automakers will need a library of similar situations (O'Flaherty, 2018).

As we look at the data requirements for AVs, three distinct data patterns emerge (O'Flaherty, 2018):

1. High-speed and scalable data are required for continuous development.
2. Rapid data identification, tracking, and access using metadata are required to manage the complexity.
3. Long-term storage is required to deliver low cost and high reliability while handling the rapid retrieval of random data.

Artificial Intelligent Infiltration

10.3.6 Five Ways Artificial Intelligence Is Impacting the Automotive Industry

When you think of AI applications in the automotive industry, you might first think of self-driving cars. But AI can do more than drive. It can keep us connected, on schedule, and safe even when we are driving ourselves. That all adds up to big business. The value of AI in automotive manufacturing and cloud services will exceed $10.73 billion by 2024. In this section, we look at five applications of AI that are impacting automakers, vehicle owners, and service providers (Novosilska, 2018).

10.3.6.1 AI Driving Features

Cars with AI offer two levels of functionality: driver assist and fully autonomous mode (Novosilska, 2018). Here are the differences (Novosilska, 2018):

10.3.6.1.1 Driver Assist

Before the automotive industry is comfortable letting AI take the wheel, it first wants to put it in the copilot's seat. AI lends itself perfectly to powering advanced safety features for connected vehicles, and that helps customers, manufacturers, and regulators get comfortable with AI as the driver before it is officially allowed to do so. By monitoring dozens of sensors, AI can identify dangerous situations. It can then alert the driver or take emergency control of the vehicle in order to avoid an accident. Emergency braking, cross traffic detectors, blind spot monitoring, and driver-assist steering can help avoid accidents and save lives in the process.

10.3.6.1.2 Driverless Automobiles

The mechanical muscle needed to control vehicle steering, braking, and acceleration has been within reach for nearly a century. The reason autonomous cars are not ready jamming the streets is their lack of a brain. The amount of processing power needed to drive a vehicle is enormous. Despite the power of modern computers, conventional computer programs simply are not up to the task. Simply stated, driving involves more than following a set of rules, or an algorithm; it involves learning. In other words, it requires AI. A number of automakers and automotive startups are working on AI applications for the automotive industry, but two companies are leading the pack in the development of truly driverless cars: Google and Tesla.

10.3.6.1.3 Google

Waymo is Google's foray into the autonomous production vehicle market, and appears to be their long-term plan to achieve market dominance. Waymo's AI software crunches data from the vehicles' LiDAR, radar, high resolution cameras, GPS, and cloud services to produce control signals that operate the vehicle. AI does more than respond to what is happening in the vehicle's vicinity. Powerful AI deep learning algorithms can accurately predict what objects in the vehicle's travel path are likely to do. A pedestrian on the sidewalk? Waymo knows they might step into the street at any moment. A vehicle stalled in the turn lane? Waymo anticipates that it might start moving again. The most valuable aspect of AI in automotive applications is that it is constantly learning and adjusting the rules it uses to navigate the road. Each vehicle makes the information it learns available to the rest of the fleet. The result is a virtual NN of self-driving vehicles that learn as they go.

10.3.6.1.4 Tesla

Tesla has succeeded in becoming a household name in the electric car market. Now it wants to do the same thing with self-driving vehicles. Eight cameras, an array of ultrasonic sensors, sonar, forward-facing radar, and GPS cull pretty much the same kind of data from the environment as Waymo. Those data are fed into an AI program that turns sensory data into vehicle control data. Tesla's Autopilot software goes beyond driving the car where you tell it to go. If you are not in the mood for talking, AutoPilot will check your calendar and drive you to your scheduled appointment. Every new Tesla comes fully equipped for autonomous driving. All that is needed is regulatory approval so the company can enable the software, putting AI in the driver's seat.

10.3.6.2 AI Cloud Services

Connected vehicles need huge amounts of data. The application of AI cloud platforms ensures that data are available when needed (Novosilska, 2018).

10.3.6.2.1 Predictive Maintenance

Unlike conventional vehicles, connected vehicles can do more than alert us with check engine lights, oil lights and low battery indicators. AI monitors hundreds of sensors and is able to detect problems before they affect vehicle operation. By monitoring thousands of data points per second, AI can spot minute changes that may indicate a pending component failure—often long before the failure could leave us stranded. In October 2018, Volkswagen and Microsoft announced a partnership, designed to transform the auto company into a digital service-driven business. By tapping the power of Azure IoT, PowerBI, and Skype, Volkswagen plans to offer customer experience, telematics, and productivity solutions for the automotive market. Volkswagen recognizes that customers need more than gimmicks to see the value in cloud-based AI solutions. That is why they intend to offer predictive maintenance and Over-the-Air (OTA) software updates for their entire brand of vehicles.

10.3.6.2.2 Individualized Marketing

From email ads that hide beneath trick subject lines to pay-per-click website ads, and social media monetization, advertising competition has become fierce. To make matters worse, sorting through raw data to target qualified prospects has become impractical. AI-based cloud platforms offer the ideal solution to accurately target a captive audience of qualified prospects. With AI's connection to big data, vehicle infotainment systems can be used to suggest products and services to drivers based on a wealth of raw data. For example, a driver whose social media posts announce wedding plans can be alerted of a sale at a bridal store just up the street. A low-fuel situation can automatically suggest the nearest gas station (which, of course, has paid for the privilege). Or the driver's eating habits can lead the system to suggest a relevant restaurant just around the corner. In short, AI has the power to know the driver's needs and wants and to know when he or she is in close proximity to businesses that can meet them.

10.3.6.3 AI Automotive Insurance

The insurance industry and AI have one thing in common both spend a great deal of time predicting the future. It is no surprise, then, that insurance has embraced the use of AI in cars to help make risk assessments in real time. Among other things, AI speeds up the process of filing claims when accidents do occur. The following sections give examples of how the insurance industry is using deep learning technology (Novosilska, 2018).

10.3.6.3.1 AI-Powered Driver Risk Assessment

AI levels the playing field for drivers by creating risk profiles based on drivers' individual risk factors found in big data. Rather than relying on driving history to set premiums, AI looks at a myriad of less-obvious factors that can predict how safe a driver is likely to be. From health issues to a recent divorce, AI can find details about a driver that may affect his or her ability to drive safely. Since such issues may be recent, they might not have affected the driver in the past but potentially could in the future.

10.3.6.3.2 DIY Auto Claims

Art Financial solidified its position in the Chinese auto market in May 2018 with the release of version 2.0 of its AI-powered video app, Dingsunbao 2.0. The app enables drivers to perform their own auto damage assessment for their insurance companies. On-screen instructions guide the user on how to video his or vehicle damage for the insurance claim. While a human insurance adjuster still must sign off on each claim, AI quickly informs the driver how to have the vehicle repaired and what will be covered by insurance. Dingsunbao is typical of the AI apps that every automotive insurance company will eventually provide to their customers. The 1 billion renminbi that the app has already saved the Chinese auto insurance industry leave no room for doubt.

10.3.6.4 AI in Car Manufacturing

AI is not only changing what a vehicle can do, it is also changing how vehicles are built. Assembly line robots are nothing new. They have helped assemble vehicles since 1960s. What is new are smart robots that work with their human counterparts, rather than just alongside of them. In early 2018, Kia Motors began working with Hyundai Vest Exoskeleton (H-VEX) to develop wearable industrial robots for their assembly lines. Having identified AI as one of its top five future growth areas, Kia assembled a robotics team to lead the way. The Vest Exoskeleton (H-VEX) and Hyundai Chairless Exoskeleton (H-CEX) wearable robots protect workers' knees, back, and neck while giving them the mobility and strength to perform their jobs. AI is helping build cars in other ways, as well. Collaborative robots use AI to sense what human workers are doing, and they adjust their motions to avoid injuring their nonmetallic co-workers. Automated guided vehicles (AVGs) are being used to move materials around automotive plants without human intervention. AI enables these autonomous delivery vehicles to identify objects in their paths and to adjust their route accordingly. Finally, painting and welding robots, when powered by AI, can do more than follow a preprogrammed routine. AI empowers them to identify defects or irregularities in materials and to adjust accordingly or to alert quality control personnel (Novosilska, 2018).

10.3.6.5 Driver Monitoring

AI enables cars to do more than watch the road, it can help them keep an eye on the driver. Israeli automotive computer vision startup eyeSight uses AI and deep learning to offer a plethora of in-car automotive solutions. Using advanced time-of-flight (TOF) cameras and IR sensors, eyeSight's AI software detects driver behavior in four key areas (Novosilska, 2018).

10.3.6.5.1 Driver Identification

eyeSight's AI software detects whether or not the driver is in the vehicle.

10.3.6.5.2 Driver Recognition

Using advanced AI facial recognition algorithms, eyeSight can detect which driver in a family is operating the vehicle. Since each driver has his or her own preferences, the system can automatically adjust the seat, mirrors, and temperature to suit the individual. eyeSight offers both stand-alone proprietary software and hardware and software bundles.

10.3.6.5.3 Driver Monitoring

By "observing" eye gaze, eye openness, and head position, eyeSight can detect distracted driving and alert the driver to keep his or her eyes on the road. Drowsiness is detected by eye openness and head position, allowing the system to wake the driver up if necessary. Contextual control allows eyeSight to tailor the content of the heads-up display (HUD) according to where the driver's eyes are focused.

Upper body detection detects the driver's posture. In case of a crash, air bags can be deployed in a manner that will reduce injury based on how the driver is sitting.

10.3.6.5.4 Infotainment Control

Just because someone is driving does not mean he or she does not need to interact with the vehicle infotainment system. eyeSight recognizes simple hand gestures to allow drivers to control the infotainment system without taking their eyes off the road.

10.3.7 How Artificial Intelligence Is Key for Autonomous Vehicle Development

Autonomous technology is set to transform the motor industry, and as the autonomous mobility industry takes shape, AI could play a much bigger role. The Society of Automotive Engineers uses a classification system of five levels of vehicle autonomy based on the amount of necessary driver intervention. Level 1, for example, requires a driver to be in control at all times but allows for automated acceleration and braking. Level 5 requires no human intervention. Most automakers are on the cusp of introducing Levels 3 and 4 in AVs. The US Self-Drive Act establishes key deadlines for how these technologies will make their way onto the streets and highways. It will allow stakeholders to put up to 100,000 AVs on the road for testing purposes by 2021. The pace at which the technology is developing has created an environment where traditional manufacturers risk being overtaken by tech firms. This is pushing cross-industry collaboration. Self-driving systems are now based on complex algorithms, cameras, and LiDAR sensors to create a digital world that helps AVs orient their position on the road and identify other vehicles, cyclists, and pedestrians. Such systems are incredibly difficult to design and produce. They must be programmed to cope with an almost limitless number of variables found on roads. As the industry looks to machine learning as the basis for autonomous systems, AI could be the next step in the development (Pawsey, 2017).

10.3.7.1 AI the Driving Force

Earlier this year, Apple Chief Executive Tim Cook described the challenge of building AVs as "the mother of all" AI projects; huge amounts of computing power are required to interpret all the data harvested from a range of sensors and then enact the correct procedures for constantly changing road conditions and traffic situations. However, deep learning systems can be "trained" to drive and develop decision-making processes like a human.

Deep learning is a set of algorithms that model high-level data concepts by using architectures of multiple nonlinear transformations. Deep learning architectures such as deep neural networks (DNN) and CNN have already been applied successfully to fields like speech recognition. Several companies, such as tech giant Nvidia and startup Drive.ai, are developing AI systems for AVs using deep learning (Pawsey, 2017).

10.3.7.2 Nvidia Drive PX

Originally a computer game chip company, Nvidia is now one of the leading tech firms in the autonomous driving field. Its Drive PX family of AI car computers enables automakers, suppliers, and startups to accelerate the production of autonomous and automated vehicles. In October, the firm introduced Pegasus, a new member of the Drive PX family. The Pegasus AI computer can compute 320 trillion deep learning operations per second. Roughly the size of a license plate, it has the AI performance of a 100-server data center and makes Nvidia the first tech company to offer a complete A/V stack for Levels 4 and 5 autonomous driving. Data are fused from multiple cameras, as well as LiDAR, radar, and ultrasonic sensors. This allows algorithms to accurately understand the full 360° environment around the car. The use of DNN for the detection and classification of objects dramatically increases the accuracy of the fused sensor data (Pawsey, 2017).

10.3.7.3 Drive.ai

Drive.ai is a Silicon Valley startup founded in 2015 by former colleagues from Stanford University's AI lab. Drive.ai says it is "building the brain of the self-driving car." Drive sees deep learning as the only viable route to fully AVs. By structuring its approach to self-driving cars around deep learning, Drive has been able to quickly scale to safely cope with the wide range of conditions and variables that AVs need to master. The difference is that while many companies follow a piecemeal approach to deep learning, using it for some components of the system but not all, Drive approaches self-driving from a holistic deep learning perspective.

The hardware used on Drive's fleet of four test vehicles consists of an array of sensors, cameras, and LiDAR retrofitted to the roof. The system also takes advantage of the car's own integrated sensors, such as radar and rear cameras. Each vehicle is continuously capturing data for map generation, deep learning, and driving tasks as it navigates roads.

One note of caution in the use of deep learning is the so-called black box. Once the system is "trained," data are fed in and an interpretation of the data is fed out. But the actual decision-making process in between is not something humans can necessarily understand. This is why many companies use deep learning for perception but are less comfortable using it for decision-making. It's one of the key issues with AI technology. If the system makes a mistake, it is important to be able to understand why in order to correct it. To counter this problem, Drive has broken the system down into parts. Understanding how each of the parts can be validated in different ways provides more confidence in how the system will behave (Pawsey, 2017).

REFERENCES

Al-Najjar B., Alsyouf I., 2003. Selecting the most efficient maintenance approach using fuzzy multiple criteria decision making. *International Journal of Production Economics*, 84(1), 85–100.

Batanov D., Nagarur N., Nitikhunkasem P., 1993. EZPERT-MM: A knowledge-based system for maintenance management. *Artificial Intelligence in Engineering*, 8(4), 283–291.

Bringsjord S., Sen A., 2016. On creative self-driving cars: Hire the computational logicians, fast. *Applied Artificial Intelligence*, 30(8), 758–86. doi:10.1080/08839514.2016.1229906.

Buntz B., 2019. Predictive maintenance and predicting industrial revolutions. April 18, 2019. IoT World Today. www.iotworldtoday.com/2019/04/18/predictive-maintenance-and-predicting-industrial-revolutions/. Viewed: July 29, 2019.

Cai J., Jia Y., Gu C., Wu W., 2011. Research of wartime equipment maintenance intelligent decision making based on case-based reasoning. *Procedia Engineering*, 15, 163–167.

CFMS, 2018a. AI for industrial inspection. Centre for Modelling & Simulation. https://cfms.org.uk/industry-themes/ai-for-industrial-inspection/. Viewed: July 29, 2019.

CFMS, 2018b. Applicability of artificial intelligence for industrial inspection. Centre for Modelling & Simulation. October 2018. https://cfms.org.uk/news-events-opinions/opinions/2018/october/applicability-of-artificial-intelligence-for-industrial-inspection/. Viewed: July 29, 2019.

Cheng Z., Jia X., Gao P., Wu B., 2008. A framework for intelligent reliability centered maintenance analysis. *Reliability Engineering System Safety*, 93(6), 806–814. doi:10.1016/j.ress.2007.03.037.

Chootinan P., Chen A., Horrocks M. R., Bolling D., 2006. A multi-year pavement maintenance program using a stochastic simulation-based genetic algorithm approach. *Transportation Research Part A: Policy and Practice*, 40(9), 725–743.

Chou J. S., 2009. Web-based CBR system applied to early cost budgeting for pavement maintenance project. *Expert Systems with Applications*, 36(2), Part 2, 2947–2960.

Christodoulou S., Deligianni A., Aslani P., Agathokleous A., 2009. Risk-based asset management of water piping networks using neurofuzzy systems. *Computers, Environment and Urban Systems*, 33(2), 138–149.

Coeckelbergh M., 2016. Responsibility and the moral phenomenology of using self-driving cars. *Applied Artificial Intelligence*, 30(8), 748–57. doi:10.1080/08839514.2016.1229759.

Cunneen M., Mullins M., Murphy F., Gaines S., 2018. Artificial driving intelligence and moral agency: Examining the decision ontology of unavoidable road traffic accidents through the prism of the trolley dilemma. *Applied Artificial Intelligence*, 33(3), 267–293.

Cunneen M., Mullins M., Murphy F., 2019. Autonomous vehicles and embedded artificial intelligence: The challenges of framing machine driving decisions, applied artificial intelligence. *Applied Artificial Intelligence*, 33(8), 706–731. doi:10.1080/08839514.2019.1600301.

Derigent W., Thomas E., Levrat E., Iung B., 2009. Opportunistic maintenance based on fuzzy modelling of component proximity. *CIRP Annals*, 58(1), 29–32.

Dhaliwal D. S., 1986. The use of AI in maintaining and operating complex engineering systems. In A. Mamdani & J. E. Pstachion (Eds.) *Expert systems and Optimisation in Process Control* (pp. 28–33). Aldershot: Gower Technical Press.

Edge 4 Industry, 2018. Drones and robots are taking over maintenance inspections ... and that's not a bad thing. November 2, 2018, by FAR-EDGE. www.edge4industry.eu/2018/11/02/drones-and-robots-are-taking-over-maintenance-inspectionsand-thats-not-a-bad-thing/. Viewed: July 29, 2019.

Floridi L., J. Cowls, M. Beltrametti, R. Chatila, P. Chazerand, V. Dignum, B. Schafer, 2018. AI4 People: An ethical framework for a good AI society: Opportunities, risks, principles, and recommendations. *Minds and Machines*, 28(4), 689–707. doi:10.1007/s11023-018-9482-5.

Franklin S., Ferkin M., 2006. An ontology for comparative cognition: A functional approach. *Comparative Cognition & Behavior Reviews*, Vol. 1, pp. 36–52.

Frewer L. J., Miles S., Marsh R., 2002. The media and genetically modified foods: Evidence in support of social amplification of risk. *Risk Analysis: An International Journal*, 22(4):701–711.

Gabbar H., Yamashita. H., Suzuki K., Shimada Y., 2003. Computer-aided RCM-based plant maintenance management system. *Robotics and Computer-Integrated Manufacturing*, 19(5), 449–458.

Gadam S., 2018. Artificial intelligence and autonomous vehicles. April 19, 2018. https://medium.com/datadriveninvestor/artificial-intelligence-and-autonomous-vehicles-ae877feb6cd2. Viewed: July 27, 2019.

Giarratana C., 2018. Artificial intelligence and the future of autonomous vehicles. Safety Resource Center. August 24, 2018. www.trafficsafetystore.com/blog/artificial-intelligence-and-the-future-of-autonomous-vehicles/. Viewed: July 28, 2019.

Goodall N. J., 2016. Away from trolley problems and toward risk management. *Applied Artificial Intelligence*, 30(8). doi:10.1080/08839514.2016.1229922.

Haupt R. L., Haupt S. E., 2004. *Practical Genetic Algorithms*, Second Edition. Hoboken, NJ: John Wiley & Sons, Inc.

Hengstler M., Enkel E., Duelli S., 2016. Applied artificial intelligence and trust — The case of autonomous vehicles and medical assistance devices. *Technological Forecasting & Social Change*, 105 105–120.

Huang S. J., 1998. A genetic-evolved fuzzy system for maintenance scheduling of generating units. *International Journal of Electrical Power & Energy Systems*, 20(3), 191–199.

Jiejuan T., Dingyuan M., Dazhi X., 2004. A genetic algorithm solution for a nuclear power plant risk–cost maintenance model. *Nuclear Engineering and Design*, 229(1), 81–89.

Johnson D. G., Verdicchio M., 2017. *Minds & Machines*, 27, 575. doi:10.1007/s11023-017-9417-6.

Khan F. I., Sadiq R., Haddara M. M., 2004. Risk-Based Inspection and Maintenance (RBIM) multi-attribute decision-making with aggregative risk analysis. *Process Safety and Environmental Protection*, 82(6), 398–411.

Kim H., Nara K., Gen M., 1994. A method for maintenance scheduling using GA combined with SA. *Computers & Industrial Engineering*, 27(1–4), 477–480.

Kobbacy K. A., 1992. The use of knowledge-based systems in evaluation and enhancement of maintenance routines. *International Journal of Production Economics*, 24, 243–248.

Kobbacy K. A. H., Proudlove N. L., Harper M. A., 1995. Towards an intelligent maintenance optimization system. *Journal of the Operational Research society*, 46, 229–240.

Kobbacy K. A., Vadera S., Rasmi M., 2007. AI and OR in management of operations: History and trends. *Journal of the Operational Research Society*, 58, 10–28.

Kobbacy K. A., Prabhakar Murthy D. N., 2008. *Complex System Maintenance Handbook. Artificial Intelligence in Maintenance* (pp. 209–231). London: Springer.

Kobbacy K. A., 2012. Application of artificial intelligence in maintenance modelling and management. In *2nd IFAC Workshop on Advanced Maintenance Engineering*, Sevilla, Spain: Services and Technology Universidad de Sevilla, 22–23 November, 2012.

Lapa C. M. F., Claudio M. N. A., Pereira C. M. N. A., Mol A. C., 2000. Maximization of a nuclear system availability through maintenance scheduling optimization using a genetic algorithm. *Nuclear Engineering and Design*, 196(2), 219–231.

Lapa C. M. F., Claudio M. N. A., Pereira C. M. N. A., Paes de Barros M., 2006. A model for preventive maintenance planning by genetic algorithms based in cost and reliability. *Reliability Engineering &System Safety*, 91(2), 233–240.

Lee J., Sanmugarasa K., Blumenstein M., Loo Y. C., 2008. Improving the reliability of a Bridge Management System (BMS) using an ANN-based Backward Prediction Model (BPM). *Article in Automation in Construction*, 17(6), 758–772. doi:10.1016/j.autcon.2008.02.008.

Levitin G., Lisnianski A., 1999. Joint redundancy and maintenance optimization for multistate series–parallel systems. *Reliability Engineering & System Safety*, 64(1), 33–42.

Levitin G., Lisnianski A., 2000. Optimization of imperfect preventive maintenance for multi-state systems. *Reliability Engineering & System Safety*, 67(2), 193–203.

Lin P., 2016. Why ethics matters for autonomous cars. In M. Maurer, J. Gerdes, B. Lenz, & H. Winner (Eds.), *Autonomes Fahren* (pp. 69–85). Berlin, Heidelberg: Springer Vieweg.

Malik H., Singh S., Kr M., Jarial R. K., 2012. UV/VIS response based fuzzy logic for health assessment of transformer oil. *Procedia Engineering*, 30, 905–912.

Marchant G. E., 2011. The growing gap between emerging technologies and the law. In G. E. Marchant, B. R. Allenby, & J. R. Heckert (Eds.), *The Growing Gap between Emerging Technologies and Legal-Ethical Oversight: The Pacing Problem* (pp. 19–33). Dordrecht: Springer.

Marseguerra M., Zio E., 2000 Optimizing maintenance and repair policies via a combination of genetic algorithms and Monte Carlo simulation. *Reliability Engineering & System Safety*, 68(1), 69–83.

Marseguerra M., Zio E., Podofillini L., 2002. Condition-based maintenance optimization by means of genetic algorithms and Monte Carlo simulation. *Reliability Engineering & System Safety*, 77(2), 151–165.

Mazhar M. I., Kara S., Kaebernick H., 2007. Remaining life estimation of used components in consumer products: Life cycle data analysis by Weibull and artificial neural networks. *Journal of Operations Management*, 25(6), 1184–1193.

Millar J., 2016. An ethics evaluation tool for automating ethical decision-making in robots and self-driving cars. *Applied Artificial Intelligence*, 30(8), 787–809. doi:10.1080/08839514.2016.1229919.

Moazami D., Behbahani H., Muniandy R., 2011. Pavement rehabilitation and maintenance prioritization of urban roads using fuzzy logic. *Expert Systems with Applications*, 38(10), 12869–12879.

Monga A., Zuo M. J., 2001. Optimal design of series parallel systems considering maintenance and salvage values. *Computers & Industrial Engineering*, 40(4), 323–337.

Moradi E., Fatemi Ghomi S. M. T., Zandieh M., 2011. Bi-objective optimization research on integrated fixed time interval preventive maintenance and production for scheduling flexible job-shop problem. *Expert Systems with Applications*, 38(6), 7169–7178.

Nahas N., Khatab A., Ait-Kadi D., Nourelfath M., 2008. Extended great deluge algorithm for the imperfect preventive maintenance optimization of multi-state systems. *Reliability Engineering & System Safety*, 93(11), 1658–1672.

National Highway Traffic Safety Administration (NHTSA), 2016. *Federal Automated Vehicles Policy: Accelerating the Next Revolution in Roadway Safety*. Washington, DC, USA: US Department of Transportation.

Nguyen D., Bagajewicz M. J. 2008. Optimization of preventive maintenance scheduling in processing plants. *Computer Aided Chemical Engineering*, 25. doi:10.1016/S1570-7946(08)80058-2.

Nourelfath M., Châtelet E., Nahas N., 2012. Joint redundancy and imperfect preventive maintenance optimization for series–parallel multi-state degraded systems. *Reliability Engineering & System Safety*, 103, 51–60.

Novosilska L., 2018. 5 ways artificial intelligence is impacting the automotive industry. Ignite Ltd. November 30, 2018. https://igniteoutsourcing.com/automotive/artificial-intelligence-in-automotive-industry/. Viewed: July 27, 2019.

O'Flaherty D., 2018. AI in action: Autonomous vehicles. IBM. October 11, 2018. www.ibm.com/blogs/systems/ai-in-action-autonomous-vehicles/. Viewed: July 27, 2019.

Otto S., 2019. Artificial intelligence for predictive maintenance. February 6, 2019. www.aerospacemanufacturinganddesign.com/article/artificial-intelligence-for-predictive-maintenance/. Viewed: July 29, 2019.

Pawsey C., 2017. How artificial intelligence is key for autonomous vehicle development. November 29, 2017 by Fueling Your Mind for the Road Ahead, @trucksdotcom. www.trucks.com/2017/11/29/artificial-intelligence-key-autonomous-vehicle-development/. Viewed: July 28, 2019.

Pidgeon N., Kasperson R. E., Slovic P., 2003. *The Social Amplification of Risk*. Cambridge, MA: Cambridge University Press.

Renn O., 2008. *Risk Governance*. London: Routledge. doi:10.4324/9781849772440.

Sasmal S., Ramanjaneyulu K., 2008. Condition evaluation of existing reinforced concrete bridges using fuzzy based analytic hierarchy approach. *Expert Systems with Applications*, 35(3), 1430–1443.

Siener M., Aurich J. C., 2011. Quality oriented maintenance scheduling. *CIRP Journal of Manufacturing Science and Technology*, 4(1), 15–23.

Singh M., Markeset T., 2009. A methodology for risk based inspection planning of oil and gas pipes based on fuzzy logic framework. *Engineering Failure Analysis*, 16(7), 2098–2113

Su K. W., Hwang S. L., Liu T. H., 2000. Knowledge architecture and framework design for preventing human error in maintenance tasks. *Expert Systems with Applications*, 19(3), 219–228.

Tan J. S., Kramer M. A., 1997. A general framework for preventive maintenance optimisation in chemical process operations. *Computers & Chemical Engineering*, 21(12), 1451–1469.

Trappl R., 2016. Ethical systems for self-driving cars: An introduction. *Applied Artificial Intelligence*, 30(8), 745–747. doi:10.1080/08839514.2016.1229737.

Tsai Y. T., Wang K. S., Teng H. Y., 2001. Optimizing preventive maintenance for mechanical components using genetic algorithms. *Reliability Engineering & System Safety*, 74(1), 89–97.

Volkanovski A., Mavko B., Boševski T., Čauševski A., Čepin M., 2008. Genetic algorithm optimisation of the maintenance scheduling of generating units in a power system. *Reliability Engineering & System Safety*, 93(6), 779–789.

Yang Z., Yang G., 2012. Optimization of aircraft maintenance plan based on genetic algorithm. *Physics Procedia*, 33, 580–586.

Yu R., Iung B., Panetto H., 2003. A multi-agents based E-maintenance system with case-based reasoning decision support. *Engineering Applications of Artificial Intelligence*, 16(4), 321–333.

Zadeh L. A., 1965. Fuzzy sets. *Information and Control*, 8(3), 338–353.

Zadeh L. A., 1999. Fuzzy logic and the calculi of fuzzy rules, fuzzy graphs, and fuzzy probabilities. *Computers & Mathematics with Applications*, 37(11–12), 35.

11

Big Data Analytics for AV Inspection and Maintenance

11.1 Big Data Analytics and Cyber-Physical Systems

11.1.1 Big Data Analytics

11.1.1.1 Big Data Analytics Definition

Big data analytics is defined as mining of pertinent knowledge and valuable insights from large amounts of stored data (Rajeshwari, 2012). The key objective of such analytics is to facilitate decision-making for researchers, such as offering dashboards, graphics, or operational reporting to monitor thresholds and key performance indicators (KPIs). This involves using mathematical and statistical methods to understand data, simulate scenarios, validate hypotheses, and make predictive forecasts for future incidents.

Data mining is a key concept in big data analytics; it consists of applying data science techniques to analyze and explore large datasets to find meaningful and useful patterns in those data. It involves complex statistical models and sophisticated algorithms, such as machine learning algorithms, mainly to perform four categories of analytics: descriptive analytics, predictive analytics, prescriptive analytics, and discovery (exploratory) analytics.

Descriptive analytics turns collected data into meaningful information for interpreting, reporting, monitoring, and visualization purposes via statistical graphical tools, such as pie charts, graphs, bar charts, and dashboards. Predictive analytics is commonly defined as data extrapolation based on available data for ensuring better decision-making. Prescriptive analytics is associated with descriptive and predictive analytics. Likewise, based on the present situation, it offers options on how to benefit from future opportunities or mitigate a future risk and details the implication of each decision option. Finally, discovery (exploratory) analytics illustrates unexpected relationships between parameters in big data (Rajeshwari, 2012). Some authors argue that the output of predictive analytics can benefit from the potential of descriptive analytics through the use of dashboards and scorecard computations (Bahri, Zoghlami, Abed, & Tavares, 2019).

11.1.1.2 Big Data Analytics

Big data analytics constitutes one of the most important arenas in big data systems, as it reveals hidden patterns, unknown correlations, and other useful information, which in turn, boosts revenue for many businesses. In this section, we present an overview techniques and tools for big data analysis (Atat et al., 2016).

    A. Data mining

        One of the interesting features of cyber-physical systems (CPSs) is automated decision-making. This means that CPS objects are supposed to be smart in sensing, identifying events, and interacting with others (Qu, Wang, & Yang, 2010). The massive data collected by CPS need to be converted into useful knowledge to uncover hidden patterns to find solutions and enhance system performance and quality of services. The process of extracting this useful information is referred to as data mining. One way to facilitate the data mining process is to reduce data complexity by allowing objects to capture only the interesting data rather than all of them. Before data mining can be applied to the data, some processing steps need to be completed, such as key feature selection, preprocessing, and transformation of data. Dimensionality reduction is one potential method to reduce the number of features of the data (Xu, Li, Shu, & Peng, 2014). Chen, Sanga, Chou, Cristini, and Edgerton (2009) used a neural network with k-means clustering via principal component analysis (PCA) to reduce the complexity and the number of dimensions of gene expression data to extract disease-related information from gene expression profiles. Knowledge discovery in databases (KDD) is also used in different CPS scenarios to find hidden patterns and unknown correlations in data so that useful information can be converted into knowledge (Fayyad, Piatetsky-Shapiro, & Smyth, 1996). One such use of KDD is in smart infrastructures, when these systems need to answer queries and make recommendations about the system operation to the facility manager (Behl & Mangharam, 2016).

Tsai, Lai, Chiang, and Yang (2014) broke down the core operations of data mining into three main operations: data scanning, rules construction, and rules update. Data scanning is selecting the needed data by the operator. Rules construction includes creating candidate rules by using selection, construction, and perturbation. Finally, candidate rules are checked by the operator, then evaluated to determine which ones will be kept for the next iteration. The process of scanning, construction, and update operations is repeated until the termination criteria are met. This data mining framework works for deterministic mining algorithms such as k-means, and the metaheuristic algorithms such as simulated annealing and the genetic algorithm (Atat et al., 2016).

Clustering, classification, and frequent pattern are different mining techniques that can be used to make CPSs smarter.

Tsai, Lai, Chiang, and Yang (2014) discussed two purposes for clustering (Atat et al., 2016):

i. Clustering in Internet of Things (IoT) infrastructure;
ii. Clustering in IoT services.

Clustering in the IoT infrastructure helps enhance system performance in terms of identification, sensing, and actuation. For example, in Kardeby (2011), nodes exchanged information to identify whether they could be grouped together depending on the needs of the IoT applications. Clustering can also help provide higher quality IoT services, such as in smart homes (Rashidi, Cook, Holder, & Schmitter-Edgecombe, 2011).

Classification does not require prior knowledge to complete the partitioning of objects into clusters, also known as unsupervised learning. Classification tools include decision trees, *k*-nearest neighbor, naive Bayesian classification, adaboost, and support vector machines. Classification can be done to improve the infrastructure and the services of IoT.

Finally, frequent pattern mining is about uncovering interesting patterns, such as which items will be purchased together with previously purchased items, or suggest items for customers to purchase based on customer's characteristics, behavior, purchase history, and so on. Figure 11.1 illustrates the CPS big data mining process for useful information extraction (Atat et al., 2016).

B. Real-time analytics

Real-time analysis is another approach to produce useful information from massive raw data. Real-time data streams are converted to structured data before being analyzed by big data tools, such as Hadoop. Many application domains such as healthcare, transportation systems, environmental monitoring, and smart cities, require real-time decision-making and control. For example, Twitter data can be real-time analyzed to enhance the prediction process and to make useful recommendations to users; terrorist incident data can be real-time analyzed to predict future incidents; big data streams in healthcare can be analyzed to help medical staff make decisions in real time, and this, in turn, can save patients' lives and improve the healthcare services provided, while reducing medical costs. A near real-time big data analysis architecture for vehicular networks proposed by Daniel, Paul, and Ahmad (2015) consists of a centralized data storage for data processing and a distributed data storage for streaming processed data in real-time analysis (Atat et al., 2016).

Big Data Analytics

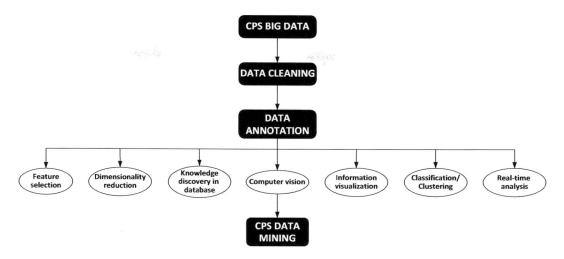

FIGURE 11.1 CPS big data mining process (Atat et al., 2016).

Wang, Mi, Xu, Shu, and Deng (2016) proposed a real-time hybrid-stream big data analytics model for big data video analysis. Zhou et al. (2017) considered online network analysis as a stream analysis issue and proposed using Spark Streaming to monitor and analyze the high-speed Internet traffic data in real time. Cao, Wachowicz, and Cha (2017) suggested mobile-edge computing nodes deployed on a transit bus, with descriptive analytics used to explore meaningful patterns of real-time data streams for transit. Trinks and Felden (2017) discussed the terms and related concepts of real-time analytics (RTA) for industry big data analytical solutions. Ali (2018) provided a framework to efficiently leverage big data technology and allow deep analysis of large and complex datasets for real-time big data warehousing (Atat et al., 2016).

Arranging the data in a representative form can provide information visualization, making the information extraction and understanding of complex large-scale systems much easier (Liu et al., 2016). Geographical Information Systems (GIS) is one important tool of visualization (Chopade, Zhan, Roy, & Flurchick, 2015), as it can help real-time analysis of many applications, such as in healthcare, urban and regional planning, transportation systems, emergency situations, public safety, and so on. Chopade, Zhan, Roy, and Flurchick (2015) proposed a large-scale system data visualization architecture called X-SimViz, which allows real-time dynamic data analytics and visualization. Computer vision is another approach to detecting security anomalies. Visualization can also be a useful tool in predicting real-time cyber attacks. For instance, Tan et al. (2015) used computer vision to transform the network traffic data into images using a multivariate correlation analysis approach based on a dissimilarity measure called Earth Mover's Distance to help detect denial-of-service attacks. A computer vision deep learning algorithm for human activity recognition was proposed by Mo, Li, Zhu, and Huang (2016). The model is capable of recognizing 12 types of human activities with high accuracy and without the need of prior knowledge, useful for security monitoring applications (Atat et al., 2016).

C. Cloud-based big data analytics

Cloud-based analysis in CPS constitutes a scalable and reliable architecture to perform analytics operations on big data streams, such as extracting, aggregating, and analyzing data of different granularities. A massive amount of data is usually stored in spreadsheets or other applications, and a cloud-based analytics service, using statistical analysis and machine learning, helps reduce the big data to a manageable size so information can be extracted, hypotheses can be tested, and conclusions can be drawn from nonnumerical data, such as photos.

Data can be imported from the cloud, and users are able to run cloud data analytics algorithms on big datasets, after which data can be stored back to the cloud. For instance, Yetis et al. (2016) used cloud computing using the MapReduce algorithm to conduct analyze on crime rates in the city of Austin using different attributes like crime type and location to help build a design that prevents future crimes and improves public safety.

Even though cloud computing is an attractive analytics tool for big data applications, it comes with some challenges, mainly concerning security, privacy, and data ownership. Tawalbeh, Mehmood, Benkhlifa, and Song (2016) extended the use of clouds to mobile cloud computing to help overcome the challenge of resource limitations, such as memory, battery life, and CPU power. A mobile cloud computing architecture was suggested for healthcare applications, along with the various big data analytic tools available.

Clemente-Castellety et al. (2015) suggested using a hybrid cloud computing consisting of public and private clouds to accelerate the analysis of massive data workloads on the MapReduce framework without requiring significant modifications to the framework. In a private cloud, cloud services delivered over the physical infrastructure are exclusively dedicated to the tenant. The hybrid cloud uses a set of virtual machines running on the private cloud, which take advantage of data locality; another set of virtual machines runs on a public cloud to run the analysis at a faster rate (Atat et al., 2016).

To optimize the utilization of cloud computing resources, predicting the expected workload and the amount of resources needed is important to reduce waste. Neves, Rose, Katrinis, and Franke (2014) developed a system that predicts the resource requirements of a MapReduce application to optimize bandwidth allocation to the application, while Islam, Keung, Lee, and Liu (2012) used linear networks, along with linear regression, to predict the future need of new resources and virtual machines (VMs). When the system falls short in predicting the right amount of resources needed, it becomes incapable of accommodating a high workload demand, leading to anomalies. Anomaly detection is an essential part of big data analytics, as it helps improve the quality of service by checking whether the measurements of the workload observed and the baseline workloads diverge by a specific margin, with the baseline workloads providing a measure of how the demand changes during a period of time based on historical records (Buyya et al., 2015).

D. Spatial-temporal analytics

Massive data obtained from widely deployed spatiotemporal sensors have caused challenges in data storage, process scalability, and retrieval efficiency. Zhong, Fang, and Zhao (2013) proposed the distributed composite spatiotemporal index approach, VegaIndexer, to efficiently process large amounts of spatiotemporal sensor data. Zheng, Ben, and Yuan (2014) investigated the big data issues in Internet of Vehicles (IOVs) applications and proposed to use cloud-based big data space-time analytics to enhance the analysis efficiency. Sinda and Liao (2017) proposed STAnD to determine anomaly patterns for potential malicious events within these spatial-temporal datasets.

The spatially distributed CPS nodes can be used to analyze location information. Ding, Tan, Wu, and Zhang (2014) proposed an efficient indoor positioning based on a new empirical propagation model using fingerprinting sensors, called the regional propagation model (RPM), based on the cluster-based propagation model theory. In another study, Ding, Tan, Wu, Zeng, and Zhang (2015) used particle swarm optimization (PSO) to estimate the location information using the Kalman filter to update the initial estimated location (Atat et al., 2016).

E. Big data analytical tools

Typical tools for big data analytics, data mining, real-time big data analytics, and cloud-based big data analytics include the following (Atat et al., 2016):

1. Tools for data mining

Hadoop is an open source managed by the Apache Software Foundation. There are two main components for Hadoop, HDFS and MapReduce. HDFS was inspired by GFS. It is a scalable and distributed storage system, an appropriate solution for data-intensive applications, such as those on a gigabyte and terabyte scale. Rather than just being a storage layer of Hadoop, HDFS adds to throughput improvement of the system and supplies efficient fault detection and automatic recovery. MapReduce is a framework used to analyze massive datasets in a distributed fashion by means of numerous machines. There are two functions in the mathematical model of MapReduce, Map and Reduce, both of which are available to be programmed. R is an open source software environment for data mining developed by AT&T Bell Labs. It is a realization of the S language used to explore data, implement statistical analysis, and draw plots. Compared with S, R is more popular and supported by a large number of database manufacturers, such as Teradata and Oracle (Atat et al., 2016).

2. Tools for real-time big data analytics

Storm is a distributed real-time computing system for big data analysis. Compared with Hadoop, Storm is easier to operate and more scalable to provide competitive and efficient services. Storm makes use of distinct topologies for different storm tasks in terms of storm clusters, which are composed of master nodes and worker nodes. The master nodes and worker nodes play two kinds of roles in the fields of big data analysis, nimbus and supervisor, respectively. The functions of these two roles are in agreement with job tracker and task tracker of the MapReduce framework. Nimbus takes charge of code distribution across the storm cluster, the schedule and assignment of worker nodes tasks, and the whole system surveillance. The supervisor compiles tasks given by nimbus. Splunk is also a real-time platform designed for big data analytics. Based on the Web interface, Splunk is available to search, monitor, and analyze machine-generated big data, and the results are exhibited in different formats including graphs, reports, alerts, and so on. Unlike other real-time analytical tools, Splunk provides various smart services for commercial operations, system problem diagnosis, and so on (Atat et al., 2016).

3. Tools for cloud-based big data analytics

The most popular tool for cloud-based big data analytics, Google's cloud computing platform consists of GFS (big data storage), BigTable (big data management), and MapReduce (cloud computing). GFS is a distributed file system, enhanced to meet the requirements of big data storage and usage demands of Google Inc. In order to deal with the commodity component failure problem, GFS facilitates continuous surveillance, error detection, and component fault tolerance. GFS adopts a clustered approach that divides data chunks into 64-KB blocks and stores a 32-bit checksum for each block. BigTable supplies highly adaptable, reliable, applicable, and dynamic control and management in the field of big data placement, representation, indexing, and clustering for enormous and distributed commodity servers; it constitutes a row, column, record tablet, and time stamp (Atat et al., 2016).

F. Summary and insights

To better extract information from big data, it is important to enhance the cloud's analytic performance. A combination of the different techniques discussed in this section can be used to optimize cloud computing resources. If VMs and cloud resources and requirements can be predicted beforehand, workloads can be efficiently processed and analyzed by taking advantage of the cloud's analytical tools. Using a hybrid cloud can further speed up the analysis of workloads, leading to reduced latency and efficient data mining (Atat et al., 2016).

11.1.1.3 Descriptive Tasks of Big Data Analytics

The descriptive task of big data analytics is to identify the common characteristics of data with the purpose of deriving patterns and relationships existing in the data. The descriptive functions of big data mining include classification analysis, clustering analysis, association analysis, and logistic regression analysis (Lee, Cao, & Ng, 2017).

- Classification analysis: Classification is a typical learning model used in big data analytics; it aims to build a model for making predictions on data features from a predefined set of classes according to certain criteria. A rule based classification is used to extract IF-THEN rules to classify different categories. Examples of classification techniques are neural networks, decision trees, and support vector machines.
- Clustering analysis: This is the process of grouping data into separate cluster of similar objects to segment data and acquire the data features. Data can be divided into subgroups according to their characteristics. Practioners may formulate appropriate strategies for different clusters. Common examples of clustering techniques are the k-means algorithm, self-organizing map, hill climbing algorithm, and density-based spatial clustering.
- Association analysis: An association model helps practitioners recognize groups of items that occur synchronously. The association algorithm is developed to search frequent sets of items with a minimum specified confidence level. The criteria support and confidence level help to identify the most important relationships among items.
- Regression analysis: Regression analysis determines the logical relationship of the historical data. The focus in regression analysis is on measuring the dependent variable given one or several independent variables; the result is a conditional estimation of the expected outcome using the regression function. Linear regression, nonlinear regression, and exponential regression are common statistical methods to measure the best fit for a set of data.

11.1.2 Cyber-Physical Systems (CPSs)

11.1.2.1 Introduction

Computing and communication capabilities will soon be embedded in all types of objects and structures in the physical environment, and applications with enormous societal impact and economic benefit will be created by harnessing these capabilities across both space and time. Such systems that bridge the cyber world of computing and communications with the physical world are referred to as cyber-physical systems (CPSs). CPSs are physical and engineered systems whose operations are monitored, coordinated, controlled, and integrated by a computing and communication core. This intimate coupling of the cyber and physical will be manifested from the nano-world to large-scale wide-area systems of systems. The Internet transformed how humans interact and communicate with one another, revolutionized how and where information is accessed, and even changed how people buy and sell products. Similarly, CPS will transform how humans interact with and control the physical world around them.

Examples of CPS include medical devices and systems, aerospace systems, transportation vehicles and intelligent highways, defense systems, robotic systems, process control, factory automation, building and environmental control, and smart spaces. CPSs interact with the physical world and must operate dependably, safely, securely, and efficiently in real time.

The World Wide Web can be considered a confluence of three core enabling technologies: hypertext, communication protocols like TCP/IP, and graphical interfaces. This integration enabled significant leaps in technology (e.g., graphics, networking, semantic Webs, multimedia interfaces and languages), infrastructure (e.g., global connectivity with increasing bandwidth, PCs for every desktop and laptop), and applications (e.g., ecommerce, auctions, entertainment, digital libraries, social networks, and online communities). Likewise, CPS can be considered to be a confluence of embedded systems, real-time systems, distributed sensor systems, and controls.

The promise of CPS is pushed by several recent trends: the proliferation of low-cost and increased-capability sensors in increasingly smaller forms; the availability of low cost, low-power, high-capacity, small form-factor computing devices; the wireless communication revolution; abundant Internet bandwidth; and continuing improvements in energy capacity, alternative energy sources, and energy harvesting. The need for CPS technologies is also being pulled by CPS vendors in sectors such as aerospace, building and environmental control, critical infrastructure, process control, factory automation, and healthcare who are increasingly finding that the technology base to build large-scale safety-critical CPS correctly, affordably, flexibly, and on schedule is seriously lacking.

CPSs bring together the discrete and powerful logic of computing to monitor and control the continuous dynamics of physical and engineered systems. The precision of computing must interface with the uncertainty and noise in the physical environment. The lack of perfect synchrony across time and space must be dealt with. The failures of components in both the cyber and physical domains must be tolerated or contained. Security and privacy requirements must be enforced. System dynamics across multiple time scales must be addressed. Scale and increasing complexity must be tamed. These needs call for the creation of innovative scientific foundations and engineering principles. Trial-and-error approaches to build computing-centric engineered systems must be replaced by rigorous methods, certified systems, and powerful tools. Analyses and mathematics must replace inefficient and testing-intensive techniques. Unexpected accidents and failures must fade, and robust system design must become an established domain. New sensors and sensor fusion technologies must be developed. Smaller and more powerful actuators must become available.

The confluence of the underlying CPS technologies enables new opportunities and poses new research challenges. CPSs will be composed of interconnected clusters of processing elements and large-scale wired and wireless networks that connect a variety of smart sensors and actuators. The coupling of the cyber and physical contexts will be driven by new demands and applications. Innovative solutions will address unprecedented security and privacy needs. New spatial-temporal constraints will be satisfied. Novel interactions among communications, computing, and control will be understood. CPS will also interface with many nontechnical users. Integration and influence across administrative boundaries will be possible.

The innovation and development of CPS will require computer scientists and network professionals to work with experts in various engineering disciplines, including control engineering, signal processing, civil engineering, mechanical engineering, and biology. This, in turn, will revolutionize how universities educate engineers and scientists. The size, composition, and competencies of industry teams that design, develop, and deploy CPS will also change dramatically. The global competitiveness of national economies that become technology leaders in CPS will improve significantly (PCAST, 2007).

The ability to interact with, and expand the capabilities of, the physical world through computation, communication, and control is a key enabler for future technology developments. Opportunities and research challenges include the design and development of next-generation airplanes and space vehicles, hybrid gas-electric vehicles, fully autonomous urban driving, and prostheses that allow brain signals to control physical objects.

Over the years, systems and control researchers have pioneered the development of powerful system science and engineering methods and tools, such as time and frequency domain methods, state space analysis, system identification, filtering, prediction, optimization, robust control, and stochastic control. At the same time, computer science researchers have made major breakthroughs in new programming languages, real-time computing techniques, visualization methods, compiler designs, embedded systems architectures and systems software, and innovative approaches to ensure computer system reliability, cyber security, and fault tolerance. Computer science researchers have also developed a variety of powerful modeling formalisms and verification tools. CPS research aims to integrate knowledge and engineering principles across the computational and engineering disciplines (networking, control, software, human interaction, learning theory, as well as electrical, mechanical, chemical, biomedical, material science, and other engineering disciplines) to develop new CPS science and supporting technology.

In industrial practice, many engineering systems have been designed by decoupling the control system design from the hardware/software implementation details. After the control system is designed and verified by extensive simulation, ad hoc tuning methods have been used to address modeling uncertainty and random disturbances. However, the integration of various subsystems, while keeping the system functional and operational, has been time consuming and costly. For example, in the automotive industry, a vehicle control system relies on system components manufactured by different vendors with their own software and hardware. A major challenge for original equipment manufacturers (OEMs) who provide parts to a supply chain is to hold down costs by developing components that can be integrated into different vehicles (Baheti & Gill, 2011).

The increasing complexity of components and the use of more advanced technologies for sensors and actuators, wireless communication, and multicore processors pose a major challenge for building next-generation vehicle control systems. Both the supplier and integrator need new science that enables reliable and cost-effective integration of independently developed system components. In particular, theory and tools are needed for developing cost-effective methods to (Baheti & Gill, 2011):

1. Design, analyze, and verify components at various levels of abstraction, including the system and software architecture levels, subject to constraints from other levels.
2. Analyze and understand interactions between the vehicle control systems and other subsystems (engine, transmission, steering, wheel, brake, and suspension).
3. Ensure safety, stability, and performance while minimizing vehicle cost to the consumer. New functionality and the cost of vehicle control systems are increasingly major differentiating factors for business viability in automobile manufacturing.

11.1.2.2 CPS Definition

A cyber-physical system is the integration of computation with physical processes. Embedded computers and networks monitor and control the physical processes, usually with feedback loops in which physical processes affect computations and vice versa. In the physical world, the passage of time is inexorable, and concurrency is intrinsic. Neither of these properties is present in today's computing and networking abstractions.

Applications of CPS arguably have the potential to dwarf the 20th-century IT revolution. They include high confidence medical devices and systems, assisted living, traffic control and safety, advanced automotive systems, process control, energy conservation, environmental control, avionics, instrumentation, critical infrastructure control (e.g., electric power, water resources, and communications systems), distributed robotics (telepresence, telemedicine), defense systems, manufacturing, and smart structures. It is easy to envision new capabilities, such as distributed micro-power generation coupled into the power grid, where timing precision and security issues loom large. Transportation systems could benefit considerably from better embedded intelligence in automobiles, as this could improve safety and efficiency. Networked autonomous vehicles (AVs) could dramatically enhance the effectiveness of our military and offer substantially more effective disaster recovery techniques. Networked building control systems (such as heating, ventilation, air conditioning or HVAC, and lighting) could significantly improve energy efficiency and demand variability, reducing our dependence on fossil fuels and our greenhouse gas emissions. In communications, cognitive radio could benefit enormously from distributed consensus on available bandwidth and distributed control technologies. Financial networks could be dramatically changed by precision timing. Large-scale service systems, leveraging of radio frequency identification (RFID) and other technologies to track goods and services could acquire the nature of distributed real-time control systems. Distributed real-time games that integrate sensors and actuators could change the (relatively passive) nature of online social interactions.

The positive economic impact of any one of these applications areas would be enormous. Today's computing and networking technologies, however, may have properties that unnecessarily impede progress towards these applications. For example, the lack of temporal semantics and adequate concurrency models in computing, and today's "best effort" networking technologies make predictable and reliable real-time performance difficult, at best. Software component technologies, including object-oriented design and service-oriented architectures, are built on abstractions that match software much better than physical systems. Many of these applications may not be achievable without substantial changes in the core abstractions (Lee, 2008).

11.1.2.3 CPS Concept

CPS is an integration of computation with physical processes; it is about the intersection, not the union of the physical and the cyber. A complex CPS definition was given by Shankar Sastry from University of California, Berkeley, in 2008: "A cyber-physical system (CPS) integrates computing, communication and storage capabilities with monitoring and/or control of entities in the physical world, and must do so dependably, safety, securely, efficiently and real-time" (Sanislav & Miclea, 2012).

CPSs are not the traditional embedded systems or real-time systems, today's sensor networks or desktop applications, but they have certain characteristics that define them, as mentioned in Huang (2008) and presented below (Sanislav & Miclea, 2012):

1. Cyber capabilities in every physical component;
2. Networked at multiple and extreme scale;
3. Dynamically reconfiguring/reorganizing;
4. High degrees of automation, with closing control loops;
5. Dependable operation and certified in some cases;
6. Cyber and physical components integrated for learning and adaptation, higher performance, self-organization, auto assembly.

CPSs, like all information and communication systems, are chosen according to certain fundamental properties (Sanislav & Miclea, 2012):

1. Functionality
2. Performance
3. Dependability and security
4. Cost.

Properties that affect dependability and security are the following (Sanislav & Miclea, 2012):

1. Input and feedback from/to the physical environment—secured communication channels.
2. Management and distributed control—a federated approach.
3. Real-time performance requirements.
4. Large geographical distribution without physical security components in various locations.
5. Very large-scale control systems (system of systems (SoS)).

11.1.2.4 Grand Challenges and Vision of CPS

The core science and technology required to support the CPS vision are essential for future economic competitiveness. Creating the scientific and technological basis for CPS can pay dividends across a wide variety of application domains, resulting in unprecedented breakthroughs in science and engineering. Groundbreaking innovations will occur because of the pervasive utility of the technology resulting in major societal and economic gains. Some possibilities are the following (Rajkumar, Lee, Sha, & Stankovic, 2010):

- Blackout-free electricity generation and distribution
- Extreme yield agriculture
- Safe and rapid evacuation in response to natural or man-made disasters
- Perpetual life assistants for busy, senior/disabled people
- Location-independent access to world-class medicine
- Near-zero automotive traffic fatalities, minimal injuries, and significantly reduced traffic congestion and delays
- Reduced testing and integration time and costs of complex CPS systems (e.g., avionics) by one to two orders of magnitude
- Energy-aware buildings and cities

- Physical critical infrastructure that calls for preventive maintenance (PvM)
- Self-correcting CPSs for "one-off" applications.

11.1.2.5 System Features of CPS

CPSs are a result of the emergence of faster computer processors, the miniaturization of electronic components, broader communication bandwidths, and the seamless integration of networked computing with everyday systems. They blend physical technologies, software and middleware technologies, and cyber technologies. Future systems will make more extensive use of synergic technologies, which integrate hardware and cyber technologies. Physical technologies enable the implementation of artifacts that can be recognized, located, operated, and/or controlled in the physical world. Cyber technologies are used for capturing, analyzing, and processing sensed signals and data produced in the physical world for decision-making. Synergic technologies enable not only a borderless interoperation between physical and cyber elements but also a holistic operation of the whole system. The design of the physical and computational aspects is becoming an integrated activity.

CPSs link the physical world with the cyber world through the use of multiple sensor and actuator networks integrated in an intelligent decision system. In other words, CPSs combine sensing and actuation with computation, networking, reasoning, decision-making, and the supervision of physical processes.

Low- and high-end implementations of CPS can be distinguished based on the extensiveness and sophistication of the resultant integrity. Low-end implementations are linearly complex, closed architecture, distributed and networked, sensing and reasoning enabled, smart and proactive (often embedded and feedback controlled) collaborative systems. High-end implementations are nonlinearly complex, open and decentralized, heterogeneous and multi-scale, intelligent and partly autonomous, and self-learning and context-aware systems.

The systems belonging to the latter class display organization without any predefined organizing principle and change their functionality, structure, and behavior by self-learning, self-adaption, or self-evolving. Complicated cyber-physical systems (C-CPSs) are low-end implementations because they are not supposed to change their functionality or architecture but to optimize their behavior, for instance, energy efficiency (e.g., due to the necessity to operate during an extended period of time), while operating under dynamically changing operating conditions or unforeseen circumstances. Some of these systems should operate in real-time applications and provide a precisely timed behavior; they should also achieve a synergic interaction between the physical and the cyber worlds by integrating computational and physical processes.

The cyber and physical parts of the systems are interconnected and affect each other through information flows. Due to this functional synergy, the overall system performance is of higher value than the total of the individual components. This synergy is particularly important for high-end CPSs, which exhibit properties such as self-organization. In general, CPSs strive towards a natural human–machine interaction that extends to the human cognitive domain. These kinds of systems are also capable of exhibiting extensive remote collaboration. Unlike linear complex systems (LCSs), CPSs work on non-dedicated networks. CPSs are often connected in a hierarchical manner, as systems of systems, in which one system monitors, coordinates, controls, and integrates the operation of other systems. For this reason, they can be considered multidimensional complex systems. Based on their functionality and characteristics, high-end CPSs can be used in areas such as transportation, healthcare, and manufacturing.

Some CPSs are mission-critical systems (MCSs) because their correct functioning is critical to ensuring the success of a mission, provisioning an essential supply, or safeguarding security and well-being. These are the systems that ensure proper and continuous operation of, for example, nuclear plants, automated robot control systems, and automatic landing systems for aircraft. Any failure in MCSs can lead to loss of human life and damage to the environment and may cause losses in terms of supply and cost. However, their operation is always characterized by the presence of uncertainty.

Big Data Analytics

This introduces challenges from the point of view of the dependability, maintenance, and repair of mission-critical nonlinear CPSs. In the long run, it is crucial to comprehensively analyze what the maintenance of these systems theoretically, methodologically, and practically means and how it can be implemented in different systems (Ruiz-Arenas, Horváth, Mejía-Gutiérrez, & Opiyo, 2014).

11.1.2.6 Application Domains of CPS

CPSs present a set of advantages: they are efficient and safe systems; they allow individual entities to work together in order to form complex systems with new capacities. Cyber-physical technology can be applied in a wide range of domains, offering numerous opportunities: critical infrastructure control, safe and efficient transport, alternative energy, environmental control, telepresence, medical devices and integrated systems, telemedicine, assisted living, social networking and gaming, manufacturing, agriculture (Huang, 2008; Lee, 2008). Critical infrastructure, i.e., assets that are essential for the functioning of a society and economy, includes facilities for: water supply (storage, treatment, transport and distribution, waste water), electricity generation, transmission and distribution, gas production, transport and distribution, oil and oil products production, transport and distribution, and telecommunication (Sanislav & Miclea, 2012).

Wan, Man, and Hughes (2010) listed some requirements that CPSs should meet according to the business sectors where they will be used, i.e., automotive, environment monitoring/protection, aviation and defense, critical infrastructure, healthcare (see Table 11.1). The physical platforms, supporting CPSs provide the following five capabilities: computing, communication, precise control, remote cooperation, and autonomy (Sanislav & Miclea, 2012).

Unlike traditional embedded systems, CPSs interface directly with the physical world, making the detection of environmental changes and system behavior adaptation the key challenges in their design (Sanislav & Miclea, 2012).

11.1.2.7 The Past, Present, and Future of CPS

A CPS is an orchestration of computers and physical systems. Embedded computers monitor and control physical processes, usually with feedback loops, where physical processes affect computations and vice versa.

Applications of CPS include automotive systems, manufacturing, medical devices, military systems, assisted living, traffic control and safety, process control, power generation and distribution,

TABLE 11.1

CPS Characteristics and Application Domains (Sanislav & Miclea, 2012)

Application Domain	CPSs Characteristics
Automotive	CPSs for the automotive industry require high computing power, due to complex traffic control algorithms that calculate, for example, the best route according to traffic situation.
Environment	CPSs for environment monitoring, distributed in a wide and varied geographical area (forests, rivers, mountains), must operate without human intervention for long periods with minimal energy consumption. In such an environment, the accurate and in-time data collection provided by the ad hoc network with low power consumption, represents a real research challenge.
Aviation, defense	CPSs for aviation and defense require precise control, high security, and high-power computing. The development of the security protocols will be the main research challenge.
Critical infrastructure	CPSs for energy control, water resources management, etc. require a precise and reliable control, leading to application software methodologies to ensure the quality of the software.
Healthcare	CPSs for healthcare and medical equipment require a new generation of analysis, synthesis, and integration technologies, leading to the development and application of interoperability algorithms.

energy conservation, HVAC, aircraft, instrumentation, water management systems, trains, physical security (access control and monitoring), asset management, and distributed robotics (telepresence, telemedicine).

As an intellectual challenge, CPS is about the intersection, not the union, of the physical and the cyber. It combines engineering models and methods from mechanical, environmental, civil, electrical, biomedical, chemical, aeronautical, and industrial engineering with the models and methods of computer science. These models and methods do not combine easily. Thus, CPS constitutes a new discipline that demands its own models and methods.

The term "cyber-physical systems" emerged around 2006, when it was coined by Helen Gill at the National Science Foundation in the United States. The related term "cyberspace" is attributed to William Gibson, who used it in the novel *Neuromancer*, but the roots of the term CPS are older and deeper. It would be more accurate to view the terms "cyberspace" and "cyber-physical systems" as stemming from the same root, "cybernetics," coined by Norbert Wiener, an American mathematician who had a huge impact on the development of control systems theory. During World War II, Wiener pioneered technology for the automatic aiming and firing of anti-aircraft guns. Although the mechanisms he used did not involve digital computers, the principles involved are similar to those used today in computer-based feedback control systems. His control logic was effectively a computation, albeit one carried out with analog circuits and mechanical parts, and, therefore, cybernetics is the conjunction of physical processes, computation, and communication. Wiener derived the term from the Greek κυβερνήτης (kybernetes), meaning helmsman, governor, pilot, or rudder.

The term CPS is sometimes confused with "cybersecurity," which concerns the confidentiality, integrity, and availability of data and has no intrinsic connection with physical processes. The term "cybersecurity" is about the security of cyberspace and only indirectly connected to cybernetics. CPS certainly involves many challenging security and privacy concerns, but these are by no means the only concerns.

CPS connects strongly to the currently popular terms IoT, Industry 4.0, the Industrial Internet, Machine-to-Machine (M2M), the Internet of Everything, TSensors (trillion sensors), and the fog (like the cloud, but closer to the ground). All of these reflect a vision of a technology that deeply connects our physical world with our information world. The term "CPS" is more foundational and durable than all of these, however, because it does not directly reference either implementation approaches (e.g., the "Internet" in IoT) or particular applications (e.g., "Industry" in Industry 4.0). It focuses instead on the fundamental intellectual problem of conjoining the engineering traditions of the cyber and the physical worlds. We can talk about a "cyber-physical systems theory" in a manner similar to "linear systems theory." Like linear systems theory, a CPS theory is all about models. Models play a central role in all scientific and engineering disciplines. However, since CPS conjoins distinct disciplines, which models should be used? Unfortunately, models that prevail in these distinct disciplines do not combine well (Lee, 2015).

11.2 Big Data Analytics in Inspection and Maintenance

11.2.1 Big Data Analytics for Predictive Maintenance Strategies

11.2.1.1 Introduction

Maintenance can be defined as all actions which are necessary to retain or restore a system and a unit to a state necessary to fulfill its intended function. The main objective of maintenance is to preserve the capability and the functionality of the system while controlling the cost induced by maintenance activities and the potential production loss. Correspondingly, failures can be defined as any change or anomaly in the system causing an unsatisfactory level of performance. Although only certain failures will cause severe risk in productivity and safety, most failures lead to disruptive, inconvenient, and expensive breakdowns and loss of quality. Maintenance plans are designed to reduce or eliminate the number of failures and the costs related to them (Lee, Cao, & Ng, 2017).

There are two broadly accepted methodologies aimed at continuously enhancing maintenance excellence, with different focuses. As a human factor management-oriented policy, total productive maintenance (TPM) involves all employees, especially the operators, in the maintenance program in order to achieve optimality in overall effectiveness and zero breakdowns. Through the operators' participation in maintenance, such as through inspections, cleaning, lubricating, and adjusting, hidden defects can be detected before service breakdown. TPM aims to diminish and eliminate six significant losses of equipment effectiveness—i.e. breakdowns, setup and adjustment, idling and stoppages, reduced speed, defects in process, and reduced yield (Jardine & Tsang, 2013).

Reliability-centered maintenance (RCM) is another approach to strengthening the system's reliability, availability, and efficiency. It focuses on design and technology. RCM program is based on systematic assessment of maintenance needs after a complete understanding of the system function and the types of failure causing function losses (Lee, Cao, & Ng, 2017).

11.2.1.1.1 Types of Maintenance

Maintenance activities can be categorized into three types (Lee, Cao, & Ng, 2017):

1. Reactive or corrective maintenance
2. Preventive maintenance
3. Predictive maintenance.

The following terms are also respectively used for the above three categories interchangeably as (Lee, Cao, & Ng, 2017):

1. Breakdown maintenance or unplanned maintenance
2. Planned maintenance
3. Condition-based maintenance (CBM) or prognostic and health management (PHM).

Reactive or corrective maintenance (CvM) follows a run-to-failure methodology, compromising repair and/or replacement work after an equipment outage has occurred. This primitive maintenance approach, which has been applied in industry for decades, is still considered the best maintenance policy for non-critical components with short repairing time in the system. However, in most cases, an equipment failure can lead to unexpected production delay and lower the production efficacy rate, or more seriously, cause severe damage to other components and/or injury to people. One goal of a proactive maintenance plan is to reduce the overall requirement for reactive maintenance and to apply preventive and predictive strategies on any feasible occasion.

Preventive maintenance (PvM) is performed based on a certain periodic interval to prevent and correct problems before breakdown without considering the actual health condition of a system. Basic preventive maintenance, including inspections, lubrication, cleaning, and adjustment, is the first step. After that, rectification or replacement can be undertaken only for components identified with defects and/or considerable risk of failure. Generally, most PvM actions can be implemented by operators with basic training (Lee, Cao, & Ng, 2017).

Predictive maintenance (PdM) is a trend-oriented policy that begins by identifying the states of each component within the equipment. PdM relies on engineering techniques and statistical tools to process the data and analyze the health condition in order to predict possible equipment failure (Lee, Ardakani, Yang, & Bagheri, 2015). The prediction of the equipment condition is based on the finding that most types of failures which occur after a certain degradation process from a normal state to abnormalities do not happen instantaneously (Fu et al., 2004). Through degradation monitoring and failure prediction, PdM reduces the uncertainty of maintenance activities and enables problems to be identified and resolved before potential damage. Condition based maintenance (CBM), the alternative term for PdM, places more emphasis on real-time inspections using RFID devices and wireless sense networks (WSNs). The three key steps of CBM are monitoring and processing, diagnosis and prognosis, and maintenance decision-making (Lee, Cao, & Ng, 2017).

11.2.1.2 Maintenance Strategies

In corrective maintenance, maintenance procedures are undertaken when a failure has occurred. Manufacturers are required to keep components inventories for maintenance, repair, and operations (MRO) to prevent disruption of the overall production by the failure of machine parts or equipment. In a preventive maintenance strategy, maintenance follows a fixed-time, interval based or condition-based schedule to avoid failure. This is a protective, process-oriented approach in which machine failure and downtime cost can be reduced by taking proper preventive measures. Decisions on maintenance schedules are based on a machine's physical properties or asset condition, with extra resources spent on non-value-added activities to estimate and measure condition (Exton & Labib, 2002). Preventive maintenance attempts to provide an empirical basis for the development of a framework design of manufacturing flexibility at machine idle periods and during maintenance activities. The assumption is that the machine failure follows the bathtub curve shown in Figure 11.2. Scheduled maintenance takes place in the wear-out phase to reduce the failure rate (Sikorska et al., 2011). The most conspicuous deficiency of the strategy is random failure within the useful life period. The impact of failure in a critical machine is a tremendous risk to downtime cost and becomes a bottleneck in production logistics operations (Lee, Cao, & Ng, 2017).

Predictive maintenance helps to remedy random machine failure in maintenance management. It is based on observation of the machine degradation process and the monitoring of its status from normal to flawed (Wu, Gebraeel, Lawley, & Yih, 2007). This is a sensor-based content-aware philosophy drawing on the "Internet of Things." The intelligent maintenance prediction support system monitors machine status by utilizing real-time sensory data (Kaiser & Gebraeel, 2009). Predictive maintenance can provide insights for maintenance scheduling to eliminate unanticipated machine breakdowns and minimize maintenance costs and downtime, before the occurrence of random machine failure (Garcia, Sanz-Bobi, & del Pico, 2006). The important factors associated with PdM in maintenance policy management (MPM) are criticality, availability of sensory data, reliability, timeliness, relevance, and knowledge-oriented strategy (Lee, Cao, & Ng, 2017).

- Criticality of failures: Predictive maintenance is heavily concentrated on real-time machine condition monitoring, using diagnostics and prognostics to reduce foreseeable machine downtime cost. Assets defined as critical must have a higher rank in priority; i.e., those with the greatest impact on production. This changes the maintenance objective from avoiding breakdown to accepting downtime and taking maintenance actions ahead of the schedule.
- Availability of sensory data: Predictive maintenance is highly dependent on extract transform load (ETL) operational data in close condition-based monitoring. The current operational status

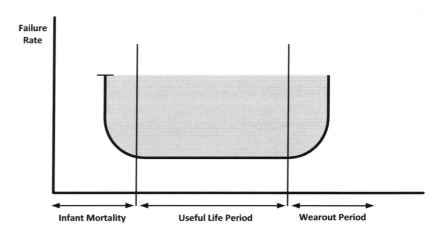

FIGURE 11.2 Classical bathtub curve (Klutke, Kiessler, & Wortman, 2003).

and abnormal performance can be assessed by equipping sensors to identify failure. Lack of sensory data may result in poor maintenance predictions.
- Reliability: Maintaining critical machine performance and leveraging the overall cost to sustain production are the major targets of PdM policy. The system must provide the correct measures and reliable performance in prediction to address feasible and foreseeable machine failure and build confidence in operation.
- Timeliness: The prediction of maintenance and the transmission of data must occur in a timely manner. The time series of the maintenance schedule and delivery of MRO should take into consideration the maintenance management required to facilitate production, with zero tolerance of equipment failure.
- Relevance: The MPM system needs to be developed based on the opinions of experts. The collected sensory information must be recorded and analyzed on a real-time basis. To improve data quality, the extraction of relevant data for maintenance decision-making is crucial. Inappropriate integration of a sensor and machine may cause poor estimation and inaccurate prediction of machine performance.
- Knowledge objective-oriented strategy: The concept of predictive maintenance involves a belief that the implicit knowledge from the collaboration of sensory information contributes to maintenance in advance of failure. The knowledge transfer system facilitates the disclosure of implicit information to maximize production efficiency and minimizes the adverse impact of idling time during maintenance and unawareness of potential failure. Predictive maintenance could be helped by using big data mining techniques to detect and defeat anomalies at an early stage.

11.2.2 The Next Step: Big Data Analytics

Predictive maintenance (PdM) is facilitated by new opportunities to capitalize on the digital revolution, more specifically on advances in decision support tools powered by big data analytics.

In our increasingly digitized world, where virtually every activity creates a digital trace, there has been an exponential growth in how much data can be used for PdM. Datasets can be obtained from both internal and external sources. Consider, for example, the vast pools of sensor data that can be collected from entire factories, transportation fleets or infrastructure networks, and distributed via IoT technology. In terms of external data, consider environmental data on temperature, humidity, and wind speeds or operator profiles or specifications of materials being processed at the time of failure. Datasets used for PdM may be structured, like spreadsheets or relational databases, but can also be unstructured, like maintenance logs or thermal images which can be "unlocked" through text mining and pattern recognition software, respectively.

We could easily drown in this sea of data. Fortunately, rapid advances in artificial intelligence (AI) techniques have enabled us to make sense of all these data. Machine learning (ML) algorithms are especially helpful. These algorithms are not constructed as a predefined set of rules, as in traditional software programming. Instead, they are self-learning. They infer rules by performing a series of trials on a set of training data and, thus, construct their own model of the world. Every subsequent amount of data is then used to refine that model and improve its predictive powers (Haarman, Mulders, & Costas Vassiliadis, 2017).

11.2.2.1 Predictive Maintenance 4.0

The application of big data analytics in maintenance represents the fourth level of maturity in predictive maintenance, as shown in Figure 11.3. We call this fourth-level Predictive Maintenance 4.0, abbreviated as PdM 4.0.

PdM 4.0 predicts future failures in assets and ultimately prescribes the most effective preventive measure by applying advanced analytic techniques to big data on technical condition, usage, environment, maintenance history, similar equipment elsewhere, and anything else that may correlate with the performance of an asset (Haarman, Mulders, & Costas Vassiliadis, 2017).

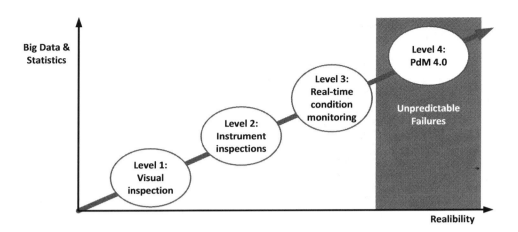

FIGURE 11.3 Four levels of maturity in PdM (Mainnovation, 2019).

The four levels of predictive maintenance can be defined as:

- Level 1, visual inspections: periodic physical inspections; conclusions are based solely on inspector's expertise.
- Level 2, instrument inspections: periodic inspections; conclusions are based on a combination of inspector's expertise and instrument readouts.
- Level 3, real-time condition monitoring: continuous real-time monitoring of assets, with alerts given based on pre-established rules or critical levels.
- Level 4 PdM 4.0: continuous real-time monitoring of assets, with alerts sent based on predictive techniques, such as regression analysis.

Real-time condition monitoring will only get to a certain level of reliability, a level with both unforeseeable and inexplicable failures. But these failures could be tackled with big data analytics. PdM 4.0 involves harnessing the power of AI to create insights and detect patterns and anomalies that escape detection by the cognitive powers of even the most gifted humans. PdM 4.0 gives us a chance to predict what was previously unpredictable. It helps us anticipate the failures and accidents increase uptime, and extend the lifetime of assets (Haarman, Mulders, & Costas Vassiliadis, 2017).

A number of success factors are critical for the implementation of PdM 4.0 (Mainnovation, 2019):

- The most important critical success factor (CSF) is the availability of data.
- Technology is the second most important success factor.
- The availability of data is critical, because one of the biggest technological challenges continues to be the collection of continuous and real-time sensor data of numerous assets.
- Multiple parts of outdated asset bases do not have the necessary sensors.
- The capacity of data networks during collection of large quantities of data is a concern especially data on fleets of trucks and trains traveling long distances and covering large areas.
- In Dangerous industrial environments require an IoT infrastructure specifically designed to meet certain safety demands.

A successful implementation of PdM 4.0 features the following steps (see Figure 11.4) (Mainnovation, 2019):

1. Asset value ranking & feasibility study.
2. Asset selection.

Big Data Analytics

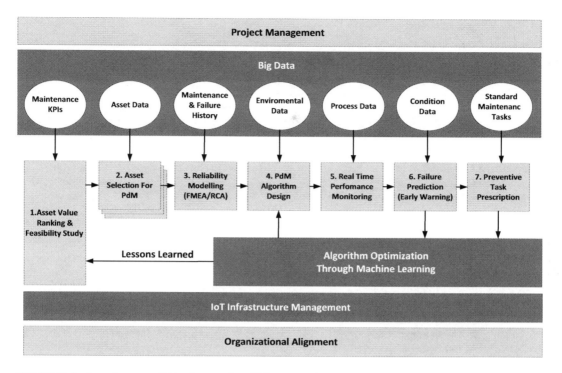

FIGURE 11.4 Steps for successful implementation of PdM 4.0 (Mainnovation, 2019).

3. Reliability modeling, i.e., root cause analysis (RCA) and fail mode effect analysis (FMEA).
4. PDM algorithm design.
5. Real-time performance monitoring.
6. Failure prediction.
7. Predictive task prescription: correcting measures are chosen from a vast array of tasks (library of tasks).

11.3 Integration of Big Data Analytics in AV Inspection and Maintenance

11.3.1 Big Data Analytics' Role in Autonomous Vehicle Development

Although various technologies have provided small steps towards automation, a fully automated vehicle has yet to be realized. However, over the last decade, with the large range of advancements in technology and the newfound use of big data, tech companies have created the necessary programming for fully automated vehicles. AVs rely on the data they receive through a variety of systems, including the Global Position System (GPS), radar, sensor technology, and cameras.

These sources provide them with the data they need to make safe driving decisions. At this point, car manufacturers are still using stores of big data to work out the kinks of the thousands of scenarios an autonomous car could find itself in, but it's only a matter of time before self-driving cars transform the automotive industry by making up the majority of cars on the road. The ability to use big data is changing industries worldwide, and one of the most essential big data analytics, deep learning, is contributing to the progress in the automotive industry towards fully AVs. Although it will still be several decades before the mass adoption of self-driving cars, the change will come. In only a few decades, we'll likely be living in a time where cars are a safer form of transportation, and accidents are tragedies that are few and far between (Potter, 2019).

11.3.2 The Rise of Autonomous Vehicles

Technology adoption by the automotive sector has mainly centered on the concept of AVs and the efforts to make them a more viable form of transport. A UK government report estimates that the connected and AV industry will be worth £28bn in the United Kingdom by 2035, making this technology a prime focus area for all industry players. The industry is already capable of producing Levels 1–3 AVs (see Chapter 1), where higher levels correspond to reduced degrees of human intervention required during driving. The capabilities that make up these levels include environmental awareness, the real-time condition monitoring of the vehicle, and weather analysis. The ultimate goal is to enable AVs to reach Levels 4 and 5, where the driver's input is minimal or the driver is made completely redundant. While this goal may only be possible in the distant future, autonomous technology has already begun to include features to make vehicles smarter and more capable of dealing with dynamic demands.

IoT and cloud integration are essential parts of the progression from "dumb" cars to smarter, intelligent vehicles that make better use of connectivity.

AVs today are integrated with sensors that constantly collect data in real time. These data are later stored in a remote cloud server for analysis. This effectively transforms the vehicle into a "black-box" recorder that can be used at any point to determine the current state of the vehicle's functioning. The analyzed data are then used to create a customized driving profile which enables the in-built software to predict future driving scenarios and devise solutions to prevent accidents. In addition to in-car components, the integration of sensors into city environments for assets such as traffic lights, lamp posts, and traffic signs will help create a larger, more detailed database and improve situational awareness at all times (Singh, 2018).

11.3.3 Keeping Big Data Analytics under Control for Autonomous Vehicles

The driverless car has been a tech dream for decades, and now that broadband connectivity, cloud computing, and AI are increasingly available, autonomous cars may go mainstream, provided certain technical and regulatory milestones are reached.

First, Autonomous cars generate a staggering amount of data; Intel CEO Brian Krzanich estimates one car generates 4 terabytes of data in 8 h of operation. Multiple image, radar/laser-illuminated detection and ranging (LiDAR), time-of-flight, accelerometer, telemetry, and gyroscope sensors generate data streams that must be analyzed to perform the calculations, inspections, and adjustments required to safely navigate a car.

Second, analysis needs to happen in real time if the car is to keep up with constantly changing driving conditions (other cars or pedestrians moving around the vehicle, changing weather, traffic signs, etc.). These real-time performance requirements mean there's no time to upload data to a central server, conduct the necessary analytics, and send instructions back to the car for execution. Thus, data that are critical to safely navigate the car must be analyzed locally by the car itself (essentially the car is an edge device in a cloud network).

Third, not only does the car need to analyze data on its own, it must also learn to pick and choose between different data streams to identify the ones best suited for analysis at any given moment to keep the car driving safely. This last requirement, the need to determine what data are required to perform an analysis, is tricky. While predefined filters can help a car's machine learning routines learn what data to use and when to use them, those filters can't be updated in real time. Accordingly, an autonomous car will need to run machine learning and analytics engines powerful enough to recognize mission-critical data requiring immediate analysis and action maintenance.

In other words, we need analytics and machine learning algorithms for autonomous cars that can:

- Identify data in all formats.
- Recognize what data are required for mission critical operations and analyze those data locally.
- Compress or aggregate noncritical data for uploading to the cloud for future use.

Big Data Analytics 383

- Schedule uploads of noncritical data from the car to the cloud when less expensive communications are available (e.g., when the car is parked overnight at home and can access the owner's Wi-Fi instead of a metered cellular network).
- Know how to call for legacy data from the cloud so the AI can use it for future analytics.

The last one is particularly important. An autonomous car manufacturer will be responsible for storing vast amounts of data generated by cars operating around the world, and many of those data will likely have no real value when initially captured. However, their value may be revealed in the future as the manufacturer's autonomous driving applications evolve and improve. Today's noncritical data can be useful for future applications, provided the data are properly stored and catalogued so they can be easily found.

Without careful cataloguing of data as they are captured, autonomous car vendors run the risk of creating a "dark data" problem. Dark data is the term used to describe data an organization collects but fails to take advantage of because they don't know how to or have forgotten they have them. This will be a significant problem for self-driving cars because of the sheer volume of data they generate. And as we see more vendors enter the autonomous driving market, the ones that will ultimately win out over others will be those vendors best prepared to analyze data at the local level and have catalogue their databases properly so future autonomous applications can find the legacy data they need, when they need them (Chala, Bayliss, & Camper, 2019).

11.3.4 Big Data Analytics—How Does It Help the Automotive Industry?

Continuous development of technologies is inevitably causing a transformation of the ecosystem of the automotive industry. The increasing demands from consumers coupled with diverse technological features, are making decision-making a tough process in the automotive industry.

However, with the availability of big data analytics, decision-making can be made easier (MARii, 2019). Ways big data analysis can help the automotive industry include the following (MARii, 2019):

1. Maintain smooth flow between links in the supply chain.
   - Auto manufacturers use huge amount of data from various systems, such as dealer management system (DMS), customer relationship management (CRM), and customer satisfaction surveys to gather metrics on numerous values ranging from customer inquiries to sales, inventory levels, customer order patterns on various models, trims, and color selection.
   - This process enables manufacturers to understand what their customers actually want in a particular vehicle.
   - In addition, the data collected allow suppliers to ensure the availability of materials needed to manufacture components that customers find appealing.
2. Enable strategic marketing planning.
   - The automotive industry is a platform for huge businesses.
   - Huge sums of money are invested in the automotive industry and it is likely to increase in value.
   - Thanks to the data collected through marketing mix analysis, comprehensive evaluation of customer responses, and internal business operations, big data analytics is helping auto manufacturers to strategically manage the use of incentives, rebates, financing deals, and other key attractions in increasing sales and return of investments (ROIs).
3. Make R&D easier.
   - The designing and engineering process of developing new vehicles involves a lot of decision-making dilemmas, often leading to the waste of time and money.
     - For example, a model in development may have a certain feature used in existing models, such as a door lock function button.
     - If the manufacturer leverages data monitoring and analytics from industry customer satisfaction studies, its dealership service departments, and its own internal research, concerns with that button based on existing models can be recognized.

- This approach allows designers and engineers to avoid repeating the same mistakes, hence saving time and money.
- The data sample can also enable the manufacturer to smooth the operational process; this allows more accurate material procurement, manpower planning, and vendor coordination.

4. Be data friendly and future-proof.
   - Data are tools to future-proof the automotive industry, significant enablers of the latest advances in the development of AVs.
   - The technology behind self-driving cars is machine learning or AI.
   - A machine "learns" to keep passengers safe through continuous research and development, facilitated by data collected from real-world driving conditions, environmental studies, and so on.
     – According to the Center for Automotive Research (CAR), one of the recent applications of big data analysis was the assessment of cost and effectiveness of powertrain technologies developed to meet global standards in fuel economy and greenhouse gas emissions.
     – The research found a discrepancy between what automakers and regulators project as the cost of meeting those standards.
   - This shows that data also have the ability to inform regulators and policy makers as they make decisions going forward, and this could benefit automakers in the long run.

11.4 Utilization of AVs in Industry 4.0 Environment

11.4.1 Understanding the Convergence of Industry 4.0 and Autonomous Vehicles

As a society, we've embraced a brave new world bursting with myriad emerging technologies so quickly that we haven't taken the time to understand the interconnections. We see new entrants like cryptocurrency and the IoT as singular, when, in fact, they are symbiotic. Together, these trends form an extensive, disruptive infrastructure.

Groundbreaking technologies and the as-a-service economy are deeply interrelated. If we take time to think about the impact of these disruptive trends, we begin to identify the power of convergence and how seemingly unique ideas are ubiquitous and closely intertwined. Convergence is the truest form of digital transformation, and it's happening all around us.

Bing Ads recently analyzed 600,000 instances of customer queries over 16 months from categories aligned with a variety of disruptive trends to better understand and quantify their momentum. The stream graph in Figure 11.5 shows six months of recent data and uses average site time as a proxy to understand user engagement (Veverka, 2018). Queries range from "FinTech" and "digital twin technology" to "predictive analytics." In the figure, colored bands rise to the right across the y-axis over time. The thickness of each band provides insight into the frequency of activity. While some bands sustain their thickness, others get thicker or dissipate. The visualization specifically denotes increased interest in technology, manufacturing, and AVs as these trends continue their journey to convergence and eventual mainstream adoption.

Convergence is most profound in the seismic revolution of Industry 4.0. While the base of Industry 4.0 is manufacturing, McKinsey defines the core DNA of this convergent disruption as the proliferation of big data, advanced analytics, human–machine interfaces, and digital-to-physical transfers providing profound opportunities for new business models and delivery systems.

As we move forward, auto, retail, finance, technology, and telecommunications will all be profoundly entrenched in Industry 4.0, and the digitalization of these verticals will demonstrate the significant impact of convergence, which we will likely witness in real time (Veverka, 2018).

Big Data Analytics

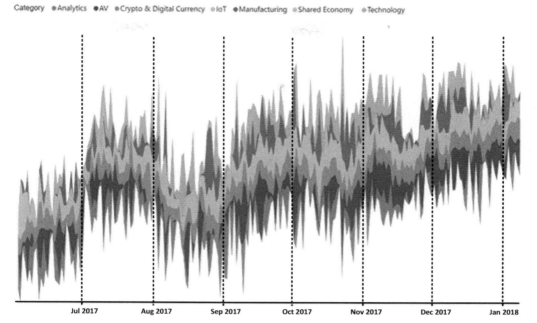

FIGURE 11.5 Recent data and used average site time as a proxy (Veverka, 2018).

11.4.2 Manufacturing and the Role of Robotics in Industry 4.0

Industry 4.0 impacts the whole manufacturing value chain—from raw materials to unfinished goods, the production shop floor, the warehouse, storage, and delivery. As information from each stage is leveraged to introduce more automation in the next, the manufacturing supply chain becomes smarter. Since automation is at the core of Industry 4.0, robots will be an essential part of manufacturing. Invariably, as smarter environments give rise to more information, robots will become more intelligent and easier to operate. Indeed, the ultimate goal of Industry 4.0 is an autonomous smart factory that can produce customizable products.

Otto Motors, a division of Clearpath Robotics, already builds self-driving vehicles for material transport in a manufacturing or industrial site. In fact, the burgeoning area of human–robot collaboration is transforming the manufacturing landscape. Robots, with their superior skills in repetitive tasks that are focused on precision and speed, provide workers with opportunities to focus on more complex tasks, such as those that involve working with large assemblies. The ideal factory of the future identifies no separation between automated and manual workstations: humans and robots will collaborate optimally without a safety fence separating them (Jain & Pai, 2019).

11.4.3 Industrial Autonomous Vehicles

Machine-to-machine (M2M) communication, along with the availability of cloud computing platforms and ubiquitous sensing, has led to the era of the Internet of Things (IoT) in industry. Combined advancements in robotics, autonomous technologies, and IoT have created an ideal environment for the adoption of connected industrial AVs across industries.

AVs have held the attention of futurists and technology enthusiasts for some time, as evidenced by the continuous research and development in AV technologies over the past two decades. Rapid advances in robotics, AI, computer vision, and edge computing capabilities are resulting in machines that can potentially think, see, hear, and move more deftly than humans. AVs in the form of self-driving cars have become the subject of both hype and intense competition among auto majors and technology companies.

Self-driving car prototypes with LiDARs, radars, cameras, ultrasonic sensors—along with heavy computational capabilities under the hood to recognize and maneuver around obstacles—are becoming a common sight in many cities. With the emergence of sophisticated AV technologies, we are on the cusp of their rapid deployment in industrial applications, and the confluence of the IoT and AV technologies is poised to remake and reimagine industries (Jain & Pai, 2019).

In manufacturing, robots traversing the shop floor can gather useful information from the production line on available inventory, tool availability, and calibration needs and identify potential hazards and near-misses to ensure worker safety. Robots can work around the clock and prepare the line for the next production shift. Robots are also connected and communicate with each other, collaboratively deciding which robot should do a particular task. All the autonomously moving robots, as well as machines on the floor, are connected to a back-end system, either on premises or cloud based, that will process the data these machines provide and accordingly perform tasks that optimize the system-level performance on the manufacturing floor. In this scenario, robots and humans collaboratively improve the operational efficiency and productivity of the manufacturing line. Similar use cases and associated benefits can be expected in other industries, such as mining and oil and gas.

Connected industrial AVs are poised to dramatically change the industrial landscape alongside the gradual but steady adoption of autonomous technologies in consumer automotive vehicles and public transport. The rise of industrial AVs signals an exciting future ahead (Jain & Pai, 2019).

11.4.4 Autonomous and Connected Vehicles: An Industry 4.0 Issue

11.4.4.1 Introduction

Industry 4.0 is characterized by innovative enabling technologies that provide the instruments through which a new collaborative environment is achieved in which humans and systems may interact, gaining advantages from self-organizing and optimizing in real time (see Figure 11.6) (Pieroni, Scarpato, & Brilli, 2018b).

The development of the processes belonging to Industry 4.0 is driven by innovations in the areas of IT, embedded systems, production, automation technique, and mechanical engineering. The aim is to create new factories able to manage much more complex systems (Laka & González Rodríguez, 2015). Smart products and smart production equipment will be connected and will overview the entire process, from the product idea to the product design, supply chain, and manufacturing. This approach will permit more efficient results in all the value chain production. Smart production also covers the delivery

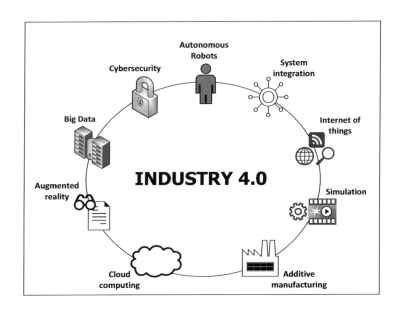

FIGURE 11.6 Industry 4.0 enabling technologies (Pieroni, Scarpato, & Brilli, 2018b).

Big Data Analytics

of products to the end users, after sales services, and product recycling. The connection of all elements within the value chain in real time is considered the basis of Internet 4.0.

The automotive industry is experiencing new challenges and frontiers with autonomous and connected vehicles which are becoming "smart" and totally connected with the rest of the world through Internet technologies. The improvements in modern technologies have allowed the comprehensive integration of essential systems and data for AV operations. Vehicles can process decisions based on defined criteria informed by actual conditions, with data provided by the following integrated systems (Pieroni, Scarpato, & Brilli, 2018b):

- GPS: Gives a satellite-based global location and time reference for accurate and constant position tracking.
- Inertial navigation system (INS): Monitors and calculates positioning, direction and speed of vehicles, assisted by sensors onboard.
- LiDAR: Used a laser detection sensors to identify surrounding objects.

All mechanisms provide decision-making data necessary for the vehicle to be aware of position, traffic conditions, and possible movements.

There are presently two different categories of smart vehicles: autonomous vehicles (AVs) that perform all driving functions with or without a "human driver," also called vehicles without drivers; connected autonomous vehicles (CAVs) which have advanced communication technologies linking them to other vehicles or infrastructure. Both will lead new investments in urban infrastructure to reinforce features (especially wireless communication) linking the vehicle and the infrastructure at the edge of the road (e.g., smart lamppost) to transmit in real time an increasing amount of bidirectional data between the vehicle and the urban infrastructure (Pieroni, Scarpato, & Brilli, 2018b).

11.4.4.2 The Technological Impact of CAVs

Several connected vehicles have been designed and developed to be tested in various markets around the globe (Kharpal, 2017). Figure 11.7 illustrates the conceptual vision of a smart, integrated, dynamic, and connected society in which the diffusion of connected AVs is growing.

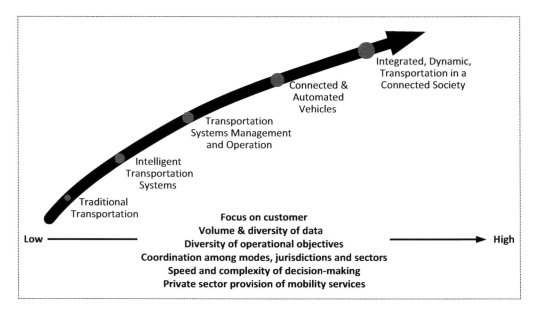

FIGURE 11.7 US Department of Transportation's (USDOT) CAV evolution vision (Pieroni, Scarpato, & Brilli, 2018b).

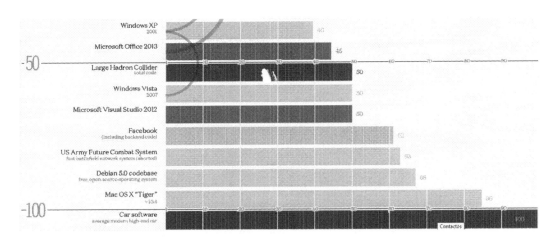

FIGURE 11.8 Complexity of modern software (Information is Beautiful, 2015).

However, the bidirectional flow of data between vehicles and infrastructure is raising concerns about the regulation about the privacy of the exchanged of data (Pieroni, Scarpato, & Brilli, 2018b).

Regulations on the privacy (Guadagni, 2015) of the collection, the preservation, and the use of data in a connected society exist, but they vary across states and countries. Several protocols have been implemented to collect data from vehicles to infrastructure (V2I) and between vehicles (V2V). The former is better defined as an infrastructure communication technology, and the latter as a cooperative communication technology (Pieroni, Scarpato, & Brilli, 2018b).

In the V2I protocol, the infrastructure plays a coordinating role by gathering global or local information on traffic and road conditions and then suggesting or imposing certain behaviors on a group of vehicles. The V2V protocol is more difficult to realize because of its decentralized structure and its aim to organize the interaction among vehicles and possibly develop collaborations (Dey, Rayamajhi, Chowdhury, Bhavsar, & Martin, 2016).

Progress in onboard computational technology has introduced "adaptive" intelligence, in other words, the vehicle changes its behavior according to environmental conditions (e.g., traffic condition). This processing must be performed in real time, considering at the same time the speed of the vehicle, the obstacles present in the roadway, and the traffic conditions, have the possibility to make appropriate changes keeping the vehicle in the lane and suggesting alternative itineraries.

Figure 11.8 illustrates the complexity (in MoC—million-of-line-code) of the software modules provided for a totally connected vehicle compared to the software onboard other types of vehicles (Pieroni, Scarpato, & Brilli, 2018b).

The software embedded in a totally connected vehicle can reach and exceed 100 millions of lines of code (Information is Beautiful, 2015). Even if the measure of line of code is not widely recognized, it can be a valid indicator of the software complexity in this type of vehicle. It is mandatory to reduce the computational time, the complexity of the embedded software while maintaining at the same time the availability and the reliability of software itself. This software complexity requires the availability of processors with increasing performance. As a consequence, the complexity of a vehicle embedded software is increasing, not only to ensure the performance of the various interactive function, but also to permit the development of an orchestration layer that allows communication between the various software modules and the on-road infrastructure (Pieroni, Scarpato, & Brilli, 2018b).

REFERENCES

Atat R., Liu L., Wu J., Li G., Ye C., Yi Y., 2016. Big data meet cyber-physical systems: A panoramic survey. *IEEE Access*, 4. doi:10.1109/ACCESS.2018.2878681.

Ali A. R., 2018. Real-time big data warehousing and analysis framework. In *Proceedings of 2018 IEEE 3rd International Conference on Big Data Analysis (ICBDA)*, Shanghai, China. March 2018, pp. 43–49.

Baheti R., Gill H., 2011. Cyber-physical Systems. *The Impact of Control Technology*, 12(1), 161–166.

Bahri S., Zoghlami N., Abed M., Tavares J. M., 2019. Big data for healthcare: A survey. *IEEE Access*. doi:10.1109/ACCESS.2018.2889180.

Behl M., Mangharam R., 2016. Interactive analytics for smart cities infrastructures, In *Proceedings of 2016 1st International Workshop on Science of Smart City Operations and Platforms Engineering (SCOPE) in partnership with Global City Teams Challenge (GCTC) (SCOPE - GCTC)*, Vienna, Austria. April 2016, pp. 1–6.

Buyya R., K. Ramamohanarao, C. Leckie, R. N. Calheiros, A. V. Dastjerdi, S. Versteeg, 2015. Big data analytics-enhanced cloud computing: Challenges, architectural elements, and future directions. In *Proceedings of 2015 IEEE 21st International Conference on Parallel and Distributed Systems (ICPADS)*, Melbourne, VIC, Australia. December 2015, pp. 75–84.

Cao H., Wachowicz M., Cha S., 2017. Developing an edge computing platform for real-time descriptive analytics. In *Proceedings of 2017 IEEE International Conference on Big Data (Big Data)*, Boston, MA, USA. December 2017, pp. 4546–4554.

Chala A., Bayliss D., Camper D., 2019. Keeping big data under control for autonomous cars. HPCC Systems. January 14, 2019. https://hpccsystems.com/blog/Keeping_Big_Data_Under_Control_with_Autonomous_Cars. Viewed: August 01, 2019.

Chen T. C., Sanga S., Chou T. Y., Cristini V., Edgerton M. E., 2009. Neural network with k-means clustering via PCA for gene expression profile analysis. In *Proceedings of 2009 WRI World Congress on Computer Science and Information Engineering*, Los Angeles, CA, USA. March 2009; Vol. 3, pp. 670–673.

Chopade P., Zhan J., Roy K., Flurchick K., 2015. Real-time large-scale big data networks analytics and visualization architecture. In *Proceedings of 2015 12th International Conference Expo on Emerging Technologies for a Smarter World (CEWIT)*, Melville, NY, USA. October 2015, pp. 1–6.

Clemente-Castellety F. J., Nicolae B., Katrinis K., Rafique M. M., Mayo R., Fernerlandez J. C., Loreti D., 2015. Enabling big data analytics in the hybrid cloud using iterative mapreduce. In *Proceedings of 2015 IEEE/ACM 8th International Conference on Utility and Cloud Computing (UCC)*, Limassol, Cyprus, December 2015. pp. 290–299.

Daniel A., Paul A., Ahmad A., 2015. Near real-time big data analysis on vehicular networks. In *Proceedings of 2015 International Conference on Soft Computing and Networks Security (ICSNS)*, Coimbatore, India, Feb 2015, pp. 1–7.

Dey K. C., Rayamajhi A., Chowdhury M., Bhavsar P., Martin J., 2016. Vehicle-to vehicle (V2V) and vehicle-to-infrastructure (V2I) communication in a heterogeneous wireless network - Performance evaluation. *Transportation Research Part C: Emerging Technologies*, 68, 168–184, 2016.

Ding G., Tan Z., Wu J., Zhang, J., 2014. Efficient indoor fingerprinting localization technique using regional propagation model. *IEICE Transactions on Communications*, E97-B(8), 1728–1741.

Ding G., Tan Z., Wu J., Zeng J., Zhang L., 2015. Indoor fingerprinting localization and tracking system using particle swarm optimization and Kalman filter. *IEICE Transactions on Communications*, E98-B(3), 502–514.

Exton T., Labib A., 2002. Spare parts decision analysis–The missing link in CMMSs (Part II). *Journal of Maintenance & Asset Management*, 17(1), 14–21.

Fayyad U. M., Piatetsky-Shapiro G., Smyth P., 1996. Advances in knowledge discovery and data mining. In U. M. Fayyad, G. Piatetsky-Shapiro, P. Smyth, & R. Uthurusamy (Eds.) *Data Mining to Knowledge Discovery: An Overview* (pp. 1–34). Menlo Park, CA: American Association for Artificial Intelligence.

Fu C., Ye L., Liu Y., Yu R., Iung B., Cheng Y., Zeng Y., 2004. Predictive maintenance in intelligent-control-maintenance-management system for hydroelectric generating unit. *IEEE Transactions on Energy Conversion*, 19(1), 179–186.

García M. C., Sanz-Bobi M. A., del Pico J., 2006. SIMAP: Intelligent system for predictive maintenance: Application to the health condition monitoring of a wind turbine gearbox. *Computers in Industry*, 57(6), 552–568. doi:10.1016/j. compind.2006.02.011.

Guadagni F., 2015. Personal and sensitive data in the e-Health-IoT universe. In *International Internet of Things Summit*, Cham: Springer, 2015, pp. 504–514.

Haarman M., Mulders M., Costas Vassiliadis C., 2017. Predictive maintenance 4.0. Predict the unpredictable. Price Waterhouse Coopers (PWC). June 2017. www.pwc.nl/nl/assets/documents/pwc-predictive-maintenance-4-0.pdf. Viewed: August 01, 2019.

Huang B., 2008. Cyber physical systems: A survey.

Information is Beautiful, 2015. Million lines of code. v0.9//September 24, 2015. www.informationisbeautiful. net/visualizations/million-lines- of-code/. Viewed: August 02, 2019.

Islam S., Keung J., Lee K., Liu A., 2012. Empirical prediction models for adaptive resource provisioning in the cloud. *Future Generation Computer Systems*, 28(1), 155–162.

Jain G., Pai M., 2019. The rise of industrial autonomous vehicles. Wipro Digital. https://ottomotors.com/blog/5-industry-4-0-technologies. Viewed: August 02, 2019.

Jardine A. K., Tsang A. H., 2013. *Maintenance, Replacement, and Reliability: Theory and Applications*. Boca Raton, FL: CRC Press.

Kaiser K. A., Gebraeel N. Z., 2009. Predictive maintenance management using sensor-based degradation models. *IEEE Transactions on Systems, Man and Cybernetics, Part A: Systems and Humans*, 39(4), 840–849.

Kardeby V., 2011. Automatic sensor clustering: Connectivity for the internet of things. *Licentiate Thesis*, Mid Sweden University, Department of Information Technology and Media, 2011.

Kharpal A., 2017. Baidu opens up driverless car tech to challenge Google, Tesla. CNBC. April 19, 2017. www.cnbc.com/2017/04/19/baidu-driverless-cars-open-source-technology.html. Viewed: August 02, 2019.

Klutke G.-A., Kiessler P. C., Wortman M., 2003. A critical look at the bathtub curve. *IEEE Transactions on Reliability*, 52(1), 125–129. doi:10.1109/TR.2002.804492.

Laka L., González Rodríguez M., 2015. "INDUSTRIA 4.0". *DYNA Ing. E Ind.*, 90(3), 16–17.

Lee E. A., 2008. Cyber physical systems: Design challenges. In *2008 11th IEEE International Symposium on Object and Component-Oriented Real-Time Distributed Computing (ISORC)*, Berkeley, CA, USA: Center for Hybrid and Embedded Software Systems, EECS. University of California.

Lee A. E., 2015. The past, present and future of cyber-physical systems: A focus on models. *Sensors*, 15, 4837–4869. doi:10.3390/s150304837.

Lee J., Ardakani H. D., Yang S., Bagheri B., 2015. Industrial big data analytics and cyber-physical systems for future maintenance & service innovation. *Procedia CIRP*, 38, 3–7. doi:10.1016/j.procir.2015.08.026.

Lee C. K. M., Cao Y., Ng K. H., 2017. Chapter 4: Big data analytics for predictive maintenance strategies. doi:10.4018/978-1-5225-0956-1.ch004.

Liu S., Yin J., Wang X., Cui W., Cao K., Pei J., 2016. Online visual analytics of text streams. *IEEE Transactions on Visualization and Computer Graphics*, 22(11), 2451–2466.

Mainnovation, 2019. PdM 4.0, Predictive maintenance. Netherlands. www.mainnovation.com/solutions/pdm4-0/. Viewed: August 01, 2019.

Malaysia Automotive Robotics and IoT Institute (MARii), 2019. Big data analytics – How does it help the automotive industry? http://marii.my/big-data-analytics-how-does-it-help-the-automotive-industry/. Viewed: August 02, 2019.

Mo L., Li F., Zhu Y., Huang A., 2016. Human physical activity recognition based on computer vision with deep learning model. In *Proceedings of 2016 IEEE International Conference on Instrumentation and Measurement Technology*, Taipei, Taiwan, May 2016, pp. 1–6.

Neves M. V., Rose C. A. F. D, Katrinis K., Franke H., 2014. Pythia: Faster big data in motion through predictive software-defined network optimization at runtime. In *Proceeding of 2014 IEEE 28th International Symposium on Parallel and Distributed Processing*, Phoenix, AZ, USA, May 2014, pp. 82–90.

Pieroni A., Scarpato N., Brilli M., 2018a. Industry 4.0 revolution in autonomous and connected vehicle a non-conventional approach to manage big data. *Journal of Theoretical and Applied Information Technology*, Vol. 96, No. 1, pp. 10–18. E-ISSN 1817-3195/ISSN 1992-8645.

Pieroni A., Scarpato N., Brilli M., 2018b. Performance study in autonomous and connected vehicles an industry 4.0 issue. *Journal of Theoretical and Applied Information Technology*, Vol. 96, No. 4, pp. 984–994. ISSN: 1992-8645.

Potter M., 2019. Big data's role in self-driving car development. Inside Big Data. April 5, 2019. https://insidebigdata.com/2019/04/05/big-datas-role-in-self-driving-car-development/. Viewed: August 02, 2019.

Qu F., Wang F. Y., Yang L., 2010. Intelligent transportation spaces: Vehicles, traffic, communications, and beyond. *IEEE Communications Magazine*, 48(11), 136–142.

Rajeshwari D., 2012. State of the art of big data analytics: A survey. *International Journal of Computer Applications*, 120(22), 39–46.

Rajkumar R., Lee I., Sha L., Stankovic J., 2010. Cyber-physical systems: The next computing revolution. In *Design Automation Conference 2010*, Anaheim, CA, USA. Copyright 2010 ACM 1-58113-000-0/00/0010.

Rashidi P., Cook D. J., Holder L. B., Schmitter-Edgecombe M., 2011. Discovering activities to recognize and track in a smart environment. *IEEE Transactions on Knowledge and Data Engineering*, 23(4), 527–539.

Report of the President's Council of Advisors on Science and Technology (PCAST), 2007. Leadership under challenge: Information technology R&D in a competitive world.

Ruiz-Arenas S., Horváth I., Mejía-Gutiérrez R., Opiyo E. Z., 2014. Towards the maintenance principles of cyber-physical systems. *Strojniški vestnik - Journal of Mechanical Engineering*, 60(12), 815–831. doi:10.5545/sv-jme.2013.1556.

Sanislav T., Miclea L., 2012. Cyber-physical systems - concept, challenges and research areas. *CEAI*, 14(2), 28–33.

Sastry S., 2008. *CPSs Definition*. Berkeley: University of California.

Sikorska J., Hodkiewicz M., Ma L., 2011. Prognostic modelling options for remaining useful life estimation by industry. *Mechanical Systems and Signal Processing*, 25(5), 1803–1836. doi:10.1016/j.ymssp.2010.11.018.

Sinda M., Liao Q., 2017. Spatial-temporal anomaly detection using security visual analytics via entropy graph and eigen matrix. In *Proceedings of 2017 IEEE 15th International Conference on Dependable, Autonomic and Secure Computing, 15th International Conference on Pervasive Intelligence and Computing, 3rd International Conference on Big Data Intelligence and Computing and Cyber Science and Technology Congress (DASC/PiCom/DataCom/CyberSciTech)*, Orlando, FL, USA, November 2017, pp. 511–518.

Singh S., 2018. Road to an autonomous future: How autonomous technologies are improving efficiency and safety. ITProPortal. August 29, 2018. www.itproportal.com/features/road-to-an-autonomous-future-how-autonomous-technologies-are-improving-efficiency-and-safety/. Viewed: August 02, 2019.

Tan Z., Jamdagni A., He X., Nanda P., Liu R. P., Hu J., 2015. Detection of denial-of-service attacks based on computer vision techniques. *IEEE Transactions on Computers*, 64(9), 2519–2533.

Tawalbeh L. A., Mehmood R., Benkhlifa E., Song H., 2016. Mobile cloud computing model and big data analysis for healthcare applications. *IEEE Access*, 4, 6171–6180.

Trinks S., Felden C., 2017. Real time analytics âĂŤ state of the art: ˇ Potentials and limitations in the smart factory. In *Proceedings of 2017 IEEE International Conference on Big Data (Big Data)*, Boston, MA, USA, December 2017, pp. 4843–4845.

Tsai C. W., Lai C. F., Chiang M. C., Yang L. T., 2014. Data mining for internet of things: A survey. *IEEE Communications Surveys Tutorials*, 16(1), 77–97.

Veverka J., 2018. Understanding the convergence of industry 4.0 and the rise of autonomous vehicles. Martech: Analytics & Data. August 17, 2018. https://martechtoday.com/understanding-the-convergence-of-industry-4-0-and-the-rise-of-autonomous-vehicles-222919. Viewed: August 02, 2019.

Wan K., Man K. L., Hughes D., 2010. Specification, analysing challenges and approaches for cyber-physical systems (CPS). *Engineering Letters*, 18(3).

Wang K., Mi J., Xu C., Shu L., Deng D.-J., 2016. Real-time big data analytics for multimedia transmission and storage. In *Proceedings of 2016 IEEE/CIC International Conference on Communications in China (ICCC)*, Chengdu, China, July 2016, pp. 1–6.

Wu S.-j., Gebraeel N., Lawley M. A., Yih Y., 2007. A neural network integrated decision support system for condition-based optimal predictive maintenance policy. *IEEE Transactions on Systems, Man and Cybernetics, Part A: Systems and Humans*, 37(2), 226–236.

Xu L., Li J., Shu Y., Peng J., 2014. Sar image denoising via clustering based principal component analysis. *IEEE Transactions on Geoscience and Remote Sensing*, 52(11), 6858–6869.

Yetis Y., Sara R. G., Erol B. A., Kaplan H., Akuzum A., Jamshidi M., 2016. Application of big data analytics via cloud computing. In *Proceedings of 2016 World Automation Congress (WAC)*, Rio Grande, Puerto Rico, July 2016, pp. 1–5.

Zheng D., Ben K., Yuan H., 2014. Research of big data space-time analytics for clouding based contexts-aware IOV applications. In *Proceedings of 2014 Second International Conference on Advanced Cloud and Big Data*, Huangshan, China, November 2014, pp. 150–156.

Zhong Y., Fang J., Zhao X., 2013. VegaIndexer: A distributed composite index scheme for big spatio-temporal sensor data on cloud. In *Proceedings of 2013 IEEE International Geoscience and Remote Sensing Symposium (IGARSS)*, Melbourne, VIC, Australia, July 2013, pp. 1713–1716.

Zhou B., Li J., Guo S., Wu J., Hu Y., Zhu L., 2017. Online internet traffic measurement and monitoring using spark streaming. In *Proceedings of IEEE Global Communications Conference (GLOBECOM)*, Singapore, Singapore, December 2017, pp. 1–6.

Index

3D Landscape visualization, 218–219
3D LiDAR, 94, 266, 268, 270
3D Visualization, 59, 121, 188, 190–192, 216, 218–221, 223, 225
3D Visualization software, 219, 223, 225

A

ACC, 1, 13–14, 116–117, 119–120, 123
Accommodation, 235, 308–309, 337
Acquisition, 40, 163, 178, 203, 210, 274, 286–287, 321–322
Active object programming, 57
Active thermography, 167
Actuators, 37, 42, 80, 82, 100, 144–145, 307, 314–315, 321, 352, 371–372
Adapted 2D views, 188
Aerial inspection, 133, 303–304, 306
AGV, 41
AHS, 2, 8–10, 151
AI, 35, 51–55, 76, 88, 122, 180–184, 297, 300, 304, 341–348, 350–363, 379–380, 382–385
Algorithms, 2–5, 14–15, 39, 48, 52–54, 56, 63, 97, 122, 131–132, 139, 166, 175, 177, 179, 188, 197, 211, 276, 282–283, 287, 296–299, 310–311, 314–315, 317, 321, 357, 368, 375, 379, 382
ALS, 156–158, 271–272, 299
Analysis, 30–31, 43, 47, 54, 63, 68, 77–79, 81, 88, 105, 154, 166–167, 187–188, 206–208, 211, 213–215, 218, 239–240, 273, 291, 299–300, 313, 330, 335, 348–350, 375, 380–384
Analyze, 30, 120
Antenna, 48, 113, 118, 132, 139, 157, 230, 232–234, 244–265
Antenna classification, 248, 253–254
Antenna types, 269
Aperture, 123, 132, 151, 206, 245, 248, 271
Artificial driving intelligence, 361
Asset data, 346
Asset inspection, 26, 28, 33, 87–88, 109, 296, 344–346
Asset maintenance, 28–29, 76, 351
Asset performance, 28–29, 33
Asset reliability, 29
Assets, 21–26, 28–29, 47, 57, 72–74, 76–77, 87, 91, 179, 288, 296–297, 345–346, 352, 375
Augmented 2D views, 188
Automatic programming, 56
Automation, 1–3, 18, 120, 180, 215, 287, 291, 299, 335, 345, 350, 353–354, 370, 373, 381
Automotive industry, 116, 132, 313, 341–342, 354, 357, 363, 371, 375, 381, 383–384, 387, 390
Automotive radar, 116, 119
Automotive sensors, 115
Autonomous behavior, 43

Autonomous decision, 343
Autonomous inspection, 73, 76–77
Autonomous maintenance, 73, 75–76
Autonomous robot, 38–39, 50, 57–59, 71–74, 109, 305, 325–326
Autonomous vehicle, 1–2, 7, 39, 80–81, 83, 111, 113, 115, 117, 120, 122–123, 125, 133, 136, 138, 141–142, 265–266, 299–300, 309, 341, 353, 372, 387
Autonomy, 4, 7, 38–39, 51, 70, 102, 105, 121, 289, 299, 327–329, 360, 375
AUV, 39–42, 62, 64, 66, 68, 102, 153, 267–269
AV, 1–3, 6–8, 13–21, 39, 41, 123, 313, 322, 324, 327–332, 342–344, 354–356, 382, 385–387
AV Inspection, 365, 381
AV security, 82, 330–332
AVRP, 116–117, 123–125
AVs, 1–7, 9–11, 13–21, 39, 53, 62, 82, 309, 314, 328–332, 353–356, 372

B

Bandwidth, 63, 119, 241–243, 245, 253, 256, 262–264, 268–270, 279, 294, 303, 356, 368, 370, 372, 374
Big data analytics, 365, 367–370, 379–381, 383, 389–391
Broadcast radio, 232
Building, 21, 23–26, 38, 40–41, 46–48, 57, 74, 87, 137, 139, 158, 160–161, 280–287, 291, 296, 298–305, 336, 342, 345–346, 360–361, 370, 372–373
Building inspection, 47, 280, 286, 296, 303, 305
Building monitoring, 287

C

Camera, 4, 42, 44, 46–48, 57, 89, 91, 93, 102, 105, 115–116, 121–124, 129–131, 133–134, 139–144, 158–159, 161, 165, 168–172, 174, 190, 272, 279–282, 284, 294, 297–299, 313–315, 317–319, 336, 342, 351, 354, 381, 386
Canals, 23, 75, 294
CAV, 387
CAVs, 299, 387
CBM, 28–29, 377
CBR, 241, 347–348, 350, 361
Cell phones, 232
Climbing robots, 90, 93, 107, 275–276
Cloud data, 141–142, 302–303, 351, 368
Collision avoidance, 35, 45, 85, 132, 134, 149, 237–239, 268–269, 336
Communication, 2, 6, 13–15, 22, 35, 37, 39–40, 59, 61–62, 79–82, 157, 180, 227–232, 241–244, 247–248, 279, 314–316, 328–333, 370–376, 385
Communication methods, 270

393

Communication systems, 15, 269, 342, 354–355, 373
Communications data rate, 241
Confined spaces, 46, 313
Confirmative analysis, 195
Connected vehicles, 341, 355, 357–358, 387, 390
Construction industry, 299–300, 304
Control, 1, 39–48, 64, 75, 80–83, 85, 111–113, 115–116, 156, 159, 162, 165, 170, 193, 210, 236–237, 272–276, 281, 284–286, 291–294, 300, 307–308, 313–317, 321–324, 328–330, 332, 334–335, 343, 359–360, 366, 369–376
Control system, 3, 43–45, 75, 90, 273, 275–276, 286, 305, 307–308, 317, 324, 330, 338, 371–374, 376
Control vehicle, 357
Controlled flight, 291
Corrective maintenance, 311, 338, 377–378
CPS, 80–82, 365–368, 370–376, 391
Crack detection, 287, 298
Criticality, 31, 330, 378

D

Data, 3–7, 28–31, 37–40, 42, 51, 61, 75–80, 89–92, 94–96, 113–115, 121–125, 131–134, 136–138, 146–147, 153–162, 166–167, 174–185, 187–188, 190–198, 205–218, 232, 239–243, 266, 276, 293–303, 312–320, 322, 326, 328, 348–358, 365–370
Data acquisition, 71, 111, 155, 157, 184, 281, 286–287, 296, 298, 352
Data analysis, 30, 78, 147, 178–179, 184, 187–188, 239, 299, 363, 365–366, 369, 383–384, 389
Data annotation, 367
Data cleaning, 367
Data collection, 77–78, 92, 101, 106, 177–178, 272, 286, 290, 299–302, 312, 345–346, 353, 375
Data flow, 354
Data integration, 210–212, 346
Data Intelligence, 180, 184
Data mining, 175, 177–180, 183, 185, 187, 191, 222, 349, 365–369, 379, 389, 391
Data post-processing, 175
Data processing, 101, 132, 157–158, 166, 183, 272, 296–297, 322, 366
Data rate, 241, 243, 268–269
Data storage, 156, 351, 366, 368–369
Data summation, 175, 177
Data transformation, 352
Data visualization, 175–176, 187, 221, 225, 367
Decision, 1, 9, 28–29, 78–80, 83, 182–183, 206, 209, 215–216, 228, 276, 284, 288, 294, 308–310, 313, 318, 321, 329, 341–344, 347, 349–350, 352–355, 365–366, 374, 383–384
Decision support, 109, 215, 221, 224, 350, 352, 379, 391
Decision-level fusion, 206
Deep learning, 296–297, 303, 341, 350, 354–361, 367, 381, 390
Detecting, 5, 41, 47, 71, 83, 88, 90, 99, 110, 116, 119, 122, 140, 153, 160, 169–171, 181, 209, 214, 224, 271, 276, 283, 298, 307–308
Diagnosis, 47, 79, 97, 100, 109, 126, 163–164, 181–182, 285, 298, 314, 317–318, 338, 347, 352, 369, 377

Digital inspection, 282
Display, 46, 71, 101, 146, 162, 166, 176, 179, 188–189, 192–200, 205, 215, 220–223, 239, 326, 359, 374
DM, 349, 383
Driver monitoring, 123
Drone inspection, 304, 311–313, 339
Drones, 39, 46–47, 70–74, 87–88, 107, 109, 240, 284, 286, 296, 300, 303–306, 312–313, 344–346, 362
DSS, 350
DTR, 241–242, 269
Dynamics, 45, 56–57, 74, 301, 337, 342, 371

E

eMaintenance, 76–80
EO Sensors, 130
Equipment, 21–23, 29–33, 122, 157, 215, 238, 240, 272, 275, 277, 299–300, 313–314, 335, 371, 375, 377–379, 386
Equipment data, 29–31
Explorative analysis, 195

F

Fail-Safe system, 338
Failure, 5–6, 12, 26, 28–31, 33, 41, 79, 88, 105, 143, 181, 238, 289, 307–311, 324, 330, 343–344, 349–352, 358, 369, 374, 377–379, 381
Failure detection, 307
Failure identification, 307
Failure prediction, 310–311, 352, 381
Failures, 5–6, 23, 29, 46, 78, 103, 132, 238, 288, 307–308, 310–311, 322, 324, 338, 348, 351–352, 371, 376–380
Fault, 34, 41, 46, 70, 79, 103, 108, 169, 181, 271, 274, 276, 286, 288, 307–309, 312–320, 330–331, 336–339, 350–352, 356, 369, 371
Fault detection, 34, 46, 70, 169, 181, 274, 308–309, 315, 317–318, 337–338, 351, 369
FDD, 318–320
FDM, 239–240, 268
FDM Data, 240
Feature-level fusion, 206
Field data, 177, 210
FL, 33–34, 71, 184, 337, 347, 350, 390–391
Flight controller, 43–44
Freight transport, 16
Frequency, 3, 28, 44, 48, 84, 111, 116–119, 125–130, 147–148, 152, 167, 172, 214, 229–232, 234, 242, 245, 247–248, 256, 259–262, 267–269, 272, 282, 321, 371–372, 384
Functionality, 82, 105, 144–146, 179–180, 190, 289, 300, 328, 332, 334, 346, 357, 372–374, 376
Fusion of multisource data, 206, 213

G

GAs, 18, 21, 38–39, 47–48, 54, 71, 73, 75, 88–92, 102, 106, 109–110, 296–297, 303, 305, 312–313, 346–350, 358, 364, 371–372, 375, 384, 386
GCS, 46–47, 284, 298

Index

GIS Data, 177, 208–209, 216
GNSS, 112, 130, 133–134, 139, 149–150, 152–154, 156–157, 160, 338
GPS, 5, 8–11, 37–38, 42–44, 67–68, 91, 105, 115, 130, 138–140, 144, 147–154, 159, 177–178, 184, 280–282, 286, 290, 293, 297, 299, 301, 310, 324–326, 335–337, 339, 356–357, 381, 387
GPS data, 67, 177–178, 210–211, 326
Graphs, 31, 189, 192, 196–197, 230, 364–365, 369

H

Hardware, 7, 10, 21–22, 37–39, 51, 57, 75, 87, 114, 131, 141, 143–145, 147, 171, 178, 181, 187, 190, 201, 204, 208, 220, 268, 284, 294, 309, 314–315, 359, 361, 371, 374
HAV, 333–336, 338
HAVs, 333–334, 336, 342
Hierarchies, 196
Hybrid systems, 22, 347, 350

I

ICT, 76, 78–79
Identification, 73, 115, 158, 164, 181, 183, 206, 222, 301, 309, 321, 337, 350–351, 356, 371–372
Image acquisition, 274, 286
Image processing, 47, 51, 74, 161–164, 184, 206, 211, 219, 287, 297–299, 305, 351
IMOS, 347, 350
Industrial assets, 21–22, 28, 33, 35, 352
Industrial autonomous vehicle, 390
Industrial AV, 385–386
Industrial equipment, 21–22
Industrial inspection, 170, 304, 350, 361
Industry 4.0, 35, 72, 109, 353, 376, 384–386, 390–391
Information visualization, 175, 187–188, 191, 195, 222–224, 367
Infrared imaging, 93, 185
Infrastructure inspection, 47, 70, 108, 132, 295–297
INS, 5, 9, 40, 105, 130, 139–140, 147, 151, 153, 156, 160, 290, 325, 339, 387
Inspection, 26–29, 33, 39, 44, 46–47, 57, 79, 87–98, 100, 102–104, 132–133, 164, 166, 169–172, 178, 267, 271–277, 279–280, 282–288, 290–291, 294–299, 311–313, 344–347, 349–351, 364, 377, 380, 382
Inspection modes, 87
Intelligence, 37, 41–42, 51–55, 61, 70–72, 76, 122, 154, 180, 184, 215, 224, 234, 297, 309, 341–344, 346, 351–352, 354, 361–363, 372, 379, 388–389
Intelligent diagnosis, 155, 179
IoT, 47, 53, 80–81, 328–329, 337, 351, 353, 358, 361, 366, 376, 379–380, 382, 384–386, 389–390
IR, 46, 91, 102, 122, 129–131, 136, 139, 165–172, 174, 184–185, 279, 283–284, 286, 321, 352, 359
IRT, 166, 168
Isolation, 34, 37, 181, 308, 317, 337

K

KBSs, 347
KDD, 178–179, 366

L

Landmark assistance, 321
Laser scanning, 107, 155–160, 184, 209, 272
LF, 248, 318–320
LiDAR, 4, 10, 36, 38, 73–74, 93–94, 105, 115, 117, 120–125, 130, 132–134, 136, 138, 141–142, 144, 151–158, 206, 208, 265–266, 268–270, 289, 314–316, 324, 333, 336, 356–357, 360–361, 382, 386–387
Line security diagnosis, 206
Linear assets, 22, 35, 57, 72–74, 76–77, 109
Local positioning systems, 321
Localization, 5, 7, 9, 38, 47, 63–64, 66, 79, 133–138, 140–143, 150–154, 266, 301, 321, 324–326, 333, 336, 338, 389

M

M2M, 376, 385
MAC, 235, 237, 239–240, 268
MAC risk monitoring, 239
Machine architectures, 48, 50
Machine learning, 30, 47, 52, 177, 179–180, 183, 188, 296–297, 300, 310, 323, 347, 349, 352–353, 356, 360, 365, 379, 382, 384
Machinery, 23–26, 32, 92, 165, 177
Maintenance, 19, 23, 28–31, 33–35, 38, 46–47, 72–80, 82, 88, 91, 100, 105, 107–109, 190–191, 286, 289, 305, 309, 311–313, 329–330, 338, 344–353, 358, 361–364, 374–380, 382, 389–391
Maintenance costs, 23, 29, 378
Maintenance inspections, 346, 362
Mapping, 7, 37–38, 43, 46, 62, 102, 105–106, 121–122, 131, 133, 135, 138–140, 142, 156, 160, 176, 196, 200, 208, 210–211, 267, 271–272, 282, 284, 289–290, 301–302, 333, 355
MAV, 280, 287, 305
MAVs, 280, 287
MF, 248, 318–320
MI, 34, 41–42, 48, 52–55, 71, 90, 151, 153–154, 183, 324, 343, 367, 391
Mission, 41–43, 46, 48, 62, 68, 70, 90, 92, 107, 130, 133, 181, 267, 269, 284–285, 305, 309, 322, 324, 334, 338, 374–375, 382
ML, 47, 52–54, 269, 379
MMS, 139, 272
Mobile, 14–15, 22, 37, 41–42, 46, 51, 59, 62, 89, 96–97, 102, 135, 139, 146–147, 155–157, 160, 191–192, 209, 220, 231, 242, 258, 263, 271–272, 294, 303, 324–325, 332, 336, 367–368
Mobile networks, 293
Model-based approach, 181, 308

Monitoring, 1, 29–31, 78–79, 87, 103, 113, 123, 146, 149–150, 159, 162, 169, 181, 187, 231, 238–239, 243, 271–272, 276, 280, 282, 284–288, 298, 300, 314–315, 324, 326, 346, 349, 355, 357–358, 365–367, 373, 375–378, 380–383
MOT, 265–266
Multi-UAV, 297, 299
Multidimensional data, 192, 194, 196, 223
Multisource data, 206, 209–213
Multisource data fusion, 209
Multisource data integration, 210–212

N

Navigation, 1, 5, 9, 33, 38, 40, 42, 57, 64, 74, 82, 90, 105, 113, 123, 126, 130, 133–135, 139, 142, 155–156, 176, 190, 235, 240, 272, 280–281, 301–302, 317, 324–327, 329–330, 332–339, 387
Navigation sensors, 281, 317
Network, 5, 16, 23, 56–58, 62–63, 78–79, 82, 90, 92, 100, 103–104, 113, 141, 152, 159, 179, 183, 210–211, 213–214, 221, 229, 232, 241–243, 247, 258, 264, 272, 283, 293–294, 308–310, 331–332, 344, 349–351, 356, 370–375, 377
Networking, 31, 81, 370–372, 374–375
NNs, 349–350, 355
Non-destructive testing, 184

O

Object detection, 121–125, 140–143, 150, 153, 206, 266, 318, 338
Object model, 57
Object-Based visualization, 202–203
Obstacle avoidance, 9, 51, 63, 83–87, 107, 109, 265, 273–274, 306, 324
Obstacle detection, 83, 141, 150, 268, 273, 276, 313
Obstacle traversing, 275
OD, 34, 203, 318–320
Offshore constructions, 312
Onboard inspection, 93
One-Dimensional data, 196
Onshore constructions, 312
Optimized production, 311
Orthophotomap, 291–293

P

PdM 4.0, 379–381, 390
Perception, 3–4, 6–7, 42, 51, 55–56, 60–61, 85, 124, 130–131, 133, 142–143, 152, 189–190, 192, 197, 215, 223, 236, 240, 295, 313, 315–319, 325, 338, 342–344, 354–355, 361
Pipelines, 23, 39, 47, 75, 90–92, 106, 285, 344
Pixel, 123–124, 131, 142, 155, 161, 167–168, 193–194, 199–201, 206–208, 222, 271, 286, 293, 319–320
Pixel-level fusion, 206–208
Planning, 3, 5, 7, 29, 39, 41–42, 45, 54, 83–85, 103, 134–135, 157–160, 162, 225, 229, 237, 265, 273, 281, 287–288, 301, 311, 322, 324–325, 329, 346, 348–349, 367

Polarization, 248, 253
Position control, 272–274
Post-processing, 47, 175, 177–179, 183–184, 208, 282, 287
Power cables, 23, 102, 107
Power line inspection, 71, 132, 272–276, 299, 304–306
Power lines, 46–47, 72, 75, 132–133, 271–276, 298, 304, 306, 322
Power supply, 92, 94, 98, 273
PP&E, 23–24
Predictive maintenance, 29–31, 33, 78, 351–353, 358, 361, 363, 377–380, 389–391
Preventive maintenance, 31, 72, 311, 348–349, 352, 363–364, 374, 377–378
Process data, 345
Programmability, 51
Programming, 56–59, 61–62, 71, 144–145, 191, 232, 349, 371, 379, 381

R

Radar, 3, 5, 8–10, 94, 113, 115–121, 123–126, 131–132, 141–142, 144, 151, 153–154, 158, 206, 231, 236, 248, 262, 266, 271, 284, 314–316, 324, 336, 354, 356–357, 360–361, 381–382, 386
Radiation, 39, 131, 134, 155, 164–165, 169–170, 230–231, 244–246, 248, 250–251, 253–256, 260–261, 263–264, 344–345
Radiation pattern, 165, 244–246, 254–256, 260–261, 264
Radio broadcasting, 231–232, 246, 269
Radio communication, 230–232, 234, 268, 270
Radio communication system, 234
Radio waves, 115–117, 126, 230–233, 244, 246–247, 314
Railway, 21, 23, 46–47, 75, 100, 104–105, 107, 288–289, 291, 293–294, 304
Railway infrastructure inspection, 47
Range measurement, 156
Real-time analytics, 366–367
Realibility, 3, 6, 28–29, 31, 53, 57, 74–75, 79, 84, 92, 128, 138, 143, 149, 182, 209, 280, 309, 321, 324, 347–349, 356, 371, 377–381, 388
Reception, 8, 126, 139, 230–231, 234, 244–245, 262
Recognition, 13, 47, 54, 70, 84, 123–124, 140, 151, 162–163, 188, 206, 208, 265, 276, 282, 297, 301, 314, 355–356, 359–360, 367, 390
Recording, 78, 93, 155, 281–282, 292, 351
Reliability, 3, 6, 28–29, 31, 35, 41, 57, 74–75, 79, 84, 92, 128, 138, 143, 149, 182, 209, 280, 309, 321, 324, 338, 347–349, 356, 361, 363, 377–381, 388
Remote sensing, 39, 43, 47, 72, 90, 131–132, 150–152, 156, 162, 168–169, 184, 206–209, 223, 225, 271, 284–285, 305
Repeated inspections, 307
Rethinking, 54
Risk assessment, 26, 28, 47, 84, 330, 358
Road inspection, 47
Road safety, 81, 333
Roads, 2–3, 5–6, 14–16, 21, 23, 46, 138, 153, 161, 299, 316, 325, 333, 341, 355, 360–361, 363
Robot mechanisms, 275

Index

Robotic, 3–5, 7, 33, 37–39, 42, 50, 53, 57, 70, 72, 75, 83, 88, 91, 94–95, 97–98, 102–103, 107–109, 134, 153, 299–300, 302–304, 321–322, 337, 359, 370, 372, 376, 385, 390
Robots, 4, 33, 37–39, 42, 51, 53, 57–59, 62, 66, 70–71, 73–76, 84, 88–90, 93–94, 102, 107–108, 275–276, 302, 305, 324, 336, 344–346, 354, 359, 362–363, 385–386
Rule-based approach, 181

S

Safety, 1–2, 6–7, 13–18, 57, 77, 81, 84–87, 89, 95, 103, 117, 123, 133, 181–182, 232, 236, 271, 299–301, 309, 311, 318, 322–324, 329–339, 342, 344–345, 367–368, 371–373, 375–376, 380, 385–386
Sampling algorithms, 71
SAR, 132, 151, 206, 208, 271, 391
Scanning, 42, 84, 94, 98, 107, 121, 155–160, 174, 184, 209, 236–237, 272, 281, 295, 366
Scene-Based visualization, 197, 205
Secure navigation, 339
Security, 6, 24–26, 37, 46, 75, 79, 81–82, 103, 107, 168, 206, 284, 286, 288, 323–324, 328–332, 337–338, 346, 367–368, 371–376, 391
Self-driving, 1, 13, 16, 34, 120, 133, 143–144, 150–151, 154, 313–314, 324, 333–334, 337, 339, 354, 356–357, 360–361, 363–364, 381, 383–386, 390
Semiautonomous flights, 291
Sensing strategies, 111
Sensor, 2–7, 11, 37–43, 46–47, 56–57, 67, 73–75, 78–84, 90–103, 105–118, 120–136, 139–147, 158–160, 162, 165, 170, 187, 206, 215, 244, 265–266, 289–290, 294, 297, 308–311, 313–322, 329–330, 342, 351–354, 356–361, 368, 376, 386–387
Sensor fusion, 34–35, 41, 114–115, 141–144, 150, 152–155, 317–318, 371
Sensor platforms, 144–146, 208
Sensors for missions, 130
SFDA, 308
SFDIA, 308
Siltation, 303
SLAM, 63, 133–135, 137–138, 142–143, 150–151, 153–154, 269, 301–302
Smart, 15, 35, 46, 53, 72, 78, 80, 109, 129, 144–146, 284, 294, 300, 314, 324, 344, 346, 351, 359, 365–366, 369–372, 374, 385–387, 389, 391
Software, 3, 6–7, 22, 29, 38–43, 46, 57–59, 74, 87, 98–99, 101, 105, 114, 145, 147, 171, 177, 179–181, 188–192, 197, 200, 204, 210, 282, 284, 345, 353, 379, 382
Software visualization, 188, 190–191, 222, 224
Stacked display, 194

T

Technologies, 1, 6–7, 10–11, 14, 16, 21, 40, 53–54, 74, 76, 78, 87, 91, 93–94, 105, 113, 120–121, 123, 133, 139–140, 166, 176, 191, 219–220, 342–343, 350–351, 353–354, 360, 370–372, 374–375, 381, 383–387

Testing, 8, 14, 16, 88, 96, 100, 106, 169, 183–184, 242, 280, 291, 302, 305, 323, 333, 338, 349, 352, 354, 356, 360, 371, 373
Text and hypertext, 196
Thermal imaging, 46, 75, 165–166, 169, 185, 283–286
Three-Dimensional MOT, 266
TPM, 377
Tracking with Multiple autonomous
vehicles, 265
Transmission, 22, 47, 74, 91, 146, 149, 171, 201, 231, 234, 237, 241–242, 244–245, 256, 271–272, 276, 303, 305, 325, 345, 372, 375, 379, 391
Transmit facility, 77

U

UAS, 46, 71–72, 103–106, 284, 286, 288–291, 293–294, 304
UAV, 43–47, 70–75, 87–88, 90–91, 93–94, 104–106, 130, 132–133, 151, 271–274, 280–281, 284, 288–292, 294–306, 308–309, 311, 338, 345–346
UAV-Driven, 300–301, 303
UAVs, 42–47, 73–75, 83, 90–94, 104–106, 130, 132–133, 208, 272, 280, 282, 284–286, 288–291, 294–297, 299–305, 307–309, 339, 345–346
Ultrasonic sensors, 73, 97, 102, 107, 124–125, 127–128, 151, 153, 299, 357, 386
UV Imaging, 169–174, 184
UV Signal, 171

V

V2I, 6, 13, 15, 81–82, 331, 388–389
V2V, 2, 6, 13–14, 81–82, 331, 388–389
Vegetation management, 132
Vehicle communication, 13, 35, 82, 331
Velocity, 5, 14, 48–50, 57, 67, 74–75, 85–86, 112–113, 115–120, 130, 139, 147, 155, 235, 314, 336
Visual data, 4, 90, 187–188, 192, 222, 300–303
Visual data exploration, 187–188, 192
Visual imaging, 155, 160
Visual Inspection, 88, 90, 93–95, 106, 164, 277, 279–280, 286, 290, 311, 350, 380
Visual scanning, 237
Visualization, 32, 47–48, 59, 121, 159–160, 162, 175–177, 187–197, 199–200, 202–205, 213–214, 216–225, 239, 287, 302–303, 345, 365, 367, 371, 384, 389–390
Visualization techniques, 187–188, 196, 222
VMT, 7, 14–17, 20–21

W

Waterway, 23, 73, 75
WSN, 92, 377

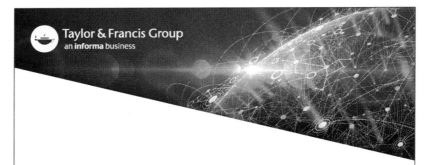

Taylor & Francis eBooks

www.taylorfrancis.com

A single destination for eBooks from Taylor & Francis with increased functionality and an improved user experience to meet the needs of our customers.

90,000+ eBooks of award-winning academic content in Humanities, Social Science, Science, Technology, Engineering, and Medical written by a global network of editors and authors.

TAYLOR & FRANCIS EBOOKS OFFERS:

- A streamlined experience for our library customers
- A single point of discovery for all of our eBook content
- Improved search and discovery of content at both book and chapter level

REQUEST A FREE TRIAL
support@taylorfrancis.com